Indira's Objective Agricultural Extension

The Authors

Renu Arya has completed her graduation and post graduation in Home science with specialization in extension education and communication management from Chandra Shekhar Azad University of Agriculture and Technology, Kanpur. She was also student of doctoral programme in Home Science in extension education and communication management in Punjab Agricultural University, Ludhiana. Presently she is working as a Subject Matter Specialist in Home Science at Krishi Vigyan Kendra, Behraich (UP) under Narendra Deo University of Agriculture and Technology, Kumarganj, Faizabad. She has published many research papers/popular articles in national and international journal of repute. She has also authored two books entitled Indira Krishi Pratiyogita Darpan in Hindi and Indira Competition Explorer, Indira's Objective in Agricultural Extension and Indira's Objective in Agriculture, Home Science and Fundaments of Horticulture. She was awarded Rajiv Gandhi Fellowship from UGC for doing Ph.D.

Dr. R.L. Arya born on 3 March 1964 at Village Jagdish Pur (Bhekhepur), Auraiya - Etawah (UP), completed his B.Sc. (Ag & A.H.) in 1984 from Chandra Shekhar Azad University of Agriculture and Technology, Kanpur, M.Sc. and Ph.D. in Agronomy from Indian Agricultural Research Institute, Pusa, New Delhi in 1986 and 1989, respectively. He served as a scientist at CRIDA, Hyderabad, IGFRI, Jhansi, IIPR, Kanpur, CTRI, Rajahmundry, CTRI Research Station, Jeelugumilli (West Godavari dist., A.P.) and CTRI Research Station, Dinhata (Cooch Behar dist, W.B.). He has good number of publication to his credit about 250 research papers/popular articles/papers presented in seminar symposia etc/book chapters in various national and international journals/megazines. Dr. Arya has vast experience (more than 20 years) in the field of dryland/rainfed agriculture, nutrient and water management, cropping system in pulses and pulse based cropping system, nutrient management and cropping system research in Jati and Motihari tobacco in West Bengal. He has also released a variety "Torsa" – an early maturing type of Motihari tobacco. He has authored several books such as Principles of Crop Production and Crops (2 Volume), Modern techniques of tobacco cultivation, Pulse production, Irrigation management, Weed management, Tobacco production in India, Indira Krishi Pratiyogita Darpan, Indira Competition Explorer, Indira Agricultural Competition Explorer, Indira's Objective in Agricultural Biotechnology, Indira's Objective in Agricultural Extension, Indira's Objective in Agronomy and Indira's Objective in Agriculture, Fundamentals of Agronomy and Soil and Plant Nutrition. He was also selected for Dr. Rajendra Prasad Puruskar by Indian Council of Agricultural Research, New Delhi and received best book award by Indian Society of Weed Management from NRC Weed Science, Jabalpur.

Janardhan Kumar obtained his M.Sc. (Ag) in Plant Biochemistry from Narendra Deo University of Agriculture and Technology, Kumar Ganj, Faizabad. He was a brilliant research Scholar in Department of Plant Biochemistry, Chandra Shekhar Azad University of Agriculture and Technology, Kanpur. He has published two book entitled Indira's Objective in Agricultural Extension and Indira's Objective in Agronomy and Indira's Objective in Agriculture. He was awarded Rajiv Gandhi Fellowship from UGC for doing Ph.D. Presently he is working as a Senior Manager in Indian Bank at Raichur (Karnataka).

Indira's Objective Agricultural Extension

(Multiple choice questions (MCQs) for competitive exams of Agricultural extension)

Renu Arya
Subject Matter Specialist
Krishi Vigyan Kendra
Behraich (U.P.)

DR. R.L. Arya
(M.Sc. & Ph.D. in Agronomy, IARI, NEW DELHI)
Ex Senior Scientist
Central Tobacco Research Institute,
Rajahmundry (A.P.)

Janardhan Kumar
Senior Manager
Indian Bank
Raichur (Karnataka)

Published by:

SCIENTIFIC PUBLISHERS (INDIA)
5A, New Pali Road, P.O. Box 91
Jodhpur 342 001, India
E-mail: info@scientificpub.com
Website: www.scientificpub.com

Print: 2020

ISBN: 978-81-72338-71-8
eISBN: 978-93-87869-69-1

Printed in India

PREFACE

India is a developing country and majority of population is dependent on agriculture. Total food grain production was 51 million tones in 1951 was increased to 257 million tones in 2012-13. There is still need to increase the total food grain production for ever increasing human population in the country with the active support and popularization of newly evolved high yielding verities/production technologies by the extension workers. Extension is becoming as the one of the most favourite and important subject for the student of agriculture, it has been also taken as an subject in undergraduate, post graduate, M.Phil and doctoral programmes is running in majority of Indian universities.

The competitive examinations have become a routine procedure of recruitment and admissions for higher position and education. Now-a-days a large number of short/objective type questions appear in the examination. These questions call for quick-answering for success within a specified short period of time. A sincere effort has been made by the authors to present them in most easy, short and understandable language for the benefit, students, farmers and those who are interested in agriculture and agricultural extension.

Indiras Objective Agricultural Extension for competitive exams in agricultural extension discipline contains 15 chapters covering all related discipline. The chapters included such as: Introduction, historical perspective in relation to agriculture and extension, extension principles, approaches and programming, extension education, teaching and methods, extension learning and evaluation, communication and communication technology, audio - visual aids, innovation, evaluation and adoption, participatory rural appraisal, rural development and panchayati raj, economic principle, sociology and social work, training manual, agricultural statistics and appendix. This book has given due importance and whole syllabus was covered as per UGC, ICAR and SAUs programmes in relation to agriculture extension. Each chapter contains multiple choice questions and total about 9000 objective questions with multiple choices have been framed and arranged sequentially for the easy understanding of the students. Recent information and development in the field of agriculture extension have been incorporated in the text. This book is primarily intended to serve as a appearing in competitive examinations of undergraduate, post graduate and doctorate programmes in agricultural extension of various universities. Thus, this book is based on the syllabus of student of agriculture stream, it may be useful not only to students but also teachers, researchers as well as extension workers. The chapters are chosen in view to cover the course contents of competitive examinations like IAS,

IFS, ARS, PCS, Banking, SAUs, UGC and to get admissions in various degree programmes of SAUs and other universities.

Finally, I have belief and hope that this book will fulfill the requirement of students of agriculture and agricultural extension stream for appearing in different competitive examinations.

Being first edition, there could be some shortcomings for which valuable and informative suggestions will be gratefully appreciated. These shortcomings if found in the text may be incorporated in forthcoming edition for the benefit of readers with your critical suggestions, so that book may be more useful to the students.

I convey my gratitude to the entire Management of Scientific Publishers, Jodhpur for taking all pains in the publication of the book. Authors shall be grateful to all those persons as well as various books, periodicals, magazines, newsletters, journals, etc., have helped me in the preparation of this book. The authors/publishers look forward to suggestions from all readers, especially for experienced researchers and scholars for further improving the subject content as well as the presentation of this book. If any suggestions from readers will be highly appreciated

2013, Kanpur

RENU ARYA
R. L. ARYA
JANARDHAN KUMAR

CONTENT

Chapter 1

INTRODUCTION

1. The word Extension is derived from:
(A) Latin (B) Greek
(C) Both (A) & (B) (D) None of these

2. The word "tentio" in extension is derived from:
(A) Greek word (B) Latin word
(C) English word (D) None of these

3. Extension literary means:
(A) Stretching out (B) Talking with
(C) Deciding for (D) Thinking about

4. The term 'Extension' have originated in:
(A) Latin (B) Persian
(C) Greek (D) Sanskrit

5. The latin roots for the word 'extension':
(A) *ex* and *tensio* (B) *ex and tension*
(C) *ex* and *luceo* (D) *ex* and *termo*

6. Extension is a/an:
(A) Informal education
(B) Out of school systems of education
(C) Both (A) & (B)
(D) Formal education

7. Extension is:
(A) Education for all village people
(B) Teaching through learning by doing and seeing is believing
(C) Two way channel
(D) All of these

8. The purpose of introducing agricultural universities in India is to improve the standard of agriculture:
(A) Extension (B) Teaching
(C) Research (D) All of these

9. Which extension method is evolved in result demonstration, posters, movies, slides and film strips?
(A) Spoken (B) Visual
(C) Written (D) None of these

10. An extension worker, while extending, scientific knowledge to the farmers, starts with:
(A) Lectures
(B) Practical demonstration
(C) Distribution of leaflet
(D) All of these

11. There are how many degrees of evaluation in Extension:
(A) 4 (B) 5 (C) 3 (D) 2

12. The extension activities was started first time in:
(A) USA (B) Russia
(C) Germany (D) India

13. The extension word was used first time in:
(A) USA (B) India
(C) France (D) USSR

14. The term 'University Extension' was first used in:
(A) 1866 (B) 1972
(C) 1901 (D) 1921

15. The person considered as father of extension:
(A) James Stewatt (B) Seaman A Knapp
(C) Edgar Dale (D) Paul Leagans

16. The term 'Extension Education' was first coined in:
(A) USA (B) UK
(C) Netherlands (D) India

17. The person considered as father of extension in India:
(A) Adivi Reddy (B) O P Dahama
(C) K N Singh (D) J B Chitambar

18. 'Vulgarisation' in French is equivalent to the word:
(A) Social taboos (B) Education
(C) Extension (D) Social change

19. Extension when aims at empowering or uplifting the poor:
(A) Emancipatory extension
(B) Formative extension
(C) Informative extension
(D) Persuasive extension

20. Extension as a science is classified under:
(A) Hard Science
(B) Soft Science
(C) Pure Science
(D) Physical Science

21. Extension is not:
(A) Voluntary
(B) Family centered
(C) Conclusion oriented
(D) Flexible

22. Extension can be considered as a:
(A) Service (B) Profession
(C) Discipline (D) All of these

23. Extension basically is a two way process between:
(A) Farmer and researcher
(B) Researcher and planner
(C) Teacher and planner
(D) Researcher and teacher

24. Extension that emphasize on human resource development:
(A) Emancipatory extension
(B) Formative extension
(C) Informative extension
(D) Persuasive extension

25. Extension when used as a policy instrument for achieving societal objective:
(A) Emancipatory extension
(B) Formative extension
(C) Informative extension
(D) Persuasive extension

26. Extension that moves beyond the traditional crop production technologies to include all land based activities:
(A) Universal extension
(B) Cyber-extension
(C) Emancipatory extension
(D) Broad based extension

27. Extension as an instrument for helping people to make well-considered choices:
(A) Emancipatory extension
(B) Formative extension
(C) Informative extension
(D) Persuasive extension

28. The instrument used by extension to bring change:
(A) Communication (B) Teaching
(C) Learning (D) Adoption

29. Extension can be effective only through:
(A) Voluntary change
(B) Democratic change
(C) Autocratic change
(D) Persuasive change

30. The main mode of extension in the farmer first model:
(A) Farmer to farmer
(B) Agent to farmer
(C) Researcher to agent
(D) Researcher to farmer

31. The satellite model concept of extension was formulated by:
(A) M S Swaminathan (B) G Perumal
(C) Curt Lewin (D) Anil Gupta

32. The fundamental objective of extension work is:
(A) Development of farmer
(B) Development of individual
(C) Development of culture
(D) Development of society

33. The ultimate objective of extension is:
(A) Increase productivity
(B) Increase income
(C) Rural development
(D) Realize ones fullest potential

34. Agricultural extension deals with:
(A) Farmers (B) Change agents
(C) Farmers and change agents
(D) None of the these

35. An extension worker help farmers to attain desirable changes in his:
(A) Temperament (B) Behaviour
(C) Health (D) Wealth

36. Leadership development in extension is:
(A) A means (B) An end
(C) Both means and end
(D) None of these

37. Most useful leader in agricultural extension is:
(A) Voluntary leader
(B) Bureaucratic leader
(C) Autocratic leader
(D) Democratic leader

38. An extension worker is a:
(A) Professional leader (B) Lay leader
(C) Local leader (D) Voluntary leader

39. An extension agent is considered to be a:
(A) Personal cosmopolite
(B) Impersonal cosmopolite
(C) Personal localite
(D) Impersonal localite

40. The full form of A.T.M.A. is:
(A) Agriculture Technology Management Association
(B) Agriculture Technology Management Agency
(C) Agriculture Technology Mission Agency
(D) None of these

41. T & V system is a good example of:
(A) Extension approach
(B) Training approach
(C) Co-operative self help approach
(D) Integrated development approach

42. To get the intelligence quotient (I.Q.) of the individual, the formula will be:

(A) $I.Q. = \frac{\text{Chronological age}}{\text{Mental age}} x100$

(B) $I.Q. = \frac{\text{Mental age}}{\text{Chronological age}} x100$

(C) $I.Q. = \frac{\text{Chronological age}}{\text{Mental age}}$

(D) $I.Q. = \frac{\text{Mental age}}{\text{Chronological age}}$

43. Which one of the step is not included in programme planning?
(A) Collect facts
(B) Analyze situation
(C) Identify problems
(D) Execute plan

44. ________ is the primary institution of society:
(A) Family
(B) Village
(C) Individual
(D) Block

45. Which of the following extension approaches has got the highest intensity of influence?
(A) Mass approach
(B) Group approach
(C) Community approach
(D) Individual approach

46. Which of the following scheme is exclusively meant for self employment for rural youths?
(A) NREP (B) TRYSEM
(C) IRDP (D) DPAP

47. First Agricultural University in India was established at:
(A) New Delhi (B) Hyderabad
(C) Bikaner (D) Pant Nagar

48. Which of the following is not the form of social interaction?
(A) Competition (B) Cooperation
(C) Coordination (D) Conflict

49. Which of the following pair is not matched:
(A) Communication : Berlo
(B) Rural weakers : Small farmers
(C) Mass media : Demonstration
(D) Helping poorest of the poor first : IRD programme

50. General meeting are broadly, the meeting of:
(A) Heterogeneous participation
(B) Homogenuos participation
(C) Community participation
(D) Society participation

51. Campaign is an:
(A) Intensive teaching
(B) Extensive teaching
(C) Invasive teaching
(D) None

52. Programme planning is a procedure of
(A) Working by people
(B) Working for the people
(C) Working with the people
(D) None of these

53. A series of still pictures on one roll is called
(A) Channel (B) Film strip
(C) Mock-up (D) Black board

54. A bulletin should contain
(A) 4-1 2 pages (B) 12-24 pages
(C) 24-48 pages (D) None

55. In India, slightly less than half the villagers are inhabited ———————persons:
(A) 500 (B) 800 (C) 1000 (D) 1500

56. About ¾ Indian villages have a human population of less than:
(A) 600 (B) 700
(C) 800 (D) 1000

57. According to CACP, sugarcane prices announced is:
(A) MSP (B) Procurement prices
(C) Statuary (D) Flat prices

58. B.R.Mehta team stimulated an active consideration of democratic bodies in India:
(A) Decentralization (B) Stabilization
(C) Centralization (D) Mobilization

59. Most people retain ——————— % of what they read:
(A) 5-10 (B) 10-15
(C) 20-25 (D) 30-35

60. Most people kept in mind ——————% of what they seen:
(A) 10-15 (B) 20-25
(C) 30-35 (D) 50-55

61. Up to ——————— % of what is taught is kept in mind by majority of people, if they participate actively, and all the senses (read, hear, and seen) are involved:
(A) 50 (B) 65
(C) 80 (D) 90

62. Symposium is a short series of lectures, usually by speaker:
(A) 1-2 (B) 2-5
(C) 5-7 (D) 7-10

63. Phillips 66 format of hurdle system is related with:
(A) Panel (B) Form
(C) Buzz section (D) Work shop

64. Small group interaction designed to en courage the free introduction of ideas on an unrestricted basis without any limitations to feasibility is known as:
(A) Conference (B) Seminar
(C) Buzz section (D) Brain storming

65. ——————— essentially follow the seminar method and the focus is on any particular subject or problem:
(A) Work shop (B) Syndicate studies
(C) Institute (D) None

66. News paper articles are included in which method of extension:
(A) Individual (B) Group
(C) Mass (D) None of these

67. "Stay at home" type of people can be contacted by which method of extension:
(A) Individual
(B) Group
(C) Mass
(D) None of these

68. —————— is essentially a recognizable imitation or replica of the original, whether workable or not:
(A) Model (B) Specimens
(C) Mock-up (D) Objects

69. A working model is known as:
(A) Objects (B) Model
(C) Mock-up (D) Poster

70. "The communication process in Rural development" is written by:
(A) Leagans (B) Berlo
(C) A.A. Reddy (D) Ray

71. —————— are expressions of the ends towards which our efforts are directed:
(A) Goal (B) Objectives
(C) Needs (D) Message

72. —————— is defined as the distance in any given direction one expects to go during a given period of time:
(A) Goal (B) Objectives
(C) Needs (D) Message

73. The ultimate objective of extension work is the full development of ——————:
(A) Individual
(B) Group
(C) Mass
(D) Village

74. The gap between the situation and objective is the area of ——————:
(A) Goal (B) Needs
(C) Interest (D) None

75. In good administration the offices of a —————— are arranged in a pyramid of authority and responsibility:
(A) Autocracy (B) Monarchy
(C) Bureaucracy (D) Democracy

76. ——————believes that if you leave workers alone the work will be done:
(A) Autocratic leader
(B) Democratic leader
(C) Laissez-faire leader
(D) None of these

77. —————— is one who has received specialized training in the field in which he work as full time and paid for his work:
(A) Operational leader
(B) Nominal leader
(C) Professional leader
(D) Prominent talent leader

78. Best method for selection of leader is
(A) Sociometry (B) Election
(C) Discussion method
(D) Key informants

79. —————— are those things which person ought or ought not to do in society
(A) Tabbo (B) Mores
(C) Folkways (D) Ritual

80. Which one of the following is a farmer agency?
(A) FCI (B) IFFCO
(C) Agro service centers
(D) Shree Ram fertilizers

81. The basic unit of development of IRDP is:
(A) Community Development Programme
(B) Village (C) District
(D) Family

82. Krishi Vigyan Kendra (KVKs) can only be under control of:
(A) Agricultural University
(B) Voluntary organisation
(C) Govt. and Voluntary organization
(D) Agricultural University and Voluntary Organisation

83. "The Lab to Land" programme also included demonstration but they were different from early demonstrations in terms of:
(A) Target groups
(B) Type of demonstration
(C) Size of demonstration
(D) Place of demonstration

84. The first Krishi Vigyan Kendra in India was established at:
(A) Bihar
(B) Rajasthan
(C) Punjab
(D) Pondichery

85. Which of the following is not a group approach device:
(A) Symposium (B) Film
(C) Demonstration (D) Workshop

86. The subject extension is a:
(A) Means (B) Discipline
(C) End (D) Branch of mean

87. The extension system used to provide precise and reliable scientific information in simple language about a simple practice is called:
(A) Leaflet (B) Newspaper
(C) Literature (D) Circular letter

88. The benefit of first hand knowledge good will and creating confidence can be developed by:
(A) Telephone
(B) Visiting home and farms
(C) Organising village fairs
(D) None of these

89. IVLP stands for:
(A) Indian Village Linkage Programme
(B) Indian Village Linkage Plan
(C) Institution Village Linkage Programme
(D) Integrated Village Lab to Land Programme

90. Audio-visual aids is:
(A) Radio (B) Wireless
(C) Telephone (D) Television

91. Which of the following country was first use the term 'extension' to imply extension knowledge from agriculture research institution to the farmers through the process of informal education?
(A) U.S.A. (B) U.S.S.R.
(C) India (D) Japan

92. Which of the following will give the clear idea of proportion?
(A) Model (B) Pie chart
(C) Radio (D) TV

93. For providing the precise and reliable scientific information in simple language the extension system used is:
(A) Charts (B) Demonstration
(C) Leaflets (D) None of these

94. CONA stands for:
(A) Comparative Oriented Need Assessment
(B) Community Oriented Need Assessment
(C) Commercial Oriented Natural Assignment
(D) None of the above

95. Which of the following visual aids are commonly used by agricultural extension workers in India?
(A) Posters, photographs, flash card and fenal graphs
(B) Slides, film-strips, models and puppets
(C) Bulletin boards, cultural programmes and black boards
(D) All of the these

96. Which of the following is not true?
(A) Programme planning is a spontaneous process
(B) Programme planning is an educational process
(C) Programme planning is a continuous process
(D) Programme planning is a coordinating process

97. Which of the following is a function of the RRB's?
(A) To develop rural economy
(B) To provide credit for agriculture and allied activities
(C) To encourage rural industries
(D) All of the these

98. Evaluation done by the agency is known as:
(A) Self evaluation
(B) Internal evaluation
(C) External evaluation
(D) None of the above

99. The all – round development of Indian rural people can be achieved by means of:
(A) Extension education
(B) Adult education
(C) More money distribution
(D) All of the these

100. Integrated Rural Development Programme (IRDP) has been conceived:
(A) Essentially to development anti hoard ing programme
(B) Essentially to development adult education facilities in the rural areas
(C) Essentially as an anti-poverty programme
(D) Essentially to develop fisheries in the area

101. Coding of the data is done to assign:
(A) Numerical to answer
(B) Symbol to answer
(C) Both (A) and (B)
(D) None of these

102. Slide projector gives the image:
(A) Direct (B) Indirect
(C) Both (A) and (B) (D) Reflected

103. What is the specific objectives of the GATT?
(A) Globilization
(B) Increase world production
(C) Complete use of resources
(D) All of these

104. High yielding Variety programme has given emphasis on:
(A) Land reform
(B) Package and practice with dwarf varieties
(C) Increase production through dry farming
(D) None of the these

105. Which of the following is the first step in developing of programme?
(A) Selection of site
(B) Collection of information
(C) Both (A) and (B)
(D) Find felt need of people

106. Kisan Mitra is an employee of:
(A) Central Govt (B) State Govt.
(C) Corporation (D) None of the above

107. Unit of the IRDP:
(A) Village
(B) Block
(C) District
(D) State

108. Puppet show in village as a method for transfer of technology is an example of:
(A) Mass media
(B) Group media
(C) Individual media
(D) Tribal media

109. The main purpose of T and V programme is to:
(A) Identify research problems
(B) Bring about co-ordination in the administration of agricultural extension
(C) Update the technical knowledge and skills of agricultural extension workers
(D) Help farmers in the adoption of latest innovations in agriculture

110. Consider the following statements: In interpersonal channels of communication
(1) Message flow tends to be one way
(2) Quantum of feedback available is very high
(3) Communication is face to face
(4) Speed of reaching a large audience is relatively slow of these statements
(A) 1, 2, 3 and 4 are correct
(B) 1 and 4 are correct
(C) 2, 3 and 4 are correct
(D) 2 and 3 are correct

111. Read the following statements and mark your answer according to the code given below:

(1) Newspaper is a popular medium of communication in rural areas.

(2) Newspaper is written form of communication

(3) Film show is audio-visual form of communication

(4) Film show is popular medium of communication in rural areas.

Answer code

(A) 1 and 2 are correct

(B) 2 and 3 are correct

(C) 3 and 4 are correct

(D) 1 and 4 are correct

112. Which one of the following is not correctly matched?

List-I	List –II
(A) Mass	Exhibition
(B) Individual	Farm visit
(C) Group	Mela
(D) Mass	Newspaper

113. Which of the following is the group contact teaching method?

(A) Farm and home visit

(B) Bulletins

(C) Result demonstration

(D) Method demonstration

114. Match list –I and list – II and select the correct answer using the code given below:

List – I (Aids)	List-II (Method of contact)
(A) Model	(1) Group contact
(B) Slide	(2) Mass contact
(C) Telephone call	(3) Personal contact
(D) Poster	(4) Mass contact

Code:

(A) A-2, B-3, C-4, D-1

(B) A-1, B-2, C-3, D-4

(C) A-3, B-2, C-2, D-1

(D) A-3, B-2, C-4, D-1

115. Read the following statements and mark your answer according to the code given below:

(1) C.D. is a programme

(2) C.D. is a cultural event

(3) C.D. is a social event

(4) C.D. is a project

Answer code:

(A) 1 and 2 are correct

(B) 1 and 4 are correct

(C) 2 and 3 are correct

(D) 2 and 4 are correct

116. The benefit of first hand knowledge, good will and create confidence can be developed by:

(A) Visiting homes and farm

(B) Telephone (C) Village fair

(D) All of the these

117. To create interest in a wider range of people and to influence people to adopt better practices the extension method generally used is:

(A) Exhibition (B) Tours

(C) Fair (D) Competition

118. The extension system used to provide precise and reliable scientific information in simple language about a simple practice is called:

(A) Leaflet (B) Literature

(C) Newspaper (D) Circular letter

119. Extension of recent agricultural knowledge to rural peoples through trained persons of the same locality is called the principle of:

(A) Panchayati Raj

(B) Centralized power

(C) Local leadership

(D) None of the these

120. Effective extension service is a process of:

(A) Helping the people which helps themselves

(B) Profiting the local leaders

(C) Helping the B.D.O.

(D) All of the these

121. Krishi Gyan Kendra's are functioning through the financial assistance of:
(A) State Govt. (B) UNICEF
(C) ICAR (D) IARI

122. While working as an extension specialist in a village one finds that the package of practice are not being followed despite the best communication methods one had employed for the purpose one finds the following reasons for non-adaptivity:
(1) 60% are laggards
(2) 20% are innovative but orthodox
(3) 20% are old and orthodox
The best remedial measure would be to:
(A) Adopt 1 and 3
(B) Adopt 2 and above
(C) Adopt 3
(D) Adopt 2 and make them team leaders

123. The concept of demonstration was given by:
(A) Lewis Jones
(B) O.P. Dahama
(C) Rudramurthy
(D) Dr. Seeman A. Knapp

124. The functions of pre-extension in the development block was:
(A) To survey block area
(B) To prepare ground for intensive stage
(C) To conduct agricultural demonstrations
(D) All of the these

125. Block samiti is guided by
(A) B.D.O. and his extension officers
(B) V.L.W.
(C) Zila Parishad
(D) Panchayat Samiti

126. The development approach of National Extension Service Block was based mainly on:
(A) Self help
(B) Govt. help
(C) State Govt. help
(D) Help by village Panchayat

127. Which of the following statement is not correct?
(A) Village Panchayat is a non-official body
(B) Village Panchayat is headed by V.L.W.
(C) The head of a block is B.D.O.
(D) All of the these

128. The responsibility of development work at village level is of:
(A) Collector (B) B.D.O.
(C) V.L.W. (D) Gram Pradhan

129. The number of V.L.W's is an I.A.D.P. block is:
(A) 5 (B) 10
(C) 15 (D) 20

130. National demonstrations are first line demonstrations which is conducted on the farmer's field by:
(A) Extension workers (B) V.L.W.
(C) Research Scientists (D) B.D.O.

131. To create interest in a wider range of people to adopt better practices, the extension method generally used is:
(A) Field trials
(B) Exhibitions
(C) Competition
(D) Fair

132. The elements of teaching learning process are:
(A) Teaching subject matter and learner
(B) Teaching equipments and physical facilities
(C) Both (A) and (B)
(D) None of the above

133. "Cone of experience" was devised by:
(A) Rogers (B) O.P. Bhatnagar
(C) A.A. Reddy
(D) Edger Dale

134. The programme for transfer of technology started by ICAR at the golden jubilee year occasion was:
(A) Operational Research Project
(B) Lab to Land Programme
(C) National Demonstration
(D) I.R.D.P.

135. Extension workers are:
(A) Professional leaders
(B) Autocratic leader
(C) Lay leaders
(D) None of these

136. Match the following

List-I (Aids)	**List-II (Method of contact)**
(a) Model	(1) Group contact
(b) Slide	(2) Mass contact
(c) Telephone call	(3) Personal contact
(d) Poster	(4) Mass contact

Code	**a**	**b**	**c**	**d**
(A)	1	2	3	4
(B)	2	3	4	1
(C)	3	4	2	1
(D)	4	1	3	2

137.

List-I (Approach)	**List-II (Technique)**
(a) Mass	(1) Exhibition
(b) Individual	(2) Farm visit
(c) Group	(3) Model
(d) Mass	(4) Newspaper

Code	**a**	**b**	**c**	**d**
(A)	2	3	4	1
(B)	1	2	3	4
(C)	4	3	2	1
(D)	2	1	3	4

138. VLW (Multi Purpose Worker) concept emerged in which project?
(A) T & V Project (B) NEP
(C) KVK
(D) Etawah Pilot Project

139. Which of the extension programmes in India advocated single line of command and strong research-extension linkages?
(A) KVKs (B) T& V
(C) NATP (D) ORP

140. SREP is:
(A) Scientific Research Extension Programmes
(B) Strategic Research Extension Plan
(C) Specific Research Extension Plan
(D) Special Resource and Enterprise Focus Programme

141. ATMA is a:
(A) Registered society at district level for decentralized planning
(B) Non-Govt. organization working very near to the poor and down trodden people
(C) Technical package of information dissemination
(D) None of these

142. Which of the following programmes is being implemented by SAUs/Institutes to strengthen institute-village linkages?
(A) ATIC (B) NATP
(C) IVLP (D) UPDASP

143. "Farmers Interest Group" concept in agricultural extension is:
(A) A group of farmers meant for cultural activities
(B) A group of farmers in the villages created to watch the interests of the farmers
(C) A group of farmers coming together voluntarily to take up some activities
(D) A group of farmers interested in employment generation for rural youths

144. The internal state which stimulate the individual to carry out certain activities is:
(A) Perception (B) Motivation
(C) Intuition (D) Learning

145. The motivational type most important for extension is:
(A) Extrinsic motivation
(B) Intrinsic motivation
(C) Situational motivation
(D) Classical motivation

146. "If I have to get that gold medal, I should work hard at studies" reflects:
(A) Determination
(B) Intrinsic motivation
(C) Industriousness
(D) Extrinsic motivation

147. Which of the following is not an approach to education:
(A) Conventional (B) Prevocational
(C) Progressive (D) Liberating

148. Effective teaching and learning is a matter of:
(A) Chance and luck
(B) Design and effect
(C) Material and non material
(D) Formal and informal

149. Education is shaped by:
(A) Society (B) Teacher
(C) Student (D) Knowledge

150. Cone of experience classify the instructional methods according to:
(A) Experiential correctness
(B)Difficulty in understanding
(C) Ease in use and application
(D) Level of abstraction in thinking

151. The cone of experience was developed by:
(A) Egder Dale
(B) Berlo C.K.
(C) Kuldeep Nair
(D) S C Parmar

152. The type of experience a farmer have when he sown wheat in lines under the supervision of extension worker:
(A) Symbolic experience
(B) Iconic experience
(C) Supervisory experience
(D) Enactive experience

153. When a person involved in iconic experiences, learning occurs through:
(A) Active participation
(B) Observation
(C) Abstract thinking
(D) Doing

154. Which among the following is an example of direct purposeful experience:
(A) Ploughing
(B) Model of a plough
(C) Blow up of a plough
(D) Diagram of a plough

155. The extension worker tells about the varieties characters to a farmer, the farmer learns by:
(A) Abstraction (B) Learning
(C) Intelligence (D) Questioning

156. Classification of extension methods into individual, group and mass is based on:
(A) Foam (B) Use
(C) Number of people contacted
(D) Periodicity

157. To get first hand knowledge on a farm situation we use:
(A) Village fairs
(B) Farm & home visit
(C) Telephone
(D) Questionnaire

158. The ratio of 'takes' to 'acceptance' is the highest in:
(A) Farm and Home visit
(B) Method demonstration
(C) Result demonstration
(D) Campaign

159. Media forum is classified under:
(A) Individual contact method
(B) Group contact method
(C) Mass contact method
(D) All of these

160. When the farmer meets the extension worker at his work place:
(A) Farmer's call
(B) Extension campaign
(C) Result demonstration
(D) Office visit

161. The extension method that act as a safety valve in reducing tension:
(A) Group discussion
(B) Method demonstration
(C) Result demonstration
(D) Campaign

162. The best extension method suited to all situation:
(A) Demonstration (B) Campaign
(C) Farm publication(D) None of these

163. Method demonstration aims more at improving ones:
(A) Cognitive skills (B) Perceptual skills
(C) Motor skills (D) Analytical skills

164. Seaman A Knapp conducted result demonstration in:
(A) Cotton (B) Maize
(C) Barley (D) Wheat

165. A method which build the confidence of the researcher, extension worker and the farmer:
(A) Minikit trial
(B) Result demonstration
(C) Method demonstration
(D) All of these

166. A method of motivating the people to adopt a new practice by showing what has actually been achieved by applying the practice under field conditions:
(A) Campaign (B) Result demonstration
(C) Field day (D) Farm bulletin

167. Maintenance of record is absolutely necessary in case of:
(A) Method demonstration
(B) Minikit trial
(C) Field day (D) Field trip

168. The method used when the farmer is not there in the field while the extension worker make a visit:
(A) Farm and home visit
(B) Method demonstration
(C) Result demonstration
(D) Flag method

169. The normal group boundaries get obliterated in:
(A) Method demonstration
(B) Farm and home visit
(C) Office calls
(D) Campaign

170. A body of general principles or laws of a field of knowledge is called:
(A) Television (B) Philosophy
(C) Mandate (D) Principle

171. The basic philosophy of extension is to teach people:
(A) How to do (B) What to do
(C) When to do (D) Where to do

172. A person, who believes in equal rights and opportunities for women and men:
(A) Communalist (B) Feminist
(C) Secularist (D) Patriachal

173. The fundamental unit of civilization according to Horton is:
(A) Home (B) Land
(C) Individual (D) Capital

174. The difference between what is and what ought to be:
(A) Need (B) Aim
(C) Goal (D) Objective

175. The condition which people after study decided needs change is:
(A) Problem (B) Programme
(C) Project (D) Solution

176. A blue print for action is:
(A) Plan (B) Programme
(C) Objective (D) Project

177. The distance in any direction one expects to go is:
(A) Aim (B) Goal
(C) Objective (D) Principle

178. Process or procedure of attaining an objective:
(A) Programme (B) Method
(C) Plan (D) Media

179. The level of objective operates at field extension work:
(A) Fundamental objective
(B) Working objective
(C) General objective
(D) None of these

180. Which of the following is not a level of objective:
(A) Basic objective
(B) Fundamental objective
(C) Working objective
(D) General objective

181. Extension work grew out of:
(A) Problem (B) Situation
(C) Condition (D) Participation

182. The need which are not realized as impor-tantce are:
(A) Felt need (B) Unfelt need
(C) Ignored need (D) Unrealized need

183. Which of the following is the most effective method in identifying real rural needs:
(A) Election method
(B) Sociometric method
(C) Workshop method
(D) Selection method

184. Measuring performance against predeter mined goals is called:
(A) Management (B) Evaluation
(C) Measurement (D) Supervision

185. The evaluation of any programme should be based on:
(A) Financial outlay
(B) Number of participation
(C) Objectives of the programme
(D) None of the these

186. The element that take time, effort and resource to move from one event to another:
(A Event
(B) Activity
(C) Critical path
(D) Slack

187. The largest possible time to perform an activity assuming everything goes well is:
(A) Optimizing time
(B) Pessimistic time
(C) Most likely time
(D) Expected time

188. IVLP uses the concept of:
(A) Agro-system analysis
(B) Agro-ecosystem analysis
(C) Eco-system analysis
(D) Socio-ecosystem analysis

189. ATMA is an important component under:
(A) NARRS (B) NATP
(C) NAEP (D) KVK

190. ATMA operated at:
(A) Block level (B) District level
(C) State level (D) National level

191. The number of ATMA developed under NATP:
(A) 22 (B) 24 (C) 25 (D) 28

192. The ATMA under NATP is a:
(A) Statutory body (B) Registered society
(C) Cooperative society
(D) Krishi Vigyan Kendra

193. The apex training institute at state level to give training support to ATMA:
(A) MANAGE (B) EEI
(C) CAPART (D) SAMETI

194. SREP is an important component in the functioning of:
(A) NARS (B) ATMA
(C) ATIC (D) KVK

195. Indian Council of Agricultural Research is a:
(A) Educational organization
(B) Research institution
(C) Registered cooperative
(D) Autonomous body

196. Extension teaching at the under graduate level was started by:
(A) Madras university
(B) Meerut university
(C) Calcutta university
(D) Mumbai university

197. The first line system of extension in India is controlled by:
(A) ICAR institutes/SAUs
(B) Ministry of Agriculture
(C) Non Government Organisations
(D) Private Extension Agencies

198. The demonstration in case of AICPND was done by:
(A) Farmer (B) Media personal
(C) Research scientist
(D) Extension person

199. The relatively non-mobile among the TOT programmes of ICAR:
(A) LLP (B) ORP
(C) AICPND (D) KVK

200. The funding of KVK is done by:
(A) ICAR (B) SAU
(C) State government
(D) Central government

201. Farm Science Centre is a synonym of:
(A) FTC (B) FDG
(C) KVK (D) Teleclub

202. The extension component of NATP:
(A) Discipline oriented research
(B) Organizational and management reforms
(C) Agro-ecosystem research
(D) Innovation in technology dissemination

203. The extension service in USA is called:
(A) National extension service
(B) US extension work
(C) Cooperative extension service
(D) Rural extension service

204. The actual physical product of source-encounter is:
(A) Receiver (B) Message
(C) Channel (D) Feedback

205. The three important dimensions of message include:
(A) Code, content, treatment
(B) Source, self, receiver
(C) Encode, decode, feedback
(D) Simplicity and clarity

206. The way in which message is handled before placing in the channel:
(A) Decoding
(B) Treatment
(C) Distortion
(D) Feedback

207. The link that connects the source with the receiver:
(A) Message
(B) Communication effect
(C) Channel (D) Treatment

208. Television is primarily a:
(A) Media mix (B) Source
(C) Medium (D) Channel

209. Trustworthiness and competency are elements of:
(A) Credibility (B) Fidelity
(C) Empathy (D) None of these

210. Acceptance and rejection are the one of the kinds of:
(A) Failure of channel
(B) Feedback
(C) Treatment of message
(D) Empathetic feelings

211. The term class media refers to:
(A) Mass media
(B) Group media
(C) Inter personal media
(D) Intra personal media

212. While communicating, the communicator can write facing the audience in:
(A) Black board (B) Flash card
(C) Slide (D) OHP

213. Which of the following statement about audio-visual aids is true:
(A) Replaces a teacher
(B) Decorates learning situation
(C) Emphasis non-verbal experience
(D) Includes text book

214. The term electronic chalkboard refers to:
(A) LCD projector (B) OHP projector
(C) VHS projector
(D) Beta chrom projector

215. The film projector commonly used in extension:
(A) 8 mm (B) 16 mm
(C) 32 mm (D) 70 mm

216. The object for projection in a slide projector is kept between:
(A) Reflector and condenser
(B) Condenser and objective
(C) Bulb and condenser
(D) Objective and screen

217. The principle made used in slide-cum-film projector:
(A) Direct projection
(B) Indirect projection
(C) Reflected projection
(D) Diffused projection

218. Which of the following instructional aids essentially involved two sense organs:
(A) Television (B) Radio
(C) Poster (D) All of the these

219. Reflected projector are used in:
(A) OHP
(B) Slide projector
(C) Lantern projector
(D) Epidiascope

220. Primary colours used in extension are:
(A) Red, Blue, Yellow
(B) Red, Blue, Green
(C) Red, Blue, Purple
(D) Red, yellow, green

221. Black and white colours are:
(A) Chromatic colours
(B) Achromatic colours
(C) Diligent colour
(D) Symbolic colours

222. The number of pages a leaflet usually will contain is:
(A) Two (B) Four
(C) Six (D) Eight

223. Non-photographic reconstruction of reality is:
(A) Illustration (B) Pictures
(C) Models (D) Diagrams

224. Exact visual recording of things is called:
(A) Illustration (B) Pictures
(C) Models (D) Photographs

225. The extension system used to provide precise and reliable scientific information in simple language about a simple practice is called:
(A) Leaflet (B) Literature
(C) News paper (D) Circular letter

226. The coverage of agricultural news in news paper is:
(A) 3 to 5 % (B) 20%
(C) 1 to 2 % (D) 50%

227. Radio is an example for:
(A) Hot media (B) Cold media
(C) Optical media (D) Fiber media

228. The effective rate of delivery for a radio talk is:
(A) 80-100 words/minute
(B) 120-140 words/minute
(C) 160-180 words/minute
(D) 200-220 words/minute

229. The best style of presentation for radio talks with the farmers:
(A) Lecture (B) Group discussion
(C) Interview (D) Songs

230. The knowledge base for a technology usually derives from:
(A) Basic research (B) Applied research
(C) Technology testing
(D) Technology development

231. Technology development process starts with:
(A) Technology testing
(B) Technology production
(C) Technology generation
(D) Technology dissemination

232. Technology generation mainly happens at:
(A) Research station
(B) Farmers fields
(C) Training centres
(D) Educational institutes

233. The function of extension starts from:
(A) Technology testing
(B) Technology production
(C) Technology integration
(D) Technology dissemination

234. Fine testing and packaging of technologies into specific recommendations is called:
(A) Technology testing
(B) Technology production
(C) Technology integration
(D) Technology dissemination

235. Linkage is the technological term used to denote communication that is:
(A) Inter-organizational
(B) Intra-organizational
(C) Inter – personal
(D) Intra-personal

236. The technology flow in the farmer first model is basically from:
(A) Agent to farmer
(B) Researcher to agent
(C) Farmer to farmer
(D) Researcher to farmer

237. The process of delivery of technology to farmers:
(A) Technology testing
(B) Technology production
(C) Technology integration
(D) Technology dissemination

238. Apart from introduction of technology, to make use of it, extension should concentrate:
(A) Adoption of indigenous ideas
(B) Mobilization of farmers
(C) Diffusion of technology
(D) Large farm demonstration

239. Evaluation of technology means:
(A) Value judgment (B) Valere
(C) Positive and negative
(D) None of these

240. A technology that represents a major im provement over the existing one embodies:
(A) Compatibility (B) Visibility
(C) Utility (D) Complexity

241. The first stage of on-farm research:
(A) Technology evaluation
(B) Technology refinement
(C) Technology verification
(D) Technology dissemination

242. The belief that technology causes changes in the society:
(A) Technological fundamentalism
(B) Technological materialism
(C) Technological structuralism
(D) Technological determinism

243. "If people utilize all the presently avail able knowledge, there will not be any hunger":
(A) Transfer of technology paradigm
(B) Emancipatory extension paradigm
(C) Informative extension paradigm
(D) Persuasive extension paradigm

244. Transfer of technology is related with:
(A) Local idea (B) Progressive idea
(C) Foreign idea (D) None of these

245. Farmers interact with:
(A) Many sources
(B) Few sources
(C) Intra personal sources
(D) Information sources

246. The collectivity of units which are functionally differentiated and engaged in joint problem solving:
(A) Social norms
(B) Social system
(C) Social organisation
(D) Social unit

247. Established behaviour patterns for the members of the social system:
(A) Norms (B) Values
(C) Attitude (D) Beliefs

248. The patternized social relationship among the members of a social system:
(A) Social structure
(B) Social system
(C) Social organization
(D) Social unit

249. The alterations in the structure and functions of the social system:
(A) Social change (B) Social movement
(C) Social alteration (D) Social interaction

250. The social change where in the source of the change is from within the system:
(A) Local change (B) Immanent change
(C) Localized change
(D) Direct change

251. Microanalysis approach in change analysis refers to:
(A) Change at individual level
(B) Change at family level
(C) Change at group level
(D) Change at social system level

252. Extension agent need to necessitate change initially on:
(A) Adoption only
(B) Knowledge and adoption
(C) Belief and value system
(D) Infrastructural facilities

253. Voluntary agencies are characterized by:
(A) Less acceptability
(B) Greater continuity
(C) Less commitment
(D) Less intensity

254. Essential characteristics of a primary group is:
(A) Face to face contact
(B) Personal and emotional relationship
(C) Interpersonal proximity
(D) Unity in diversity

255. A person who actually initiates the action in the community is:
(A) Professional leader
(B) Assured leader
(C) Operational leader
(D) Popular leader

256. The leader who is well trained and paid:
(A) Autocratic leader (B) Democratic leader
(C) Lay leader (D) Professional leader

257. Contact farmer can be considered as:
(A) Opinion leader (B) Professional leader
(C) Official leader (D) Lay leader

258. Which among the following find the greatest application in desire stage of extension teaching:
(A) Farm publication (B) Radio
(C) Exhibition (D) Mass media

259. Poster is most effective at which stage:
(A) Awareness (B) Evaluation
(C) Affection (D) Persuasion

260. The terminology that does to relates with the most innovative member of a social system:
(A) Drones (B) Ultra adopters
(C) Progressists (D) Light houses

261. The chairman of District Rural Development Agency:
(A) District Collector
(B) District Planning Officer
(C) District Panchayat President
(D) Block Development Officer

262. The authority responsible for planning and implementation of rural development programme at state level:
(A) Planning commission
(B) State Development Committee
(C) Ministry of Rural Development
(D) Chief Minister

263. The chairman of State Development Committee:
(A) Governor (B) Chief Minister
(C) Chief Secretary
(D) Minister of Rural Development

264. The Gandhian approach of rural development have its basic emphasis on:
(A) Village industries
(B) Development of inner man
(C) Local self government
(D) Production improvement

265. CDP differs from IADP as the former have:
(A) Area based approach
(B) Decentralized administration
(C) Degree of investment
(D) Multi purpose approach

266. The extension approach used under IADP was:
(A) Proscriptive (B) Perscriptive
(C) Participatory (D) Commercial

267. The base level extension functionary in T & V system:
(A) BDO
(B) VLW
(C) AEO
(D) SDAO

268. The new name of VLW is:
(A) VDO (B) VEO
(C) CDO (D) VDW

269. The ratio of extension agent to farmer in a typical T&V system is:
(A) 1-250 (B) 1-500
(C) 1-800 (D) 1-1500

270. Single line of command is one of the important feature of:
(A) T&V (B) CDP
(C) IAAP (D) IRDP

271. ABC of posters means:
(A) Attractive, Brief and Complete
(B) Attractive, Brief and Clear
(C) Attractive, Brief and Creative
(D) None of these

272. A blue print for action is:
(A) Plan (B) Programme
(C) Project (D) None of these

273. A codified system of mores is generally called:
(A) Norms (B) Laws
(C) Customs (D) Values

274. A critical activity must have:
(A) Positive float (B) Negative float
(C) Zero float (D) None of these

275. A five minute radio talk has:
(A) 400 words (B) 300 words
(C) 600 words (D) 800 words

276. A new story technology is can be published in:
(A) Extension bulletin
(B) Folder
(C) Newspaper
(D) Sports magazine

277. A poster has letter height of 22 inches. It can be clearly viewed up to the distance of:
(A) 10 feet (B) 20 feet
(C) 60 feet (D) 80 feet

278. A predetermined course of action is called:
(A) Objective (B) Project
(C) Plan (D) Management

279. A recognizable imitation or replica of an object whether workable or not is called:
(A) Chart (B) Specimen
(C) Mock-up (D) Model

280. A three dimentional recognizable imitation of an object is:
(A) Illustration (B) Model
(C) Mock up (D) Specimen

281. A working model is known as:
(A) Objects (B) Model
(C) Mockup (D) Poster

282. A visual aid has caption whose letter height is 0.5 inch. It can be visible clearly from:
(A) 10 Feet (B) 20 Feet
(C) 40 Feet (D) 60 Feet

283. Affective domain relates to:
(A) Knowledge (B) Speech
(C) Attitude (D) None of these

284. After a suitable extension teaching, the villagers attain desirable changes in his:
(A) Behaviour (B) Health
(C) Habits (D) All of these

285. Action research programme is considered as an example for:
(A) Formal training
(B) Non formal training
(C) Direct training
(D) Distant training

286. Action-reaction interdependence in communication is called:
(A) Credibility (B) Feed back
(C) Message treatment
(D) Fidelity

287. Age, sex and caste indicates:
(A) Social status (B) Class status
(C) Ascribed status (D) Achieved status

288. An extension worker while extending scientific knowledge to the farmers starts with:
(A) Practical demonstration
(B) Lectures
(C) Leaflets
(D) All of these

289. An innovation has relatively high degree of complexity, its adoption will be:
(A) Quick
(B) Fast
(C) Slow
(D) Cannot be said anything

290. Aristotle model of communication contains how many ingredients:
(A) Five (B) Four
(C) Three (D) Two

291. Any decision and/or action to be realistic has to be based on:
(A) Facts (B) Opinion
(C) Value judgement (D) Moral grounds

292. Andragogy deals with:
(A) Change
(B) Adult learning
(C) Society (D) Ancient civilization

293. An individual letter written by the extension worker to a farmer in connection with extension work is called:
(A) Circular (B) News letter
(C) Personal letter (D) None of these

294. At which level, communication is fact and problem oriented:
(A) Explorative level (B) Participative level
(C) Convenient level (D) None of these

295. At which of following level farmers information and advisory centres under ATMA operates:
(A) District level (B) Block level
(C) Tehsil level (D) Village level

296. At which of following level Jawahar Gram Samridhi Yojna is implemented:
(A) District panchayat
(B) Intermediate panchayat
(C) Village panchayat
(D) Nager panchayat

297. ATMA Governing board is a/an:
(A) Advisory body
(B) Statutory body
(C) Autonomous body
(D) Financial body

298. ATMA management committee consists of:
(A) 6 members (B) 12 members
(C) 18 members (D) 24 members

299. ATMA management meets:
(A) Twice in a month (B) Once in a month
(C) Once in a crop season
(D) Once in a year

300. Behaviour of an individual comprises of:
(A) Knowledge and Attitude
(B) Attitude and Skill
(C) Skill and Knowledge
(D) Knowledge, Attitude and Skill

301. Behavioral changes in human beings in extension education include:
(A) Changes in knowledge
(B) Changes in attitude
(C) Changes in skill (D) All of these

302. Campaign is an:
(A) Intensive teaching
(B) Extensive teaching
(C) Invasive teaching
(D) None of these

303. Communication behaviour essentially consists of:
(A) Information seeking
(B) Information processing
(C) Information dissemination
(D) All of these

304. Folkways which deals with moral aspects are:
(A) Taboos (B) Laws
(C) Mores (D) Norms

305. Folkways have:
(A) Moral validity
(B) Conventional validity
(C) Legal validity (D) All of these

306. In National Programme of Doordarshan, Krishi Darshan is broadcasted at:
(A) 7:00 p.m. to 7:30 p.m.
(B) 7:00 a.m. to 7:30 a.m.
(C) 7:30 a.m. to 8:00 a.m.
(D) 6:30 p.m. to 7:00 p.m.

307. Systems approach is also known as :
(A) Contingency theory
(B) Integrated theory
(C) Modern theory
(D) Ecological theory

308. System approach to training has been developed by:
(A) S. Tyson (B) Alfred York
(C) B. Richard (D) Sameul Paul

309. The ability to use the most effective means to reach a given goal is:
(A) Rationality (B) Objectivity
(C) Competency (D) Relativity

310. The accuracy or precision of a measuring instrument is:
(A) Validity (B) Measurement
(C) Reliability (D) Readability

311. The amount of information that can be transmitted per unit of time is called:
(A) Channel capacity
(B) Rate of transmission
(C) Transmission power
(D) Transmission ratio

312. The approach to leadership which assumes that " leaders are born and not made" is:
(A) Innate approach (B) Trait approach
(C) Contingency approach
(D) Behavioral approach

313. The communication of innovations:
(A) Involves a time factor
(B) Does not involve time factor
(C) May or may not involve time factor
(D) Has no relevance

314. The component of message is/are:
(A) Code (B) Content
(C) Treatment (D) All of above

315. The concept of feedback in communication is generally used to reflect:
(A) Resource orientation
(B) Receiver orientation
(C) Process orientation
(D) All of these

316. The essence of the scalar principle is:
(A) Unity of command (B) Span of control
(C) Communication (D) Coordination

317. The first line extension system mainly comprises of:
(A) ICAR and SAUs (B) NGO
(C) Ministry of Agriculture and Cooperation
(D) Ministry of Rural Development

318. The shortest possible time to perform an activity, assuming that everything goes well is:
(A) Optimistic time (B) Pessimistic time
(C) Most likely time (D) None of these

319. The social network model ideally suited or imparting knowledge change:
(A) Star (B) Wheel
(C) Circle (D) Slash

320. There are how many basic social processes:
(A) 5 (B) 4
(C) 3 (D) 2

321. DASP stands for:
(A) Development Agriculture Service Project
(B) Diversified Agriculture Support Project
(C) Diversified Agriculture Service Project
(D) Development Agriculture Support Project

322. Farmers and parliament is published from:
(A) Pantnagar (B) New Delhi
(C) Hisar (D) Lucknow

323. Integrated crop management is a combination of:
(1) Integrated nutrient management
(2) Integrated water management
(3) Integrated Weed management
(4) Integrated pest management

Correct answer is:
(A)1 and 2 (B) 2 and 3
(C)1, 3 and 4 (D) 1, 2, 3, and 4

324. Haryali yojna is related to:
(A) Food grain production
(B) Fertilizer production
(C) Irrigation
(D) Grass management

325. New name of ATMA is:
(A) National Agricultural Extension and Technology Mission
(B) National Food Security Mission
(C) Agriculture Technology Management Agency
(D) Integrated Scheme for Oilseeds, pulses, palmoil and maize

---

True or False

326. In India community development projects were launched in 1952.
327. IRDI was launched in 1980
328. Under NREP, the state government was its share 40%.
329. National extension service in India was established in 1953.
330. Programme planning is an educational process.
331. ICAR was launched in 1934.
332. Etawah programme was organized by Albert Mayer in 1945.
333. Radio is a media of mass communication.
334. Television includes visual device only.
335. The rate of an extension worker in extension education is to create an effective awareness.
336. The SFDA and MFAL programme were started in 1965.
337. The Taccavi loan is given by Govt.
338. Agriculture Minister is the chairman of SFDA.
339. Zamidari system was introduce by Mughals.
340. Gurgaon Programme was started by R.N. Tagore.
341. Shriniketan 1921 programme was started by Mahatma Gandhi.
342. Major 14 banks were nationalized in the year 1975.
343. The T and V system was proposed by Dr. Daniel Benor.
344. Lead Bank scheme was introduced in the year 1969.
345. Krishi Vigyan Kendra's are sponsored by IARI.
346. Integrated Rural Development Programme was launched in 1978.
347. The purpose of introducing agricultural universities in India is to improve the standard of agriculture through extension, teaching and research.
348. In NERP the share of centre and state will be 50:50.
349. The first agricultural university of India was established in 1960.
350. A poster as a visual aid is always used alone.
351. National Demonstration Programme was started in 1958 In India.
352. Training of Rural Youth for Self Employment (TRYSEM) was launched in 1979.
353. First KVK was started at Pondichery in 1974.
354. Demonstration means showing by doing.
355. Method demonstration is long type demonstration.
356. The father of method demonstration was Dr. Seeman A Knapp.

357. Result demonstration is long type demonstration.
358. Puppet, drama, cinema is include in audio-aids.
359. Teaching and learning is the process of communication.
360. Communication is derived from Greek word.
361. Production of desirable changes in human behaviour through gaining desirable knowledge, attitude and mental and manual skill is called education.
362. Extension activity was started first time in USA.
363 Planning is decision making and involved selecting and integrating the courses of action that an organization will follow to attain its objectives.
364. Bulletins, leaflets, posters and radio are the media of group communication method.
365. A extension service is only concerned with the transfer of technology.
366. Dryland farming is the key to stability of agricultural production in India.
367. Education is the production of desirable changes in human behaviour.
368. Panchayati Raj System was established in India in the year 1957.
369. "Cone of experience" deviced by A. Reddy.
370. Democratic Decentralization recommended by a committee headed by B.R. Mehta.
371. National Extension service was started in the year 1958.
372. National demonstration was started in the year 1964
373. The pilot project for rural development was started in 1948.
374. IRDP was launched in 1978.
375. The scheme of Regional Rural Banks was launched in India on Oct. 1965.
376. Programme planning is a spontaneous process.
377. A good extension programme should be rigid.
378. Rural Development Programme should be formed to meet long term objectives.
379. Orissa has no Panchayati Raj.
380. B.D.O. is responsible for village women education about rural development programme.
381. Farm and home visit are the method of individual communication.
382. The purpose of introducing agricultural universities in India is to improve the standard of agricultural extension.
383. Basic operational unit for rural development in India is Block.
384. Function of pre extension stage block was to conduct agricultural demonstration.
385. The period of stage II of a community development block was 5 year.
386. In India the use of television as a medium of communication of rural development was started in 1961.
387. In India there are about 175600 village panchayats.
388. Programme planning is a co-ordinating process.
389. Village Panchayat is Headed by V.L.W.
390. The head of the Block is B.D.O.
391. Village Panchayat is non-official body.
392. National demonstration were started in 1965.
393. Block Samiti are guided by Zila Parishads.
394. In the normal community development block the number of V.L.W. to V.D.O.

Fill in the Blanks:

395. Co-operative movement were started in the year——.
396. Etawah Project was started in the year of————-
397. Firka development scheme was started in the year ——
398. Nilokheri Project was started in the year———— by————
399. Etawah Programme was organized by————
400. Bhoodan movement was started in the year————
401. IRDP was launched in————
402. Basic operational unit for rural development in India is————
403. Antyodaya Programme was started in the year————
404. Command Area Development Programme were started in ———-
405. The SFDA and MFAL Programme were started in————
406. T and V system is introduced in———— in————
407. Demonstration means showing by————
408. The father of method demonstrations was————
409. Method demonstration is———— demonstration.
410. Result demonstration is———— demonstration.
411. National demonstration was first time started in 1965 by————
412. T and V system is also called——— and was evolved by————
413. The extension activity was started first time in————
414. The word "Learning by doing" was coined by———— in————
415. Adoption is an————
416. Elements of adoption process————
417. Bulletins, leaflets, posters and radio are the media of————
418. ——— effective communication.
419. Radio belong to the———— word is a means of communication.
420. Written form of extension teaching methods does not include————
421. The basic principle of demonstration is learning by————
422. ———— is a composite of both method and result demonstration
423. Teaching and learning is the process of————
424. "Communication" is derived from————.
425. The extension work must be based on————
426. A good extension service is an process of————
427. The fundamental objective of extension education——
428. The technical message are carefully selected and are relayed during the visit to the——— ————
429. Extension workers are————
430. IRDP was launched in———
431. Krishi Vigyan Kendra's are sponsored by————
432. "Panchayati Raj" was started in the year————
433. A good extension programme should be————

434. An extension education job is ———
435. The method of extension education for effectively impressing the people about a particular recommended practice suitable for a particular area is ———
436. The chairman of SFDA is ————————
437. 20 Point programme was started by ——— in the year ——
438. High yielding variety programme was started in the year ———— by the Govt. of India.
439. Panchayati Raj was first started at ———— in the year ————
440. The non-projected aid can be divided into ———— and ————
441. The District Rural Development Agency has a governing body headed by ——
442. National Agricultural Research Project was launched in January ——— and is funded by the ————
443. Aim of extension work is to bring about the changes———
444. The National Agricultural Extension Project is being launched by the ———
445. Sevagram attempt was started under guidance of ————
446. ———————— is energy mobilization towards the attainment of the goal.
447. The advantage of the —— projector is that a leader can always face the class.
448. Under TRYSEM the proposal is to train about ———— youth every year in various skills.
449. The new name of community development programme is ———
450. Gurgaon experiment was first started by F.L.——
451. A good extension programme should be——————-
452. The Intensive agricultural District programme as popularly known as a ————
453. The —— and ————————— programmes wee started in 1970.
454. The Taccavi loan is given by ——
455. The ninth five year plan was started in ————
456. The principle "one village one society" was suggested by ——————
457. Chairman of planning commission is ————
458. Central co-operative bank operates normally at ———— level.
459. ——— is the degree of positive or negative effect associated with some physiological object.
460. M.E.S. was established in India in the year————
461. The effective method for the success of extension education in rural areas depends on ——
——
462. There are ——— degrees of evaluation.
463. Desert Development Programme was started in ————
464. In a normal community development block, the number of V.L.W. is ———
465. Prestige is the evaluation of the ————
466. The new name of V.L.W. is ——
467. ———— is the gap between what is and what ought to be.
468. To prove the worth of a nine practice suited to a particular are, the best method of extension education ————
469. Radio is a media of ————————
470. T and V system includes————————

471. ——— is the statement of situation objectives problems and solution.
472. NABARD was set up in ————
473. Caste is a gained by ————
474. "Cone of experience" was devised by ————
475. ———— is an assignment of values to alternatives.
476. Family is included under ———

View points:

477. Rural participatory appraisal is related with agriculture extension.
478. Success of rural projects depend upon partnership of beneficiaries.
479. The comilla project approach was first used by the academy of rural development at comilla in Bangladesh
480. The principle of "one village one society" was suggested by Meelogen committee.
481. The rural system research idea was motivated by Dr. M.S. Swaminathan in 1988. A round table discussion is called panel.
482. ATMA (Agriculture Technology Management Agency) is a registered society associated with IVLP (Institutional Village Linkage Programme).
483. The basic unit of development under I.R.D. programme is family.
484. Extension workers should be well versed with economic principles without it they cannot give effective advice from business point of view.
485. The commission recommending the minimum support prices for farmers is known as com mission for agriculture costs and prices.
486. In India, community development projects were launched in 1952.
487. The T and V system was proposed by Dr. Daniel Benor.
488. TRYSEM scheme was started in 15th August, 1979.
489. Integrated Rural Development Programme was started in 2nd October, 1980.
490. The extension education aim at changing knowledge, skills and attitudes.
491. Extension education is different from formal education in a sense that we start from practicals and end up with theory.
492. Krishi Vigyan Kendra (KVKs) can only be under control of agricultural university and voluntary organization.
493. The 'Lab to Land' programme also included demonstration but they were different from early demonstrations in terms of size of demonstration.
494. District credit plan is prepared by Lead Bank.
495. UPDASP is sponsored by World Bank.
496. Kisan mitra is an employee of State Govt.
497. Puppet is a traditional and simple aid of communication
498. August Compte is the father of sociology.
499. Scope of sociology is studied community.
500. The term "human communication" means exchange of ideas.

501. The term "inter-communication" means communication on any issue, communication be tween two persons, communication between various groups.
502. Evaluation done by the agency is known as external evaluation
503. In extension education exhibition is helpful for mass communication.
504. The method of establishing "mass contracts" is Bulletin.
505. Extension education is a educational process.
506. Knowledge, skill, attitude is the element of behaviour.
507. Family is included under primary group.
508. The process of initiating a conscious and purposeful action is known as motivation.
509. Kisan mandal system under T and V system was initiated in the year of 1993.
510. Rural Development programme at district level is being looked by DRDA.
511. In extension education communication is a expression of thoughts.
512. Method demonstration is the most effective approach of the extension for the group of people.
513. The Firka development project was first time introduced in the state of Tamilnadu.
514. Training and Visit programme was first time introduced in Rajasthan and MP
515. The word 'tensio' in extension is derived from Latin word.
516. J. Paul Leagons is a father of extension education.
517. First model of communication was given by Berlo.
518. Jawahar Rozgar Yojna was introduced in April, 1989.
519. National Agriculture Research Project was launched by DCAR for location specific research in the year Ist January1979.
520. T and V system was introduced with the assistance of world bank.
521. Socio-metric method is best suited for selection of leader.
522. The main purpose of T and V programme is to help farmers in the adoption of latest innovation in agriculture
523. Trait is made of behaviour.
524. Democratic decentralization was recommended by B,.R. Mehta.
525. The fundamental objective of community development is overall development of the people.
526. Extension is helping the people who help themselves.
527. The goal of teaching is to bring about the desire change in the behaviour of the "Audience".
528. Social culture is composed of group.
529. Motivation is known as oal directed or goal seeking behaviour.
530. The state of being moved or stirred up feeling in one way of the other is called motion.

Chapter 2

HISTORICAL PERSPECTIVES OF AGRICULTURE AND EXTENSION

1. Which one of the following is the correct match?
(A) Etawah pilot project -Albert Mayer (1948)
(B) Grow more food campaign - J.L. Nehru (1947)
(C) Co-operative movement- Nicholas (1904)
(D) Sewagram project- M.K.G andhi (1927)

2. The first KVK was established in 1974 at:
(A) Nagpur (B) Nilokheri
(B) Ludhiana (D) Pondichery

3. Which of the following rural development project was launched before independence?
(A) Etawah pilot project
(B) Co-operative movement
(C) Community development
(D) National Extension service

4. Lab to Land programme was stared by:
(A) NBSSLUP (B) Govt. of U.P.
(C) Smt. Indira Gandhi
(D) ICAR

5. National Academy of Agriculture Research Management (NAARM) is located at:
(A) New Delhi (B) Hyderabad
(C) Bangalore (D) Cuttack

6. Community development project was started in India in:
(A) 1947 (B) 1949
(C) 1952 (D) 1953

7. T and V system of extension was started by:
(A) D.Benor (B) M.Jackson
(C) M.Anderson (D) O.P.Dhama

8. High yielding variety programme was started in the year:
(A) 1960 (B) 1963
(C) 1966 (D) 1968

9. "Kissan Bharti" periodical is published from:
(A) Delhi (B) Lucknow
(C) Hisar (D) Pant Nagar

10. Gurgaon project was organized by:
(A) Mahatma Gandhi
(B) F.L. Bryne
(C) Hatch (D) Albert Mayer

11. Sevagram attempt was started under the supervision of:
(A) F.L. Bryne (B) R.N. Tagore
(C) J.L. Nehru (D) M.K. Gandhi

12. Shriniketan attempt was started by R.N. Tagore in collaboration with Elmhirst in:
(A) Uttar Pradesh (B) Bengal
(C) Travancore (D) Etawah

13. Marthandum attempt was started by:
(A) Howard (B) Hatch
(C) Albert Mayer (D) F.L. Bryne

14. Which one of the following pairs is not correct:

Attempt	Year of start
(A) Gurgaon	1920
(B) Sevagram	1920
(C) Marthandum	1921
(D) Etawah	1950

15. NREP was started in the year:
(A) 1977 (B) 1978
(C) 1979 (D) 1982

16. The government sponsored Firka development scheme of Madras state was launched in:
(A) 1956 (B) 1946-47
(C) 1976 (D) 1980

17. The first KVK, Pondicherry was established in 1974 related with:
(A) RAU, Pusa
(B) TAU, Coimbatore
(C) AAU, Hyderabad
(D) ICAR

18. Lab to land programme was launched by the ICAR as a part of its————jubilee celebration in 1979:
(A) Centenary (B) Platinum
(C) Golden (D) Silver

19. The Rural System Research idea was motivated by M.S.Swaminathan in:
(A) 1978 (B) 1988
(C) 1998 (D) 2004

20. The government of India set up Planning Commission is:
(A) 1948 (B) 1950
(C) 1952 (D) 1954

21. The Chairman of Planning Commission is:
(A) President (B) Prime Minister
(C) Finance Minister
(D) Speaker of Lok Sabha

22. Operation Research Projects (ORP) were initiated in:
(A) 1960 – 61 (B) 1966 – 67
(C) 1970 – 71 (D) 1975 – 76

23. State Agricultural Universities in India were setup on the pattern of Land Grant College of:
(A) U.K. (B) France
(C) Italy (D) U.S.A.

24. Potato famine of Ireland occurred in:
(A) 1645 (B) 1745
(C) 1845 (D) 1854

25. Potato famine of Ireland was caused by the attack of fungus
(A) *Phytophthora palmivora*
(B) *Plasmopora viticola*
(C) *Puccinia graminis*
(D) *Phytophthora infestans*

26. Natural resistance to late blight of potato is present in:
(A) *Solanum tuberosum*
(B) *Solanum acaule*
(C) *S. demissum*
(D) *S. stoloniferum*

27. Gene bank are part of:
(A) Ex situ conservation
(B) In situ conservation
(C) Both ex situ and in situ conservation
(D) Tribal diet

28. *In situ* conservation are carried out through:
(A) Natural parks
(B) Wildlife santuaries
(C) Biosphere reserves
(D) All the these

29. *Ex situ*conservation is carried out through:
(A) Funneling the threatened species into trade
(B) Offsite collection
(C) Gene banks (D) All the these

30. Gene banks comprise:
(A) Seed banks, orchards, tissue culture and cryopreservation
(B) Seed banks, tissue culture and cryopreservation
(C) Seed banks and cryopreservation
(D) Tissue culture and cryopreservation

31. Orthodox seeds can be preserved for long periods at a temperature of:
(A) -196 ^{0}C (B) -10 to -20 ^{0}C
(C) 0 – 4 ^{0}C (D) 10 to 15 ^{0}C

32. Dwarfing gene of wheat is:
(A) Pal -1 (B) Atomita – 1
(C) Norin – 10 (D) Pelita -2

33. Jojoba yields wax for cosmetics from its:
(A) Stem (B) Leaves
(C) Seeds (D) Fruit pulp

34. IR 36 was developed through breeding of:
(A) Six rice varieties and *Oryza nivara*
(B) 13 rice varieties and *Oryza nivara*
(C) *Oryza sativa* and *O. indica*
(D) *Oryza indica* and *O. nivara*

35. Which one is beverage plant:
(A) *Ilex paraguriensis*
(B) *Caltha edulis*
(C) *Poullinia cupana*
(D) All of these

36. The three top most major crops of the world in the order of total production are:
(A) Wheat>Rice > Maize
(B) Rice>Wheat>Maize
(C) Wheat>Maize>Rice
(D) Rice>Maize>Wheat

37. Potential food plant is:
(A) Winged bean (Psophocarpus tetragonolobus)
(B) Jajoba (*Simmondsia chinensis)*
(C) Guayule (*Parthenium argentatum)*
(D) *Leucaena*

38. A potential source of rubber is:
(A) Guayule (*Parthenium argentatum*)
(B) *Leucaena leucocephala*
(C) *Cucurbita foetidissima*
(D) *Citrullus colocynthis*

39. Cola, the common beverage, is obtained from:
(A) Seeds (B) Flowers
(C) Leaves (D) Stems

40. Maximum genetic diversity of crop plants occurs where agriculture is:
(A) Run on commercial lines
(B) Advanced
(C) Rainfed
(D) Primitive

41. Which is more suitable for germplasm collection:
(A) Centre of production
(B) Homelands
(C) Tropical areas
(D) Temperate areas

42. Which one will be scarched for obtaining maximum genetic diversity:
(A) Recent varieties
(B) Old varieties
(C) Wild relatives
(D) All the these

43. Paraguay Tea is got from:
(A) Leaves of *Ilex*
(B) Seeds of *Ilex*
(C) Leaves of *Paullinia*
(D) Paste of crushed kernels of *Paullinia*

44. Arabian Tea is:
(A) Leaves and buds of *Paullinia*
(B) Seed of Ilex
(C) Leaves and buds of *Catha edulis*
(D) Seeds of Catha

45. Cryopreservation is carried out at:
(A) -10 to -20 ^{0} (B) -50 to -60 ^{0}C
(C) -196 ^{0}C (D) -100 ^{0}C

46. Co-varieties of sugarcane obtained red rot resistance from:
(A) *Saccharum munja*
(B) *S. spontaneum*
(C) *S. arundinacearum*
(D) *S. edula*

47. A calcitrant seed will be killed in:
(A) Low temperature (B) Dry air
(C) Further dehydration
(D) All the these

48. A gift from new world to old world is:
(A) Barley (B) Rice
(C) Wheat (D) Maize

49. Which plant was introduced from old to new world:
(A) Potato (B) Rice
(C) Corn (D) Sweet potato

50. Dwarf wheat varieties brought into India from Mexico were:
(A) Sonora-64 and Sonalika
(B) Sharbati Sonora and Pusa Lerma
(C) Sonalika and Sharbati Sonora
(D) Sonora-64 and Lerma Rojo-64

51. Vavilov centres of origin of crop plants are located in:
(A) Tropical plains
(B) Mountains of tropical areas
(C) Mountains of temperate areas
(D) Both (B) and (C)

52. An under utilized oil crop is:
(A) *Brassica campestris*
(B) *Cocos nucifera*
(C) *Helianthus annus*
(D) *Azadiracta indica*

53. Liquid wax is got from:
(A) Jajoba (B) Winged Bean
(C) Guayule (D) Leucaena

54. Wheat varieties collected by Vavilov were:
(A) 26000 (B) 2600
(C) 4600 (D) 260

55. Hevea Rubber belongs to:
(A) Malayasia (B) Peruvial Andes
(C) Brazil (D) South east Asia

56. Germplasm is contained in:
(A) Somatic cells
(B) Growing points
(C) Reproductive cells
(D) Both (A) and (C)

57. Which one will be best for germplasm collection:
(A) Britain (B) France
(C) Portugal (D) Peru

58. Dwarf wheat were evolved by:
(A) Swaminathan (B) Vavilov
(C) Khush (D) Borlaug

59. In a gene bank, genetic material is stored as:
(A) Genes (B) Seeds
(C) Vegetative material/tissue culture
(D) Both (B) and (C)

60. Genetic erosion is due to:
(A) Deforestation
(B) Shifting cultivation
(C) Adoption of genetically uniform varieties
(D)All the above

61. Exchange of germplasm is carried mostly through shoot tips because it is:
(A) Cheap (B) Small and handy
(C) Virus free (D) Genetically stable

62. Tomato and potato belong to:
(A) North America
(B) South America
(C) Asia minor
(D) Ethiopia

63. A plant introduced from new world to old world is:
(A) Rice (B) Wheat
(C) Sugarcane (D) Potato

64. National Botanical Research Institute is located at:
(A) Dehradun (B) Delhi
(C) Lucknow (D) Mumbai

65. In India, the largest herbarium is located at:
(A) Dehradun (B) Delhi
(C) Lucknow (D) Calcutta

66. Which of the following countries is the centre of origin of rice:
(A) India (B) China
(C) Indo-Malayan region
(D) Russia

67. Central Sugarcane breeding Research In stitute is situated at:
(A) Coimbatore (B) Lucknow
(C) Delhi (D) Bhopal

68. A new world spice that has become an essential part of Indian cuisine is:
(A) Red pepper (B) Black Pepper
(C) Cardamom (D) Ginger

69. A social forest tree is:
(A) Teak (B) Sandal wood
(C) Subabul (D) Rosewood

70. A plant species threatened with extinction can be preserved in:
(A) Herbarium (B) National park
(C) Botanical Garden
(D) Gene Bank

71. A living fossil is:
(A) Gnetum (B) Ginkgo
(C) Riccia (D) Pinus

72. Jojoba is:
(A) Crotolaria (B) Leucaena
(C) Simmondsia (D) Parthenium

73. A place where endangered genetic materials of plants are kept is:
(A) Gene library (B) Gene bank
(C) Gene pool (D) Herbarium

74. Wild varieties of plants must be conserved to:
(A) Feeding wild animals
(B) Maintain ecosystem
(C) Future evolution
(D) Incorporate useful traits in future crop varieties

75. The energy required daily by hunter gatherer humans was:
(A) 2000-4000 kcal (B) 6000-10000 kcal
(C) 10000-70000 kcal (D) 10000-12000 kcal

76. Per capita daily energy consumption in nineteenth century industrialized society was:
(A) 6000 k cal (B) 10000 kcal
(C) 30000 kcal (D) 70000 kcal

77, Per capita daily energy consumption of USA is:
(A) 10000 kcal (B) 30000 kcal
(C) 70000 k cal (D) 250000 kcal

78. Which is the most important crop in the world:
(A) Maize (B) Sorghum
(C) Barley (D) Wheat

79. The great popularity of wheat is on accounts of:
(A) Well balanced carbohydrates and proteins
(B) Inability for complete mechanized production
(C) Short storage period of grain and flour storage
(D) Limited to narrow range of climate

80. According to Vavilov , the common bread wheat originated in :
(A) The Caucasus-Turkey-Iraq area
(B) Temperate zone of America
(C) Naturally fertile regions of France
(D) Cool areas of China

81. *Acacia senegal* and *Ahlagi camelorum* are the xerophytes belonging to:
(A) Psammophytes (B) Psychrophytes
(C) Lithophytes (D) Halophytes

82. *Ceina parviflora* is a xerophyte that has the succulent:
(A) Stems (B) Leaves
(C) Roots (D) Fruits

83. "The dominant species with maximum canopy coverage of two or more species" represents a community character known as:
(A) Society (B) Location
(C) Consociation (D) Association

84. In —— the buds (life forms) are invariably buried in the soil or the substratum as the bulbs and rhizomes:
(A) Chamaephytes (B) Crypto/geophytes
(C) Hemicryptophytes (D) Therophytes

85. Collective farming system has originated from:
(A) Bulgaria (B) Poland
(C) Soviet Union (D) France

86. Commercial plantation farming comprises an:
(A) Import-oriented system
(B) Export-oriented system
(C) Indigenous use-oriented system
(D) None of these

87. Agricultural growth is very soon likely to be hampered by scarcity of:
(A) Natural resources
(B) Man-made resources
(C) International cooperation
(D) Local coordination

88. A set of shared, similar characteristics (as applied to logically deduced and symbolically expressed models and observations) may be called as:
(A) Negative isomorphism
(B) Positive isomorphism
(C) Neutral isomorphism
(D) None of these

89. Rubber has probably the —— range of tolerance for temperature:
(A) Narrow west (B) Widest
(C) Medium (D) None of these

90. The requirement and tolerance of a species may be considered as:
(A) Supplementary (B) Complementary
(C) Antagonistic (D) None of these

91. The cactus from North America and the jack rabbit from Europe were introduced into:
(A) Holland (B) Poland
(C) Austria (D) Australia

92. When competition effect is small for both the plant species so that the interspecific-effects are ——— the intraspecific effects, both species may be able to live together:
(A) More than (B) Equal to
(C) Less than (D) None of these

93. The self-crowding effect represent the ———— competition:
(A) Interspecific (B) Intraspecific
(C) Intergeneric (D) Intrageneric

94. The interplant competition drastically affects the ——— of the plant:
(A) Size and Shape
(B) Number and Size
(C) Number and Shape
(D) None of these

95. Banana (*Musa sapientum*) plant is an herbaceous ———— which may grow to a height of about 30 feet:
(A) Annual (B) Biennial
(C) Perennial (D) None of these

96. ———— has been the world's leading producer of cane sugar:
(A) India (B) Indonesia
(C) Hawaii (D) Cuba

97. Cotton has been grown in India as a crop of fibre for perhaps ——— years and was known in Peru as early as 1000 BC:
(A) 3000 (B) 4000
(C) 5000 (D) 6000

98. When one batch species react with the environment to facilitate the entry of a new set of species, such a model is known as:
(A) Inhibition model
(B) Tolerance model
(C) Facilitation model
(D) Elucidation model

99. The centre of origin of wheat is:
(A) South western Asia
(B) South-eastern Asia
(C) India (D) China

100. High yielding dwarf varieties of wheat were developed by:
(A) Dr. B.P. Pal
(B) Dr. Norman E. Borlaug
(C) Dr. E.W. Burton
(D) Dr. M.S. Swaminathan

101. The maximum area under wheat is in:
(A) USSR (east while) (B) India
(C) USA (D) China

102. The production of wheat is the highest in:
(A) USSR (east while) (B) India
(C) China (D) USA

103. The *Triticum aestivum* is a wheat:
(A) Diploid (B) Tetraploid
(C) Hexaploid (D) Monoploid

104. Maize protein is deficient in:
(A) Betadine (B) Tryptophane and Lysine
(C) Lysine (D) Tryptophane

105. The country ranking first in respect of area under maize (corn) is:
(A) India (B) China
(C) Brazil (D) USA

106. The largest maize producing country in the world is:
(A) USA (B) Mexico
(C) India (D) Canada

107. The dent corn is also known as:
(A) *Zea mays indentata*
(B) *Zea mays indurate*
(C) *Zea mays tunicate*
(D) *Zea mays everta*

108. All India Co-ordinated Maize Improvement Project was started from:
(A) 1957 (B) 1955
(C) 1965 (D) 1967

109. The place of origin of urdbean is:
(A) India (B) Burma
(C) Bangladesh (D) Ceylon

110. The scientific name of early maturing bold and black seeded urdbean varieties is:
(A) *Vigna mungo var viridis*
(B) *Vigna mungo var niger*
(C) *Vigna mungo var macropode*
(D) *Vigna mungo var micropode*

111. Soybean besides good quality protein also contains about———— per cent of oil:
(A) 10 (B) 20
(C) 30 (D) 40

112. The protein content of soybean is about:
(A) 40 % (B) 20%
(C) 30% (D) 10%

113. The grain legume which is also grown as an oilseed is:
(A) Chickpea (B) Urdbean
(C) Lentil (D) Soybean

114. The ————is considered to be a progenitor of soybean:
(A) *Glycine max*
(B) *Glycine sinensis*
(C) *Glycine ussuriensis*
(D) *Glycine usitattisimum*

115. Among the oilseed crops, groundnuts ranks ———— in India:
(A) First (B) Second
(C) Third (D) Fourth

116. Groundnut seed normally contains percent protein and oil:
(A) 25 and 45 (B) 22 and 43
(C) 45 and 26 (D) 43 and 22

117. Quadricarpellatum capsule of sesamum contains how many carpels:
(A) Two (B) Three
(C) Four (D) Six

118. Sesamum belongs to the family of:
(A) Leguminoceae (B) Papilionaceae
(C) Pedaliaceae (D) Chinipideaceae

119. In south India sesamum is grown in:
(A) Rabi season only
(B) Kharif season only
(C) Rabi and kharif season
(D) Summer season only

120. Castor belongs to family:
(A) Compositeae (B) Euphorbiaceae
(C) Legiminoceae (D) Gramineae

121. The largest producer of rapeseed and mustard in the world is:
(A) China (B) Japan
(C) India (D) Canada

122. Rapeseed and mustard belongs to the family:
(A) Leguminoceaea (B) Compositeae
(C) Euphorbiaceae (D) Crucifereae

123. The 2n chromosome number in Rugosa is:
(A) 20 (B) 24 (C) 32 (D) 36

124. The fruit of rapeseed and mustard is known as:
(A) Pod (B) Silique (C) Grain (D) Caryopsis

125. The scientific name of erect type of ground nut is:
(A) *Arachis hypogaea* subspecies *fastigiata*
(B) *Arachis hypogeae* subspecies *procumbens*
(C) *Arachis hypogeae* subspecies *oleifera*
(D) *Arachis hypogeae* subspecies *gigantean*

126. The gynophore of groundnut is commonly referred to as the:
(A) Perinth (B) Peg
(C) Ovate (D) Monadelph

127. Linseed can be used as a crop:
(A) Oilseed (B) Fibre
(C) Oilseed and fibre (D) None of these

128. The protein content in linseed cake has been estimated at:
(A) 12% (B) 24%
(C) 36% (D) 48%

129. Linseed crop belongs to the family:
(A) Compositeae (B) Euphorbiaceae
(C) Leguminoceae (D) Linaceae

130. The fruit of linseed is known as:
(A) Pod (B) Canum
(C) Seedball (D) Dutum

131. Sunflower as an oilseed crop was introduced into India in:
(A) 1965 (B) 1969
(C) 1972 (D) 1985

132. Sunflower is a :
(A) Self pollinated crop
(B) Cross-pollinated crop
(C) Self and cross pollinated crop
(D) Parthenocarpous

133. Cotton belongs to the family of:
(A) Malvaceae (B) Compositeae
(C) Cricifereae (D) Lineaceae

134. The number of known species in genus Gossypium have been estimated at:
(A) 10 (B) 15
(C) 20 (D) 25

135. Jute belongs to the family of:
(A) Malvaceae (B) Compositeae
(C) Lineaceae (D) Tiliaceae

136. Tobacco has been classified as a crop:
(A) Cash (B) Catch
(C) Inter (D) Relay

137. *Nicotiana rustica* has been derived as an amphidiploid from a cross between:
(A) *Nicotiana tabacum* x *Nicotiana paniculata*
(B) *Nicotiana paniculata* x *Nicotiana sylvestris*
(C) *Nicotiana undulata* x *Nicotiana sylvestris*
(D) *Nicotiana undulata* x *Nicotiana paniculata*

138. Tobacco belongs to the family:
(A) Compositeae (B) Solanaceae
(C) Crucifereae (D) Leguminaceae

139. Nicotine content of the leaves of *Nicotiana rustica* ranges between:
(A) 2.0 – 5.5 (B) 3.0-4.5
(C) 3.5-8.0 (D) 5.0 – 10.5

140. Nicotine content of the leaves of *Nicotiana tabacum* ranges between:
(A) 2.0 – 5.5 (B) 3.0-4.5
(C) 3.5-8.0 (D) 5.0 – 10.5

141. Out of the following ———— is commonly cultivated potato:
(A) *Solanum demissum*
(B) *Solanum stenotonum*
(C) *Solanum tuberosum*
(D) *Solanum andigenum*

142. The family of sugarbeet is a:
(A) Chenopodiaceae
(B) Crucifereae
(C) Euphorbiaceae
(D) Compositeae

143. A tuber represents:
(A) Enlarged underground root
(B) Enlarged underground stem
(C) Enlarged underground nodule
(D) None of these

144. Potato belongs to the family:
(A) Solanaceae (B) Compositeae
(C) Leguminaceae (D) Euphorbiaceae

145. Department of Revenue, agriculture and Commerce was set up in India on:
(A) April 17, 1871 (B) April 27, 1871
(C) April 27, 1872 (D) April 17, 1872

146. Agriculture colleges were opened in India first of all in year:
(A) 1905 (B) 1900
(C) 1907 (D) 1911

147. The Department of Revenue, Agriculture and Commerce was set up during the reign of:
(A) Lord Curzon (B) Lord Mayo
(C) Lord Wellington (D) None of these

148. Department of agriculture in provinces were set up in:
(A) 1871 (B) 1881
(C) 1891 (D) 1911

149. First Inspector General of Agriculture was appointed in year:
(A) 1901 (B) 1902
(C) 1900 (D) 1907

150. Imperial Agricultural Research Institute was set up in Pusa, Bihar in year:
(A) 1905 (B) 1901
(C) 1871 (D) 1872

151. IARI was shifted to New Delhi in year:
(A) 1925 (B) 1926
(C) 1936 (D) 1947

152. Imperial Council of Agricultural Research was set up in a year of:
(A) 1901 (B) 1926
(C) 1929 (D) 1936

153. Who was the first president of ICAR:
(A) Dr. B.P. Pal
(B) Khan Bahadur Sir Mod. Habibullah
(C) Dr. M.S. Swaminathan
(D) Dr. Panjab Singh

154. Indian Council of Agricultural Research came into existence in:
(A) 1929 (B) 1936
(C) 1946 (D) 1947

155. Total net sown area of India is:
(A) 120 mha (B) 142 mha
(C) 180 mha (D) 95 mha

156. Total cropped area of India is:
(A) 120 mha (B) 142 mha
(C) 180 mha (D) 95 mha

157. The headquarter of National Bureau of Soil Survey and Land Use Planning is situated at:
(A) Nagpur
(B) Pune
(C) New Delhi
(D Hyderabad

158. The headquarter of National Bureau of Plant Genetic Resources is situated at:
(A) Nagpur (B) Pune
(C) New Delhi (D) Hyderabad

159. Match the name of ICAR institute and their headquarters:

Institutes	**Headquarter**
1. IGFRI	i. Jhansi
2. CAZRI	ii. Cuttack
3. CRRI	iii. Jodhpur
4. IISR	iv. Lucknow

Which of the match is correct?
(A) 1. – i 2. – ii 3. – iii 4. - iv
(B) 1. – i 2. – ii 3. – iii 4. – iv
(C) 1. – i 2. – iii 3. – ii 4. – iv
(D) 1. – i 2. – iv 3. – iii 4. – ii

160. IIHR stands for:
(A) Indian Institute of Horticultural Research
(B) Indian Inland Horticulture Research
(C) Institute of Inland Horticulture Research
(D) None of these

161. Which one of the following is matched correctly
(A) CIAE – Bhopal
(B) CSWCTI – Jodhpur
(C) CTCRI – Dehradun
(D) NAARM – New Delhi

162. CRIDA is situated at:
(A) Hyderabad
(B) New Delhi
(C) Bangalore
(D) Lucknow

163. The ICAR institutes in 'A' and its head quarters in 'B' are given below:

A	B
1. CTRI	i. Trivendrum
2. CTCRI	ii. Rajahmundry
3. IASRI	iii. Nagpur
4. CICR	iv. New Delhi

Which of the match is correct?
(A) 1. – i 2. – ii 3. – iii 4. - iv
(B) 1. – i 2. – iv 3. – iii 4. – ii
(C) 1. – ii 2. – i 3. – iv 4. – iii
(D) 1. – iii 2. – iv 3. – i 4. – ii

164. CIMMYT is working mainly for:
(A) Wheat (B) Maize and Wheat
(C) Rice (D) Cotton

165. The name of International Research Institutes and its head quarters are given below:

Institutes	Head quarters
1. ICRISAT	Hyderabad
2. IRRI	Philippines
3. ICARDA	Syria
4. IBPGR	Rome

Which of the above pairs is matched correctly?
(A) 1 (B) 3 (C) 4 (D) All of the above

166. Zein is known as:
(A) Rice protein (B) Wheat protein
(C) Maize protein (D) Millet protein

167. Maize protein is deficient in:
(A) Tryptophane (B) Lysine
(C) Both (A) and (B) (D) None of these

168. Origin of maize is:
(A) Mexico (B) India
(C) Africa (D) China

169. Pulses are rich source of:
(A) Fat (B) Carbohydrate
(C) Protein (D) Vitamin

170. Who introduce tobacco cultivation in India?
(A) ITC (B) Portugeese merchants
(C) Chinese (D) Persians

171. Central Tobacco Research Institute is situated in:
(A) UP (B) MP
(C) AP (D) Karnataka

172. Removal of terminal bud in tobacco is called:
(A) Topping (B) Desuckering
(C) Nipping (D) Curing

173. Jute (*Corchorus spp*) belongs to the family:
(A) Malvaceae (B) Tiliaceae
(C) Pedaliaceae (D) Solanaceae

174. Jute Agricultural Research Institute is situated in:
(A) Karnataka (B) Andhra Pradesh
(C) Tamilnadu (D) WB

175. The process by which the fibre in the bark get loosened and separated from the woody stalk is called as:
(A) Retting (B) Extraction
(C) (A) and (B) (D) None of these

176. Wheat protein is called as:
(A) Zein (B) Gluten
(C) Dhurin (D) Ricin

177. Which of the following states has the highest area of groundnut?
(A) Maharastra (B) Karnataka
(C) Punjab (D) Gujarat

178. The rice variety containing 'Dee-geo-woo-gen' gene is:
(A) Indrasan (B) Barmoti
(C) Tilak (D) IR 8

179. The origin place of soybean is:
(A) India (B) China
(C) America (D) Brazil

180. Soybean is a:
(A) Day – neutral plant
(B) Long day plant
(C) Short day plant
(D) None of these

181. Origin of berseem crop is:
(A) USSR (B) Egypt
(C) India (D) China

182. *Phalaris minor* is also known as:
(A) Barmuda grass (B) Canery grass
(C) Water Hyacinth (D) Kasni

183. How much sugar can be prepared from 1000 quintals of sugar cane?
(A) 125 q (B) 100 q
(C) 80 q (D) 50 q

184. The crop in which maximum pesticides are used?
(A) Tomato (B) Potato
(C) Sugarcane (D) Cotton

185. Central Rice Research Institute is situated at:
(A) Cuttuck (B) Karnal
(C) Mumbai (D) Manila

186. One acre is equivalent to:
(A) 1.8 ha (B) 0.652 ha
(C) 0.404 ha (D) 0.038 ha

187. Origin of urd bean is:
(A) Burma (B) Pakistan
(C) India (D) China

188. Yellow revolution indicates the production of:
(A) Oilseeds production
(B) Sugar production
(C) Fruits production
(D) Pulses production

189. First Director General of ICAR was:
(A) Dr. M.S. Swaminathan
(B) Dr. N.S. Randhawa
(C) Dr. G.S. Randhawa
(D) Dr. B.P. Pal

190. Total food grain production in India in 2008-09 is:
(A) 207 mt (B) 275 mt
(C) 300 mt (D) 234 mt

191. White revolution indicates the production of:
(A) Eggs production
(B) Sugar production
(C) Fruits production
(D) Pulses production

192. Green revolution indicates the production of:
(A) Foodgrain production
(B) Sugar production
(C) Fruits production
(D) Pulses production

193. Blue revolution indicates the production of:
(A) Fish production (B) Sugar production
(C) Fruits production(D) Pulses production

194. First State Agricultural University of India is:
(A) Gujarat Agricultural University
(B) Punjab Agricultural University
(C) G. B. Pant University of Agricultural & Tech.
(D) Hisar Agricultural University

195. Inflorescence of sugarcane is known as:
(A) Arrow (B) Tillers
(C) Silk (D) Tassel

196. The highest productivity of wheat during is 2011-12
(A) Haryana (B) Punjab
(C) UP (D) Bihar

197. The highest production of wheat is in:
(A) Haryana (B) Punjab
(C) UP (D) Bihar

198. The principle male sterile line introduced for hybrid pearlmillet production in India was:
(A) Tift-23 A (B) BJ-104
(C) MHB-118 (D) CO-2

199. Essential oils are used in:
(A) Cosmetics and Toiletries
(B) Food beverage
(C) Paint industry
(D) All of these

200. First Face Pack was developed in:
(A) France (B) Rome
(C) Greece (D) None of these

201. Oil content in fresh herbage in menthol-mint is:
(A) 0.1-0.5 % (B) 0.5-0.8 %
(C) 1.0-1.5 % (D) 1.5 – 2.0 %

202. Major constituent of *Occimum cannum*is:
(A) Methyl chavical (B) Thymal
(C) Linalool (D) Engenol

203. Origin of cow pea crop is:
(A) India (B) Pakistan
(C) China (D) Japan

204. Sunflower crop belongs to the family of:
(A) Cruciferеae (B) Compositeae
(C) Chenopodiaceae (D) Liliaceae

205. What is the optimum pulse requirement per capita per day?
(A) 20 g (B) 40 g
(C) 80 g (D) 100 g

206. Origin of French bean is:
(A) America (B) India
(C) China (D) Pakistan

207. Protein content in rajmash is:
(A) 10.9 % (B) 22.9 %
(C) 25.9 % (D) 32.9 %

208. Carbohydrate content in rajmash is:
(A) 50.6 % (B) 60.6 %
(C) 70.6 % (D) 80.6 %

209. Protein content in pea is:
(A) 10 % (B) 22 %
(C) 25 % (D) 32 %

210. Carbohydrate content in potato is:
(A) 40.3 % (B) 70.3 %
(C) 81.3 % (D) 91.3 %

211. India occupies position in vegetable production in world:
(A) First (B) Second
(C) Third (D) Fourth

212. India occupies position in production of onion in world:
(A) First (B) Second
(C) Third (D) Fourth

213. Total cattle population in India during 2006-07 is:
(A) 181 million
(B) 196 million
(C) 201 million
(D) 210 million

214. Total buffalo population in India during 2006-07 is:
(A) 80 million (B) 90 million
(C) 95 million (D) 99 million

215. Total goat population in India during 2006-07 is:
(A) 100 million (B) 115 million
(C) 125 million (D) 135 million

216. Total sheep population in India during 2006-07 is:
(A) 35 million (B) 51 million
(C) 58 million (D) 63 million

217. Total camel population in India during 2006-07 is:
(A) 0.63 million (B) 0.75 million
(C) 0.76 million (D) 0.63 million

218. Total milk production in India during 2006-07 is:
(A) 77.7 million tonnes
(B) 80.6 million tonnes
(C) 85.8 million tonnes
(D) 100.9 million tonnes

219. The contribution of livestock to total agricultural output is:
(A) 20 % (B) 25 %
(C) 30 % (D) 35 %

220. The contribution of livestock sector to total GDP is:
(A) 1 % (B) 3 %
(C) 6 % (D) 8 %

221. India's share in total milk production of world is:
(A) 10 % (B) 13 %
(C) 15 % (D) 21 %

222. The average productivity of Indian cow is:
(A) 810 kg/lactation
(B) 987 kg/lactation
(C) 1200 kg/lactation
(D) 1500 kg/lactation

223. Dairy sector provides employment to the total work force of:
(A) 5 % (B) 10 % (C) 15 % (D) 20 %

224. India's position in total cattle population is:
(A) First (B) Second
(C) Third (D) Fourth

225. Central Sheep and Wool Research Institute is situated at:
(A) Avikanagar (Raj)
(B) Bangalore (KN)
(C) Bikaner (Raj)
(D) Pusa (New Delhi)

226. Which of the following state has highest sheep production in India:
(A) Rajasthan (B) Gujarat
(C) Madhya Pradesh (D) Uttar Pradesh

227. India's share in goat population in the world is:
(A) 15% (B) 20%
(C) 25% (D) 30%

228. Which of the following state is leading in goat and sheep population?
(A) Madhya Pradesh (B) Rajasthan
(C) J & K (D) Gujarat

229. The meat contribution by goat to total meat production in India is:
(A) 20 % (B) 30 % (C) 37 % (D) 48 %

230. The highest percentage of total goat production is found in the region:
(A) Western and semi-arid region
(B) South humid tropical region
(C) Eastern region
(D) North temperate region

231. The contribution of goat in total milk production of India is only:
(A) 1.7% (B) 3.2% (C) 5.7% (D) 6.1%

232. The mean rate of goat slaughtering is:
(A) 25% (B) 30% (C)36% (D) 41%

233. Which of the state is leading in poultry production?
(A) UP (B) AP
(C) Karnataka (D) Maharastra

234. The total egg production in India during is 2010-11
(A) 20.2 million (B) 35.1 million
(C) 63.0 million (D) 40.3 million

235. The highest poultry production in India is:
(A) UP (B) AP
(C) MP (D) WB

236. Which of the country has the highest population of poultry birds?
(A) USA (B) UK
(C) China (D) India

237. Per capita availability of animal protein in Indian diet is:
(A) 10 g/day (B) 15 g/day
(C) 18 g/day (D) 20 g/day

238. The protein per cent in egg yolk is:
(A) 5% (B) 11% (C) 16% (D)18%

239. The highest producing state of duck is:
(A) UP (B) WB
(C) Bihar (D) Assam

240. Turkey production is highest in the world is:
(A) USA (B) India
(C) Brazil (D) China

241. Per capita availability of milk in India is:
(A) 200 g/day (B) 226 g/day
(C) 230 g/day (D) 240 g/day

242. India's rank in world egg production is:
(A) Ist (B) IInd
(C) IIIrd (D) Vth

243. Total wool production in India during 2006-07is:
(A) 40 million kg (B) 50 million kg
(C) 45.1 million kg
(D) 60 million kg

244. Total fish production in India during 2006-07 is:
(A) 68.69 million tones
(B) 70.12 million tones
(C) 75.63 million tones
(D) 80.06 million tones

245. Highest inland fish production during 2006-07 is
(A) 40.25 million tones
(B) 38.45 million tones
(C) 42.68 million tones
(D) 45.69 million tones

246. Highest marine fish production during 2006-07 is:
(A) 21.36 million tones
(B) 25.36 million tones
(C) 30.24 million tones
(D) 32.69 million tones

247. The number of milk recording centres at national level are:
(A) 30 (B) 50
(C) 94 (D) 115

248. The number of Central Cattle Breeding Centres in India are:
(A) 3 (B) 5 (C) 7 (D) 9

249. Central Cattle Breeding Centre is situated at:
(A) Suratgarh (Rajasthan)
(B) Andeshnagar (UP)
(C) Alamadi (TN)
(D) All of these

250. Central Frozen Semen Production and Training Institute is located at:
(A) Hisar (B) Suratgarh
(C) Hessarghatta (D) Bangalore

251. Central Sheep Breeding Farm is located at:
(A) Hisar (B) Suratgarh
(C) Hesarghatta (D) Bangalore

252. How many pig breeding farms have been established in India?
(A) 101 (B) 158
(C) 260 (D) 275

253. Central Fodder Seed Production Farm is located at:
(A) Jhansi (B) Hessarghatta
(C) Hyderbad (D) Hisar

254. India's share in world buffalo population is:
(A) 20 % (B) 40 %
(C) 57 % (D) 75 %

255. India's share in world cattle population is:
(A) 10% (B) 12%
(C) 15% (D) 18%

256. Central Institute of Fisheries, Nautical and Engineering Training is located at:
(A) Kochi (B) Mumbai
(C) Vishakhapattanam (D) Kolkata

257. Central Institute of Coastal Engineering for Fisheries is located at:
(A) Kochi (B) Mumbai
(C) Bangalore (D) Kolkata

258. The number of major fishing harbours in India is:
(A) 2 (B) 4 (C) 6 (D) 8

259. Total honey production in India is:
(A) 7500 tonnes (B) 10200 tonnes
(C) 13000 tonnes (D) 15000 tonnes

260. Which country has highest milk production is:
(A) India (B) Denmark
(C) Switzerland (D) USA

261. India's share in world bovine population is:
(A) 17% (B) 19%
(C) 21% (D) 23%

262. Buffalo constitute the total bovine population of:
(A) 35% (B) 40%
(C) 45% (D) 47%

263. What is the contribution of dairy sector to the employment:
(A) 5% of total work force
(B) 8% of total work force
(C) 9% of total work force
(D) 12% of total work force

264. Milk production in India during 2006-07 is:
(A) 72 m tonnes (B) 77 m tonnes
(C) 100.9 m tonnes (D) 100 m tonnes

265. Largest milk colony of Asia is:
(A) Aere milk colony
(B) Allahabad dairy farm
(C) Rajasthan dairy farm
(D) Kanpur diary farm

266. First Dairy farm in India was opened in year of:
(A) 1870 (B) 1877
(C) 1889 (D) 1890

267. First Dairy farm in India was opened at:
(A) Anand (B) Allahabad
(C) Karnal (D) New Delhi

268. Contribution of dairy sector to the total agricultural output is nearly:
(A) 21 % (B) 25 %
(C) 29 % (D) 31 %

269. Contribution of dairy sector to the GDP is around:
(A) 4 % (B) 6 % (C) 8 % (D) 10 %

270. Before 1951, India's milk production was:
(A) Less than 20 mt
(B) 20-30 mt
(C) 30-40 mt
(D) More than 40 mt

271. India's share in world milk production is:
(A) 10% (B) 13%
(C) 15% (D) 21%

272. Indian Dairy Corporation was setup mainly for:
(A) Success of Operation of Flood Program
(B) Implementation of Operation Flood Program
(C) Trade exchange related to Operation Flood Program
(D) All of these

273. The plan investment in Animal Husbandry and Dairying is about:
(A) 4% (B) 5%
(C) 6% (D) 8%

274. India's share in Asian bovine population is:
(A) 27% (B) 38%
(C) 51% (D) 60%

275. According to ICAR, minimum requirement of milk/day/head is:
(A) 200 g (B) 280 g
(C) 290 g (D) 300 g

276. National Agricultural Insurance Scheme launched in India in year:
(A) 1972 (B) 1985
(C) 1999 (D) 2000

277. Comprehensive Crop Insurance Scheme was launched in year:
(A) 1975 (B) 1985
(C) 1995 (D) 2001

278. Central Institute of Agricultural Engineering is situated at:
(A) Kanpur (B) Bhopal
(C) Ludhiana (D) Nagpur

279. The intermediaries came into Indian Agriculture in which year:
(A) 1793 (B) 1783
(C) 1773 (D) 1763

280. The first bill for abolition of intermediaries or zamidari introduced in:
(A) 1946, UP (B) 1948, Bihar
(C) 1971, TN (D) 1967, Gujarat

281. Central Farm Machinery Training and Testing Institute is situated at:
(A) Budni, MP (B) Ludhiana, Punjab
(C) Kanpur, UP (D) Jabalpur, MP

282. Zamidary system was introduced in year:
(A) 1792 (B) 1793
(C) 1794 (D) 1795

283. Rayatwari system was introduced by:
(A) Thomas Munro and Captain Read
(B) Lord Powel
(C) Lord Curzon
(D Lord Cornwalish

284. Land Reform Implementation Committee was set up in 1963 under the chairmanship is:
(A) G.L. Nanda (B) Prof. Dantwola
(C) Arvind Ghosh (D) M.K. Gandhi

285. The minimum wages Act (Labour) was enacted during:
(A) 1932 (B) 1936
(C) 1948 (D) 1952

286. The Employees Provident Fund and Family Pension Act was enforced in:
(A) 1952 (B) 1962
(C) 1972 (D) 1982

287. The payment of Bonus Act was implemented in:
(A) 1948 (B) 1958
(C) 1965 (D) 1968

288. The payment of Gratuity Act was implemented in:
(A) 1952 (B) 1962
(C) 1972 (D) 1982

289. The Industrial Disputes Act was enacted in:
(A) 1937 (B) 1947
(C) 1957 (D) 1967

290. The Trade Union Act was enacted in:
(A) 1926 (B) 1936
(C) 1946 (D) 1956

291. The Bonded Labour System (Abolition) ordinance came into force in:
(A) 1971 (B) 1973
(C) 1975 (D) 1978

292. The Livestock Importation Act was implemented in:
(A) 1862 (B) 1872
(C) 1898 (D) 1903

293. The Agriculture Produce (grading) and Marketing Act was passed in:
(A) 1927 (B) 1937
(C) 1947 (D) 1957

294. The Cattle Trespass Act was implemented in:
(A) 1781 (B) 1881
(C) 1981 (D) 1991

295. The prevention of Cruelty to animals Act, 1950 was implemented in:
(A) 1857 (B) 1891
(C) 1907 (D) 1950

296. Dangerous Machines (Regulation) Act was passed in:
(A) 1971 (B) 1981
(C) 1983 (D) 1993

297. The Land Improvement Act was initiated in:
(A) 1869 (B) 1870
(C) 1871 (D) 1872

298. The Land Improvement Loan Act was started in:
(A) 1871 (B) 1876
(C) 1883 (D) 1891

299. The Agriculturists Loans Act was implemented in:
(A) 1881 (B) 1882
(C) 1883 (D) 1884

300. The National Co-operative Development Corporation Act was started in:
(A) 1947 (B) 1959
(C) 1962 (D) 1971

301. The Agricultural Produce (Development and Warehousing) Corporation Act was implemented in:
(A) 1947 (B) 1956
(C) 1962 (D) 1971

302. The National Co-operative Development Corporation was set up in:
(A) 1952 (B) 1962
(C) 1963 (D) 1964

303. The Agricultural Refinance and Development Corporation Act came into existence in:
(A) 1961 (B) 1962
(C) 1963 (D) 1964

304. The Warehouse Corporation Act was implemented in:
(A) 1960 (B) 1961
(C) 1962 (D) 1963

305. The Regional Rural Banks Act was enforced in India in:
(A) 1971 (B) 1974
(C) 1976 (D) 1981

306. The Multi-Unit Co-operative Societies Act was implemented in:
(A) 1940 (B) 1941
(C) 1942 (D) 1943

307. The All India Rural Credit Survey Committee was constituted by RBI in:
(A) 1931 (B) 1941
(C) 1951 (D) 1961

308. National Development Council Resolution was passed in :
(A) 1949 (B) 1959
(C) 1969 (D) 1979

309. The New Co-operative Policy Resolution was adopted in:
(A) 1975 (B) 1976
(C) 1977 (D) 1978

310. National Cooperative Union of India was set up in:
(A) 1951 (B) 1953
(C) 1957 (D) 1961

311. Agricultural Finance Corporation was set up in:
(A) April 10, 1965 (B) April 10, 1966
(C) April 10, 1968 (D) April 10, 1969

312. Agricultural Refinance and Development Corporation (ARDC) was setup in:
(A) 1961 (B) 1962
(C) 1963 (D) 1964

313. Tribal Area Development Project was launched in:
(A) 1971-72 (B) 1973-74
(C) 1976-77 (D) 1978-79

314. State Trading Corporation Ltd was constituted in:
(A) 1956 (B) 1957
(C) 1958 (D) 1959

315. Indian Standard Institute was established in:
(A) 1945 (B) 1946
(C) 1947 (D) 1948

316. Marketing and Designing centre is situated at:
(A) Pune (B) Faridabad
(C) Nagpur (D) New Delhi

317. Central Agmark Laboratory is situated at:
(A) New Delhi (B) Nagpur
(C) Bangalore (D) Kanpur

318. Indian Standard Institute is situated at:
(A) Kolkata (B) New Delhi
(C) Madras (D) Hyderabad

319. Agricultural Produce (Grading and Marketing) Act was established in:
(A) 1932 (B) 1937
(C) 1939 (D) 1940

320. Central Marketing department was opened in:
(A) 1932 (B) 1035
(C) 1937 (D) 1941

321. The per capita availability of land has been decreased from 0.5 ha in 1950-51 to 0.14 ha in 2000 and will further reduced by 2020:
(A) 0.08 ha (B) 1.08 ha
(C) 1.18 ha (D) 1.28 ha

322. First Agricultural Marketing Advisor was appointed on:
(A) January 1, 1935
(B) January 26, 1935
(C) October 2, 1935
(D) August 15, 1935

323. Agricultural Price Commission was constitutes in:
(A) 1955 (B) 1965
(C) 1975 (D) 1985

324. Essential Commodities Act was passed on:
(A) 1954 (B) 1955
(C) 1956 (D) 1957

325. Cold Storage Order came into existence in:
(A) 1962 (B) 1963
(C) 1964 (D) 1965

326. Meat Food Products order was passed on:
(A) 1971 (B) 1972
(C) 1973 (D) 1974

327. Export (Quality Control and Inspection) Act was passed on:
(A) 1965 (B) 1968
(C) 1969 (D) 1971

328. Head office of Directorate of Marketing and Inspection is situated at:
(A) New Delhi (B) Faridabad
(C) Gurgaon (D) Ghaziabad

329. Food Corporation of India (FCI) was came into existence on:
(A) 1965 (B) 1966
(C) 1967 (D) 1968

330. A expert committee to review the Cooperative Marketng was set up in 1964 under the chairmanship of:
(A) Dr. M.L. Dantewala
(B) R.N. Mirdha
(C) Raj Gopalacharya
(D) R.L. Mehta

331. The Agricultural Finance sub committee was constituted under the chairmanship of:
(A) V.L. Mehta
(B) D.R. Gadgil
(C) Sardar Ballabh Bhai Patel
(D) J.L. Nehru

332. Annual loss due to pests, diseases and weeds annually is estimated nearly:
(A) >50000 crores
(B) <50000 crores
(C) > 1.0 crores
(D) < 1.0 crores

333. India is using biopesticides presently:
(A) 1% of total pesticide
(B) 5 % to total pesticide
(C) 10 % of total pesticide
(D) 15 % of total pesticide

334. Average biopesticide use globally is:
(A) 5 % of total pesticide
(B) 10 % of total pesticide
(C) 12 % of total pesticide
(D) 15 % of total pesticide

335. Total pesticide consumption in India during 2006-07 is :
(A) 35.65 million tonnes
(B) 37.95 million tonnes
(C) 39.26 million tonnes
(D) 42.56 million tonnes

336. Destructive Insects and Pest Act was launched in India in:
(A) 1907 (B) 1914
(C) 1921 (D) 1946

337. Directorate of Plant Protection Quarantine and Storage was started in:
(A) 1907 (B) 1914
(C) 1921 (D) 1946

338. Which of the airport does not have plant quarantine station?
(A) Lucknow (B) Kolkata
(C) New Delhi (D) Madras

339. Headquarter of Central Silk Board is:
(A) Lucknow (B) Bangalore
(C) Hyderabad (D) Behrampur

340. Central Sericulture Research Institute and Technology is situated at:
(A) New Delhi (B) Kolkata
(C) Behrampur (D) Madras

341. Headquarter of Tobacco Board is:
(A) Rajahmundry (B) Guntur
(C) Vijaywada
(D) Vishakhapatannam

342. What is origin place of cauliflower:
(A) Canada (B) Italy
(C) India (D) China

343. What is origin place of cabbage:
(A) West Europe (B) West Africa
(C) West China (D) East India

344. What is origin place of bottle guard:
(A) Canada (B) USA
(C) India (D) China

345. What is origin place of pumpkin:
(A) Canada (B) America
(C) India (D) China

346. What is origin place of carrot:
(A) Canada (B) USA
(C) Asia (D) China

347. What is origin place of Sweet potato:
(A) North America
(B) South America
(C) Australia
(D) China

348. What is origin place of chilli:
(A) Brazil (B) Chile
(C) Gutemala (D) China

349. APEDA refers to:
(A) Agricultural and Processed Food Products Export Development Authority
(B) Agricultural Production and Export Development Authority
(C) Agricultural Production and Export Development Agency
(D) Asian Policy and Export Development Association

350. India is leading in production of:
(A) Turmeric
(B) Black pepper
(C) Ginger
(D) All of these

351. Total production of vegetable in India during 2006-07 is:
(A) 93.92 mt (B) 96.54 mt
(C) 102.11 mt (D) 120.71 mt

352. What is the contribution of India in world vegetable production:
(A) 10 % (B) 15 %
(C) 20 % (D) 23 %

353. Which of the following vegetable is produced maximum in India is:
(A) Cauliflower (B) Cabbage
(C) Tomato (D) Brinjal

354. Which of the following spices is most important in India is:
(A) Chilli (B) Turmeric
(C) Garlic (D) All of these

355. India's share in chilli production is:
(A) 27 % (B) 33.7 %
(C) 45 % (D) 51.2 %

356. Minimum requirement of fruit per head per day for balanced diet is:
(A) 60 gm (B) 80 gm
(C) 85 gm (D) 120 gm

357. CFTRI is situated at:
(A) Mysore (B) Kolkata
(C) New Delhi (D) Bangalore

358. First fruit product research laboratory was opened in India at:
(A) Kodour (B) Pune
(C) Lucknow (D) Bhopal

359. The origin place of banana is:
(A) South-East Asia
(B) South-West Asia
(C) South Africa
(D) South America

360. The origin place of loquat is:
(A) China (B) India
(C) South Africa (D) South America

361. The origin place of jackfruit is:
(A) China (B) Mexico
(C) India (D) South America

362. The origin place of grape is:
(A) Asia (B) China
(C) South Africa (D) South America

363. The origin place of peach is:
(A) Europe (B) China
(C) South Africa (D) South America

364. The origin place of pear is:
(A) Iran (B) Asia
(C) Africa (D) America

365. The origin place of gladiolus is:
(A) China (B) Egypt
(C) South Africa (D) South America

366. India's place in fruit production in world is:
(A) First (B) Second
(C) Third (D) Fourth

367. The fruit production in India is:
(A) 26 mt (B) 40 mt
(C) >45 mt (D) 80 mt

368. The total area under fruit production in India is:
(A) 1.5 m ha (B) 2.5 m ha
(C) 3.8 m ha (D) 8.5 m ha

369. What is the total production of horticultural crops in India during 2001-02 is:
(A) 152.5 mt (B) 155.65 mt
(C) 171.0 mt (D) 185.13 mt

370. What is the share of India in total world fruit production?
(A) 5 % (B) 10 % (C) 15 % (D) 20 %

371. Total honey production in India is:
(A) 2.2 thousand tones
(B) 8.7 thousand tones
(C) 10.2 thousand tones
(D) 12.5 thousand tones

372. The area under herbal garden in India is:
(A) 200 ha (B) 226 ha
(C) 310 ha (D) 510 ha

373. The fruits are arranged in ascending order in respect of its area. Which of the following sequence is correct:
(A) Mango>Citrus>Banana>Apple
(B) Mango>Apple>Banana>Citrus
(C) Guava>Mango>Banana>Citrus
(D) Apple>Mango>Citrus>Litchi

374. What is the share of India in total mango production of world?
(A) 23% (B) 42%
(C) 54% (D) 60%

375. Presently the total production of cut flowers in India during 2008-09 is:
(A) 550 million (B) 780 million
(C) 1952 million (D) 910 million

376. Which of the following fruit has its highest productivity in India?
(A) Grape (B) Mango
(C) Banana (D) Guava

377. The practice of raising plants and animals is:
(A) Agriculture (B) Horticulture
(C) Agronomy (D) Silviculture

378. Agriculture originated about
(A) 2300-3000 years back
(B) 20000-30000 years back
(C) 7000-13000 years back
(D) 90000-100000 years back

379. Kung tribe is
(A) Primitive agriculturist
(B) Hunter-gatherer
(C) Advanced agriculturist
(D) Completely dependent on horticulture

380. In various parts of the world, agriculture
(A) Came from Asia Minor
(B) Has spread from Yellow Rive Valley
(C) Originated independently
(D) Has developed to various degree

381. Fire helped in the cultivation of
(A) Fruit plants (B) Fibre Plants
(C) Drug yielding plants
(D) Grain yielding plants

382. Early human civilization was based on
(A) Development of fibre crops
(B) Fire
(C) Agriculture
(D) Fire and Agriculture

383. A nonagricultural society living today is
(A) Kung tribe (B) Urban Society
(C) Hintza tribe (D) Kuntizha tribe

384. Early man preferred plants
(A) Grain yielding (B) Fruit yielding
(C) Multipurpose (D) Fibre yielding

385. A multipurpose plant of India is
(A) Walnut (B) Almond
(C) Apple (D) Coconut

386. A grain producing plant where spathe covering helped in domestication
(A) Wheat (B) Maize
(C) Rice (D) Barley

387. The main source of irrigation in India is:
(A) Canals (B) Tubewells
(C) Wells (D) Tanks

388. The total food grain production in India during 2007-08 was:
(A) 202 million tones
(B) 220 million tones
(C) 225 million tones
(D) 230 million tones

389. Agriculture became state subject in year:
(A) 1910 (B) 1919
(C) 1947 (D) 1952

390. Indian Agricultural Services ceased to function in the year:
(A) 1919 (B) 1921
(C) 1924 (D) 1947

391. ICAR was born in:
(A) 1908 (B) 1910
(C) 1929 (D) 1949

392. What was the name of ICAR in initial years of its establishment?
(A) Indian Council of Agricultural Research
(B) Imperial Council of Agricultural Research
(C) Imperial Centre of agricultural Research
(D) Indian Centre of Agricultural Research

393. The establishment of ICAR is based on the recommendation of which commission?
(A) Royal Commission on Agriculture
(B) Famine Commission
(C) Grow More Food Commission
(D) Balwant Rai Mehta Commission

394. ICAR was renamed Indian Council of Agricultural Research in which year?
(A) 1929 (B) 1946
(C) 1949 (D) 1952

395. Who was the first Director General of ICAR and Secretary to Ministry of Agriculture?
(A) Dr. B.P. Pal
(B) Dr. M.S. Swaminathan
(C) Dr. O.P. Gautam
(D) Dr. R.S. Paroda

396. When did ARS service start?
(A) 1947 (B) 1952
(C) 1973 (D 1982

397. What is the main mandate of ICAR?
(A) To plan aid, undertake, promote and coordinate agriculture research in the country
(B) To plan, aid undertake, promote and coordinate agricultural research in the country
(C) To plan, aid undertake, promote and coordinate agricultural education and research in the country
(D) To plan, aid undertake, promote and coordinate agricultural research, education and extension in the country

398. How many Deputy Director Generals are working with ICAR?
(A) 05 (B) 06 (C) 07 (D) 08

399. Who works as President of ICAR?
(A) The Prime Minister of India
(B) The Union Agriculture Minister
(C) The State Minister of Agriculture
(D) Director General, ICAR

400. The growth of ICAR started with he establishment of which of the following institutes, being the first one?
(A) Imperial Bacteriological Laboratory, Pune
(B) Agricultural Research Institute, Pusa, Bihar
(C) Indian Agricultural Research Institute, New Delhi
(D) Agriculture College, Kanpur

401. How many National Institutes are working in the ICAR system?
(A) 04 (B) 05 (C) 06 (D) 07

402. Which of the following was the first institute/university to be establishment only land Grant College System in 1958?
(A) Indian Agricultural Research Institute, New Delhi
(B) G.B. Pant University of Agriculture & technology, Pantnagar
(C) College of Agriculture, Coimbatore
(D) College of Agriculture, Kanpur

403. NAARM (National Academy of Agricultural Research Management), Hyderabad was established to train scientists in which year?
(A) 1949 (B) 1973
(C) 1976 (D) 1986

404. AICRP stands for?
(A) All India Crop Research Programme
(B) All India Coordinated Research Project
(C) All India Cropping System Research Project
(D) None of these

405. MANAGE, Hyderabad is an institution working in the area of:
(A) Agricultural research management
(B) Agricultural extension management
(C) Agricultural Education Management
(D) Agricultural research, education and extension management

406. What for ASRB stands?
(A) Agricultural Services Recruitment Board
(B) Agricultural Scientist Recruitment Board
(C) Agricultural Science Research Bureau
(D) Agricultural Scientist Research Bureau

407. How many Trainers Training Centres (TTCs) are in India?
(A) 05 (B) 08
(C) 13 (D) 21

408. IVLP (Institution Village Linkage Programme) centres are functioning at how many places?
(A) 70 (B) 79 (C) 83 (D) 89

409. How many Zonal Research Stations have been remandated to function as KVKs also?
(A) 50 (B) 53 (C) 57 (D) 59

410. How many central institutes of ICAR are working in agricultural science?
(A) 10 (B) 36 (C) 73 (D) 89

411. How many Bureaus have been established in India?
(A) 01 (B) 02 (C) 05 (D) 11

412. National Bureau of Plant Genetic Resources is established at which place?
(A) New Delhi (B) Modipuram
(C) Coimbatore (D) Bangalore

413. National Bureau of Soil Survey & Land Use Planning is established at which place?
(A) Jorhat (B) Nagpur
(C) New Delhi (D) Kochin

414. National Research Centre for Women in Agriculture is established at which place?
(A) Calcutta (B) Bhubaneshwar
(C) Jorhat (D) New Delhi

415. National Centre of Agricultural Economics & Policy Research is established at which place?
(A) New Delhi (B) Hyderabad
(C) Bangalore (D) Madras

416. How many agricultural universities are functional in India?
(A) 25 (B) 30 (C) 29 (D) 54

417. How many central agricultural universities are established in India?
(A) 01 (B) 02 (C) 03 (D) 04

418. How many ICAR institutes have been upgraded to the status of Deemed universities?
(A) 02 (B) 04
(C) 05 (D) 12

419. CIMMYT- an international organization supports research activity on which of the following crops?
(A) Paddy (B) Maize
(C) Wheat (D) Maize & Wheat

420. ICRISAT stands for:
(A) International Crops Research Institute for the Semi Arid Tropics
(B) International Centre for Research in Sugarcane & Tobacco
(C) Indian Crop Research Institute for Semi Arid Tropics

421. Agriculture is the subject of which of the following?
(A) State Govt.
(B) Central Govt.
(C) State & Central Govt.
(D) None of these

422. ICAR/SAUs conduct what kind of extension?
(A) Public extension
(B) Frontline extension
(C) Private extension
(D) None of these

423. What is the main mandate of ICAR/SAUs in the area of agricultural extension?
(A) Technology transfer
(B) Technology assessment on refinement
(C) Technology assessment, refinement
(D) None of these

424. Extension Education Institutes (EEIs) impart training in the area of:
(A) Subject matter
(B) Training & Communication methodology
(C) Subject matter, training and communication methodology
(D) None of these

425. The most serious economic problem of India are:
(A) Poverty and Unemployment
(B) Unemployment and stagnation
(C) Underdevelopment and Poverty
(D) Corruption and Unemployment

426. The percentage of people below poverty line in India is:
(A) 25.1 (B) 27.8
(C) 20.3 (D) 21.45

427. By eradicating poverty, who wished to "wipe out every tear from every eye":
(A) Jawahar Lal Nehru
(B) Mahatma Gandhi
(C) Rabindra Nath Tagore
(D) Vinoba Bhave

428. The poverty lines in rural area on the basis of an average daily intake:
(A) 2100 calories (B) 2300 calories
(C) 2400 calories (D) 3400 calories

429. The 'Garibhi hatao' slogan was coined by:
(A) Jawahar Lal Nehru
(B) Mahatma Gandhi
(C) Rabindra Nath Tagore
(D) Indira Gandhi

430. The proportion of Indian population living in the villages:
(A) One-third (B) Half
(C) Two-third (D) Three-fourth

431. Who said 'If the villages perish, India will perish:
(A) Jawahar Lal Nehru
(B) Mahatma Gandhi
(C) Rabindra Nath Tagore
(D) Indira Gandhi

432. Rural development can be defined as a strategy designated to improve:
(A) Economic and social life of rural poor
(B) Condition of all section of people
(C) Economic condition of rural society
(D) Physical infrastructure

433. The basic guide, philosopher and friend in rural development:
(A) BDO (B) VDO
(C) Pradhan (D) Pramukh

434. Rural development programmes are related to:
(A) Self employment
(B) Employment generation
(C) Poverty alleviation (D) All of these

435. The body responsible for rural development at base level:
(A) Panchayat (B) Block
(C) District (D) State

436. The basic operational unit for rural development in India is:
(A) Village (B) District
(C) Tehsil (D) Block

437. Who coined the term Green revolution:
(A) Willianm Gadd
(B) M.S. Swaminathan
(C) Norman E. Borlaoug
(D) C. Subramanian

438. Agricultural Scientists Recruitment Board (ASRB) was set up based on the recommendations of the committee headed by:
(A) Mohan Sinha Mehta
(B) M S Swaminathan
(C) Gajendra Gadkar
(D) B P Sinha

439. Agricultural Scientists Recruitment Board (ASRB) was set up in the year:
(A) 1973 (B) 1975
(C) 1977 (D) 1979

440. Agriculture and livestock was published by:
(A) Ministry of Agriculture and Cooperation
(B) Ministry of Rural Development
(C) Indian Council of Agricultural Research
(D) Ministry of Animal Husbandry

441. The concept of agro-climatic zones were introduced during:
(A) NATP (B) NARP
(C) NREP (D) NAIP

442. The concept of agro-eco system analysis was developed during:
(A) Late 60s (B) Late 70s
(C) Late 80s (D) Late 90s

443. The concept of agro-eco system analysis was developed in:
(A) USA (B) Philippines
(C) Thailand (D) Israel

444. The concept of agro-eco system was basically originated at University of Chiang Mai by:
(A) Tonnis Fermetty
(B) Adam Groves
(C) Levis Carol (D) Gordon Conway

445. The first farm magazine in India is:
(A) Kheti
(B) Phool Phul
(C) Krishi Chayanika
(D) Yojana

446. The general micro-unit of an agro-ecosystem is:
(A) District (B) Village
(C) Farmer (D) Household

447. The headquarter of Food and Agricultural Organization are at:
(A) New York (B) New Delhi
(C) Geneva (D) Rome

448. The ideal types of social norms are:
(A) Traditional (B) Modern
(C) Both of these (D) None of these

449. The II most important hygiene factor according to Heirzberg is:
(A) Policy and administration
(B) Achievement
(C) Salary
(D) Working conditions

450. The impact of green revolution was largely noted in:
(A) Rice only (B) Wheat only
(C) Rice and Wheat
(D) Rice, Wheat and Pulses

451. The National Steering Committee (NSC) is chaired by:
(A) Director General, ICAR
(B) Union Minister of Agriculture
(C) Minister of State for Agriculture
(D) Principle Secretary Agriculture

452. The president of ICAR is:
(A) Prime Minister
(B) Agriculture Minister
(C) Director General
(D) Finance Minister

453. Scientific name of Flint corn is:
(A) *Zea mays indurata*
(B) *Zea mays tunicata*
(C) *Zea mays saccharata*
(D) *Zea mays ceratina*

454. Scientific name of Pod corn is:
(A) *Zea mays indurata*
(B) *Zea mays tunicata*
(C) *Zea mays saccharata*
(D) *Zea mays ceratina*

455. Scientific name of Sweet corn is:
(A) *Zea mays indurata*
(B) *Zea mays tunicata*
(C) *Zea mays saccharata*
(D) *Zea mays ceratina*

456. Scientific name of Waxy corn is:
(A) *Zea mays indurata*
(B) *Zea mays tunicata*
(C) *Zea mays saccharata*
(D) *Zea mays ceratina*

457. Scientific name of Pop corn is:
(A) *Zea mays indurata*
(B) *Zea mays tunicata*
(C) *Zea mays everta*
(D) *Zea mays ceratina*

458. Scientific name of Soft or flour corn is:
(A) *Zea mays indurata*
(B) *Zea mays amylacea*
(C) *Zea mays saccharata*
(D) *Zea mays ceratina*

459. In America which crop is known as yellow gold:
(A) Maize (B) Rice
(C) Soybean (D) Wheat

460. Scientific name of green varieties of black gram is:
(A) *Vigna mungo var viridis*
(B) *Vigna mungi var niger*
(C) *Vigna mungo var macropode*
(D) *Vigna mungi var micropode*

461. Scientific name of bunch type of ground nut is:
(A) *Arachis hypogea sp fastigiata*
(B) *Arachis hypodea sp procumbens*
(C) *Arachis hypogea sp viridis*
(D) *Arachis hypogea sp niger*

462. Scientific name of spreading type of ground nut is:
(A) *Arachis hypogea sp fastigua*
(B) *Arachis hypogea sp procumbens*
(C) *Arachis hypogea sp viridis*
(D) *Arachis hypogea sp niger*

463. Scientific name of desi cotton is:
(A) *Gossypium herbacium*
(B) *Gossypium arboreum*
(C) Both (A) and (B) (D) None of these

464. Scientific name of American cotton is:
(A) *Gossypium barbedensii*
(B) *Gossypium hirsutum*
(C) Both (A) and (B)
(D) None of these

465. Diploid species of cotton are:
(A) Desi (B) American
(C) Both (A) and (B) (D) None of these

466. Tetraploid species of cotton are:
(A) Desi (B) American
(C) Both (A) and (B) (D) None of these

467. Poor man's timber is:
(A) Mango wood (B) Bamboo
(C) Acacia wood (D) Eucalyptus wood

468. Tendu/Kendu leaves used as wrappers for bidis are got from:
(A) *Diospyros ebenum*
(B) *Dipterocarpus indicus*
(C) *Dipterocarpus tuberculatus*
(D) *Diospyros exsculpta*

469. Conservation is:
(A) Protection of natural resources
(B) Management of natural resources
(C) Proper use of natural resources
(D) All of these

470. MAB is:
(A) Man and Botany
(B) Man and Biosphere
(C) Man and Biotic community
(D) Man, Antibiotics and Bacteria

471. Silent valley has been preserved as it has:
(A) Recreational value
(B) Valuable timber plants
(C) Natural forest
(D) Rare plant and animals

472. Silent valley having rare plants and animals is located in:
(A) Kerala
(B) Karnataka
(C) Jammu and Kashimir
(D) Andhra Pradesh

473. Which one is connected with conservation of forests?
(A) Gir (B) Silent valley
(C) Kaziranga (D) Bharatpur

474. New approach to conservation is establishment of:
(A) Reserve forests
(B) Santuries
(C) Biosphere reserves
(D) National parks

475. Social forestry is:
(A) Management of forests by cooperative societies
(B) Growing and management of useful plants on common lands
(C) Growing one type of trees
(D) Growing different types of plants together

476. Decrease in species diversity in tropical countries is mainly due to:
(A) Urbanisation (B) Pollution
(C) Deforestation (D) Soil erosion

477. The key village scheme was initiated in August 1952 comprising first systematic attempt to improve the:
(A) Quality and productivity of buffaloes and cattle in India
(B) Quality and productivity of sheep and goat in India
(C) Quality of pigs
(D) All of these

478. Farm demonstration work began in —— —— in USA by Dr. Seaman:
(A) 1903 (B) 1908
(C) 1955 (D) 1960

479. Extension work in Japan was first started by:
(A) Government organization
(B) Farmer's organization
(C) Labour's organization
(D) Business's organization

480. Which of the following is not correct:
(A) Department of revenue, agriculture and commerce – 1871
(B) The famine commission _ 1880
(C) The famine commission - 1901
(D) None of these

481. The Royal commission's report came in:
(A) 1920 (B) 1924
(C) 1928 (D) 1931

482. "Grow more food campaign" was started in:
(A) 1944 (B) 1942
(C) 1952 (D) 1953

483. Intensive Agricultural Development Programme (IADP) popularly known as package programme started originally in 7 districts in:
(A) 1959 (B) 1960-61
(C) 1964 (D) 1970

484. Intensive Agriculture Area Programme came into operation in:
(A) March, 1964 (B) March, 1969
(C) October, 1969 (D) None of these

485. Which Five year plan aimed at attaining self sufficiency in food?
(A) 3rd (B) 4th
(C) 5th (D) 6th

486. For social justice government started new programme:
(A) Small Farmers Development Agency (SFDA)
(B) Marginal Farmers and Agriculture Labours Scheme (MFALS)
(C) Drought Prone Area Programme
(D) All of these

487. The leader of the Boodan movement was:
(A) Achary Vinoba Bhave
(B) Spencer Hatch
(C) F.L. Bryne
(D) T. Prakasam

488. Young Farmers Association was formed in India under the guidance of Dr. P.S. Deshmukh in:
(A) April, 1956 (B) April, 1966
(C) April, 1976 (D) April, 1980

489. National commission on farmers was set up in India in:
(A) March, 1999 (B) March, 2002
(C) February, 2004 (D) June, 2004

490. Who was the first chairman of National Commission on Farmers:
(A) M.S. Swaminathan
(B) Sompal
(C) Balram Jakhar
(D) V.L. Chopra

491. National Agricultural Science Museum was inaugurated on 3rd Nov, 2004 by:
(A) Dr. Man Mohan Singh
(B) Dr. A.P.J. Abdul Kalam
(C) A.B. Vajpayee
(D) Sonia Gandhi

492. National Agricultural Science Museum located at:
(A) New Delhi (B) Bangalore
(C) Kolkata (D) Chennai

493. National Food for work programme was launched on:
(A) November14, 2000
(B) November 14, 2004
(C) October 2, 2000 (D) October 2, 2004

494. National Food for work programme implemented as a:
(A) 100% centrally sponsored scheme
(B) 80:20 center state sponsored scheme
(C) 50:50 centre state sponsored scheme
(D) None of these

495. Swaranjayanti Gram Swarozgar Yojana (SGSY) was launched in:
(A) April, 1999 (B) October, 1999
(C) January, 2000 (D) January, 2002

496. Which one of the following pairs not correct:
(A) Sompoorna Grameen Rozgar Yojana - 2001
(B) Pradhan Mantri Gramodaya Yoajna - 2000-01
(C) Rural Employment Generation Programme-1995
(D) Pradhan Mantri Gram Sadak Yojana - 1999

497. The government sponsored Firka development scheme was launched under the guidance of
(A) T. Prakasam (B) Jayachandran
(C) Spencer Hatch (D) Murlidharan

498. In India ———— of the total population lives in villages
(A) One third (B) Half
(C) Two third (D) Three fourth

499. National Agriculture Technology Project was started in:
(A) 1995-96
(B) 1998-99
(C) 2000-01
(D) 2002-03

500. NATP was funded from
(A) USA (B) World Bank
(C) GOI (D) ICAR

501. National Institute of Agriculture Marketing (NIAM) is located at:
(A) New Delhi (B) Jaipur
(C) Hisar (D) Meerut

502. Under NRDP, the state government has its share:
(A) 20% (B) 75%
(C) 80% (D) 50%

503. National Demonstration was started in the year of:
(A) 1964 (B) 1965
(C) 1970 (D) 1975

504. In India, Community Development Projects were launched in:
(A) 1950 (B) 1951
(C) 1952 (D) 1962

505. The High Yielding Variety Programme started in the year of:
(A) 1966 (B) 1964
(C) 1971 (D) 1976

506. I.R.D.P. was launched in:
(A) 1980 (B) 1960
(C) 1970 (D) 1978

507. Which of the following state has no panchayati raj:
(A) Rajasthan (B) Meghalaya
(C) Bihar (D) Orissa

508. The first Agriculture University of India:
(A) P.A.U., Ludhiana (Punjab)
(B) H.A.U., Hissar (Haryana)
(C) G.B.P.U.A. & T., Pantnagar (U.A.)
(D) A.P.A.U., Hyderabad (A.P.)

509. What does the term DPAP denotes?
(A) Dry Project Area Programme
(B) Drought Prone Area Programme
(C) Development Project Area Programme
(D) None of these

510. Lead Bank scheme was started in the year:
(A) 1965 (B) 1969
(C) 1975 (D) 1964

511. The scheme of Regional Rural Banks was launched in India on:
(A) 1stOct.1975 (B) 15thAug. 1978
(C) 2ndOct. 1975 (D) 1stJan. 1970

512. Fourth five year plan was started in the year:
(A) 1966 (B) 1969
(C) 1967 (D) 1960

513. National Agricultural Extension Project was launched by:
(A) ICAR (B) IARI
(C) Govt. of India (D) State Govt.

514. There are of evaluation:
(A) Four (B) Three
(C) Five (D) Two

515. The S.F.D.A. and M.F.A.L. Programme were started in:
(A) 1970 (B) 1965
(C) 1967 (D) 1969

516. Command Area Development Programme was started in:
(A) 1962 (B) 1965
(C) 1978 (D) 1974

517. Who is the chairman of S.F.D.A.?
(A) Agriculture Minister
(B) Governor
(C) Chief Minister (D) Collector

518. The ninth five year plan was started in:
(A) 1995-96 (B) 1996-97
(C) 1997-98 (D) 1998-99

519. The Taccavi loan is given by:
(A) The bank (B) The government
(C) The co-operative society
(D) None of these

520. The principle "one village one society" was suggested by:
(A) Ashok Mehta Committee
(B) Mec Lagen Committee
(C) Wanchoo Committee
(D) Rajamannan Committee

521. T and V system of extension was developed by:
(A) M. Jackson (B) D. Benor
(C) M. Anderson (D) O.P. Dhama

522. Chairman of planning commission is:
(A) President (B) Governor
(C) Finance Minister
(D) Prime Minister

523. The major 14 banks were nationalized in the year:
(A) 1969 (B) 1970
(C) 1973 (D) 1975

524. I.R.D.P. was started in the year:
(A) 1965-66 (B) 1978-79
(C) 1964-65 (D) 1980-81

525. Which of the following guides the Block samiti:
(A) Block Pramukh
(B) V.D.O.
(C) B.D.O. and other extension officers
(D) None of these

526. Desert Development programme was started in:
(A) 1977-78 (B) 1980-81
(C) 1981-82 (D) 1983-84

527. Which of the following method is the best for selection of a leader?
(A) Statistic method
(B) Psycho-metric method
(C) Socio- metric method
(D) Bio-metric method

528. Albert Mayer of USA started a project in September, 1948 known as:
(A) Gurgaon experiment
(B) Marthandam Project
(C) Etawah pilot project
(D) None of these

529. The Gurgaon experiment was started by:
(A) R.N. Tagore (B) Vinoba Bhava
(C) F.L. Brayne (D) Mahatama Gandhi

530. The effective media of self-dependent of village of Rajasthan is:
(A) Formulation of village sponsored economic plans
(B) Extension of cities
(C) Extension of village education
(D) Service to rural unemployment in cities

531. First KVK was established in Rajasthan was:
(A) Badgaon (1952)
(B) Fathehpur (1976)
(C) Beechwal (1972)
(D) Banswara (1975)

532. National Bank for Agriculture and Rur Development came into existence on:
(A) July 22, 1984
(B) July 12, 1982
(C) October 15, 1986
(D) August 10, 1983

533. The T & V system of Agricultural extension was first introduced in the year:
(A) 1964 (B) 1954
(C) 1984 (D) 1974

534. The programme which have been merged under IRDP were:
(A) SFDA, MFAL, DPAP and CADP
(B) MFAL, SFDA, IAAP and IADP
(C) SFDA, MFAL and Lab to Land
(D) DRDA, MFAL and TV

535. The first five year plan was started in:
(A) 1947-48 (B) 1950-51
(C) 1951-52 (D) 1952-53

536. In India the use of television as a medium of communication for rural development was started in:
(A) 1964 (B) 1965
(C) 1966 (D) 1967

537. National extension service was started in the year:
(A) 1951 (B) 1952
(C) 1953 (D) 1954

538. In India major commercial banks were nationalised in:
(A) 1965 (B) 1967
(C) 1969 (D) 1971

539. Kisan Mandal System under T and V system was initiated in the year of:
(A) 1996 (B) 1995
(C) 1994 (D) 1993

540. Fore runner of C.D. programme is:
(A) Etawah project
(B) Mrthundum project
(C) Gurgaon project
(D) Sevagram

541. Which of the following is the special scheme launched by the ICAR for transfer of technology to the farmers?
(A) All India Co-ordinated Research Project
(B) Operational Research Project
(C) Lab to Land Programme
(D) National Demonstration Programme

542. Which of the following rural development project was launched before independence:
(A) Etawah Pilot Project
(B) Co-operative Project
(C) Community Development
(D) National Extension Service

543. The new name of community development programme is:
(A) Intensive Agriculture Programme
(B) C.A.D.P.
(C) Rural Development Programme
(D) None of the above

544. For demonstration the value of a new extension programme to rural peoples of the locality, the programme should be sponsored by group of:
(A) Rural people of locality
(B) Urban people
(C) Local leaders
(D) Gram Pradhan and V.L.W.s

545. The Etawah pilot project for rural development was started in:
(A) 1971 (B) 1978
(C) 1952 (D) 1948

546. The national rural development programme was replaced by food and work programme during:
(A) 1982 (B) 1980
(C) 1979 (D) 1976

547. In India, the use of television as a medium of communication for rural development was started in:
(A) 1960 (B) 1965
(C) 1966 (D) 1967

548. National demonstration was started in:
(A) 1960 (B) 1962
(C) 1950 (D) 1965

549. **List-I (Programme)** **List-II (Starting year)**
(a) Gurgaon project (1) 1920
(b) Marthandom Project (2) 1948
(c) Rural development (3) 1935
(d) F.A.O. (4) 1928

Code

	a	b	c	d
(A)	2	3	4	1
(B)	3	4	2	1
(C)	1	4	3	2
(D)	1	2	3	4

550. **List-I (Programme)** **List-II (Worker)**
(a) Indian Village Service (1) S.K. Dubey
(b) Etawah Pilot Project (2) Spencer Hatch
(c) Nilokheri Project (3) W.H. Wisher
(d) Marthandom Project (4) Albert Mayer

Code	a	b	c	d
(A)	3	4	1	2
(B)	3	2	1	4
(C)	4	3	2	1
(D)	1	2	4	3

551. **List-I (Programme)** **List-II (Founder)**
(a) Planning Commission (1) Nicholos
(b) Bhoodan Movement (2) Madras Govt.
(c) Firka development Scheme (3) Viinoba Bhave
(d) Co-operative Movement (4) Govt. of India

Code	a	b	c	d
(A)	4	3	2	1
(B)	3	4	2	1
(C)	3	2	1	4
(D)	1	2	3	4

552. **List-I (Programme)** **List-II (Starting year)**
(a) Multiple cropping scheme (1) 1950
(b) Rural development (2) 1908
(c) Sriniketan (3) 1960
(d) Sarvodaya Programme (4) 1935

Code	a	b	c	d
(A)	2	3	1	4
(B)	3	4	2	1
(C)	4	3	2	1
(D)	1	2	3	4

553. **List-I** **List-II**
(a) IIHR (1) Syria
(b) CTRL (2) Bangalore
(c) IRRI (3) Matunga
(d) ICARDA (4) Philippines

Code	a	b	c	d
(A)	1	4	3	2
(B)	4	3	2	1
(C)	3	2	1	4
(D)	2	3	4	1

554. **List-I** **List-II**
(a) NABARD (1) 1935
(b) RBI (2) 1982
(c) CACO (3) 1975
(d) RRB (4) 1965

Code	a	b	c	d
(A)	4	3	2	1
(B)	3	4	2	1
(C)	2	1	4	3
(D)	1	2	3	4

555. **List-I** **List-II**

(a) AGMARK (1) 1953
(b) ICAR (2) 1973
(c) NES (3) 1937
(d) T and V system (4) 1929

Code	a	b	c	d
(A)	1	4	3	2
(B)	4	3	2	1
(C)	3	4	1	2
(D)	2	3	4	1

556. **List-I** **List-II**

(a) First KVK of India (1) Sultanpur
(b) Total KVK in India (2) Pondichery
(c) First KVK in Rajasthan (3) 261
(d) First KVK in U.P. (4) Fatehpur

Code	a	b	c	d
(A)	2	3	4	1
(B)	3	2	1	4
(C)	3	1	2	4
(D)	1	2	3	4

557. **List-I** **List-II**

(a) CPCRI (1) Manila
(b) CTCRI (2) Jodhpur
(c) IRRI (3) Trivendrum
(d) CAZRI (4) Kesargod

Code	a	b	c	d
(A)	1	2	4	3
(B)	4	3	1	2
(C)	3	2	4	1
(D)	2	3	4	1

558. Name of the project in which 'Gaon Sathi' worked:
(A) Etawah Pilot Project
(B) Sriniketan Project
(C) Yamuna par Punarnirvan Project
(D) Marthandam Project

559. Who aspired for 'Villagers to be self-reli ant and self–sufficient'?
(A) Indira Gandhi (B) M.K. Gandhi
(C) Lal Bahadur Shastri
(D) Jawahar Lal Nehru

560. Balwant Rai Mehta Committee recommended:
(A) Three Tier Panchayati Raj System
(B) T & V System
(C) Small Farmers Development Agency
(D) None of these

561. Community Development Project started with the agreement between?
(A) Indo-Japan Govt.
(B) Indo-US Govt.
(C) Indo-British Govt.
(D) Indo-Netherland Govt.

562. "Agricultural Administrative Committee" was constituted in 1958 to review the CD programmes under whose chairmanship?
(A) R.N. Tagore
(B) Balwant Rai Mehta
(C) Raja Sunder Singh Nalagarh
(D) F.L. Brayne

563. Ford Foundation team was invited to help in:
(A) The study of agricultural scenario of India
(B) To support grow more food campaign
(C) To support food crises by way of ensuring supplies
(D) None of these

564. Ford Foundation Committee recommended:
(A) Efforts in specific areas and specific-crops
(B) Research in major crops
(C) Opening of institutions for agricultural research and education
(D) Major emphasis on extension

565. Post graduate programme in extension was started in:
(A) 1950 (B) 1952
(C) 1955 (D) 1960

566. The first extension education institute in India is established in:
(A) Nilokheri (B) Anand
(C) Hyderabad (D) Jorhat

567. NATP was launched in the year:
(A) 1998 (B) 1999
(C) 2000 (D) 2001

568. NATP is funded by:
(A) ICAR (B) FAO
(C) World Bank (D) Reserve Bank

569. Agriculture was made a state subject through:
(A) Government of India Act (1919)
(B) Royal Commission Report (1928)
(C) Ford Foundation Report (1959)
(D) GMF Enquiry Report (1955)

570. Bharti Balika was a component of:
(A) Gurgaon experiment
(B) Sevagram (C) Sriniketan
(D) Etawah project

571. The place where Gandhiji started his sewagram project:
(A) Wardha (B) Pohri
(C) Gwalior (D) Rajkot

572. The concept of village Guide was introduced by:
(A) R N Tagore (B) B Pant
(C) M K Gandhi (D) F L Brayne

573. Bombay plan was formulated in:
(A) 1943 (B) 1947
(C) 1950 (D) 1951

574. People's plan was envisaged by:
(A) Jawaharlal Nehru
(B) Viswaswrayya
(C) M N Roy (D) Mahalanobis

575. Indian Village Service was started in:
(A) Wardha (B) Pohri
(C) Gwalior (D) Lucknow

576. The person actively involved with Indian Village Service:
(A) Albert Mayer (B) F L Brayne
(C) S K Dey (D) A T Mosher

577. The concept of multipurpose village level worker was evolved from:
(A) Etawah Pilot Project
(B) Community Development Project
(C) National Extension Service
(D) Gurgaon Project

578. Etawah Pilot Project received financial assistance from:
(A) US Point 4 programme
(B) US Point 7 programme
(C) US Point 20 programme
(D) World Bank

579. The project started to rehabilitate the refugees from Pakistan during partition:
(A) Gurgaon project
(B) Nilokheri project
(C) Sriniketan project
(D) Sewagram project

580. 'Betting on the strong horse' is a slogan associated with:
(A) IRDP (B) HYVP
(C) IADP (D) SFMA

581. The IADP was popularly known as:
(A) Package programme
(B) Training programme
(C) Community programme
(D) Agriculture programme

582. Linking farm plan with credit is a characteristics feature of:
(A) IRDP (B) IADP
(C) CDP (D) Etawah project

583. The first state in India to introduce Training and Visit system was:
(A) Punjab (B) Rajasthan
(C) Tamilnadu (D) Bihar

584. National Extension Service (NES) in India was initiated on:
(A) October 2, 1952
(B) October 2, 1953
(C) January 26, 1950
(D) August 15, 1947

585. Ethopia is the origin home of:
(A) Coffee (B) Maize
(C) Rice (D) Cabbage

586. Cabbage originated in:
(A) Mexico and Central America
(B) Mediterranean
(C) Peruvian Andes
(D) South west Asia

587. Maize evolved in:
(A) Brazil
(B) Mexico and Central America
(C) USA
(D) Asia minor and Afganistan

588. The centre of origin of Almond and Apple is:
(A) South West Asia
(B) South East Asia (C) China
(D) Asia minor and Afganistan

589. The origin home of rice is:
(A) South West Asia
(B) South East Asia
(C) Mediterranean
(D) China

590. Peruvian Andes are believed to be centre of origin of:
(A) Tomato (B) Potato
(C) Tomato and Potato
(D) Tomato, Potato and Chillies

591. The origin of sunflower is believed to be in:
(A) Mexico and Central America
(B) Brazil (C) USA
(D) Peruvian Andes

592. South East Asia is thought to be centre for the origin of:
(A) Rice
(B) Rice and Sugarcane
(C) Rice, Sugarcane and Mango
(D) Rice, Sugarcane, Mango and Banana

593. The famous cultivated plant was developed in China is:
(A) Tea (B) Coffee
(C) Cotton (D) Orange

594. Production centre of Hevea Rubber is:
(A) Malayasia (B) Brazil
(C) Peruvian Andes (D) South West Asia

595. The centre of origin of Wheat is:
(A) South West Asia
(B) South East Asia
(C) Asia minor and Afganistan
(D) Mexico

596. Which is not a new world crop:
(A) Tapioca (B) Peanut
(C) Cashewnut (D) Poppy

597. Which one belong to old world:
(A) Squash (B) Pumpkin
(C) Water Melon (D) Guava

598. Select a crop which belongs to new world:
(A) Jackfruit (B) Papaya
(C) Spinach (D) Pear

599. Which crop belong to new world:
(A) Red Pepper (B) Black pepper
(C) Onion (D) Garlic

600. Which is the crop which originated in Peru and Brazil but is now mainly grown in India is:
(A) Groundnut (B) Potato
(C) Maize (D) None of these

601. The crop which originated in Peru but is most extensively grown in East Europe is:
(A) Coffee (B) Potato
(C) Rubber (D) Maize

602. Name the crop which had its birthplace in Tropical America but the centre of production is now Midwest USA:
(A) Pineapple (B) Oil palm
(C) Cocoa (D) Maize

603. Christopher Columbus landed in island of Bahmas (Carribean sea) in:
(A) 1692 (B) 1592
(C) 1492 (D) 1792

604. Tropical forests are being lost at the annual rate of:
(A) 11 million ha/yr
(B) 110 million ha/yr
(C) 1 million ha/yr
(D) 1100 million ha/yr

605. Post Vavilov megadiversity centre is:
(A) Australia (B) Siberia
(C) Andamana (D) Tibet

606. The number of centres of origin of crop plants recognized by Vavilo is:
(A) 8 (B) 11
(C) 7 (D) 21

607. Guarana is:
(A) Fodder plant (B) Oil plant
(C) Beverage plant (D) Food plant

608. The Central Marketing Department was established in India in:
(A) 1930 (B) 1947
(C) 1935 (D) 1955

609. The Food Corporation in India was born in:
(A) 1952 (B) 1960
(C) 1965 (D) 1971

610. Cotton Corporation of India is a ideal agency of:
(A) Cereal
(B) Pulses and Oilseeds
(C) Cotton (D) Jute

611. Central Warehousing Corporation was set up in:
(A) 1930 (B) 1957
(C) 1966 (D) 1970

612. Indian Standard Institute was set up in:
(A) 1935 (B) 1942
(C) 1947 (D) 1960

613. Internatinal Organisation for Standardization (I.S.O.) came into existence in:
(A) 1947 (B) 1950
(C) 1958 (D) 1965

614. The Commission or Agricultural Costs and Prices was setup in:
(A) 1950 (B) 1960
(C) 1965 (D) 1970

615. Council of State Agricultural Marketing Board was established in:
(A) 1970 (B) 1975
(C) 1988 (D) 1990

616. National Agricultural Co-operative Marketing Federation (NAFED) was established is:
(A) 1955 (B) 1958
(C) 1965 (D) 1968

617. Cultivation of hybrid rice is most popular in :
(A) China (B) India
(C) Phillipines (D) Canada

618. Central Upland Rice Research Station is located in:
(A) Masodha (B) Orissa
(C) Hyderabad (D) Bihar

619. Green revolution in wheat basically attributed to:
(A) Greater use of seeds of high yielding varieties
(B) Higher use of N, P and K fertilized
(C) Greater availability of irrigation
(D) Greater use of plant protection chemicals

620. The land/man ratio in India at present is:
(A) 0.10 ha (B) 0.15 ha
(C) 0.20 ha (D) 0.25 ha

621. A severe drought occurred during the last decade is:
(A) 1985 (B) 1986
(C) 1987 (D) 1988

622. Total flood prone area in India is approximately:
(A) 30 mha (B) 35 mha
(C) 40 mha (D) 45 mha

623. Area sown more than once is maximum in the state of :
(A) UP (B) MP
(C) Rajasthan (D) Gujarat

624. National Cooperative Development Corporation (NCDC) was established in:
(A) 1960 (B) 1961
(C) 1962 (D) 1963

625. When the quantitative restriction (QR) was over in W.T.O.?
(A) 1951 (B) 1994
(C) 1999 (D) 2001

626. National Agricultural Policy was declared in:
(A) 1951 (B) 1961
(C) 1971 (D) 1981

627. National Seed Corporation (NSC) was established in:
(A) 1960 (B) 1963
(C) 1965 (D) 1968

628. Directorate of Marketing an Inspection is situated in:
(A) Kanpur (UP)
(B) Faridabad (Haryana)
(C) Kolkata (WB) (D) Hyderabad (AP)

629. Out of the total cropped area of India, forest occupy only:
(A) 26% (B) 33% (C) 35% (D) 19.5%

630. The first Director General of Indian Council of Agricultural Research was:
(A) Dr. M.S. Swaminathan
(B) Dr N.S. Randhawa
(C) Dr. B.P Pal
(D) Dr. O.P. Gautam

631. Indian Institute of Subtropical Horticulture is situated in:
(A) Bangalore (B) Lucknow
(C) Mysore (D) New Delhi

632. The agro-climatic zones in India is divided into:
(A) Ten (B) Twelve
(C) Fifteen (D) Twenty

633. The thrust of the agricultural programmes during the Eight Plan is on:
(A) Increasing production of rice
(B) Increasing production of wheat
(C) Increasing production of oilseeds, fruits, vegetables and pulses
(D) Increasing production of cotton

634. On October 2, 1952, was launched:
(A) National Extension Service
(B) Integrated Rural Development Programme
(C) Training and Visit System
(D) Community Development Programme

635. Cooperative movement in India was started in:
(A) 1914 (B) 1947
(C) 1904 (D) 1935

636. When was the 'Minimum Wages Act' for agricultural laboures passed?
(A) 1947 (B) 1948
(C) 1951 (D) 1956

637. Training and Visit system of extension started in UP in:
(A) 1970 (B) 1980
(C) 1985 (D) 1990

638. Which one is the latest programme in India?
(A) KVK (B) NARP
(C) NAEP (D) NATP

639. Which one is the latest programme in India?
(A) ATMA (B) NATP
(C) RKVY (D) ISOPALM

640. National Extension Services in India was established in:
(A) 1952 (B) 1953
(C) 1990 (D) 1995

641. The first hybrid variety of Bajra was developed in:
(A) 1956 (B) 1965
(C) 1970 (D) 1976

642. The Bengal famine of 1942 was due to failure of paddy crop caused by:
(A) *Helminthosporium oryzae*
(B) *Xanthomonas compestris pv oryzae*
(C) *Pyricularia oryzae*
(D) *Rhizoctonia solani*

643. Highest number of sugar mills in India is located in:
(A) Maharastra (B) Tamilnadu
(C) Bihar (D) UP

644. Sustainable agriculture can be achieved by:
(A) More resources
(B) More subsidies
(C) More technological support
(D) Better prices of the produce

645. The cause of famous 'Irish famine' was:
(A) Early blight disease of potato
(B) Late blight disease of potato
(C) Blast disease of paddy
(D) Rust disease of wheat

646. Contribution of agriculture sector to GDP in India is:
(A) 10% (B) 20% (C) 30% (D) 40%

647. High yielding variety programme was started in the year :
(A) 1965 (B) 1966
(C) 1967 (D) 1968

648. N.E. Borlaug was awarded Nobel Prize in:
(A) Peace (B) Economic
(C) Agriculture (D) Physics

649. NATP as sponsored by:
(A) Planning commission
(B) Government of India
(C) ICAR (D) World Bank

650. IVLP stands for:
(A) Institution Village Linkage Programme
(B) Integrated Village Lab to Land Programme
(C) India Village Linkage Programme
(D) Indian Village Linkage Plan

651. Kisak Bharti periodical is published from:
(A) New Delhi (B) Lucknow
(C) Pantnagar (D) Hisar

652. Directorate of Knowledge management in Agriculture is situated at:
(A) Pusa, New Delhi
(B) Pusa, Samasti pur
(C) Hyderabad, AP
(D) Hisar, Haryana

653. Fakhurruddin Ali Ahmad award relate to:
(A) Best research work on Tribal people upliftment
(B) Team work in agricultural development
(C) Best technical book in hindi
(D) None of these

654. Hari Om Trust Award relate to:
(A) Best research work on Tribal people upliftment
(B) Team work in agricultural development
(C) Best technical book in hindi
(D) None of these

655. Dr. Rajendra Prasad award relate to:
(A) Best research work on Tribal people upliftment
(B) Team work in agricultural development
(C) Best technical book in hindi
(D) None of these

656. Evergreen revolution is related to:
(A) Potato (B) Food grain production
(C) Milk production
(D) Increasing production in all together

657. Rainbow revolution is related to:
(A) Potato
(B) Food grain production
(C) Milk production
(D) Increasing production in all together

658. Green revolution is related to:
(A) Potato
(B) Food grain production
(C) Milk production
(D) Increasing production in all together

659. Brown /Black revolution is related to:
(A) Potato
(B) Food grain production
(C) Milk production
(D) Non conventional energy

660. Round revolution is related to:
(A) Potato
(B) Food grain production
(C) Milk production
(D) Increasing production in all together

661. White revolution is related to:
(A) Potato
(B) Food grain production
(C) Milk production
(D) Increasing production in all together

662. Blue revolution is related to:
(A) Potato (B) Food grain production
(C) Milk production
(D) Fish production

663. Grey revolution is related to:
(A) Fertilizer
(B) Food grain production
(C) Milk production
(D) Increasing production in all together

664. Red revolution is related to:
(A) Potato
(B) Meat/tomato production
(C) Milk production
(D) Increasing production in all together

665. Golden revolution is related to:
(A) Potato
(B) Food grain production
(C) Horticultural crop
(D) Increasing production in all together

666. Pink revolution is related to:
(A) Potato
(B) Food grain production
(C) Milk production
(D) Prawan production

667. Silver revolution is related to:
(A) Chicken/eggs production
(B) Food grain production
(C) Milk production
(D) Increasing production in all together

668. Yellow revolution is related to:
(A) Potato
(B) Oilseed production
(C) Milk production
(D) Increasing production in all together

669. The agricultural Price Commission was set up to evolve balanced and integrated price structure in the year:
(A) 1962 (B) 1963
(C) 1964 (D) 1965

670. In Uttar Pradesh Krishi Utpadan Mandi Adhiniyam was passed in:
(A) 1962 (B) 1963
(C) 1964 (D) 1965

671. International financing agency to support National Agricultural Technology Project in India is:
(A) FAO (B) World Bank
(C) Asian Bank (D) UNDP

672. Which year is declared as International year of rice:
(A) 2004 (B) 2005
(C) 2006 (D) 2007

673. Which year is declared as International year of parthenium/neem patented year:
(A) 2004 (B) 2005
(C) 2006 (D) 2007

674. Which year is declared as International year of Desert and desertification:
(A) 2004 (B) 2005
(C) 2006 (D) 2007

675. Which year is declared as International year of water:
(A) 2004 (B) 2005
(C) 2006 (D) 2007

676. Which year is declared as International year of potato:
(A) 2008 (B) 2005
(C) 2006 (D) 2007

677. Which year is declared as International year of Astronomy:
(A) 2008 (B) 2009
(C) 2010 (D) 2011

678. Which year is declared as International year of bio-diversity:
(A) 2008 (B) 2009
(C) 2010 (D) 2011

679. Which year is declared as International year of forest:
(A) 2008 (B) 2009
(C) 2010 (D) 2011

680. Who says "Everything else can wait, but not agriculture"?
(A) Shri. J.L. Nehru
(B) Dr. A.P.J. Abul Kalam
(C) Dr. H. Rembold
(D) Mahatma Gandhi

681. Who says "Agriculture is the culture of the times as/or all culture"?
(A) Shri. J.L. Nehru
(B) Dr. A.P.J. Abul Kalam
(C) Dr. H. Rembold
(D) Mahatma Gandhi

682. Who says "We have to have the second green revolution for ever increasing population of the country"?
(A) Shri. J.L. Nehru
(B) Dr. A.P.J. Abul Kalam
(C) Dr. H. Rembold
(D) Mahatma Gandhi

683. Who says "India is sitting on gold mines will all its Neem trees"?
(A) Shri. J.L. Nehru
(B) Dr. A.P.J. Abul Kalam
(C) Dr. H. Rembold
(D) Mahatma Gandhi

684. Total food grain production in India during 2011- 12 was:
(A) 235.00 mt (B) 250.55 mt
(C) 257.44 mt (D) 267.00 mt

685. Total pulses production in India during 2011- 12 was:
(A) 17.2 mt (B) 18.2 mt
(C) 19.2 mt (D) 20.2 mt

686. Total Oilseeds production in India during 2011- 12 was:
(A) 23.5 mt (B) 30.0 mt
(C) 35.1 mt (D) 35.5 mt

687. Total wheat production in India during 2011- 12 was:
(A) 93.90 mt (B) 95.54 mt
(C) 95.98 mt (D) 96.31 mt

688. Total rice production in India during 2011- 12 was:
(A) 85.00 mt (B) 90.54 mt
(C) 95.98 mt (D) 104.3 mt

689. Food grain production in India during 2011- 12 increased in comparison to 2010- 11 was:
(A) 12.95 mt (B) 13.95 mt
(C) 14.95 mt (D) 15.95 mt

690. Total pulses production in India during 2010-11was:
(A) 17.24 mt (B) 18.24 mt
(C) 19.24 mt (D) 20.24 mt

691. Total oilseed production in India during 2010-11was:
(A) 23.5 mt (B) 25.0 mt
(C) 32.5 mt (D) 35.5 mt

692. Wheat production in India during 2011- 12 increased in comparison to 2010- 11was:
(A) 4.03 mt (B) 5.03 mt
(C) 6.03 mt (D) 7.03 mt

693. Rice production in India during 2011- 12 increased in comparison to 2010-11was:
(A) 8.34 mt
(B) 9.34 mt
(C) 10.34 mt
(D) 11.34 mt

694. Area under total food grain in India during 2011- 12 was:
(A) 115.05 mha (B) 125.03 mha
(C) 128.04 mha (D) 130.28 mha

695. Area under total pulses in India during 2011- 12 was:
(A) 23.28 mha (B) 26.40 mha
(C) 24.78 mha (D) 25.85 mha

696. Area under total oilseeds in India during 2011- 12 was:
(A) 27.56 mha (B) 25.96 mha
(C) 27.22 mha (D) 26.44 mha

697. Area under wheat in India during 2011- 12 was:
(A) 29.90 mha (B) 29.07 mha
(C) 28.46 mha (D) 27.75 mha

698. Area under rice in India during 2011- 12 was:
(A) 35.85 mha (B) 40.50 mha
(C) 43.97 mha (D) 45.85 mha

699. The Central AGMARK lab is located at:
(A) Lucknow (B) Nagpur
(C) Bhopal (D) Hyderabad

700. Cultivation of duram wheat is primarily confined to:
(A) Central India
(B) Southern India
(C) Central & Southern India
(D) UP hills

701. Tick the correct agro-climatic zone of UP which has highest average population density per km^2:
(A) Bundelkhand zone
(B) Central plains zone
(C) Hills zone
(D) North-eastern plains zone

702. National Food Security Mission started in which five year plan:
(A) 9th (B) 10th
(C) 11th (D) 12th

703. Bharat Ratna Dr. C. Subramanian Award is related to:
(A) Best agriculture teacher
(B) Best technical book
(C) Tribal people upliftment
(D) Best research in agriculture

704. In India, during 1960 milk production was 2 crore metric tones, it was rose in 2010-11 to:
(A) 10.2 crore metric tones
(B) 11.2 crore metric tones
(C) 12.2 crore metric tones
(D) 13.2 crore metric tones

705. Father of White Revolution in India is:
(A) Dr. M.S. Swaminathan
(B) Dr. Vergeese Kurien
(C) Dr. R.S. Paroda
(D) Dr. B.P. Pal

706. National Dairy Development Board was established in:
(A) 1965 (B) 1966
(C) 1967 (D) 1968

707. Operation of Flood was launched in:
(A) 1969 (B) 1970
(C) 1971 (D) 1972

708. Gujarat Cooperative Milk Marketing Federation (GCMMF) was established in:
(A) 1971 (B) 1972
(C) 1973 (D) 1074

709. India Rank in milk production during 2010-11 was:
(A) First (B) Second
(C) Third (D) Fourth

710. Marketing of AMUL is to be done by:
(A) AMUL it self (B) GCMMF
(C) Government of India
(D) Gujarat State

711. Who is known as milk man of India and Amul man is?
(A) Dr. M.S. Swaminathan
(B) Dr. Vergeese Kurien
(C) Dr. R.S. Paroda
(D) Dr. B.P. Pal

Chapter 3

EXTENSION PRINCIPALS, APPROACHES AND PROGRAMMING

1. Discouraging the application of chemical fertilizers and advocating organic farming:
(A) Principle of transfer
(B) Principle of set
(C) Principle of disassociation
(D) Principle of association

2. The extension scientific knowledge to villagers by specialists of the same area:
(A) Principle of local leadership
(B) Principle of Panchayati Raj
(C) Principle of grass root organisation
(D) Principle of self activity

3. Learning to use a sprayer correctly requires repeated trying of the instrument:
(A) Principle of motivation
(B) Principle of attitude
(C) Principle of practice
(D) Principle of satisfaction

4. If a farmer has learned the technique of water management in a particular crop, they should be able to use this method in other crops as well:
(A) Principle of achievement
(B) Principle of transfer
(C) Principle of disassociation
(D) Principle of association

5. Issue of advertisement regarding public health and sanitation at the one set of an epidemic outbreak is a classical example of:
(A) Principle of set (B)Principle of time
(C) Principle of association
(D) Principle of transfer

6. When planting a crop in lines gives better yield, the farmers may be advised not to practice broadcasting:
(A) Principle of achievement
(B) Principle of transfer
(C) Principle of disassociation
(D) Principle of association

7. When farmers are ready to cooperate, with good guidance, they may be able to form a cooperative society:
(A) Principle of motivation
(B) Principle of attitude
(C) Principle of readiness
(D) Principle of satisfaction

8. When a farmer develop a favourable set towards scientific method of cultivation, they shall learn more on the topic:
(A) Principle of motivation
(B) Principle of attitude
(C) Principle of readiness
(D) Principle of satisfaction

9. The attainment of perfection demands that undesirable and useless movements are replaced by desirable and useful movements:
(A) Principle of motivation
(B) Principle of attitude
(C) Principle of readiness
(D) Principle of satisfaction

10. Favourable experience of planting trees motivate tribal farmers to collect samplings from the forest nursery:
(A) Principle of motivation
(B) Principle of attitude
(C) Principle of readiness
(D) Principle of satisfaction

11. When insect have appeared or are likely to appear in crops, farmers shall learn about plant protection:
(A) Principle of timing
(B) Principle of adaptation
(C) Principle of maximum response
(D) Principle of satisfaction

12. When villagers assist the extension worker at different levels in implementing a new programme:
(A) Principle of cooperation
(B) Principle of satisfaction
(C) Principle of implementation
(D) Principle of interest

13. When farmers use their crop loan for the very purpose for which the loan is sanctioned:
(A) Principle of achievement
(B) Principle of completion
(C) Principle of clarity of objectives
(D) Principle of satisfaction

14. The principle under which custom religion is findout?
(A) Principle of social change
(B) Principle of cultural differences
(C) Principle of adaptability
(D) Principle of whole family

15. The principle under which children and family are included:
(A) Principle of social change
(B) Principle of cultural differences
(C) Principle of adaptability
(D) Principle of whole family

16. The basis of any extension approach is:
(A) Individual is the end
(B) Individual is a means to an end
(C) Rural development
(D) Transfer of technology

17. Extension approach in India generally follows the norms of:
(A) Cooperation
(B) Participation
(C) Prescription
(D) Mobilization

18. The most preferred approach in extension work is:
(A) Autocratic (B) Democratic
(C) Lessiz faire (D) None of these

19. The interactive socio-political process motivated by desire to meet an individuals needs through collective actions on mutual benefit basis:
(A) Participation (B) Cooperation
(C) Extension (D) Collaboration

20. People participate by taking initiatives independent of external institutions to change systems:
(A) Self-mobilization
(B) Interactive participation
(C) Functional participation
(D) Passive participation

21. People participation by being told what is going to happen or has already happened:
(A) Self-mobilization
(B) Interactive participation
(C) Functional participation
(D) Passive participation

22. The participatory approach of extension recognizes the importance of:
(A) Exploitation of local resources
(B) Farm planning
(C) Democratic way of problem solving
(D) Learning from farmers

23. Participatory development approach in otherwards is called:
(A) Bottom up approach
(B) Top down approach
(C) Circle centered approach
(D) Centralized approach

24. The extension approach wherein short-run change is the measure of success:
(A) Project approach
(B) Commodity specialized approach
(C) Cost sharing approach
(D) General agricultural extension approach

25. The extension approaches wherein success is measured in terms of farm people's willingness and ability to share some of the cost:
(A) Project approach
(B) Commodity specialized approach
(C) Cost sharing approach
(D) General agricultural extension approach

26. The privatization approach wherein government grants private entity authority to provide a particular service within a specific geographical area:
(A) Contracting (B) Fanchise
(C) Vouchers (D) Self help

27. The privatization approach where in the public pays and private delivers:
(A) Service within a specific geographical area:
(A) Contracting (B) Fanchise
(C) Vouchers (D) Self help

28. The privatization approach most commonly used in sugarcane production sector:
(A) Contracting (B) Fanchise
(C) Vouchers (D) Self help

29. The extension approaches wherein mea sure of success is the farming people's attendance at and participation in the school's agricultural extension activities:
(A) Project approach
(B) Commodity specialized approach
(C) Educational institution approach
(D) General agricultural extension approach

30. Extension is fairly centralized and is oriented towards one commodity or crop and the agent has many functions:
(A) Project approach
(B) Commodity specialized approach
(C) Educational institution approach
(D) General agricultural extension approach

31. The extension approaches wherein the success is measured in terms of production increases of the particular crops covered by the programme:
(A) Training and visit approach
(B) Commodity specialized approach
(C) Educational institution approach
(D) General agricultural extension approach

32. The extension approaches wherein success is measured by the extent to which farming people adopt the technologies developed by the programme and continue using them over time:
(A) Commodity specialized approach
(B) General agricultural extension approach
(C) Cost sharing approach
(D) Farming system development approach

33. The farming system development approach essentially aims at:
(A) Development of technology which fits the need
(B) Development of farmer's innovation
(C) Increasing farm production and productivity
(D) Integrated development of the farm

34. "How do I help people achieve what they themselves want to achieve":
(A) "DO TO" strategy
(B) "DO FOR" strategy
(C) "DO WITH" strategy
(D) "DO AGAINST" strategy

35. The communication strategy embedded in the transfer of technology paradigm:
(A) Persuasive + Paternalistic
(B) Persuasive + Participatory
(C) Educational + Paternalistic
(D) Educational + Participatory

36. The strategy that often lead to the so called progressive farmer approach in agricultural extension:
(A) "DO TO" strategy
(B) "DO FOR" strategy
(C) "DO WITH" strategy
(D) "DO AGAINST" strategy

37. "How do I develop an offering which my target clients want" relates with:
(A) "DO TO" strategy
(B) "DO FOR" strategy
(C) "DO WITH" strategy
(D) "DO AGAINST" strategy

38. The most important factor in FSR/E is:
(A) Farmer (B) Researcher
(C) Extensionist (D) All of the there

39. Which of the followings is not a factor contributing to the limited application is FSR/E:
(A) Action based approach
(B) Multidisciplinary approach
(C) Resource poor clientale
(D) Limited research output

40. The extension approach wherein success is measured in terms of the rate of take-up of the recommendations, and increases in national production:
(A) Project approach
(B) Commodity specialized approach
(C) Cost sharing approach
(D) General agricultural extension approach

41. Under subsistent farming situation agricultural development is characterized by:
(A) Commercial orientation
(B) Subsidy orientation
(C) Information orientation
(D) Companion-mission orientation

42. Which among the following is a bottom up approach:
(A) Farming system approach
(B) Commodity specialized approach
(C) Training and visit approach
(D) Cost sharing approach

43. ———— is a statement of situation, objectives, problems and solutions
(A) A plan of work
(B) Extension programme
(C) A calendar of work
(D) A project

44. ———— is a plan of work arranged chronologically.
(A) A project (B) A calendar of work
(C) A plan of work
(D) Programme planning

45. A sound extension programme building based on
(A) Analysis of the facts in the situation
(B) Select problems based on needs
(C) Determine objective and solutions
(D) All of these

46. First step for making a programme planning includes
(A) Evaluation
(B) Analysis of the situation and determining problems
(C) Deciding an objectives (D) Teaching

47. The main objective of evaluation in extension is
(A) To facilitate effective decision making directly goes to conclusion
(B) To facilitate effective decision making without jumping to conclusion
(C) To see weak points in programme
(D) None of these

48. When an extension programme is successfully completed and the farmer and his family members are satisfied by solving the problem and getting a new idea, it is called:
(A) The principle of achievement
(B) The principles of completion
(C) The principles of maximum response
(D) The principles of satisfaction

49. The strategy for supporting the rapidly expanding agricultural programme requires:
(A) That cultivators are educated in favour of modern farming
(B) That efforts are made to manufacture more fertilizers
(C) That efforts are made to provide more irrigational facilities
(D) All of the these

50. A good extension programme should be:
(A) Rigid (B) Flexible
(C) Both (A) and (B) (D) None of these

51. Determination of extension programme consists of:
(A) Definition and finding solutions
(B) Preparing plan of work and carry out the plan of work
(C) Collection and analysis of data's determination of needs and objectives
(D) Both (A) and (B)

52. ATIC has been initiated with:
(A) Single window concept of technology dissemination
(B) Diversification of agricultural technology
(C) Intensification of technologies
(D) Intensification and diversification of technologies

53. PARA extension workers are considered to be:
(A) The workers working with NGOs
(B) Rural youths trained to take up small extension works on payment basis
(C) Extension workers having specialization in rural areas and enterprises
(D) Workers with knowledge of all the enterprises attached to farm house hold

54. Which are the main reasons for the limited success of T & V in India:
(A) Heavy infrastructure investment
(B) Assumptions at different levels
(C) Top to down approach
(D) All the these

55. Participating extension management envisages:
(A) Participation of all the stakeholders in the project
(B) Participation of farmers in the project
(C) Technology transfer by extension agencies
(D) None of these

56. What is "Cyber extension":
(A) Extension through group approach
(B) Extension through use of television and other mass media
(C) Extension through computers
(D) Extension through NGOs

57. Gender development in extension programmes is related to:
(A) Empowering farm women
(B) Development of both genders on equal basis
(C) Development of households with emphasis on gender
(D) None of these

58. People's participate in an extension programme is significant when:
(A) Local leaders participate
(B) Literate section of villagers participate
(C) Gram panchayat members participate
(D) Majority of villagers participate

59 ....is that of working with rural people through out of school education along with their current interest and needs for getting overall development of rural families:
(A) Extension education
(B) Extension process
(C) Extension jobs
(D) Extension service

60. The extension programme is a statement of:
(A) Situation and objectives
(B) Solution and problems
(C) Problems and situation
(D) All of the these

61. Who, what, when and where are part of:
(A) Plan of work
(B) Calendar of work
(C) Programme evaluation
(D) Decision making

62. The plan of work when arranged chronologically is:
(A) Calendar of work
(B) Calendar of operation (C) Objective
(D) Programme

63. A sound programme must have:
(A) Permanence with rigidity
(B) Permanence with flexibility
(C) Temporarity with rigidity
(D) Temporarity with flexibility

64. Extension programme planning is a:
(A) Educational process
(B) Administrative process
(C) Effective process
(D) Uniform process

65. The first step in extension programme planning at local level is:
(A) Determination of objectives
(B) Analysis of existing situation
(C) Decision of extension teaching method
(D) Determination of target

66. The participation of people in extension programme is:
(A) Compulsory (B) Voluntary
(C) Involuntary (D) None of these

67. The pilot study in extension programme represent:
(A) Post evaluation
(B) Concurrent evaluation
(C) Pre evaluation
(D) Monitoring

68. Last stage of programme planning process is:
(A) Evaluation (B) Auditing
(C) Supervision (D) Reconsideration

69. Under T & V system what is mainly responsible to make regular contact with farmers?
(A) Deputy Director (Extension)
(B) Kisan Sahayak/Sewak
(C) SDAEO
(D) BDO

70. The extension work must be based on needs and interests of:
(A) People (B) Leader
(C) Officers (D) Government

71. Need based and skill oriented vocational training to practicing farmers and employed extension personnel is provided through:
(A) Training and Visit
(B) Krishi Vigyan Kendra
(C) National Demonstration
(D) Lab to Land programme

72. Which of the following scheme have been merged under Jawahar Rozgar Yojna:
(1) NREP (2) RLEGP
(3) IRDP (4) TRYSEM
(A) 1 & 2 (B) 1 & 3
(C) 3 & 4 (D) All of these

73. Operational flood programme relate to:
(A) Conserving soil
(B) Controlling floods
(C) Boosting milk supply
(D) Funds for flood victims

74. The first step in extension programme planning is:
(A) Analysis of situation
(B) Fixing objectives
(C) Collection of facts
(D) Determination of goals

75. The best method to convince people is:
(A) Video
(B) Demonstration
(C) Classroom teaching
(D) Group discussion

76. I.B.R.D. is synonymous to:
(A) World Bank
(B) Asian Development Bank
(C) International Monetary Fund
(D) World Trade Organisation

77. Lab to Land Programme was started by:
(A) Government of India
(B) Government of UP
(C) Smt. Indira Gandhi
(D) ICAR

78. Read the following statement and mark your answer according to code given below:
Principle of extension is to:
I. Plan for Gram Panchayat members
II. Plan for whole village
III. Plan for youth, women and farmers
IV. Plan for village leaders
Answer code:
(A) I and II are correct
(B) II and III are correct
(C) III and IV are correct
(D) IV and IV are correct

79. The components of programme planning are:
(A) Simple (B) Complex
(C) Easy (D) None of these

80. Components of programme planning is/are
(A) People their needs
(B) Useful technology
(C) The educational process
(D) Analysing situation
(E) All of these

81. Principles of programme planning is/are:
(A) It is based on an analysis of the facts of the situation
(B) Based on the needs and interests of the local people
(C) It determines priorities, define objectives and solution which differ satisfaction
(D) All of these

82. A plan of economic development of India is a:
(A) The Bombay plan
(B) Peoples plan
(C) Visvesvaraya plan
(D) Gandhian plan

83. Bombay plan was prepared by eight business man of Bombay in:
(A) 1944
(B) 1945
(C) 1946
(D) 1947

84. The main objective of Bombay plan was:
(A) To double up the per capita income within the period of 5 years
(B) To double up the per capita income within the period of 10 years
(C) To double up the per capita income within the period of 15 years
(D) To double up the per capita income within the period of 20 years

85. How many stages of Bombay plan:
(A) The people plan
(B) The Visvesvaraya plant
(C) The Gandhian plan (D) All of these

86. People's plan was prepared by:
(A) M.N. Roy (B) M. Visvesvaraya
(C) S.N Agarwal (D) None of these

87. Visvesvaraya plan was prepared by:
(A) M.N. Roy (B) M. Visvesvaraya
(C) S.N Agarwal (D) None of these

88. Gandhian plan was prepared by:
(A) M.N. Roy (B) M. Visvesvaraya
(C) S.N Agarwal (D) None of these

89. Basic principle of people's plan was:
(A) Production for use s distinct from production for exchange
(B) To raise the country's economic efficiency to a level at which the average citizen would be able to find employment and to earn a livelihood
(C) To raise the material as well as the cultural level of the masses to a basic standard of living within a period of 10 years.
(D) All of these

90. Basic principle of Visvesvaraya plan was:
(A) Production for use s distinct from production for exchange
(B) To raise the country's economic efficiency to a level at which the average citizen would be able to find employment and to earn a livelihood
(C) To raise the material as well as the cultural level of the masses to a basic standard of living within a period of 10 years.
(D) All of these

91. Basic principle of Ganghian plan was:
(A) Production for use s distinct from production for exchange
(B) To raise the country's economic efficiency to a level at which the average citizen would be able to find employment and to earn a livelihood
(C) To raise the material as well as the cultural level of the masses to a basic standard of living within a period of 10 years.
(D) All of these

92. The concept of T and V system was evolved in:
(A) 1971 (B) 1972
(C) 1973 (D) 1974

93. The concept of T and V system was evolved in 1973 by:
(A) Daniel Benor
(B) Albert Mayer
(C) F. L. Bryne
(D) Arthur T.Mosher

94. T and V system firstly evolved in:
(A) France
(B) Turkey
(C) Nigeria
(D) USA

95. Government sponsored rural development prograamme is:
(A) Shantineketan programme
(B) Sarvodaya programme
(C) T and V
(D) None of these

96. T and V system introduced in India in:
(A) 1973 (B) 1974
(C) 1075 (D) 1976

97. T and V system is introduced in India in 1974 in:
(A) Rajasthan canal area
(B) Chambal command area in MP
(C) Both (A) and (B)
(D) None of these

98. The basic objective of T and V system is/are:
(A) To improve the effectiveness of agricultural extension
(B) Transfer of technical literacy increases agricultural production
(C) Transfer of technical literacy increases agricultural production
(D) To provide training to youth for agricultural and home development

99. Training and visit programme also called:
(A) Intensive agricultural Extension
(B) Sarvodaya programme
(C) Indian Village Scheme
(D) Shatineketan programme

100. The credit for the success of KVK is goes to:
(A) Dr. MS. Swaminathan
(B) Dr. Vergees Kurien
(C) Dr. Chandrika Prasad
(D) Dr. B.P. Pal

101. In the panchayati raj system, village panchayat is the:
(A) Lowest unit (B) Middle unit
(C) Higher unit (D) Highest unit

102. General body of panchayat is:
(A) Zila sabha
(B) Tehsil sabha
(C) Block sabha
(D) Gram sabha

103. The basic idea and concept of community development propgramme was borrowed from:
(A) England (B) USA
(C) Canada (D) China

104. 4H-Club was established in:
(A) 1900 (B) 1910
(C) 1920 (D 1930

105. 4 H – Club was established in:
(A) Russia
(B) Canada
(C) America
(D) Japan

106. Basic aim of 4H-Club is:
(A) To improve training at the kendras or out of the centre
(B) Transfer of technical literacy increases agricultural production
(C) To provide training to youth for agricultural and home development
(D) To improve the effectiveness of agricultural extension

107. 4H-Club means Heart, Head, Hand and Health were explained by:
(A) Dr. MS. Swaminathan
(B) Dr. Vergees Kurien
(C) Dr. Chandrika Prasad
(D) A.B. Graham

108. Panchayati raj was first started at:
(A) Nagaur (Rajasthan)
(B) Andhra Pradesh
(C) Both (A) and (B)
(D) None of these

109. Panchayati raj was started in:

(A) 1958	(B) 1959
(C) 1960	(D) 1961

110. The main objective of KVK is:
(A) To improve training at the kendras or out of the centre
(B) Transfer of technical literacy increases agricultural production
(C) To provide training to youth for agricultural and home development
(D) To improve the effectiveness of agricultural extension

111. First KVK was started at Pandicherry in 1974 with the aim of:
(A) To improve training at the kendras or out of the centre
(B) Transfer of technical literacy increases agricultural production
(C) To provide training to youth for agricultural and home development
(D) To improve the effectiveness of agricultural extension

Chapter 4

EXTENSION EDUCATION, TEACHING AND METHODS

1. The goal of Extension Education is to promote:
(A) Income of the farmers
(B) Area of the new crop
(C) Scientific outlook
(D) Production of the crops

2. The basic element of the core of Extension Education is:
(A) Men himself
(B) Man and environment
(C) Man created devices
(D) All of these

3. Principle of extension education is:
(A) Learning by doing
(B) Learning by seeing
(C) Learning by reading
(D) Learning by hearing

4. Goal of extension education is:
(A) To promote income of farmers
(B) To promote production of the crops
(C) To promote new crops
(D) To promote scientific out look

5. Purpose of extension evaluation is:
(A) To identify the weak points
(B) To identify the strong points
(C) To identify the gaps and errors
(D) To identify all the above

6. The fundamental objective of extension education is:
(A) To provide the farmer knowledge and help that will enable him to increase his income
(B) To encourage the farmer to grow own food, eat well and live well
(C) The development of the people
(D) None of these

7. Extension education is:
(A) A discipline (B) A profession
(C) Discipline and profession both
(D) None of these

8. The objectives of education are decided by:
(A) Social sciences (B) Physical sciences
(C) Normative sciences
(D) Biological sciences

9. Less intensive and less effective method of extension is:
(A) Individual (B) Group
(C) Mass (D) Both (A) & (B)

10. The goal of extension education is to promote:
(A) Income of the farmers
(B) Area of the new crop
(C) Scientific outlook
(D) Production of the crops

11. Extension education is different from formal education in a sense that:
(A) We start from practical and end up with theory
(B) We start from theory and end up with practical
(C) We teach practical only
(D) We teach theory only

12. Extension education aims at:
(A) Increasing agricultural production
(B) Increasing Milk production
(C) Changing knowledge, skills and attitudes
(D) Changing administrative system

13. In extension education communication is a:
(A) Process (B) Method
(C) Idea
(D) Expression of thoughts

14. To induce emotional participation of the local community and to create a conductive psychological climate for adoption of a new practice, the method of extension generally applied is:
(A) Competition (B) Exhibitions
(C) Fair (D) Field trails

15. The method of extension education for effectively impressing the people about a particular recommended practice suitable for a particular area is:
(A) Radio (B) Tours
(C) Charts (D) Demonstrations

16. The objective of education are decided by:
(A) Biological science
(B) Physical science
(C) Normative science
(D) Social science

17. The extension of scientific knowledge to villagers by specialists of the same area is known as the principle of:
(A) Panchayat Raj
(B) Local leadership
(C) Modernization
(D) None of them

18 In extension we prefer:
(A) Laiser fair approach
(B) Democrative approach
(C) Automatic approach
(D) None of the these

19. The two important principles of extension education are:
(A) Participation and leadership
(B) Participation and class room teaching
(C) Leadership and classroom teaching
(D) Leadership and subject matter specialist

20. Which of the following extension programmes linked farm plan with credit for farmers?
(A) Integrated Rural Development Programme
(B) Community Development Programme
(C) Intensive Agricultural Development Programme
(D) Etawah Pilot Project

21. Which one of the following categories the discipline of extension education belong?
(A) Pure science
(B) Applied biological science
(C) Applied behaviour science
(D) Art

22. An extension worker practically explains Japanese method of paddy cultivation to a group of farmers and also show its superiority over the traditional method. In the process, which of the following communication methods have been used by them?
(1) Method demonstration
(2) Result demonstration
(3) Field visit
(4) Study tour
Select the correct answer using the codes given below:
(A) 1, 2, 3 and 4
(B) 1 and 3
(C) 2 and 3
(D) 1, 2 and 4

23. Consider the following statements: Extension in the long run means change in:
(A) Knowledge (2) Skill
(3) Work (4) Attitude
Of these statements:
(A) 1, 2 and 3 are correct
(B) 1, 3 and 4 are correct
(C) 1, 2 and 4 are correct
(D) 2, 3 and 4 are correct

24. Extension workers should be well versed with economic principles:
(A) Never
(B) Partially
(C) Both 'A' and 'B'
(D) Without it they can't give effective advice from business point of view

25. Extension education focuses upon development of:
(A) An individual (B) A family
(C) A social group (D) A community

26. Aim of extension work is to bring about the changes:
(A) Toward betterment
(B) In the heart of peoples
(C) In rural areas
(D) None of these

27. The objective of TRYSEM is to provide technical skills to the Rural youth from families below the poverty line. The target group comprises rural youth between the age of:
(A) 20 and 25 (B) 18 and 35
(C) 14 and 35 (D) 14 and 3

28. An extension education job is:
(A) To educate the people
(B) To advice the Urban people
(C) To advice the Rural people
(D) All of these

29. Extension work must be based on:
(A) The interest of Gram Pradhan
(B) The interest of BDO
(C) The needs and interest of the people
(D) None of these

30. In India, the development of poorers from all the sides is possible only by:
(A) More money supply
(B) Making strong rules
(C) Extension education
(D) All of these

31. In India the development of poorer sections from all the sides possible only by:
(A) Extension education
(B) Making strong rules
(C) More money supply
(D) All of these

32. The basic element at the core of extension education is:
(A) Man itself
(B) Man and Environment
(C) Institutions (D) Farmers

33. Extension education is:
(A) Formal education
(B) Informal education
(C) Non formal education
(D) All of these

34. The basic objective of extension education is to create opportunities for:
(A) Greater awareness
(B) Effective learning
(C) Promoting knowledge
(D) Feeling aloof

35. The basic logic applicable in extension education:
(A) Horizontal logic
(B) Deductive logic
(C) Inductive logic
(D) Lateral logic

36. Education is a process of:
(A) Interaction
(B) Learning
(C) Socialization
(D) Institutionalization

37. Subject and target group specificity are characteristics feature of:
(A) Formal education
(B) Informal education
(C) Non formal education
(D) Elementary education

38. The objectives of education are decided by:
(A) Biological sciences
(B) Physical science
(C) Normative science
(D) Social science

39. The last stage in the extension education process is:
(A) Evaluation (B) Reconsideration
(C) Adoption (D) Teaching

40. Education is essentially considered as production of desirable change in:
(A) Knowledge (B) Attitude
(C) Practice (D) Behaviour

41. The two important principles of extension education are:
(A) Participation and leadership
(B) Participation and classroom teaching
(C) Leadership and classroom teaching
(D) Leadership and subject knowledge

42. The method of extension education for effectively impressing the people about a particular recommended practice suitable for a particular area is:
(A) Radio (B) Demonstration
(C) Tours (D) None of these

43. To prove the worth of a new practice suited to a particular area, the best method of extension education is:
(A) Method demonstration
(B) Result demonstration
(C) Field trails
(D) Farm visits

44. The steps in extension teaching was put forwarded by:
(A) Paul Leagans
(B) Curl Lewin
(C) Ensminger
(D) Wilson & Gallup

45. For effective Extension teaching, one must use:
(A) Combination of appropriate aids
(B) Only visual aids
(C) Only audio aids
(D) Printed literature

46. Correct sequence of steps in extension teaching is:
(A) Attention – Interest – Desire – Conviction – Action – Satisfaction
(B) Desire – Interest – Attention – Conviction – Action – Satisfaction
(C) Interest – Attention – Desire – Conviction – Action – Satisfaction
(D) Action – Attention – Interest – Conviction – Desire – Satisfaction

47. Which methods of extension teaching is a way of showing people the value of an improved practice:
(A) Result demonstration
(B) Method demonstration
(C) Newspaper
(D) Circular letter

48. Leaflet is a single sheet of paper folded to make a ———— page piece of printed matter:
(A) 4 (B) 6 (C) 8 (D) 10

49. For one talk, how many flash cards should be used?
(A) 10 – 12 (B) 12 – 14
(C) 14 – 18 (D) 20 – 22

50. Steps of extension teaching are:
(A) Source, Message, Channel, Recommendation etc.
(B) Awareness, Interest, Trial, Evaluation
(C) Attention, Interest, Desire, Conviction, Action and Satisfaction
(D) None of these

51. The process of arranging situation in which the things to be learned are called to the attention of the learners, their interest developed, desire aroused and action promoted:
(A) Communication
(B) Training
(C) Teaching
(D) Learning

52. Steps in teaching is shortly called as:
(A) AIETA (B) AIDCAS
(C) KPDC (D) AIDACS

53. In extension teaching the teacher should first know:
(A) Subject matter
(B) Teaching aids
(C) Place of teaching
(D) Attitude of the learner

54. Effective teaching is:
(A) For informing people
(B) For transforming people
(C) Like filling a bucket
(D) Done by trial and error

55. When an extension worker approaches large number of people, spread over large areas for disseminating new improved production technology, it is called:
(A) Mass Communication
(B) Undirectional communication
(C) Public communication
(D) Group communication

56. Written form of extension teaching methods does not include:
(A) Bulletins (B) News articles
(C) Circular letters (D) None of these

57. To show the relative worth of a new practice, the extension method best suited is:
(A) Result demonstration
(B) Method demonstration
(C) Campaign (D) Group discussion

58. Result demonstration is conducted by:
(A) Researcher
(B) Extension worker
(C) Farmer
(D) All of these

59. Method demonstration works on the principle of:
(A) Learning is believing
(B) Learning by teaching
(C) Seeing and doing
(D) Learning by doing

60. Result demonstration is done in:
(A) Research station
(B) Farmer's field
(C) District farm
(D) None of these

61. Result demonstration works on the principle of:
(A) Learning is believing
(B) Learning by teaching
(C) Seeing is believing
(D) Learning by doing

62. An extension method that can also be used as a group teaching method:
(A) Campaign
(B) Method demonstration
(C) Office calls
(D) Farm and home visits

63. Education cause:
(A) Change in knowledge
(B) Change in attitude
(C) Change in skills
(D) All of these

64. Which extension method is evolved in result demonstration, posters and movies?
(A) Spoken (B) Visual
(C) Written (D) None of these

65. The term extension was first used by:
(A) USA (B) England
(C) India (D) China

66. The basic unit of I.R.D.P. programme is:
(A) A village
(B) A community development block
(C) A family
(D) A district

67. Extension education is:
(A) A discipline (B) A profession
(C) Discipline an profession both
(D) None of these

68. Principles of extension education is:
(A) Learning by hearing
(B) Learning by seeing
(C) Learning by reading
(D) Learning by doing

69. Goal of extension education is:
(A) To promote income of the farmers
(B) To promote production of the crops
(C) To promote new crops
(D) To promote scientific out look

70. Extension education aims at:
(A) Increasing agricultural production
(B) Increasing milk production
(C) Changing knowledge, skills and attitudes of the participants
(D) Changing administrative system

71. The work at Shantineketan with the objectives of:
(A) To create a real interest in people for rural improvement
(B) To study rural problems
(C) To improve village sanitation
(D) All of these

72. Sh. Rabindra Nath Tagore established a rural reconstruction institute in Shantineketan in:
(A) 1921 (B) 1922
(C) 1923 (D) 1924

73. The main principles of extension education is:
(A) Principles of interest and need
(B) Principles of cultural change
(C) Principles of learning by doing
(D) Principles of leadership
(E) All of these

74. Rural welfare movement on a mass scale was:
(A) Work at Shantineketan
(B) Gurgaon experiment
(C) Marthndam Programme
(D) Indian Village Service

75. Self help with internate expert counsel is the technique of:
(A) Work at Shantineketan
(B) Gurgaon experiment
(C) Marthandam Programme
(D) Indian Village Service

76. The aim of Etawah Pilot Project was:
(A) Introducing intensive work on the rural recontraction front
(B) Promotion of cottage industries and spread of literacy
(C) The development of village panchayat
(D) None of these

77. Who is the founder of Indian Village Service (IVS):
(A) Arthor T. Mosher
(B) B.N. Gupta
(C) Both (A) and (B)
(D) None of these

78. The main features of Sarvodaya Programme were:
(A) Simplicity and non violence
(B) Holiness of laboures
(C) Reconstruction of human value
(D) All of these

79. The main aims of Sarvodaya Programme were:
(A) Raising the standard of living
(B) Scientific development of agriculture
(C) Promotion of cottage industries and spread of literacy
D) Spread of medical and health facilities
(E) The development of village panchayat
(F) All of these

80. The intensive Agricultural District Programme was popularly known as:
(A) Package programme
(B) Knowledge programme
(C) Innovative Programme
(D) None of these

81. Government sponsored and aimed adoptment of the Gandhian ideas of Gram Swaraj is:
(A) Work at Shantineketan
(B) Gurgaon experiment
(C) Firka Development Scheme
(D) Indian Village Service

82. The main aims of Firka development scheme of Madras were:
(A) Educational and economic improvements
(B) Sanitary improvement
(C) Making the people self confident
(D) All of these

83. The package programme name was given of I.A.D.P. due to:
(A) Collective application of all improved practices
(B) Simultaneous application of all improved practices
(C) Collective and simultaneous application of all improved practices
(D) None of these

84. In which scheme, the Collector, who was primarily responsible for the successful working of the scheme in the district:
(A) Work at Shantineketan
(B) Gurgaon experiment
(C) Firka Development Scheme
(D) Indian Village Service

85. IADP programme was started in 1960 in the month of:
(A) July (B) August
(C) September (D) October

86. Etawah Pilot Project was started in 1948 in which month:
(A) July (B) August
(C) September (D) October

87. Extension education centres around:
(A) Farmers or rural peoples
(B) Innovation or inventions
(C) Extension workers
(D) All of these

88. The technology development of education can be related to various field of extension like:
(A) Agriculture (B) Home Science
(C) Agriculture and Home Science
(D) None of these

89. Agricultural extension is concerned with agricultural education assisting farmers to bring about continuous improvement in their:
(A) Physical (B) Economic
(C) Social
(D) Physical, economic and social

90. Agricultural extension makes available to the farming community to solve the problems of agriculture including:
(A) Animal husbandry
(B) Gardening
(C) Agricultural engineering
(D) All of these

Chapter 5

EXTENSION LEARNING AND EVALUATION

1. Learning is motivated by:
(A) Self (B) Group
(C) Environment (D) All of these

2. Learning is an internal process mainly controlled by:
(A) The learner (B) The teacher
(C) The subject matter
(D) The learning situation

3. Learning the rope refers to:
(A) Training (B) Socialization
(C) Both (A) and (B) (D) None of these

4. Learning become the most rapid by maximizing concentration on:
(A) One seasons at a time
(B) Two seasons at a time
(C) Three seasons at a time
(D) All the seasons simultaneously

5. Learning is a process which continues:
(A) Upto college education
(B) Upto employment
(C) Life long
(D) Occasionally

6. Which is not a principle of learning:
(A) Learning is the growth like and continuous
(B) Learning should be meaningful
(C) Learning should be challenging and satisfying
(D) Learning should not develop functional understanding of learners

7. Which one of the following is not an element of good learning situation:
(A) Physical facilities
(B) Extension worker
(C) Subject matter (D) Transfer process

8. The significant aspect of non formal learners:
(A) Undirectional flow of knowledge
(B) Flexible curriculum
(B) Hierarchically structure system
(D) Homogenous learners

9. Learning is a internal process mainly controlled by:
(A) Teacher (B) Learner
(C) Subject matter (D)Teaching materials

10. The process of acquiring or imparting the ability to perform a behavioral pattern through experience and practice:
(A) Communication (B) Learning
(C) Perception (D) Adoption

11. Learning become the most rapid by maximizing concentration on:
(A) One sense at a time
(B) Two sense at a time
(C) Three sense at a time
(D) All the sense simultaneously

12. "People learn more rapidly and permanently when the learning experience is pleasant or enjoyable" relate to:
(A) Law of readiness (B) Law of exercise
(C) Law of effect (D)Law of belonging

13. To Do, By Doing and For Doing are the fundamental objectives of:
(A) Learning (B) Teaching
(C) Training (D) Perception

14. The learning curve in teaching follows a:
(A) S-Shape (B) U-Shape
(C) V-Shape (D) Ω Shape

15. The experimental technique of learning emphasis on:
(A) Use of psychoanalysis
(B) Use of simulation games
(C) Use of psychiatry
(D) Use of learner groups

16. Progressive behaviour adaptation of an individual to a stimulus is called:
(A) Learning (B) Teaching
(C) Perceiving (D) Listening

17. Learning that suddenly appears when a reward or incentives for performance is given:
(A) Discovery learning
(B) Individualized learning
(C) Rote learning (D) Latent learning

18. The general approach in progressive mode of learning is:
(A) Authoritarian (B) Paternalistic
(C) Humanitarian (D) Democratic

19. New learning must be related to the old:
(A) Principle of readiness
(B) Principle of association
(C) Principle of practice
(D) Principle of timing

20. Life long process of learning happening through daily experience and exposure:
(A) Informal education
(B) Non formal education
(C) Formal education
(D) Professional education

21. Learning must result in functional understanding:
(A) Principle of readiness
(B) Principle of association
(C) Principle of practice
(D) Principle of timing

22. Learning must make sense to the learner:
(A) Principle of readiness
(B) Principle of clarity
(C) Principle of practice
(D) Principle of timing

23. A relatively enduring change in overt and covert response as a result of perceived stimulus is called:
(A) Endurance (B) Attribute
(C) Learning (D) Perception

24. Adult learning is:
(A) Problem centered
(B) Subject centered
(C) Instructor centered
(D) Institute centered

25. Adult literacy programme can be considered:
(A) Formal education
(B) Informal education
(C) Non-formal education
(D) All of the these

26. Which of the following statements represents true concepts of extension evaluation?
(A) Extension evaluation begins before the start of an extension programme
(B) Extension evaluation begins at the mid step of the programme
(C) Extension evaluation begins at the last step of the programme
(D) Extension evaluation begins inbuilt in each step of programme

27. Important element of evolution process is:
(A) Observations, criteria to observations and making some decisions
(B) Observations and criteria to observations
(C) Observations and collection of some information and final decision
(D) All of these

28. The science that deals with adult learning is:
(A) Androgogy (B) Pedagogy
(C) Pedology (D) Edaohology

29. The emphasis of androgogical method in comparison with pedogological method is on:
(A) Content and Process
(B) Content
(C) Process
(D) None of these

30. Androgogy literally means:
(A) Adult leading
(B) Adult learning
(C) Child leading
(D) Child learning

31. Adult learning is:
(A) Problem centered
(B) Subject centered
(C) Instructor centered
(D) Institute centered

32. The central problem in adult learning is:
(A) Age (B) Literacy
(C) Sex difference (D) Motivation

33. Adult literacy programme can be considered as:
(A) Formal education
(B) Informal education
(C) Non-formal education
(D) All of these

34. "Learning the rope" refers to:
(A) Training (B) Socialization
(C) Both (A) and (B) (D) None of these

35. Learning is facilitated mostly by:
(A) Taste (B) Hearing
(C) Touch (D) Sight

Chapter 6

COMMUNICATION AND COMMUNICATION TECHNOLOGY

1. Communication is more meaningly:
(A) Extension approach
(B) Subject of broadcasting
(C) Exchange of ideas, facts, feelings etc
(D) Transference of meaning

2. Communication is sharing together and is for:
(A) Apathy (B) Commonness
(C) Differences (D) Popularity

3. Communicator is the characteristic element in the model of:
(A) Leagans (B) Lezwell
(C) Berlo (D) Rogers

4. Leagans model of communication contains:
(A) 6 elements (B) 5 elements
(C) 4 elements (D) 3 elements

5. Leagans model of communication process is:
(A) Communication- Message- Channel- Treatment- Audience- Audience response
(B) Sender- Encoding- Channel- Decoding- Receiver
(C) Communication source- Encoder- Message- Channel- Decoder-Communication receiver
(D) None of these

6. Linear model of communication was given by:
(A) Barlo (B) Weaver
(C) Sanane (D) Leagans

7. Real object taken out of their natural settings is called as:
(A) Objects (B) Specimens
(C) Model (D) Mock-up

8. Sound can be recorded in ————- ways:
(A) 1 (B) 2
(C) 3 (D) 4

9. The disk recording is made by the ———— process:
(A) Mechanical (B) Magnetic
(C) Optical (D) None of these

10. Tape and wire recording are made by ————process:
(A) Magnetic (B) Mechanical
(C) Optical (D) None of these

11. The movie film recording is made by———— process:
(A) Magnetic (B) Mechanical
(C) Optical (D) None of these

12. Which one of the following is not a example of still pictures:
(A) Drawing (B) Painting
(C) Etching (D) None of these

13. Flash cards are a series of illustrated cards when flashed or presented before a group in proper sequence tell a:
(A) Part of story
(B) Complete story
(C) Incomplete story
(D) None of these

14. A B C of posters means:
(A) Attractive, Brief and Complete
(B) Attractive, Brief and Clear
(C) Attractive, Brief and Creative
(D) None of these

15. ____________ consist of a series of individual charts which are bound together and hung on a supporting stand:
(A) Bar charts (B) Job charts
(C) Flip charts (D) Tree charts

16. Linear model of communication was given by:
(A) Barlo (B) Weaver
(C) Sanane (D) Leagons

17. Which cf the following scientist has given the theory of information of communication?
(A) Shannon and Weaver (B) Leagon
(C) Berlo (D) Aristotle

18. Which one is not the part of communication module?
(A) Communicator (B) Message
(C) Advice (D) Audience

19. The number of elements in the communication model of Berlo:
(A) 4 elements (B) 5 elements
(C) 6 elements (D) 7 elements

20. The smallest meaningful unit of speech i called:
(A) Heuristic (B) Productive
(C) Morphemes (D) Phonemes

21. The communication model better applicable for a person reading newspaper:
(A) Shannon-Weaver model
(B) Leagan's model
(C) Berlo's model
(D) Barker Wiseman model

22. The basic elements in Aristotle's model of communication is:
(A) Speaker (B) Speech
(C) Listener or Audience
(D) All of these

23. The S-M-C-R model of communication is given by:
(A) Berlo (B) Laswell
(C) Rogers (D) Leagans

24. The influence of mass media according to hypodermic model of communication is:
(A) Direct (B) Indirect
(C) Secondary (D) No effect

25. Communication IQ was studied by:
(A) Bhanja (B) Paul Vegans
(C) Hovland (D) None of these

26. Communication competency concept was stressed by:
(A) Wiemann and Backlund
(B) Hymes
(C) Sannob and Weaver
(D) Thayer

27. Source-transmitter-signal-receiver-destination is the model of communication given by:
(A) Aristotle
(B) Shannon and Weaver
(C) Laswel
(D) Westley and Machean

28. The first basic persuasive communication model:
(A) Aristotle
(B) Shannon and Weaver
(C) Laswell
(D) Berlo

29. The communication model that matches well with political symbolism:
(A) Aristotle (B) Shannon and Weaver
(C) Laswell (D) Berlo

30. Mathematical theory of communication was put forwarded by:
(A) Aristotle
(B) Shannon and Weaver
(C) Laswell (D) Berlo

31. The amount of information that can be transmitted per unit of time:
(A) Redundancy (B) Channel capacity
(C) Coupling (D) Rhetoric

32. A communication model developed based on behavioral research:
(A) EM Rogers (B) Lazerfield
(C) D Berlo (D) Stevens

33. Which of the following extension 4 sources of communication do you consider most credible for non progressive/ remote village situation:
(A) Radio (B) Demonstration
(C) Print media (D) Exhibition

34. Which one is not the mode of radio broadcast?
(A) Straight talks (B) Interviews
(C) Songs (D) Lecture

35. Which is the traditional media of communication?
(A) Print (B) Tape recorder
(C) Radio (D) Ram leela

36. Which of the following is a traditional and simple aid of communication?
(A) Model (B) Puppet
(C) Radio (D) TV

37. Attitude of communication source is important from the point of view of:
(A) Code, content, treatment
(B) Source, self, receiver
(C) Encode, decode, feedback
(D) Self, audience, content

38. The degree to which individuals involved in a communication act differ in certain characteristics:
(A) Homophily (B) Credibility
(C) Empathy (D) Heterophily

39. The communication skills that get emphasized on listening and reading:
(A) Attitude (B) Effect
(C) Encoding (D) Decoding

40. Speaking and writing involve which communication skills:
(A) Encoding (B) Decoding
(C) Feedback (D) Knowledge

41. The process of selecting information that is relevant to a problem while ignoring distractions:
(A) Selective encoding
(B) Selective decoding
(C) Selective comparison
(D) None of these

42. Communication behaviour essentially consist of:
(A) Information seeking
(B) Information processing
(C) Information disseminating
(D) All of these

43. Communication barriers include:
(A) Social (B) Semantic
(C) Technical (D) All of these

44. The communication barrier wherein people attach different meaning to things:
(A) Semantic barrier
(B) Personal barrier
(C) Technical barrier
(D) None of these

45. Problem of homogeneity is the communication barrier related to:
(A) Source (B) Message
(C) Channel (D) Receiver

46. Providing too much information is a communication barrier related to:
(A) Communicator (B) Message
(C) Treatment (D) Receiver

47. Which among the following calls for effective communication:
(A) Listening (B) Learning
(C) Training (D) Looking

48. Indirect and inferential feedback is available in:
(A) Group media (B) Mass media
(C) Audio media (D) Visual media

49. The communication type best apply to largely illiterate audience:
(A) Written communication
(B) Informal communication
(C) Oral communication
(D) Formal communication

50. Reading newspaper is:
(A) Intra personal communication
(B) Inter personal communication
(C) Mass communication
(D) Dyadic communication

51. Constant feedback between the sender and the receiver is the characteristic feature of:
(A) Inter personal communication
(B) Mass communication
(C) Written communication
(D) Vertical communication

52. The main problem with mass communication is:
(A) Costly
(B) No local dialect
(C) Same information
(D) Cumbersome

53. Mass communication is generally identified with modern mass media like:
(A) Television and radio
(B) Cinema and news paper
(C) Farm and home visit
(D) All of these

54. An informal communication type involved in the spread of runner:
(A) Horizontal communication
(B) Vertical communication
(C) Diagonal communication
(D) Grapevine communication

55. Cybernetics is the science of:
(A) Decision making
(B) Organizations
(C) Ecology
(D) Communication

56. Communication breakdown is a common feature of:
(A) Ulterior transaction
(B) Cross transaction
(C) Straight transaction
(D) Vertical transaction

57. The art of communication is:
(A) Learning (B) Listening
(C) Feeling (D) None of these

58. Which of the following are the elements of communication process?
(A) Source – Message – Channel – Receiver
(B) Production process – Farmers – Farm – Farm business
(C) Research – Education – Extension – Training
(D) Information – Interest – Evaluation – Persuasion

59. ______a unit of two or more people in reciprocal communication and interaction with each other:
(A) Family (B) Group
(C) Society (D) Social stratification

60. The term community development appears to have originated from in England in 1946:
(A) London
(B) France
(C) Cambridge
(D) Belfast

61. Among the community, voluntary agencies are characterized by:
(A) Less intensity (B) Greater continuity
(C) Greater commitment
(D) Less accountability

62. ________ is the process by which two or more people exchange ideas, facts, feelings or impressions.
(A) Adoption
(B) Teaching
(C) Learning
(D) Communication

63. Leagans model of communication process is:
(A) Communication-message-channel-treatment-Audience-Audience response
(B) Sender-encoding-channel-decoding-receiver
(C) Communication source-Encoding-Message-Channel-Decoder-Communication receiver
(D) None of these

64. The term 'intercommunication' means:
(A) Communication on any issue
(B) Communication between two persons
(C) Communication between various groups
(D) All of these

65. The term 'human communication' means:
(A) Exchange of service
(B) Exchange of liabilities
(C) Exchange of ideas
(D) Exchange of ideas with commodity

66. Which one of the sources of the communication is most credible for non-progressive/ remote village situation?
(A) Farm radio (B) Demonstration
(C) Printed material (D) Exhibition

67. In extension education exhibition is helpful for:
(A) Individual learning
(B) Group communication
(C) Mass communication
(D) None of these

68. In extension education communication is a:
(A) Process
(B) Method
(C) Idea
(D) Expression of thoughts

69. The authority of PACS is known as:
(A) President (B) Secretary
(C) General body (D) Executive body

70. Action-reaction inter-dependable communication is termed as:
(A) Barrier (B) Feed back
(C) Fidelity (D) Treatment

71.

List-I	List-II
(a) Demonstration means	(1) Short type demonstration
(b) Method demonstration is	(2) Showing by doing
(c) Result demonstration is	(3) Long type demonstration
(d) National Demonstration is	(4) Front line extension project

Code	a	b	c	d
(A)	2	1	3	4
(B)	3	4	2	1
(C)	4	3	2	1
(D)	1	2	3	4

72. Under limited resource of a manpower, time and money, we should select:
(A) Individual contact methods
(B) Group contact methods
(C) Mass contact methods
(D) All of these

73. An extension method that can be considered both spoken and visual:
(A) Posters (B) Radio
(C) Campaign
(D) Agricultural clinics

74. Flag method is classified under:
(A) Spoken method
(B) Written method
(C) Group method
(D) Mass method

75. Agricultural clinic is classified under:
(A) Group contact method
(B) Individual contact method
(C) Mass contact method
(D) All of these

76. Educational television is classified under:
(A) Broadcast media
(B) Screen media
(C) Printed media
(D) None of these

77. Communication is the process by which message are transferred from source to:
(A) Channel (B) Message
(C) Receiver (D) Effect

78. Which of the following is not a method of group communication?
(A) Sabarmati (B) Sevagram
(C) Shriniketan (D) Bhatinda

79. Which of the following is not a method of group communication?
(A) Conference (B) Field trips
(C) National demonstration
(D) Farm and home visits

80. The most basic function of communication:
(A) Information (B) Instruction
(C) Integration (D) Influence

81. Communication is more meaningfull:
(A) Extension approach
(B) Subset of broadcasting
(C) Exchange of facts and feeling
(D) None of these

82. Level of communication are:
(A) Conventional
(B) Exploratory
(C) Participative
(D) All of these

83. On which level, communication is fact and problem oriented:
(A) Explorative level
(B) Participative level
(C) Conventional level
(D) None of these

84. The essence of communication is:
(A) Sharing information
(B) Imparting knowledge
(C) Sharing understanding
(D) Transmitting information

85. The art of communication is:
(A) Learning (B) Listening
(C) Feeling (D) None of these

86. The most important element in the communication process:
(A) Source (B) Receiver
(C) Channel (D) Message

87. Action-reaction inter-dependence in communication process is:
(A) Feedback (B) Homophily
(C) Competence (D) Fidelity

88. Which of the following is not communication process:
(A) Source (B) Encoding
(C) Transmission (D) Meetings

89. Communication process is stabilized by:
(A) Feedback (B) Credibility
(C) Channel (D) Empathy

90. The type of audience suited for communication process is called:
(A) Active audience
(B) Potential audience
(C) Available audience
(D) None of these

91. The communication channels most active in diffusion process:
(A) Mass media channel
(B) Interpersonal channel
(C) Both (A) and (B)
(D) None of these

92. The communication channel most suited to inform the audience about the existence of an innovation:
(A) Mass media channel
(B) Interpersonal channel
(C) Both (A) and (B)
(D) None of these

93. Mass communication is intended to be consumed immediately hence it is called:
(A) Rapid (B) Transient
(C) General (D) None of these

94. Feedback is inferential in case of communication made through:
(A) Village Level Worker
(B) Group meeting
(C) Radio (D) Method Demonstration

95. Mass media had direct, immediate and powerful effects on a mass audience:
(A) Two step flow model
(B) One step flow model
(C) Hypodermic needle model
(D) None of these

96. The controlling of the flow of information through a communication channel:
(A) Gate keeping (B) Gate locking
(C) Gate security (D) Gate guard

97. A system of decision centres interconnected by patterned flows of information:
(A) Communication techniques
(B) Communication network
(C) Communication level
(D) Communication channel

98. The number of communication network links possible in a social system of 100 members:
(A) 1000 (B) 10000
(C) 9876 (D) 4950

99. Two linked individuals in a network having personal communication networks that overlap:
(A) Communication proximity
(B) Communication preference
(C) Communication acceptance
(D) None of these

100. The communication network ideally suited for imparting knowledge change:
(A) Star network (B) Wheel network
(C) Circle network (D) Slash network

101. Which of the following is not a decentralized communication networks:
(A) Star (B) Chain
(C) Circle (D) Both (A) and (B)

102. ATIC under NATP functions under:
(A) State Department of Agriculture
(B) Ministry of Agriculture, GOI
(C) Ministry of RD, GOI
(D) ICAR

103. AKIS expands to:
(A) Agricultural Knowledge and Information Syndrome
(B) Agricultural Knowledge and Information Service
(C) Agricultural Knowledge and Information Science
(D) Agricultural Knowledge and Information System

104. Agricultural and rural Industries Information System Network to strengthen Small and Micro Enterprises (SMEs):
(A) AFPINET (B) ARISNET
(C) AGRISNET (D) VISTARNET

105. Land Information System network linking all institutions involved in land and water management for agricultural productivity and production systems:
(A) LISNET (B) ARISNET
(C) AGRISNET (D) VISTARNET

106. Agricultural Research Information System Network:
(A) HORTNET (B) ARISNET
(C) AGRISNET (D) VISTARNET

107. Agricultural Extension Information System Network:
(A) COOPNET (B) FERTNET
(C) AGRISNET (D) VISTARNET

108. A network that connect agricultural produce wholesale markets and rural markets:
(A) AFPINET (B) AGMARKNET
(C) HORTNET (D) VISTARNET

109. An infrastructure network upto block level agricultural offices facilitating agricultural extension services and agribusiness activities to user in rural prosperity:
(A) AGRISNET (B) ARISNET
(C) HORTNET (D) VISTARNET

110. The toll free telephone number on kisan call centre is:
(A) 1551 (B) 1331
(C) 5115 (D) 3113

111. When started kisan call centre:
(A) 1970 (B) 1971
(C) 1972 (D) 1973

112. Warna village network project was started in the state of:
(A) Uttar Pradesh (B) Maharastra
(C) Tamilnadu (D) Madhya Pradesh

113. The e-choupal concept in extension delivery is promoted by:
(A) ITC (B) MSSRF
(C) KINFRA (D) CAPART

114. "Rice Doctor" an interactive pest diagnostic expert system is developed by:
(A) ICAR (B) IRRI
(C) ICRISAT (D) INIBAP

115. Krishi Darshan programme in India is telecasted from:
(A) Mumbai (B) Kolkata
(C) New Delhi (D) Bangalore

116. Prasar Bharti came into existence on:
(A) 15th October, 1995
(B) 18th January, 1996
(C) 20th December, 1996
(D) 23rd January, 1997

117. The 4-H club colours are:
(1) Green (2) Yellow
(3) White (4) Red
(5) Blue
Select the right answer using the codes given below:
(A) 1 and 2 (B) 3 and 4
(C) 1 and 3 (D) 1 and 5

118. The 4-H club motto is:
(A) To make the better best
(B) To make the best better
(C) To make the good better
(D) To make the good best

119. The 4-H clubs are functioning in USA, 4 –H implies:
(A) Head, Hand, Heart, Health
(B) Head, Home, Health, Hands
(C) Hand, Health, Hope, Heart
(D) Home, Heart, Health, Head

120. The 4 H of communication are:
(A) Informal, Instruction, Influence and Integration
(B) Intermediation, Information, Interest and Instruction
(C) Information, Instruction, Influence and Integration
(D) Information, Integration, Influence and Instruction

121. The abbreviation COIK in COIK fallacy means:
(A) Clear Only If Known
(B) Clarity Of Informational Knowledge
(C) Capacity Of Indigenous Know-how
(D) Concept of Imaging Knowledge

122. The COIK fallacy is associated with which of the following:
(A) Communication channels
(B) Audio visual aids
(C) Communicator (D) Receiver

123. The 4-H club was forwarded by:
(A) O.H. Benson (B) A.B. Graham
(C) Seaman A. Knapp (D)A.B. Ross

124. The motto of 4 H club is to make:
(A) The poorer rich (B) Better best
(C) Adult motivated (D) Child educated

125. Which one is not the part of communication module:
(A) Communicator (B) Message
(C) Advice (D) Audience

126. Round table discussion is called:
(A) Panel (B) Symposium
(C) Forum (D) Workshop

127. Intelligence Quotient determines:
(A) Capacity age
(B) School going age
(C) Chronological age
(D) Mental age

128. The extension work must be based on needs and interests of:
(A) People (B) Leader
(C) Worker/Official (D) Government

129. Which of the following is not the element of communication process?
(A) Message (B) Feedback
(C) Channel (D) Fidelity

130. A person receiving favourable response from other is called:
(A) Authoritative (B) Psychological support
(C) Physical support (D) Social support

131. The most conspicuous development in the globalization context has been the:
(A) Information technology influx
(B) Communication technology influx
(C) Information and communication technology influx
(D) None of these

132. Developing world is gradually catching up with the technological advancement to solve its:
(A) Sociological problem
(B) Economical problem
(C) Both (A) and (B)
(D) None of these

133. The set of activities that facilitates, by electronic means the capturing, storage, processing, transmission and display of information is:
(A) Communication
(B) Information and communication technology
(C) Agricultural extension
(D) Telecommunication technology

134. The objective of Information and Communication Technology is/are:
(A) To elucidate the role of some of the major ICT initiatives in rural development
(B) To study the impact of ICT efforts on agriculture, health, women empowerment and panchayat raj
(C) To print out the impediments for the effective implementation of ICT based project
(D) All of these

135. ICT tackle key problems in mainly:
(A) Agriculture and cooperation sector
(B) Health and education
(C) Panchayat raj and women empowerment
(D) All of these

136. Which programme is designed to assist their development and develop their information, education and communication base of rural people is:
(A) PRA (B) ICT
(C) AV AIDS (D) None of these

137. ICT programme came into existence in India is:
(A) 1982 (B) 1983
(C) 1984 (D) 1985

138. ICT programme was given a vent in 1984 by the:
(A) Congress Govt. (B) BJP Govt
(C) Janta Party Govt. (D) BSP Govt

139. ICT programme was given a vent in 1984 by Rajiv Gandhi and the technologist:
(A) R. S. Paroda (B) Sam Pitroda
(C) Panjab Singh (D) B.P. Pal

140. A high power national task force on information and software development was set up in:
(A) 1996 (B) 1997
(C) 1998 (D) 1999

141. A high power national task force on information technology and software development was set up in 1998 by the:
(A) Congress Govt. (B) BJP Govt
(C) Janta Party Govt. (D) BSP Govt

142. The national task force on information technology prepared the blue print for making the adoption of IT or a national movement by establishing a wide network of empowered task force at:
(A) Government level
(B) Nongovernment level
(C) Both (A) and (B)
(D) None of these

143. The new telecommunication policy was enacted in :
(A) 1999 (B) 2000
(C) 2001 (D) 2002

144. The new telecommunication Policy 1999 was stipulated targets in terms of establishing a telecom network with a view to achieving teledensity by 2005:
(A) 5 per cent (B) 7 per cent
(C) 10 per cent (D) 15 per cent

145. The new telecommunication Policy 1999 was stipulated targets in terms of establishing a telecom network with a view to achieving teledensity by 2010:
(A) 5 per cent (B) 7 per cent
(C) 10 per cent (D) 15 per cent

146. The concept of knowledge society means that it will incorporate new knowledge will take over the role of capital and labour in the economy further leading to:
(A) Structural changes
(B) Functional changes
(C) Both (A) and B)
(D) None of these

147. Sponsoring agency of Cyber Extension Project is:
(A) MANAGE, Hyderabad
(B) ICAR, New Delhi
(C) NAARM, Hyderabad
(D) CSAUA&T, Kanpur

148. Sponsoring agency of Agriculture Technology Information Centre is:
(A) MANAGE, Hyderabad
(B) ICAR, New Delhi
(C) NAARM, Hyderabad
(D) CSAUA&T, Kanpur

149. Sponsoring agency of Agricultural Gateway of India is:
(A) MANAGE, Hyderabad
(B) ICAR, New Delhi
(C) NAARM, Hyderabad
(D) CSAUA&T, Kanpur

150. Sponsoring agency of Helpline Service is:
(A) MANAGE, Hyderabad
(B) ICAR, New Delhi
(C) NAARM, Hyderabad
(D) CSAUA&T, Kanpur

151. Sponsoring agency of Warna Wired Village Project is:
(A) NIC
(B) Warna Cooperative Complex, Warn district Kolhapur (MH)
(C) Both (A) and (B)
(D) CSAUA&T, Kanpur

152. Sponsoring agency of Gyandoot Project is:
(A) MANAGE, Hyderabad
(B) ICAR, New Delhi
(C) CSAUA&T, Kanpur
(D) Gram Panchyat Dhar Dist. MP

153. Sponsoring agency of Interactive Touch Screen Information , KIOSK is:
(A) Dept. of International Development, Rajiv Gandhi College of Animal Sciences (Pondicherry)
(B) University of Reading (UK)
(C) Both (A) and (B)
(D) CSAUA&T, Kanpur

154. Sponsoring agency of Lokmitra Project is:
(A) National Information Centre
(B) Gram Panchayat Hamirpur dist HP
(C) Both (A) and (B)
(D) CSAUA&T, Kanpur

155. Sponsoring agency of E. choupal is:
(A) Imperial Tobacco Company
(B) EID Parry (Tamilnadu)
(C) Nagarjuna Group
(D) MS Swaminathan Research Foundation, Chennai

156. Sponsoring agency of Rural Information Kiosks is:
(A) Imperial Tobacco Company
(B) EID Parry (Tamilnadu)
(C) Nagarjuna Group
(D) MS Swaminathan Research Foundation, Chennai

157. Sponsoring agency of I-Kisan Portal is:
(A) Imperial Tobacco Company
(B) EID Parry (Tamilnadu)
(C) Nagarjuna Group
(D) MS Swaminathan Research Foundation, Chennai

158. Sponsoring agency of Village knowledge Centre is:
(A) Imperial Tobacco Company
(B) EID Parry (Tamilnadu)
(C) Nagarjuna Group
(D) MS Swaminathan Research Foundation, Chennai

159. Sponsoring agency of Computer of Wheels is:
(A) Imperial Tobacco Company
(B) Pingali Rajeswari Enterpreneur, A.P.
(C) Nagarjuna Group
(D) MS Swaminathan Research Foundation, Chennai

160. Sponsoring agency of www.dctoranywhere.com is:
(A) Imperial Tobacco Company
(B) EID Parry (Tamilnadu)
(C) Tata Council for Community Initiatives (TCCI), Pune (MS)
(D) MS Swaminathan Research Foundation, Chennai

161. Important features of Cyber Extension are:
(A) Village Information Centres
(B) Institutional support to other ICT projects
(C) Both (A) and (B)
(D) None of these

162. Important features of Agriculture Technology Information Centre is:
(A) Call centre
(B) Expert system
(C) Researchers and farmers linkage
(D) All of these

163. Important features of Agricultural Gateway of India are:
(A) Helpline service through telephones
(B) Researchers-farmers linkage
(C) Both (A) and (B)
(D) None of these

164. Important features of Helpline Service are:
(A) Helpline service through telephones
(B) Researchers-farmers linkage
(C) Both (A) and (B)
(D) None of these

165. Important features of Warna Wired Village Project are:
(A) Wide Area Network
(B) Computer booths
(C) Both (A and (B)
(D)None of these

166. Important features of Gyandoot Project are:
(A) Soochanalayas
(B) Portal partial recovery
(C) Panchayat community partnership
(D) All of these

167. Important features of Interactive Touch Screen Information Kiosk are:
(A) Information Kiosks text
(B) Picture with sound backup
(C) Both (A) and (B)
(D) None of these

168. Important features of Lokmitra Project are:
(A) Lokmitra Soochana Kendra in 25 panchayat
(B) Panchayat Community partnership
(C) Researcher-farmers linkage
(D) None of these

169. Important features of E-choupal are:
(A) e-choupal
(B) Crop specific internet
(C) Both (A) and (B)
(D) None of these

170. Important features of Rural Information Kiosks are:
(A) Farmers managed Kiosks portal
(B) Public private partnership in applying ICT improvations and content development
(C) Both (A) and (B)
(D) None of these

171. Important features of I-kisan portal are:
(A) e-choupal
(B) Portal in regional language
(C) Both (A) and (B)
(D) None of these

172. Important features of Village Knowledge Centre are:
(A) Knowledge centre
(B) Pro-poor
(C) Pronature prowomen
(D) Community ownership
(E) All of these

173. Important features of computer on wheels are:
(A) Internet on motor bikes (B) Portals
(C) Remote area
(D) Resource poor
(E) All of these

174. Important features of www.doctor .com are:
(A) Telemedicine Service through websites
(B) Telemedicine Service through computer
(C) Both (A) and (B) (D) None of these

175. Major thrust or areas or level of cyber extension project are:
(A) Post harvest processing
(B) Women and child welfare
(C) Both (A) and (B)
(D) None of these

176. Major thrust or areas or level of Agricultural Technology Information Centre project are:
(A) Agronomic practices
(B) Technologies inputs
(C) Both (A) and (B)
(D) None of these

177. Major thrust or areas or level of Agricultural Gateway of India are:
(A) Cultivation practices
(B) Plant protection
(C) New technologies
(D) All of these

178. Major thrust or areas or level of Helpline service are:
(A) Cultivation practices
(B) Plant protection
(C) New technologies
(D) All of these

179. Major thrust or areas or level of Warna Wired Village Project are:
(A) Cultivation practices
(B) Pest and diseases
(C) Marketing (D) All of these

180. Major thrust or areas or level of Gyandoot Project are:
(A) Agricultural produce
(B) Market intelligence
(C) Auction rates
(D) Land records
(E) All of these

181. Major thrust or areas or level of Interactive Touch Screen Information Kiosk are:
(A) Information about important cattle disease
(B) Management of cattle
(C) Both (A) and (B)
(D) None of these

182. Major thrust or areas or level of Lokmitra Project are:
(A) Classified components
(B) Job vacancies
(C) Market rates (D) Tenders
(E) All of these

183. Major thrust or areas or level of E-choupal are:
(A) Market price
(B) Cultivation practices
(C) Weather data (D) All of these

184. Major thrust or areas or level of Rural Information kiosks are:
(A) Farm practices (B) Farm advisory
(C) Market intelligence
(D) Weather (E) All of these

185. Major thrust or areas or level of I - Kisan portal are:
(A) Agricultural practices
(B) Plant protection
(C) Animal husbandry
(D) Weather (E) All of these

186. Major thrust or areas or level of Village Knowledge Centre are:
(A) Agronomic practices (B) Cattle feed
(C) Weather (D) All of these

187. Major thrust or areas or level of Computer on wheels are:
(A) Market (B) Weather
(C) Plant protection (D) Animal health
(E) All of these

188. Major thrust or areas or level of www.doctoranywhere.com are:
(A) Medical council
(B) Medical treatment in PHC
(C) Both (A) and (B) (D) None of these

189. Which ICT project is related to cattle diseases and management of cattle is:
(A) Interactive Touch Screen Information kiosks
(B) E - choupal
(C) Lokmitra project
(D) Gyandoot project

190. Which ICT project is related to medical treatment in PHC is:
(A) Interactive Touch Screen Information kiosks
(B) www.doctoranywhere.com
(C) Lokmitra project
(D) Gyandoot project

191. E-choupal launched by ITC in:
(A) June, 1998 (B) June, 1999
(C) June, 2000 (D) June, 2001

192. ITC's E - choupal is a unique example of using ICT for:
(A) Agricultural development
(B) Animal management
(C) Human health improvement
(D) All of these

193. Which ITC's has already become the largest initiatives among all internet based interventions in rural India is:
(A) E-choupal (B) I-kisan portal
(C) Lokmitra project
(D) Gyandoot project

194. E - choupal services render in which crops:
(A) Soybean and Coffee
(B) Wheat and rice
(C) Pulses (D) All of these

195. E - choupal covers across the:
(A) MP and Karnataka
(B) AP and UP
(C) Maharashtra and Rajasthan
(D) All of these

196. Disseminating various farming information to the farmers at various places throu_h ICT centres by:
(A) I- Kisan portal (B) E – choupal
(C) Gyandoot project
(D) Lokmitra project

197. Which ICT's provide strong effective coverage and dissemination of livestock out reach information to the farming community, e – mail conferencing system are:
(A) goat-net-cirg.nic.in
(B) livestock-net@cirg.nic.in
(C) fishnet@cirg.nic.in (D) All of these

198. Under the animal health project funded by DFID Rajiv Gandhi College of Veterinary and Animal Science, Pondicherry in collaboration with University of Reading, UK has installed a:
(A) Interactive Touch Screen Information, Kiosk
(B) Agricultural Technology Information System
(C) Warna Wired Village Project
(D) Agricultural Gateway of India

199. Specific features of Interactive Touch Screen Information, Kiosk are:
(A) Cattle diseases
(B) Management of cattle
(C) Method of acquiring information
(D) All of these

200. Apollo hospital started telemedicine project on:
(A) 1999 (B) 2000
(C) 2001 (D) 2002

201. The ISRO and EDCU has initiated several telemedicine project for human health in:
(A) Remote areas (B) Rural areas
(D) Urban areas (D) Tribal areas

202. Sankara Nethralaya is a super-specility eye hospital in Chennai launched a telemedicine project through satellite connectivity is:
(A) goat-net-cirg.nic.in
(B) livestock-net@cirg.nic.in
(C) fishnet@cirg.nic.in
(D) Tele-ophthalmology

203. www.doctoranywhere.com is a telemedicine service project with aim that headquarters connected with the PHC's with atleast how many doctors always present at headquarter to give advice and medicine to the rural people:
(A) Only one doctor
(B) Team of two doctors
(C) Team of 5-6 doctors
(D) Team of many doctors

204. Lokmitra project launched in:
(A) May, 2000 (B) May, 2001
(C) May, 2002 (D) May, 2003

205. For Panchayat raj ICT project are:
(A) Lokmitra
(B) Gyandoot and Saytu
(C) E-Gram Suvidha (D) All of these

206. Lokmitra Soochana Kendras were initially set up in how many panchayats:
(A) 25 (B) 30 (C) 35 (D) 40

207. Who provide the Lokmitra software and necessary training to the soochaks (operators):
(A) National Information Centre (NIC)
(B) Agricultural Technological Information Centre (ATIC)
(B) Both (A) and (B)
(D) None of these

208. ICT for panchayat raj initiated at Dhar district of MP is:
(A) Lokmitra (B) E-choupal
(C) Gayandoot (D) None of these

209. ICT for panchayat raj was successfully implemented in Maharashra is:
(A) SAYTU (B) E-Gram Suvidha
(C) E-choupal (D) I-Kisan portal

210. ICT is Geographical Information System (GIS) based on management information system is:
(A) SAYTU (B) E-Gram Suvidha
(C) E-choupal (D) I-Kisan portal

211. E-gram suvidha is currently operated in five district of MP:
(A) MP (B) AP
(C) WB (D) UP

212. Name of website of E-Gram Suvidha is:
(A) goat-net-cirg.nic.in
(B) livestock-net@cirg.nic.in
(C) f shnet@cirg.nic.in (D) www.nic.in

213. ICT programmee is related to women's empowerment is:
(A) Lokmitra
(B) Village Knowledge Centre
(C) Gayandoot (D) E-choupal

214. Village Knowledge Centre provided:
(A) Proper training to the women
(B) Imparting skills to the women
(C) Proper training and Imparting skills to the women
(D) None of these

215. The ICT for co-operative development is:
(A) Dairy cooperative
(B) Cooperative societies
(C) Household good cooperative
(D) None of these

216. ICT project 'Dairy cooperative' developed by:
(A) AMUL (B) NDDB
(C) PARAG (D) None of these

217. Anand Milk Union Limited also introduced a ICT programme is:
(A) Dairy cooperative
(B) Cooperative societies
(C) Household good cooperative
(D) Amul cooperative

218. Amul cooperative are:
(A) Milk procurement only
(B) Milk distribution only
(C) Milk procurement and distribution both
(D None of these

219. Important features of dairy cooperative are:
(A) Measure butter fat content in milk
(B) Test the quality of the milk
(C) Help to make payment to the farmers
(D) All of these

220. Warna Wired Village Project also introduced a ICT programme is:
(A) Dairy cooperative
(B) Cooperative societies
(C) Sugar cooperative
(D) Amul cooperative

221. ICT programme sugar cooperative on project of Warna Wired Village Project covering 70 villages of:
(A) Kolhapur district of Maharashra
(B) Sangli district of Maharashtra
(C) Kolhapur and Sangli districts of Maharashtra
(D) None of these

222. Village Information Centre of sugar cooperative has providing indepth knowledge and information about:
(A) Sugarcane cultivation
(B) Marketing to the farmers
(C) Both (A) and (B) (D) None of these

223. ICT programmes have significant inputs in the development paradigm are:
(A) Reduction in transition cost
(B) Time saving
(C) Awareness and utilization by entitlement
(D) All of these

224. Through ICT enabled services, a farmer could save this travelling time and transaction costs by more than:
(A) 25-30% (B) 40-50%
(C) 60-70% (D) 80-90%

225. By using e-choupal (soya choupal), the farmers could save transaction costs by:
(A) 25-30% (B) 40-50%
(C) 67-68% (D) 80-90%

226. Soya choupal saves charges towards freight to providing factory by:
(A) 35-36% (B) 40-50%
(C) 60-70% (D) 80-90%

227. Farmers increases in the sale value for their Soyabean produce by ———— due to the premium price offered by soyachoupal for their quality farm produced:
(A) 1% (B) 2% (C) 3% (D) 4%

228. Soyachoupal is more beneficial mainly for:
(A) Small holding farmers
(B) Medium holding farmer
(C) Large holding farmers
(D) None of these

229. Farmers could place their requirement of inputs to the ————, who place bulk demand with the ICT:
(A) Sanchalak (B) Scientist
(C) Leader (D) Soochanalayas

230. Gyandoot kiosks, farmers could get the land records in village itself and save transaction costs by:
(A) 25-30% (B) 40-50%
(C) 66-67% (D) 80-90%

231. Through gyandoot, the farmer particularly getting higher returns on the basis of the timely access to market information in:
(A) Small holding farmers
(B) Medium holding farmer
(C) Large holding farmers
(D) None of these

232. E-choupal has:
(A) LAN connectivity
(B) WAN connectivity
(C) VSAT connectivity
(D) WLL connectivity

233. COW has:
(A) LAN connectivity
(B) WAN connectivity
(C) VSAT connectivity
(D) GPRS connectivity

234. KIOSKs is internet connectivity through technological device such as:
(A) LAN connectivity
(B) WAN connectivity
(C) VSAT connectivity
(D) WLL connectivity (E) All of these

235. It is widely expected that ICT's are playing role in the:
(A) Developmental process
(B) Management process
(C) Transactional process (D) All of these

236. KIOSKs is internet connectivity for small towns and villages, that is just similar to the:
(A) PCO booths (B) STD booths
(C) Both (A) and (B) (D) None of these

237. Farmer-owned Amul Cooperative in Anand has shown that since the:
(A) 1950s (B) 1960s
(C) 1970s (D) 1980s

238. Which ICT project become a model for most dairy development project in India:
(A) AMUL (B) PARAG
(C) VISAKHA (D) Namestey India

239. India is a developing nation that is making remarkable progress toward development through the:
(A) Information strategy
(B) Communication strategy
(C) Both (A) and (B)
(D) None of these

240. The aims of the task force was to formulate the draft of a national policy on Information to enable India to emerge as in Information Technology super power within the next:
(A) 1 year (B) 2 years
(C) 5 years (D) 10 years

241. Indigenous communication channels are available in many forms such as:
(A) Folk media (B) Social gathering
(C) Deliberate instruction
(D) Record and unstructured channels
(E) All of these

242. Traditional communication channels such as:
(A) Puppet show and folk drama
(B) Community tea house and open market
(C) Story telling and village meeting
(D) All of these

243. Traditional communication media are deeply rooted in the socio-cultural moves of the society are:
(A) Folk songs (B) Folk drama
(C) Both (A) and (B) (D) None of these

244. Folk media may constitute a part of the:
(A) Indigenous communication system
(B) Traditional communication system
(C) Both (A) and (B) (D) None of these

245. In India, the puppetry styles are:
(A) Sutradharika
(B) Rod puppets shadow puppets
(C) Hand puppets (D) All of these

246. Traditonal media, folk theatre forms in Maharashtra are:
(A) Tamasha (B) Powada or Pawada
(C) Dashavatar (D) All of these

247. Traditional media, Folk theatre forms in Gujarat are:
(A) Bhavai (B) Dangi tamasha
(C) Both (A) and (B) (D) None of these

248. Traditional media, Folk theatre forms in UP are:
(A) Burrakatha (B) Yakshagana
(C) Ramleela (D) Noutanki

249. Traditional media, Folk theatre forms in North India are:
(A) Burrakatha (B) Yakshagana
(C) Ramleela (D) Noutanki

250. Traditional media, Folk theatre forms in Karmataka are:
(A) Burrakatha (B) Yakshagana
(C) Ramleela (D) Noutanki

251. Traditional media, Folk theatre forms in AP are:
(A) Burrakatha (B) Yakshagana
(C) Ramleela (D) Noutanki

252. Traditional media, Folk theatre forms in Bihar are:
(A) Bidesia (B) Jaljatin
(C) Both (A) and (B) (D) Noutanki

253. Traditional media, Folk theatre forms in Tamilnadu are:
(A) Therukoothu
(B) Yakshagana
(C) Burrakatha
(D) Noutanki

254. Traditional media, Folk theatre forms in Bengal and Orissa are:
(A) Burrakatha (B) Yakshagana
(C) Ramleela (D) Jatra

255. Traditional media, keertana in Maharashtra are:
(A) The Naraditya Harikatha
(B) The Warkari Haarikatha
(C) The National Harikatha
(D) All of these

256. Traditional media, folksongs in UP are:
(A) Chaiti (B) Kaijri
(C) Both (A) and (B) (D) None of these

257. Traditional media, folksongs/folk verse/ballads/folk saying and riddles in Kerala is:
(A) Kolkali pattus (B) Bihu
(C) Mnad and Panihari
(D) Rauf and Chhakri

258. Traditional media, folksongs/folk verse/ballads/folk saying and riddles in Assam is:
(A) Kolkali pattus (B) Bihu
(C) Mnad and Panihari
(D) Rauf and Chhakri

259. Traditional media, folksongs/folk verse/ballads/folk saying and riddles in Rajasthan is:
(A) Kolkali pattus (B) Bihu
(C) Mnad and Panihari
(D) Rauf and Chhakri

260. Traditional media, folksongs/folk verse/ballads/folk saying and riddles in Kashmir is:
(A) Kolkali pattus (B) Bihu
(C) Mnad and Panihari
(D) Rauf and Chhakri

261. Traditional media, folksongs/folk verse/ballads/folk saying and riddles in Madhya Pradesh is:
(A) Sua and Dadaria
(B) Mando and Dhalo
(C) Boli (D) Lavani

262. Traditional media, folksongs/folk verse/ballads/folk saying and riddles in Goa is:
(A) Sua and Dadaria
(B) Mando and Dhalo
(C) Boli (D) Lavani

263. Traditional media, folksongs/folk verse/ballads/folk saying and riddles in Eastern Punjab is:
(A) Sua and Dadaria
(B) Mando and Dhalo
(C) Boli (D) Lavani

264. Traditional media, folksongs/folk verse/ballads/folk saying and riddles in Maharashtra is:
(A) Sua and Dadaria
(B) Mando and Dhalo
(C) Boli (D) Lavani

265. Indigenous communication , folk media are:
(A) Story telling (B) Poetry
(C) Plays and Puppet shows
(D) All of these

266. Indigenous communication, Indigenous organization and social gatherings are:
(A) Religious groups
(B) Village meetings
(C) Irrigation associations
(D Mothers clubs
(E) Loan associations (F) All of these

267. Indigenous communication, deliberate instruction are:
(A) Parents teach children
(B) Craft people instruct apprentices
(C) Elder guide young people
(D) Adolescents undergo initiation rites
(E) All of these

268. Indigenous communication, many societies keep records in:
(A) Written (B) Carved
(C) Painted (D) Memorized forms
(E) All of these

269. Unstrumented channels of communication media are:
(A) Talk at home and at the well
(B) Talk in the field and on the road
(C) Talk in tea house and coffee house
(D) Talk in chief's house and at the market
(E) All of these

270. Communication does not have to be intentional is:
(A) Unstructured channels
(B) Structured channels
(C) Both (A) and (B)
(D) None of these

271. The traditional media are the closest to the:
(A) Hearts (B) Minds
(C) Both of these (D) None of these

272. The traditional media appeals is at a:
(A) Personal (B) Intimate levels
(C) Both of these (D) None of these

273. The colloquial dialects used make for clarity in communication are:
(A) Familiar format (B) Content
(C) Bothe of these (D) None of these

274. The traditional media are:
(A) Credible
(B) Culturally compatible
(C) Both (A) and (B) (D) None of these

275. The traditional media are:
(A) Inexpensive (B) Informal
(C) Both (A) and (B) (D) None of these

276. The greatest advantage of the traditional media over the electronic media in accommodating new themes is:
(A) Flexibility (B) Rigidness
(C) Unaffected (D) None of these

277. The traditional media have:
(A) Entertainment (B) Informative value
(C) Both (A) and (B)(D) None of these

278. Indian government established a Song and Drama Division in:
(A) 1954 (B) 1955
(C) 1956 (D) 1957

279. All India Radio also used folk media in its rural broadcasting in the form of a :
(A) Daily programme
(B) Weekly programme
(C) Monthly programme
(D) None of these

280. Traditional media are important channels of:
(A) Innovation (B) Communication
(C) PRA (D) Learning

281. Several traditional methods can be used in secondary school for teaching:
(A) Extension (B) Biotechnology
(C) Agriculture (D) Home Science

282. Among different folk media, story telling, folk songs and dances are outstanding to augment in imparting education in an:
(A) Interesting manner
(B) Effective manner
(C) Interesting and effective manner both
(D) None of these

283. Folk song can be both effective and efficient as a communication medium or a society which has a:
(A) Folk theatre forms
(B) Strong oral tradition
(C) Both (A) and (B) (D) None of these

284. The pioneering institution to draw upon folk music is:
(A) All India Radio (B) Doordarshan
(C) Akashwani (D) Star Plus

285. The radio, doordarshan, drama and theatre have integrated various:
(A) Folk theatre forms
(B) Strong oral tradition
(C) Both (A) and (B)
(D) None of these

286. Traditional media have remarkable impact on the rural society because of their:
(A) Acceptable idioms
(B) Functional significance
(C) Emotional touch
(D) Entertainment component
(E) All of these

287. Traditional media can more effective in rural areas because:
(A) The techniques are simple and easy
(B) The techniques are tough and complex
(C) The techniques are no understandable
(D) The techniques are politically accepted

288. Traditional media can more effectively stir and felling and affect the attitude of:
(A) Literate masses (B) Illiterate masses
(C) Leaders (D) Servicemen

289. Folk arts laying a meaningful role in rural areas in educating the:
(A) Urban people (B) Rural people
(C) Tribal people (D) Student

290. Traditional communication forum such as the market place, tea or coffee house pubs, transportation system and village meetings should be used to reach:
(A) Urban audience (B) Tribal audience
(C) Rural audience (D) None of these

291. Midwife is the best communicator in many parts of:
(A) Asia (B) Africa
(C) America (D) Russia

292. Which type of market plays an important role in the life of the tribal people is:
(A) Daily market (B) Seasonal market
(C) Weekly market (D) Monthly market

293. The tribal market brings together people from different ethnic group for not only economic in the tribal region but also:
(A) Secular activities
(B) Religious activities
(C) Secular and religious activities
(D) None of these

294. The market is the most powerful channel of communication in the:
(A) Tribal region (B) Urban region
(C) Rural region (D) None of these

295. The impact of weekly market on their traditional life has shown an attitude of accepting:
(A) Diffusion (B) Communication
(C) Innovation (D) All of these

296. Africa has a number of unique traditional communication technique, among them are the:
(A) Taking Drums (B) Taking horns
(C) Both of these (D) None of these

297. The Comila project in Bangladesh was built into traditional communication approaches such as:
(A) Village meeting (B) Block meeting
(C) Taluka meeting (D) 'District meeting

298. The important source of information to village women is:
(A) Barber's wife (B) Dai
(C) Washer women (D) Bangle sellers
(D) All of these

299. Traditional media may compute for attention will modern media such as:
(A) Radio (B) Television
(C) Cassettes or video (D) All of these

300. Which bank used puppetry show in Pantnagar Farmer's fair to arouse the interest of the visitors in bank savings and loans?
(A) Punjab Bank
(B) State Bank of India
(C) Central Bank (D) Allahabad Bank

301. Traditional channel of communication (Munadi) is more effective than ———— by KVK worker and newspaper in disseminating message of local importance in the tribal villages:
(A) Interpersonal contact
(B) Intra personal contact
(C) Official contact (D) None of these

302. The process of effecting an interchange of understanding between two or more people is called:
(A) Communication (B) Innovation
(C) Diffusion (D) None of these

303. The process of communication is:
(A) Teaching (B) Learning
(C) Teaching and learning
(D) None of these

304. Teaching act as:
(A) Analyst
(B) Catalyst
(C) Analyst and catalyst
(D) All of these

305. Communication is a :
(A) One way process
(B) Two way process
(C) Three way process
(D) Multi way process

306. Communication involves:
(A) Source and his message
(B) Receiver and his response
(C) Both (A) and (B)
(D) None of these

307. The chief emphasis in communication on:
(A) Target audience
(B) Target communicator
(C) Target receiver
(D) Target channel

308. In which communication system of agencies works:
(A) Knowledge, Generating and System
(B) Knowledge, Disseminating and System
(C) Knowledge, Consuming and System
(D) None of these

309. In which communication system of mass media works:
(A) Knowledge, Generating and System
(B) Knowledge, Disseminating and System
(C) Knowledge, Consuming and System
(D) None of these

310. In which communication system of farmers works:
(A) Knowledge, Generating and System
(B) Knowledge, Disseminating and System
(C) Knowledge, Consuming and System
(D) None of these

311. Individual approaches in communication and extensions are:
(A) Personal visits (B) Personal letters
(C) Both (A) and (B)(D) None of these

312. Group approaches in communication and extension are:
(A) Demonstration
(B) Training the leaders
(C) Discussion meetings
(D) Field days and tours
(E) All of these

313. Mass approaches in communication and extensions are:
(A) Film and slides
(B) Flash cards, printed material and Newspaper
(C) Radio, television and recorded talks
(D) Charts, diagrams and circular letters
(E) Campaign (F) All of these

314. Projected audio-visual aids are:
(A) Film projector (B) Slide projector
(C) Overhead projector
(D) Opaque projector (E) All of these

315. Non-projected audio-visual aids are:
(A) Flash cards and groups
(B) Charts and leaflets
(C) Folders and black boards
(D) All of these

316. Audio aids are:
(A) Radio (B) Tape recorder
(C) Both (A) and (B) (D) None of these

317. Visual aids are:
(A) Flash cards (B) Overhead projector
(C) Slides (D) All of these

318. Audio-visual aids are:
(A) Puppet (B) Drama
(C) Television (D) Cinema
(E) All of these

319. The basic principles of demonstration is learning by:
(A) Seeing only (B) Doing only
(C) Seeing and doing both
(D) All of these

320. Discussion meeting approach of communication are:
(A) Group discussion and Panel discussion
(B) Symposium, workshop and Forum
(C) Group interview and dialogue or public conversation
(D) All of these

321. The single practice demonstrations is:
(A) Method demonstration
(B) Result demonstration
(C) Frontline demonstration
(D) Single line demonstration

322. Father of method demonstration is:
(A) Mr Arthur T. Mosher
(B) F. L. Bryne
(C) Dr. Seeman A Knapp
(D) Mr Albert Mayer

323. In which demonstration process, extension worker may take help of audio-visual aids:
(A) Method demonstration
(B) Result demonstration
(C) Frontline demonstration
(D) Single line demonstration

324. In which demonstration practices are shown in comparison to existing practices in:
(A) Method demonstration
(B) Result demonstration
(C) Frontline demonstration
(D) Single line demonstration

325. In which demonstration cannot compare between old and new techniques or skill:
(A) Method demonstration
(B) Result demonstration
(C) Frontline demonstration
(D) Single line demonstration

326. Which is correct:
(A) Short type demonstration Result demonstration
(B) Comparison to existing Result demonstration practices can be done
(C) Comparison between old Result demonstration and new techniques cannot be done
(D) May take help of audio-visual aids Method demonstration

327. Which is correct:
(1) Method demonstration is the short type demonstration
(2) In method demonstration cannot compare between old and new techniques
(3) The result demonstration practices are shown in comparison to existing practices
(4) In result demonstration process, the extension worker may take the help of audio-visual aids

Options
(A) 1, 2 and 4 (B) 1, 3 and 4
(C) 1, 3 and 4 (D) 1, 2, 3 and 4

328. The important points of result demonstration is/are:
(A) Never try to discover new truths
(B) Do not repeat demonstration repeatedly
(C) Use audio-visual aids to support the result demonstration
(D) Better to conduct more than one demonstration in the same village on the same subject
(E) All of these

329. There are methods of recording the sound:
(A) Mechanical process
(B) Magnetic process
(C) Optical process (D) All of these

330. Disc recording is done by:
(A) Mechanical process
(B) Magnetic process
(C) Optical process (D) All of these

331. Tape and wire recording is done by:
(A) Mechanical process
(B) Magnetic process
(C) Optical process (D) All of these

332. Recording of sound on movie film by the:
(A) Mechanical process
(B) Magnetic process
(C) Optical process
(D) All of these

333. In India and England, projector is also called as Epidiascope:
(A) Opaque projector
(B) Slide projector
(C) Overhead projector
(D) Movie projector

334. The projector are based on principle of direct projection:
(A) Opaque projector
(B) Slide projector
(C) Overhead projector
(D) Movie projector

335. The projector are based on principle of indirect projection:
(A) Opaque projector
(B) Slide projector
(C) Overhead projector
(D) Movie projector

336. Using the principle of magnetic process recording can be done on tapes made of:
(A) Paper (B) Plastic
(C) Metal (D) All of these

337. The disc or gramophone records, the disc is made of:
(A) Lac (B) Plastic
(C) Both (A) and (B) (D) None of these

338. Leaflets contain number of pages:
(A) 4-12 (B) 12-24
(C) 24-48 (D) Unlimited

339. Pamphlets contain number of pages:
(A) 4-12 (B) 12-24
(C) 24-48 (D) Unlimited

340. Bulletin contain number of pages:
(A) 4-12 (B) 12-24
(C) 24-48 (D) Unlimited

341. Quick and cheap method of written communication and proves very effective if done well is:
(A) Leaflet (B) Pamphlet
(C) Bulletin (D) Circular letter

342. Which written method of communication can teach villagers easily:
(A) Circular letters (B) Leaf lets
(C) Pamphlets (D) Bulletin

343. Brainstorming is basically a:
(A) Creative technique
(B) Motivation technique
(C) Evaluation technique
(D) Reasoning technique

344. Which type of communication is called as public relations:
(A) Across communication
(B) Interpersonal communication
(C) Internal communication
(D) External communication

345. An official communication either oral or written is called:
(A) Informal communication
(B) Formal communication
(C) Both (A) and (B) (D) None of these

346. A formal communication may be:
(A) Mandatory (B) Indicative
(C) Both (A) and (B) (D) None of these

347. A mandatory communication is mostly:
(A) Vertical (B) Horizontal
(C) Both (A) and (B) (D) None of these

348. The indicative or explanatory communication may exist:
(A) Diagonal (B) Horizontal
(C) Vertical (D) All of these

349. The communication type which provides instructions, informal and clarification is called:
(A) Informal communication
(B) Formal communication
(C) Administrative communication
(D) Both (A) and (B)

350. The communication that take place between two or more functionaries at the same level under the same supervisors:
(A) Downward communication
(B) Upward communication
(C) Horizontal communication
(D) Grapevine communication

351. Which one of the following is not an element of communication:
(A) Consistency (B) Uniformity
(C) Rigidity (D) Adequacy

352. A large number of communication models exist because of:
(A) Contextual variation
(B) Behavioural variation
(C) Sequential variation
(D) Geographical variation

353. 'Communicator' is the characteristic element in the model of:
(A) Leagans (B) Lazwell
(C) Berlo (D) Rogers

Chapter 7

AUDIO-VISUAL AIDS

1. Flag method is classified under:
(A) Written (B) Spoken
(C) Visual (D) Audio-visual

2. Flash cards, pull charts , slides and film strips are the type of aids:
(A) Audio (B) Visual
(C) Audio-visual (D) None of these

3. Flash cards are a series of illustrated cards when flashed or presented before a group in proper sequence tell a:
(A) Part of story (B) Complete story
(C) Incomplete story (D) None of these

4. Television broadcast for rural development in India, started in:
(A) 1947 (B) 1957
(C) 1967 (D) 1977

5. Radio mass medium is characterized by:
(A) One way with immediate feedback
(B) One way without instant audience response
(C) One way and colourful
(D) One way and timeless

6. Which one of the following is not correctly matched?

Approach	Technique
(A) Mass	Exhibition
(B) Individual	Farm visit
(C) Group	Mela
(D) Mass	Newspaper

7. Read the following statements and mark your answers according to the code given below:
(a) Newspaper is popular medium of communication in rural areas.
(b) Newspaper is written form of communication.
(c) Film show is audio visual form of communication.
(d) Film show is popular medium of communication in rural areas.

Answer code:
(A) A and C are correct
(B) B and C are correct
(C) C and D are correct
(D) A and D are correct

8. Consider the following statements:
Assertion (A): Film is not effective medium in rural areas.
Reason (R): Film can not be projected with low voltage power.
Select the correct answer from the code given below:
(A) A is true but R is false
(B) R is true but A is false
(C) Both A and R are correct and R is correct explanation of A
(D) Both A and R are correct but R is not the correct explanation

9. Match list I with list II and select the correct answer using the code given below.

List I (Form)	**List II (Technique)**
(a) Spoken	1. Folder
(b) Written	2. Tape recorder
(c) Projected	3. Film
(d) Non projected	4. Puppet

Code:	a	b	c	d
(A)	2	1	3	4
(B)	1	2	3	4
(C)	2	3	1	4
(D)	4	3	2	1

10. Match list I with list II and select the correct answer using the code given below:

List I (Aids)	**List II (Methods of contact)**
(a) Model	1. Group contact
(b) Slide	2. Mass contact
(c) Telephone	3. Personal contact
(d) Poster	4. Mass contact

Code:	a	b	c	d
(A)	2	3	4	1
(B)	1	2	3	4
(C)	3	4	2	1
(D)	3	2	4	1

11. Which of the following is not a method of group communication?
(A) Demonstration (B) Symposium
(C) Flannel graph (D) Circular letter

12. Which one of the following is not correctly matched?

Approach	**Technique**
(A) Mass	Television
(B) Mass	Film
(C) Mass	Tape recorder
(D) Mass	Demonstration

13. Which source of interview is the best in a rural society?
(A) Key informant (B) Group
(C) Focused group (D) None of these

14. Method of mass communication is:
(A) Demonstration (B) Farm visit
(C) Group discussion (D) News paper

15. Which one is not the part of communication module?
(A) Communicator (B) Message
(C) Advice (D) Audience

16. Which one of the following is not a method of mass contacts:
(A) Circular letters (B) Television
(C) Leaflets
(D) Result demonstration

17. Which one of the following is a method of contact group:
(A) Farms and home visits
(B) Office calls
(C) Personal letters (D) Tours

18. Which one of the following is an audio – visual method of extension teaching:
(A) Drama (B) Recordings
(C) Flash board (D) Tape recorders

19. Which one is a non – projected aid:
(A) Cinema (B) Slide
(C) Over head projector (D) Pictures

20. Puppet shows are included in:
(A) Audio aids (B) Visual aids
(C) Audio – visual aids (D) None of these

21. Flash cards, pull charts , slides and films strips are the type of aids:
(A) Audio (B) Visual
(C) Audio – visual aids (D) None of these

22. Visual teaching aid flannel graph is also known as:
(A) Chalk board (B) Bulletin board
(C) Khadder graph (D) None of these

23. Television, Fairs, Radios and Exhibitions are media of:
(A) Group communication
(B) Individual communication
(C) Community communication
(D) Mass communication

24. If we want to impress the people about a particular recommended practice suitable for a specific area, then effective extension method is:
(A) TV (B) Radio
(C) Demonstration (D) Pumphlets

25. Radio is the medium of mass communication in which, use is made up of:
(A) Audio device (B) Visual device
(C) both (A) and (B) (D) None of these

26. Bulletins, leaflets, circular letters, posters, fairs, exhibitions, radio and television are the media of:
(A) Public communication
(B) Official communication
(C) Group communication
(D) Mass communication

27. Puppet, Drama, Television and Cinema are example of:
(A) Audio-aids (B) Visual – aids
(C) Audio-visual aids (D) None of these

28. Radio and tape recorder is the examples of:
(A) Audio-aids (B) Visual – aids
(C) Audio-visual aids (D) All of these

29. **List-I** **List-II**
(a) Mass contact (1) Bulletin
(b) Group contact (2) Method demonstration
(c) Individual contact (3) Symposium
(d) Individual contact (4) Result demonstration

Code	a	b	c	d
(A)	1	3	4	2
(B)	2	3	4	1
(C)	4	3	2	1
(D)	1	2	3	4

30. **List-I** **List-II**
(a) Audio (1) Flash cards
(b) Visual (2) Radio
(c) Audio (3) Drama
(d) Audio-visual (4) Tape recorder

Code	a	b	c	d
(A)	2	1	4	3
(B)	3	2	1	4
(C)	4	3	2	1
(D)	1	2	3	4

31. **List-I** **List-II**
(a) Personal contact (1) Mass method
(b) Public address equipment (2) Mass method
(c) Circular letter (3) Individual method
(d) Television (4) Group method

Code	a	b	c	d
(A)	3	4	1	2
(B)	4	3	2	1
(C)	1	2	3	4
(D)	2	1	3	4

32. The fastest method of delivering of message in the village is:
(A) Newspaper (B) Radio
(C) Television (D) Group discussion

33. Television radio, fairs and exhibitions are the media of:
(A) Group communication
(B) Individual communication
(C) Community communication
(D) Mass communication

34. Which one of the following is not correctly matched?

Approach	Technique
(A) Mass	Television
(B) Mass	Demonstration
(C) Mass	Tape recorder
(D) Mass	Film

35. Which one is related to mass approach method?
(A) Film, Demonstration and Radio
(B) Radio, TV and Lecture
(C) Phone call, Radio and TV
(D) Exhibition, Newspaper and Radio

36. The best method for showing the value of new project is:
(A) Result demonstration
(B) Exhibition
(C) Farmers fair (D) Meeting

37. Which one of the sources of communication is most credible for non-progressive/ remote village situation?
(A) Farm radio (B) Demonstration
(C) Printed material (D) Exhibition

38. Method of mass communication is:
(A) Demonstration (B) Farm visit
(C) Group discussion (D) Newspaper

39. Best method for showing the value of new practice is:
(A) Result demonstration (B) Exhibition
(C) Farmer's fair (D) Meeting

40. In a good chart facts are presented:
(A) Only in words
(B) Only in figures
(C) Only in photographs
(D) Both in figures and photographs

41. Non-institutional communication method is:
(A) Exhibition
(B) Family member
(C) Extension literature
(D) Radio

42. Which one of the following aids involves two sensory organs simultaneously?
(A) Poster (B) Model
(C) Radio (D) Television

43. Identify the projected audio-visual aid:
(A) Puppet (B) Film
(C) Drama (D) Folk song

44. Which one is not the example of nonverbal communication?
(A) Ram Leela
(B) Movements of arms
(C) Facial expression
(D) Gestures

45. We learn maximum through:
(A) Taste (B) Touch
(C) Hearing (D) Sight

46. The extension service is called " Cooperative Extension Service" in:
(A) USA (B) India
(C) Mexico (D) China

47. The main functions of an agricultural university in India are:
(A) Technology generation, technology transfer and technology adoption
(B) Drawing attention, creating interest and taking action
(C) Teaching, research and extension
(D) Preparation, deliberation and action

48. General meeting are broadly the meeting of:
(A) Heterogeneous participation
(B) Homogeneous participation
(C) Community participation
(D) Social participation

49. A photograph is called:
(A) A diagnostic representation of facts or ideas
(B) A meaningful exchange of ideas
(C) An exact visual recording of a thing
(D) An act of influencing others

50. Identify one which is not correctly matched in the following pairs:
(A) Method demonstration : Group approach
(B) Circular letter : Mass approach
(C) Radio : Individual approach
(D) Campaign : Mass approach

51. Non projected teaching method is:
(A) Cinema (B) Slides
(C) Puppets (D) 3-D film

52. Which one is not the group approach device?
(A) Demonstration (B) Symposium
(C) Workshop (D) Film

53. Television, fairs, radios and exhibition are the media of:
(A) Group communication
(B) Individual communication
(C) Commodity communication
(D) Mass communication

54. Powerful audio-visual media for group communication is:
(A) Television (B) Film
(C) Puppet (D) None of these

55. Puppet can be made out of easily available and cheap material in the community like:
(A) Paper mach and cloth
(B) Cardboard and mud
(C) Saw dust and hay
(D) All of these

56. Puppetry in an 'INDIAN' medium of communication which attracts people of:
(A) Young (B) Old
(C) Children (D) All age group

57. Video tapes can be used for promoting audio-visual literacy for:
(A) Motivation and attitudinal change
(B) Behaviour reinforcement
(C) Community participation and entertainment
(D) All of these

58. Indian folk theatre is full of exploitative stories of kings and priests tuned into Gods, picked up from different Hindu epics like:
(A) Ramayana
(B) Mahabharat
(C) Ramayana and Mahabharat
(D) None of these

59. Like video, similar results even with more clarity and immediacy can be now used through:
(A) Computer system
(B) Television system
(C) Gramophone system
(D) All of these

Chapter 8

INNOVATION, EVALUATION AND ADOPTION

1. These are how many degree of evaluation:
(A) 4 (B) 3 (C) 5 (D) 2

2. Diffusion is the subset of:
(A) Communication
(B) Teaching
(C) Learning (D) Motivation

3. Diffusion has:
(A) 3 elements (B) 4 elements
(C) 6 elements (D) 8 elements

4. Diffusion and adoption of a new technology are respectively decided by:
(A) Society and society
(B) Individual and Individual
(C) Individual and Society
(D) Society and Individual

5. First step for making a programme planning includes:
(A) Evaluation
(B) Analysis of the situation and determining problems
(C) Deciding on objectives
(D) Teaching

6. At which of the stage of adoption group discussion is most effective:
(A) Awareness (B) Evaluation
(C) Testing (D) Persuasion stage

7. The correct sequence of innovation – decision process or adoption process is:
(A) Interest – awareness – evaluation – trial – adoption
(B) Evaluation – awareness — interest - trial – adoption
(C) Awareness -Interest — evaluation – adoption - trial
(D) Awareness – interest – evaluation – trial – adoption

8. ______________ are the first farmers to adopt a new idea:
(A) Innovators (B) Early adopters
(C) Early majority (D) Laggards

9. ______________ are the last to adopt new idea:
(A) Innovators (B) Early adopters
(C) Laggards (D) Early majority

10. Which one of the following pair is not correctly matched:

Adopter Category	Percentage
(A) Innovators	2.5
(B) Early adopters	13.5
(C) Early majority	40
(D) Laggards	16

11. Normal rate of adoption requires how many years from the introduction of the innovation to its adoption throughout the community:
(A) 2 – 6 (B) 6 – 10
(C) 10 – 14 (D) 14 – 18

12. In case of diffusion of innovation, the overt behavior means:
(A) Adoption or rejection
(B) Remembering or forgetting
(C) Fooding or lodging
(D) Mental weighing or field trial

13. At the adoption stage, information sources is :
(A) Village level change agent and to some extent mass media
(B) Extension agency and other farmers
(C) Informal personal sources including family members
(D) Self-experience gained at the trial stage

14. Adopter category which respect and regarded by many others in the social system as a role model are :
(A) Innovators (B) Early adopters
(C) Late majority (D) Laggards

15. The spread of technology in a social system:
(A) Diffusion (B) Teaching
(C) Learning (D) Adoption

16. Diffusion is essentially concerned with messages that are:
(A) Profitable (B) New ideas
(C) Vital (D) Widely known

17. The elements of diffusion includes all except:
(A) Innovation (B) Social system
(C) Adoption (D) Time

18. The oldest research tradition in diffusion research is from:
(A) Communication (B) Sociology
(C) Education (D) Anthropology

19. The major drawback of past diffusion studies was their inability to take up:
(A) Source research (B) Channel research
(C) Message research (D) Audience research

20. Largest number of diffusion studies are conducted in the field of:
(A) Communication (B) Sociology
(C) Education (D) Anthropology

21. The model of diffusion as proposed by Rogers assumes:
(A) The homogeneity of population
(B) The progressiveness of farmers
(C) Availability of infrastructural facilities
(D) The differential farming situation

22. Diffusion and adoption of a new technology are respectively decided by:
(A) Society and Society
(B) Individual and Individual
(C) Individual and Society
(D) Society and Individual

23. Diffusion can be considered as a subject of:
(A) Communication (B) Teaching
(C) Learning (D) Motivation

24. The degree to which the results of an innovation may be diffused to others:
(A) Credibility (B) Compatibility
(C) Communicability
(D) Relative advantage

25. Homophilous diffusion pattern with in a system helps in the spread of ideas:
(A) Horizontally (B) Vertically
(C) Diagonally (D) None of these

26. Which of the following is the first in motivating the farmers:
(A) Food (B) Money
(C) Farm literature (D) All of these

27. CONA expands to:
(A) Career Oriented Need Assessment
(B) Community Oriented Need Assessment
(C) Creative Organizational Need Assessment
(D) Career Oriented Need Analysis

28. The idea which is perceived as new is termed as:
(A) Information (B) Innovation
(C) Perception (D) Invention

29. An innovation may have in general:
(A) One component
(B) Two component
(C) Three component
(D) Four component

30. The material aspect of an innovation is:
(A) Idea component (B) Object component
(C) Product component
(D) Relative component

31. Symbolic adoption take place in case of innovation with:
(A) Idea component
(B) Object component
(C) Product component
(D) Relative component

32. To be called as an innovation, an idea, practice or object must be:
(A) Objectively new
(B) Subjectively new
(C) All the above
(D) None of these

33. The characteristic of an innovation important from extension point of view:
(A) Intrinsic (B) As people see them
(C) Extraordinary (D) Typological

34. An innovation can be considered to have originated from:
(A) Farmer research
(B) Extension programme
(C) Research trials (D) All of these

35. An innovation include all except:
(A) Material products(B) Ideological belief
(C) Social movement
(D) Traditional technologies

36. An innovation is liked or disliked based on:
(A) Innovation effects
(B) Innovation decision
(C) Perceived attributes
(D) Traditional makeup

37. The perceived newness of the innovation can be expressed in terms of:
(A) Knowledge (B) Decision to adopt
(C) Persuasion (D) All of these

38. The perceived attribute of an innovation that reduces the risk component:
(A) Complexability
(B) Compatibility
(C) Trialability
(D) Absorbability

39. Providing subsidy to farmers on introduction of farm machinery is an example for:
(A) Relative advantage (B) Complexity
(C) Compatibility (D) Observability

40. An innovation with low relative advantage may have:
(A) Slow rate of adoption
(B) High rate of discontinuance
(C) Low return on investment
(D) All of these

41. Compatibility is the degree to which an innovation is perceived consistent with:
(A) Clients needs for innovation
(B) Previously introduced needs
(C) Socio-cultural values and benefits
(D) All of these

42. The luxuriant growth of the plant on the application of urea relates:
(A) Compatibility (B) Trialability
(C) Observability (D) Complexity

43. 'Preventive innovations' have less adoption rate mainly due to:
(A) Low relative advantage
(B) High complexity
(C) Low trialability
(D) Low observability

44. Providing free seed samples to the farmers is closely associated with:
(A) Relative advantage
(B) Complexity
(C) Trialability
(D) Observability

45. The perceived attribute of an innovation that shows a negative correlation with adoption:
(A) Complexability (B) Compatibility
(C) Trialability (D) Observability

46. A compactable innovation offers the adopter:
(A) Greater risk (B) Less security
(C) All of these (D) None of these

47. The attributes of the innovation as matters by whom is most important:
(A) Research scholars (B) Manufactures
(C) Potential adopters (D) Change agents

48. Time is very important consideration in the process of:
(A) Communication (B) Adoption
(C) Diffusion (D) Innovation

49. Individual adopters in a social system is described in terms of his:
(A) Economic conditions
(B) Time of adoption
(C) Frequency of adoption
(D) Concurrent ability

50. Frequency of adoptors when plotted over time follows a:
(A) Normal curve (B) Skewed curve
(C) Linear curve (D) Curvilinear shape

51. The rate of adoption of an innovation in a social system over time follows:
(A) Normal curve (B) Skewed curve
(C) Linear curve (D) S- shaped curve

52. The S shape adoption curve increases at an increasing rate for:
(A) Innovators (B) Early adopters
(C) Early majority (D) All of these

53. The length of time required to pass through the innovation decision process:
(A) Innovation decision matrix
(B) Innovation decision period
(C) Innovation decision index
(D) Innovation decision time

54. Innovation decision period is actually a:
(A) Gestation period (B) Production period
(C) Mental period (D) None of these

55. The innovation decision process can be shortly called as:
(A) KPDIC (B) AIDTCA
(C) NAIDTEA (D) AIETA

56. The chief concern of farmer at the persuasion stage is:
(A) How to information
(B) How do information
(C) How much information
(D) None of these

57. The individual become more psychologically involved with the innovation at:
(A) Knowledge stage
(B) Persuasion stage
(C) Decision stage
(D) Confirmation stage

58. The outcome of persuasion leads to subsequent change in:
(A) Covert behaviour
(B) Overt behaviour
(C) Significant behaviour
(D) None of these

59. Persuasion function of innovation decision process is mainly:
(A) Cognitive (B) Affective
(C) Behavioural (D) Normative

60. The reinforcement of innovation decision making occurs at:
(A) Persuasion stage
(B) Knowledge stage
(C) Confirmation stage
(D) Decision stage

61. Conflicting information should be strictly avoided during:
(A) Readiness stage
(B) Knowledge stage
(C) Decision stage
(D) Confirmation stage

62. The decision to adopt or reject made after the prior innovation decision is called:
(A) Optimal decision
(B) Contingent decision
(C) Collective decision
(D) Authority decision

63. The basic unit of decision making in the optimal innovation decision model is:
(A) Change agent (B) Social system
(C) Researcher (D) Individual

64. The type of innovation decisions that resulted in the emergence of classical diffusion model:
(A) Optimal innovation decisions
(B) Collective innovation decisions
(C) Authority innovation decisions
(D) None of these

65. Generally the fastest rate of adoption of innovation results from:
(A) Optimal innovation decisions
(B) Collective innovation decisions
(C) Authority innovation decisions
(D) Contingent innovation decisions

66. The unit of decision making and unit of adoption are different in case of:
(A) Optimal decision
(B) Collective decision
(C) Authority decision
(D) Mental decision

67. In collective decision making who screens the idea for approving or puts sanction on it:
(A) Stimulators (B) Initiators
(C) Actors (D) Legitimizers

68. Which of the following is not an element in the model of innovation decision process:
(A) Antecedents (B) Variability
(C) Consequences (D) Processes

69. Dissonance does not occur at which of the stages of innovation decision process:
(A) Knowledge (B) Persuasion
(C) Decision (D) Confirmation

70. The degree to which an innovation is changed or modifies by a user in the process of its adoption:
(A) Re-invention (B) Re-engineering
(C) Re-modeling (D) Re-making

71. An agricultural innovator is a:
(A) Traditional leader
(B) Opinion leader
(C) Functional leader
(D) Formal leader

72. The decision to make use of a innovation as best course of action:
(A) Adoption (B) Diffusion
(C) Innovation (D) Technology

73. Adoption is basically a:
(A) Mental process
(B) Social process
(C) Psychological process
(D) Physical process

74. The scientist first to recognize adoption of innovation occurs in stages:
(A) Wilkening
(B) Rogers and Shoemaker
(C) Johnson and Haver
(D) Ryan and Gross

75. The stages of adoption are dynamic and not static according to:
(A) Ryan and Gross
(B) Wilkening
(C) Johnson
(D) Y P Singh

76. Who for the first time reported that adoption process involves decision:
(A) Pederson (B) Wilkening
(C) Holmberg (D) Mason

77. The first stage in the process of adoption according to Rogers is:
(A) Attention (B) Interest
(C) Desire (D) Satisfaction

78. The stage wherein the individual makes full use of the innovation:
(A) Evaluation (B) Adoption
(C) Awareness (D) Trial

79. The stage wherein the individual decides to continue the full use of the innovation:
(A) Evaluation (B) Adoption
(C) Awareness (D) Trial

80. The stage wherein an individual is exposed to the innovation but lacks complete information about it:
(A) Attention
(B) Interest
(C) Desire
(D) Satisfaction

81. The most important source of information at the attention stage of adoption:
(A) Mass media (B) Group media
(C) Extension worker(D) Personal localities

82. The stage of adoption wherein group discussion is most effective:
(A) Awareness (B) Evaluation
(C) Testing (D) Persuasion

83. Unfreezing existing behaviour and motivating people to change is expressed at:
(A) Desire (B) Interest
(C) Conviction (D) Attention

84. The extension activity that need to be emphasized in the interest stages of adoption:
(A) Limited trial is good
(B) Rapport building is needed
(C) Group training is needed
(D) Specialists are needed

85. Symbolic adoption is an example for:
(A) Covert behaviour
(B) Overt behaviour
(C) Affection
(D) Perception

86. Adopter categories are the classification of the members of the social system on the basis of:
(A) Innovativeness
(B) Adoption process
(C) Diffusion of innovation
(D) Innovation decision period

87. The time of adoption of an innovation by all the members in a social system:
(A) Always same (B) Always vary
(C) May vary (D) All of these

88. The classification of adopters into various categories is based on:
(A) Mean and variance
(B) Mean and standard deviation
(C) Mean and median
(D) Median and standard deviation

89. The adopter category considered as deviant by other members of the society:
(A) Innovators (B) Early majority
(C) Early adopters (D) Laggards

90. The adopter category considered as friends and neighbours by other members in the social system:
(A) Innovators (B) Early majority
(C) Early adopters (D) Laggards

91. The adopter category considered by many as the men to "check with":
(A) Innovators (B) Early majority
(C) Early adopters (D) Laggards

92. Opinion leaders belong to the category of:
(A) Innovators (B) Early majority
(C) Early adopters (D) Late majority

93. The point of reference for laggards is:
(A) Past (B) Present
(C) Future (D) All of these

94. The adopter categories who is also considered as the custodians of indigenous knowledge:
(A) Innovators (B) Early majority
(C) Early adopters (D) Laggards

95. The person who adopt an innovation before every average member of the society does it:
(A) Innovators (B) Early majority
(C) Early adopters (D) Late adopters

96. Agricultural innovators are generally viewed as:
(A) Opinion leader (B) Social deviants
(C) Skeptical (D) Traditional

97. 'Legitimization' is done by:
(A) Innovators (B) Early adopters
(C) Early majority (D) Late majority

98. Who among the following is more concerned about triability of the innovation:
(A) Early adopters (B) Early majority
(C) Innovators (D) Laggards

99. "Be not the last to lay the old aside nor be the first by which the new is tried" is the motto of:
(A) Innovators (B) Early majority
(C) Early adopters (D) Laggards

100. The percentage of population constituted by the innovators:
(A) 2.5 per cent (B) 7.5 per cent
(C) 16.5 per cent (D) 33 per cent

101. The character best represent an innovator:
(A) Venturesome (B) Skeptical
(C) Traditional (D) Respectfulness

102. The adoption of innovation rather vigourously when experts feel that it should be stopped is called:
(A) Excessive adoption
(B) Over adoption
(C) Vigorous adoption
(D) Overt adoption

103. A type of rejection which consists of considering adoption of the innovation but then deciding not to adopt it:
(A) Active rejection
(B) Passive rejection
(C) Pro-active rejection
(D) None of these

104. Who incorporates the innovations into a specific plan of action adopted to needs of the system:
(A) Stimulators (B) Initiators
(C) Actors (D) Legitimizers

105. 'Complementary adoption' is usually associated with:
(A) Agricultural adoption
(B) Technological adoption
(C) Industrial adoption
(D) Both (B) and (C)

106. The phenomena wherein adoption of one innovation requires the previous adoption of another innovation:
(A) Interaction effect (B) Diffusion effect
(C) Technology effect (D) Adoption effect

107. The individuals in the same adopter category have:
(A) Varied degree of innovativeness
(B) Multitude of innovativeness
(C) Similar degree of innovativeness
(D) None of these

108. Consequences are the changes that occur is a social system as a result of:
(A) Adoption of innovation
(B) Rejection of innovation
(C) Both (A) and (B)
(D) None of these

109. Discontinuance occurs only after an innovation has been:
(A) Fully adopted
(B) Partially adopted
(C) Not at all adopted
(D) None of the above

110. A decision to reject an idea as a result of dissatisfaction with it's performance:
(A) Disenchantment discontinuance
(B) Replacement discontinuance
(C) Betterment discontinuance
(D) Distinguishable discontinuance

111. The decision to reject an idea in order to adopt a better idea:
(A) Disenchantment discontinuance
(B) Replacement discontinuance
(C) Betterment discontinuance
(D) Distinguishable discontinuance

112. The dysfunctional effect of an innovation to an individual or to a social system:
(A) Direct consequence
(B) Indirect consequence
(C) Desirable consequence
(D) Undesirable consequence

113. The change to an individual or a social system that occur in immediate response to an innovation:
(A) Direct consequence
(B) Indirect consequence
(C) Desirable consequence
(D) Undesirable consequence

114. The changes from an innovation that is neither intended nor recognized by the members of a social system:
(A) Direct consequence
(B) Indirect consequence
(C) Desirable consequence
(D) Unanticipated consequence

115. Acceptance and rejection are the one of the kinds of:
(A) Failure of channel
(B) Feed back
(C) Treatment of message
(D) Empathetic feelings

116. Adoption is a/an:
(A) Social process
(B) Mental process
(C) Psychological process
(D) Individual process

117. Serialise the following stages of adoption process:
(1) Interest (2) Evaluation
(3) Trial (4) Awareness
Pick out the answer from the following
(A) 1, 2, 3, 4 (B) 3, 1, 2, 4
(C) 4, 1, 3, 2 (D) 3, 2, 1, 4

118. Purpose of extension evaluation is:
(A) To identify the weak points
(B) To identify the strong points
(C) To identify the gaps and errors
(D) To identify all the above

119. Read the following statements and mark your answer according to the code given below:
I. Extension evaluation is a planned process
II. Extension evaluation is spontaneous
III. Extension evaluation is indivisible portion of teaching
IV. Extension evaluation is beginning process
Answer code:
(A) I and III are correct
(B) II and III are correct
(C) I and IV are correct
(D) II and IV are correct

120. In case of diffusion of innovation, the overt behavior means:
(A) Adoption or rejection
(B) Remembering or forgetting
(C) Fooding or loading
(D) Mental weighing or field trial

Chapter 9

PARTICIPATORY RURAL APPRAISAL

1. The collection of data in PRA is done by:
(A) Local Extension agents
(B) Progressive farmers
(C) Research assistants
(D) A multidisciplinary team

2. For a study of farming system, the best PRA exercise is:
(A) Transect walk (B) Resource mapping
(C) Time line (D) Chapatti diagram

3. PRA is known for:
(A) People Resource Appraisal
(B) Public Resource Appraisal
(C) Participatory Rural Appraisal
(D) People Rural Agency

4. The concept of PRA was given by:
(A) Henry Ford (B) Robert Chambers
(C) Neils Roling (D) Nila Mukerjee

5. The person considered as father of PRA:
(A) Neils Rolling (B) Robert Chambers
(C) Van den Ban (D) Paul Leagans

6. The term PRA was first used in:
(A) Kenya (B) USA
(C) UK (D) Israel

7. PRA was introduced in India by:
(A) Government departments
(B) ICAR (C) NGO's
(D) Ford foundation

8. PRA had it's origin during late:
(A) Sixties (B) Seventies
(C) Eighties (D) Nineties

9. The main innovation in PRA is:
(A) Methods (B) Behaviour
(C) Process (D) Actors

10. The role of extension personnel's in PRA is:
(A) Elective (B) Extractive
(C) Participatory (D) None of these

11. Which of the following is not the characteristic of PRA:
(A) Quick (B) Rigid
(C) Interactive (D) Collective

12. The collection of data in PRA is based on:
(A) Extension workers
(B) Multi disciplinary team
(C) Research scientists
(D) Progressive farmers

13. The type of PRA used to understand broad patterns and trends in a rural society:
(A) Exploratory (B) Topical
(C) Participatory (D) Monitoring

14. Which among the following is not a principle of PRA:
(A) Optimal ignorance
(B) Setting uniformity
(C) Offsetting bias (D) Triangulation

15. The PRA technique that shows cause effect relationship:
(A) Venn diagrams (B) Flow diagram
(C) Social mapping (D) Livelihood analysis

16. The PRA technique wherein rural elders narrates life histories:
(A) Time line (B) Village transits
(C) History sheets (D) Ranking

17. Social calendar is a tool most commonly used in:
(A) PRA (B) RRA
(C) LFA (D) FSR

18. Which of the following is not a tool for conducting PRA:
(A) Transect walk (B) Social map
(C) Time line analysis
(D) Role play

19. The PRA technique that represent the degree of importance of individuals and institutions in decision making:
(A) Social mapping (B) Venn diagram
(C) Semi structured interview
(D) None of these

20. People's participation in an extension programme is significant when:
(A) Literate section off villagers participate
(B) Gram panchayat members participate
(C) Majority of villagers participate
(D) Local leader participate

21. Consider the following statements:
Assertion (A): Farmers posses keen interest in training programme at block headquarter
Reason (R): Farmers enjoy free meal during training period.
Select the correct answer from the code given below
(A) A is true but R is false
(B) R is true but A is false
(C) Both A and R are correct and R is the correct explanation of A
(D) Both A and R are true but R is not the correct explanation of A

22. Participatory rural appraisal is directly related to:
(A) Soil science
(B) Entomology
(C) Agricultural economics
(D) Agricultural extension

23. Systematic process of involvement of people for the analysis of their own situations and finding out solution for improving their present condition is called:
(A) Participatory research
(B) Explorative research
(C) Amiliorative research
(D) None of these

24. An effort to introduce innovation not known to social system is called:
(A) Participatory research
(B) Explorative research
(C) Amiliorative research
(D) None of these

25. It emphasizes in the improvement of what people are doing (at present). It is an effort to make the good better (Compatible) is:
(A) Participatory research
(B) Explorative research
(C) Amiliorative research
(D) None of these

26. The characteristics of participatory research is/are:
(A) An activity carried out by group of people to understand their own situation for their own benefit
(B) It is a collective effort to bring desirable results taking into consideration the various aspects of problem in a collective manner
(C) It is an analysis that lead to action
(D) Understanding of personal realities and larger social realities it will facilitate awareness raising about the connections between micro-aspects of their life and the real macro-aspects of society at large
(E) All of these

27. Participatory research is a:
(A) Cyclic process (B) Noncyclic process
(C) Both of these (D) None of these

28. The limitations of the cyclic process of participatory research may be the:
(A) More participation of practioners
(B) Least participation of practioners
(C) Either (A) or (B) (D) None of these

29. Steps in conducting a participatory research are:
(A) To know the problem
(B) Solution is needed or not
(C) To know the present status and additional information needed for the problem to be tackled
(D) All of these

30. A well known method of data collection is to carry out a:
(A) Big story (B) Medium story
(C) Small story (D) None of these

31. A small number of people (10-25) come together to start experiences and decide for any action through:
(A) Individual discussion
(B) Group discussion
(C) Individual interview
(D) Group interview

32. A well known method of data collection is to carry out a:
(A) Small survey (B) Big survey
(C) Reconscence survey (D) None of these

33. A small number of people (10-25) come together to store experiences and decide for any action through:
(A) Individual discussion
(B) Group discussion
(C) Individual interview
(D) Group interview

34. Sometimes the group engaged in participatory research may find itself uncomfortable with:
(A) Written method (B) Oral method
(C) Written and oral method both
(D) None of these

35. Women can be engaged to draw their perceptions of the problem on:
(A) Blackboard (B) Paper or cloth
(C) Black and white photographs
(D) Cutting from old magazines
(E) All of these

36. A group discussion or individual interview can be taped and then can be:
(A) Replayed for analysis
(B) Data collection
(C) Questionnaire (D) None of these

37. A group of people can investigate a particular problem by using ———— to known the opinions of a larger number of people:
(A) Discussion (B) Survey
(C) Questionnaire (D) None of these

38. The questionnaire should be single have only questions?
(A) 4-6 (B) 6-8
(C) 8-10 (D) 10-12

39. Each question of questionnaire should be:
(A) Short simple and straight forward
(B) Long, simple and straight forward
(C) Short complex and straight forward
(D) Long, complex and straight forward

40. A questionnaire of self survey should possess:
(A) Should have few (8-10) questions
(B) Should be short simple and straight forward
(C) Should have easy in conversation
(D) All of these

41. Group discussion is one of the most popular and easy method of combining:
(A) Data collection (B) Analysis
(C) Action planning (D) All of these

42. The important consideration of group discussion is to insure the:
(A) Uncomfortable atmosphere
(B) Comfortable atmosphere
(C) Positive atmosphere
(D) None of these

43. Whcih method can easily be combined with other methods is/are:
(A) Individual discussion
(B) Group discussion
(C) Individual interview
(D) Group interview

44. The important characteristics of peoples camp is/are:
(A) May be one or two days programme
(B) Common lodging ad boarding arrangement
(C) Peoples comes from longer distance
(D) All of these

45. Which method can provide an informal and free atmosphere:
(A) People's camp (B) Exhibition
(C) Campaign (D) Seminar

46. In a people's camp how many peoples can participate easy:
(A) 30-40 (B) 50-100
(C) 100-200 (D) Unlimited

47. It is always beneficial to visit an area where a similar problem has been solved for which the group is interested is:
(A) Demonstrations
(B) Group discussion
(C) Exhibition (D) Campaign

48. Indian folk theatre has a important role to play in the process of:
(A) Social changes
(B) Political changes
(C) Economic changes
(D) Socio-economic changes both

49. Folk theatre media are:
(A) Low cost (B) Indigenous
(C) May provide better for focusing and solving rural problems
(D) All of these

50. Rural poor can communicate better and convince people by making their:
(A) Own theatre
(B) Own house
(C) Leased house
(D) None of these

51. Any inanimate thing animated on stage to communicate any subject is called:
(A) Puppets (B) Drama
(C) Kisan mela (D) All of these

52. All the instincts basic to human learning are stimulated and get satisfied through puppet show:
(A) Communicative dramatic
(B) Artistic musical
(C) Inquisitive constructive
(D) All of these

53. Groups of local people analyzing their own conditions and choosing their own means of improving them is:
(A) Participatory Rural Appraisal
(B) Rapid Rural Appraisal
(C) Both (A) and (B) (D) None of these

54. The technique developed for analyzing the needs of rural community quickly, cost effectively and with little destruction to every day life is:
(A) Participatory Rural Appraisal
(B) Rapid Rural Appraisal
(C) Both (A) and (B) (D) None of these

55. The techniques are a range of survey methods to be used principally in the extraction of qualitative data is:
(A) Participatory Rural Appraisal
(B) Rapid Rural Appraisal
(C) Both (A) and (B)
(D) None of these

56. Rapid Rural Appraisal (RRA) evolved in the :
(A) 1979 (B) 1980
(C) 1981 (D) 1982

57. PRA which began in the late:
(A) 1970s (B) 1975s
(C) 1980s (D) 1990s

58. Which method emphasized multi-disciplinary teams, careful observation, semi-structured interviewing and focus group in:
(A)Participatory Rural Appraisal
(B) Rapid Rural Appraisal
(C) Both (A) and (B) (D) None of these

59. The method which involve group rather than individuals and visual representations rather than solely verbal communication in:
(A) Participatory Rural Appraisal
(B) Rapid Rural Appraisal
(C) Both (A) and (B) (D) None of these

60. Key issue of Participatory Rural Appraisal is:
(A) To highlight the different PRA techniques available for understanding the access strengths and constraints of the rural poor
(B) To describe the best condition for use of PRA techniques
(C) To highlights the constraints of using PRA techniques
(D) All of these

61. Assumptions of PRA are:
(A) The environmental problem is a common concern, the solution of which requires the active and responsible involvement of the entire community
(B) The collective action is possible only when a stakeholders of a community develop a clear common understanding about the issue
(C) Both (A) and (B)
(D) None of these

62. Rapid Rural Appraisal is a:
(A) Extractive in nature (B) Elicited
(C) Information owned by outsiders
(D) All of these

63. Participatory Rural Appraisal is a:
(A) Learning with local people
(B) Information owned by rural people but shared with outsiders
(C) Both (A) and (B) (D) None of these

64. Principles of PRA is/are:
(A) Using optimal ignorance
(B) Offsetting biases (C) Triangulation
(D) Learning from and with rural people
(E) Learning rapidly and progressively
(F) All of these

65. Good features of PRA is/are:
(A) Iterative (B) Innovative
(C) Interactive (D) Informal
(E) In the community
(F) All of these

66. The menu of PRA is/are:
(A) Secondary data review
(B) Semi structured interviews
(C) Analytical games
(D) Stories ad portraits
(E) Diagrams (F) Workshop
(G) All of these

67. Helpers of PRA is/are:
(A) What and Why (B) Who and How
(C) Where and When (D) All of these

68. Important tips of PRA is/are:
(A) Look (B) Listen
(C) Learn (D) All of these

69. Participatory approaches are not substitutes for , but are rather a integral part of:
(A) Long term dialogue
(B) Sustained interaction
(C) Both (A) and (B) (D) None of these

70. A single, brief participatory exercise with a group of local people will not lead to:
(A) Positive and lasting change
(B) Negative and lasting change
(C) Positive and integral change
(D) Negative and integral change

71. PRA are not a panacea of:
(A) Qualitative surveying
(B) Quantitative surveying
(C) Either (A) or (B)
(D) Neither (A) or (B)

72. PRA work must effectively where they are carried out over a sufficient:
(A) Length of time (B) Money
(C) Research (D) None of these

73. Primary importance of PRA is:
(A) Behaviour (B) Attitudes
(C) Longevity (D) All of these

74. A method of linking different survey methods is order to cross-check the information collected from each method is:
(A) Triangulation
(B) Pilot experiment
(C) Action research
(D) Survey design

75. Triangulation is of particular use when employing:
(A) Mapping
(B) Ranking and scoring
(C) Flow and venn diagrams
(D) All of these

76. One of the strengths of PRA is that many of the methods are visual and therefore, accessible to a:
(A) Longer group of peoples
(B) Small group of peoples
(C) Peoples are not required
(D) None of these

77. Participatory methods have contributed to adjusting the interview to make it more conversational, while still:
(A) Controlled
(B) Structured
(C) Controlled and structured both
(D) None of these

78. What are the characteristics of informal conversational interview:
(A) Questions emerge from the immediate context and are asked in the natural course of things
(B) There is no predetermination of question topics or wording
(C) Either (A) or (B) (D) Both (A) and (B)

79. What are the strengths of informal conversational interview:
(A) Increases the salience and relevance of questions
(B) Interview the built on and emerge from observations
(C) The interview can be matched to individuals and circumstances
(D) All of these

80. Key interview methods in PRA system are:
(A) Informal conversational interview
(B) Interview guide approach
(C) Standardised open-ended interview
(D) Closed quantitative interviews
(E) All of these

81. What are the weaknesses of informal conversational interview are:
(A) Different information collected from different people with different questions
(B) Less systematic and comprehensive
(C) Data organization and analysis can be quite difficult
(D) Requires maximum attention by interviewer
(E) All of these

82. What are important characteristics of interview guide approach:
(A) Topics and issues to be covered are specified in advance, in out line form
(B) Interviewer decide sequence and working of questions in the course of the interview
(C) Both of these
(D) None of these

83. What are the strengths of Interview guide approach:
(A) The outline increases the comprehensiveness of the data
(B) Makes data collection systematic for each respondent
(C) Interviews remain fairly conversational and situational
(D) All of these

84. What are the weaknesses of Interview guide approach:
(A) Important and salient topics may be inadventently omitted
(B) Interviewer flexibility in working questions can result in incomparatively of responses
(C) Both (A) and (B)
(D) None of these

85. Important characteristics of Standardised open-ended interview:
(A) The exact wording and sequence of questions are determined in advance
(B) Al interviewees are asked the same basic questions in the same order
(C) Both (A) and (B) (D) None of these

86. Important strengths of standardized open-ended interview:
(A) Respondents answer the same questions, thus increasing comparability of response
(B) Reduces interviewer bias when several interviewers are used
(C) Facilitates organization and analyses of data
(D) All of these

87. What are the weakness of standardized open - ended interview:
(A) Little flexibility in rilching the interview to particular individuals and circumstances
(B) Requires maximum attention by interviewer
(C) Data organization and analysis can be quite difficult
(D) All of these

88. Important characteristics of closed quantitative interviews:
(A) Questions and response categories are determined in advance
(B) Responses are fixed
(C) Respondent chooses from among these fixed responses
(D) All of these

89. Important strengths of closed quantitative interviews:"
(A) Data analysis is simple
(B) Responses can be directly compared and easily aggregate
(C) Man questions can be asked in a short time
(D) All of these

90. Important weaknesses of closed quantitative interviews:
(A) Respondents must fit their experience and fallings into the researchers categories
(B) May be perceived as impersonal, irrelevant and mechanistic
(C) Can distort what respondents really mean or experienced
(D) All of these

91. Participatory methods tend to make use if the:
(A) Informal conversational interview
(B) Inter view guide approach
(C) Both (A) and (B) (D) None of these

92. Informal conversational interview and interview guide approach are:
(A) Semi-structured interviews
(B) Long-structured interviews
(C) Short-structured interviews
(D) All of these

93. Ranking and scoring have long been used to assess:
(A) People's expectations
(B) Beliefs and attitudes
(C) Preferences and opinions
(D) All of these

94. Which is useful tool to be used in generating basic information which helps to focus further questioning is:
(A) Ranking (B) Scoring
(C) Ranking and scoring
(D) None of these

95. What are the important categories of ranking:
(A) Preference ranking
(B) Wealth ranking
(C) Matrix ranking (D) All of these

96. The way of quickly identifying main preferences as experienced by individual villager or village groups:
(A) Preference ranking
(B) Matrix ranking
(C) Wealth ranking (D) None of these

97. Ranking can be used to express other issues, where a two dimensional comparison is useful:
(A) Preference ranking
(B) Matrix ranking
(C) Wealth ranking
(D) None of these

98. The most common technique for ranking is:
(A) Preference ranking
(B) Matrix ranking
(C) Wealth ranking
(D) Card sorting

99. Closed scoring can be done in:
(A) Each 'box' (element for that criteria) can be scored between eg. 0-5, this is repeated over the whole matrix
(B) A fixed number of 'points' can be awarded for each criteria (row or column), and these are distributed between the elements in that two or column in relation to their perceived importance
(C) A fixed number of points is allowed for the entire matrix, and there must be distributed between both elements and criteria as participants decide
(D) All of these

100. Ranking enables villagers to divide households in the community according to economic and other well being categories is:
(A) Preference ranking
(B) Matrix ranking
(C) Wealth ranking
(D) None of these

101. Maps can be used to identify the comparative location and importance of different resources within an area is called:
(A) Participatory mapping
(B) Social mapping
(C) Contour mapping
(D) Flow diagram

102. Diagrams are used for the systematic analysis of a wide range of issues whereby a whole stories of cause and effect relationships are examined is known as:
(A) Participatory mapping
(B) Social mapping
(C) Contour mapping
(D) Flow diagram

103. Maps can be used to locate houses services and infrastructure within an area is called:
(A) Participatory mapping
(B) Social mapping
(C) Contour mapping
(D) Flow diagram

104. Flow diagram can act as a basis for discussing the relationship between:
(A) Different groups
(B) Individuals or issues
(C) Can demonstrate potential multiplier effects
(D) All of these

105. The diagrams are used to depict key institutions, organizations and individuals, and their interaction with the local community is called:
(A) Venn diagram (B) Flow diagram
(C) Both of these (D) None of these

106. In the rural transport context, venn diagram can be used to demonstrate the interaction between:
(A) Local villagers
(B) Transport operators
(C) Local government (D) All of these

107. The process of making the community inventory is sometimes called:
(A) Semi-structured interviewing
(B) High-structured interviewing
(C) Both of these (D) None of these

108. The techniques widely used in recent years in most of the developmental programme:
(A) Participatory Rural Appraisal
(B) Rapid Rural Appraisal
(C) Participatory Resource Appraisal
(D) All of these

109. A process that empowers local people to change their own condition and situation:
(A) Participatory Rural Appraisal
(B) Rapid Rural Appraisal
(C) Participatory Resource Appraisal
(D) All of these

110. A way of learning for local community about their natural resources is called:
(A) Participatory Rural Appraisal
(B) Rapid Rural Appraisal
(C) Participatory Resource Appraisal
(D) Participatory Appraisal of Natural Resources

111. The important principles which PRA is governed are:
(A) Listening and learning
(B) Offsetting biases
(C) Proper utilization of precious community time
(D) Seeking diversity
(E) Cross checking
(F) All of these

112. Components of PRA are:
(A) Attitude and Behaviour
(B) Process of participation
(C) Methods (D) Sharing
(E) Facilitation (F) All of these

113. PRA is a:
(A) Unidimensional approach
(B) Multidimensional approach
(C) Straight dimensional approach
(D) All of these

114. Sharing of information and knowledge is integral part in the process of :
(A) Participation (B) Communication
(C) Learning (D) Adoption

115. Attitude and behavior of the agent constitute the base for approaching and establishing links with:
(A) Local communities
(B) Outside communities
(C) Urban communities
(D) None of these

116. PRA is an essential component of:
(A) Rural developmental projects
(B) Rural developmental schemes
(C) Both (A) and (B) (D) None of these

117. Advantages of PRA is/are:
(A) Empowering the poor and weak
(B) Diversity
(C) Research priority
(D) Organizational changes
(E) All of these

118. PRA has extensive range of application in research and planning for:
(A) Land tenure (B) Urban planning
(C) Natural resource management and domestic violence
(D) All of these

119. Sharing of information and knowledge has magic effort in the process of :
(A) Participation
(B) Communication
(C) Learning
(D) Adoption

120. Successful training requires encouraging new ways of learning:
(A) Within oragnisation
(B) Outside the organization
(C) Both of these (D) None of these

121. Training courses are always part of the:
(A) Questions (B) Answer
(C) Questionnaire (D) None of these

122. PRA, a field based methodology which has evolved and spread in the:
(A) 1980s (B) 1985s
(C) 1990s (D) 2000s

123. The primary focus of PANR is an local natural resources from:
(A) Local community
(B) Urban community
(C) Rural community
(D) None of these

124. The conventional techniques help the experts to arrive at numerical value for appraising:
(A) Manmade resources
(B) Artificial resources
(C) Natural resources
(D) None of these

125. PRA is a technique for involving people to work for their:
(A) Other development
(B) Own development
(C) Rural development
(D) Urban development

126. All principles of PRA are framed on the:
(A) NGO involvement
(B) Government involvement
(C) People's involvement
(D) None of these

Chapter 10

RURAL DEVELOPMENT AND PANCHAYATI RAJ

1. Success in rural development project depends upon:
(A) Regular training of workers
(B) Regular contacts of workers
(C) Amount of subsidy
(D) Participation of beneficiaries

2. Role of different agencies for village development is included in
(A) Resource mapping
(B) Time line
(C) Transect walk
(D) Chapati diagram

3. Rural development depends on:
(A) Research
(B) Research – Extension
(C) Research – Teaching – Extension
(D) None of these

4. The basic unit of development under IRD program is:
(A) A village
(B) A community development block
(C) A family (D) A district

5. From among the following identify one which is three – tier system of the local self – government:
(A) Gram panchayat – Khand samiti – Zila parishad
(B) Gram panchayat –Village school– Village cooperative
(C) IRD programme – TRYSEM – NREP
(D) Lab to land – KVK – KGK

6. The University Education Commission was headed by Dr. S. Radha Krishanan recommended the establishment of:
(A) Urban Universities
(B) Rural Universities
(C) Both (A &) (B)
(D) None of these

7. There are ———— development blocks in India:
(A) 5000 (B) 6000
(C) 7000 (D) 8000

8. ———— is a dominant social institution permeating social and economic relations
(A) Society (B) Creed
(C) Caste (D) Customs

9. Rural community differ from urban community in relation to
(A) Population density (B) Culture
(C) Environment (D) All of these

10. Which of the following statements about "joint family system of rural areas" is correct?
(A) Joint families are always a cause
(B) Joint families are both an impediment and a factor or rural development
(C) There are no joint families in the rural areas
(D) Joint families are the only factors of rural development

11. Rural development programme at district level is being looked after by:
(A) DRDA
(B) MP of that constituency
(C) Additional S.P. (D) None of the above

12. Family is included under:
(A) Secondary group (B) Primary group
(C) Tertiary group
(D) Depends upon the situation

13. Integrated Rural Development Programme was started in:
(A) 1978 (B) 1974
(C) 1975 (D) 1979

14. Co-operative society is basic institutions for:
(A) Political growth of the villagers
(B) Social growth of the villagers
(C) Cultural growth of the villagers
(D) Socio-economic growth of the villagers

15. According to Balwantrai Mehta committee, Panchayati Raj should be a
(A) Two tier structure of local self government bodies from village to district
(B) Three tier structure of local self government bodies from village to district
(C) Two tier structure of local self government bodies from village to block
(D) Three tier structure of local self government bodies from village to block

16. Panchayati Raj was first time started on 2nd October, 1959 in:
(A) Bombay (Maharashtra)
(B) Nagaur (Rajasthan)
(C) Shimla (Himachal Pradesh)
(D) Patna (Bihar)

17. The village Panchayati Act came into existence in:
(A) 1958 (B) 1959
(C) 1960 (D) 1961

18. Panchayati Raj came after
(A) Pre-determined programme
(B) Self determined programme
(C) Fact determined programme
(D) None of these

19. Panchayati Raj in India was first recommended by:
(A) Balwant Rai Mehta Committee
(B) Mahatma Gandhi
(C) Kriplani committee
(D) Pandit Jawaharlal Nehru

20. Panchayati Raj was first time introduced in which of the following state:
(A) Rajasthan (B) Punjab
(C) Andhra Pradesh (D) Orissa

21. Panchayati Raj was first started at:
(A) Nagpur (B) Nagaur
(C) Nasik (D) Narendrea Nagar

22. Zamidari system was introduced by:
(A) Mugals (B) British
(C) Dutch (D) Frenches

23. **List-I** **List-II**
(a) Theory of rent (1) Edward H.
(b) Quasi rent (2) Adam Smith
(c) Monopolistic competition (3) Alfred Marshall
(d) Theory of Wealth of Nation (4) Ricardio

Code	**a**	**b**	**c**	**d**
(A)	3	4	2	1
(B)	4	3	1	2
(C)	4	2	1	3
(D)	2	3	4	1

24. The establishment of Panchayati Raj bodies at various level fulfils:
(A) Principle of cooperation
(B) Principle of leadership
(C) Principle of participation
(D) Principle of grass root

25 Co-operative society is basic institution for:
(A) Political growth of the villagers
(B) Social growth of the villagers
(C) Cultural growth of the villagers
(D) Socio-economic growth of the villagers

26. The new Panchayat Raj Act (73rd Constitution amendment) was enacted all over the country from:
(A) 24th April, 1993 (B) 24th April, 1994
(C) 24th April, 1995 (D) None of these

27. Which of the following rural development project was launched before independence:
(A) Etawah Pilot Project
(B) Cooperative movement
(C) Community development
(D) National Extension Service

28. Cooperative society is basic institution for:
(A) Political growth of the villagers
(B) Social growth of the villagers
(C) Cultural growth of the villagers
(D) Socio-economic growth of the villagers

29. Success of rural projects depend upon:
(A) Regular training of staff
(B) Constant visit of staff
(C) Quantum of subsidies
(D) Partnership of beneficiaries

30. Programme related to elimination of rural poverty is:
(A) JRY (B) RLEGP
(C) IRDP (D) All of these

31. Name the 'Poverty Eradication Programme' in which bank play a dominant role:
(A) MNP (B) IRDP
(C) RBY (D) JRY

32. When National Development Council was formed?
(A) August 1947 (B) August 1950
(C) August 1952 (D) August 1960

33. Which one of the following scheme is exclusively meant for self-employment of rural youth?
(A) NREP (B) TRYSEM
(C) IRDP (D) DPAP

34. Which one of the following is not the example of social system?
(A) Village (B) Communicator
(C) Block (D) District

35. The first cooperative societies established in India were:
(A) Credit societies
(B) Marketing societies
(C) Agricultural societies
(D) Housing societies

36. In distinguished unemployment is:
(A) Marginal physical productivity becomes zero
(B) Average physical productivity becomes zero
(C) Total physical productivity becomes zero
(D) None of these

37. Social changes refer to changes in:
(A) Social structure of the society
(B) Culture of the society
(C) Adoption behavior of the people
(D) Communication network

38. Capital is the result of savings and its formation depends upon:
(A) Capacity to save (B) Desire to save
(C) Facility to save (D) All of these

39. Selecting 'poorest among poor first' is the approach under:
(A) Jawahar Rojgar Yojna
(B) Training and Visit
(C) Integrated Rural Development Programme
(D) National Agricultural Technology Programme

40. Which one of the following does directly affect the division of labour?
(A) Capital
(B) Climate
(C) Demand of labour
(D) Nature of work

41. The basic unit of development under IRD programme is:
(A) Village Community
(B) Development block
(C) Family (D) District

42. An effective facilitator in the development of any society is:
(A) Communication technology
(B) Information and communication technology
(C) Information technology
(D) Rural Development

43. Rural India reflects the very essence of:
(A) Indian culture (B)Tradition
(C) Both (A) and (B) (D) None of these

44. The most serious economic problem of India are:
(A) Poverty and unemployment
(B) Unemployment and stagnation
(C) Underdevelopment and poverty
(D) Corruption and unemployment

45. The percentage of people below poverty line in India is:
(A) 25.1 (B) 27.8
(C) 29.3 (D) 21.5

46. The National Rural Employment Guarantee Act was enacted initially to cover:
(A) 175 districts (B) 200 districts
(C) 257 districts (D) 304 districts

47. "Attacking Rural Poverty" is a book written by:
(A) Paul leagans
(B) A.T. Mosher
(C) Coomb and Ahamed
(D) Henry Parker

48. The important philosophy behind community development is:
(A) Work based on felt needs
(B) Peoples participation
(C) Philosophy of self help
(D) Emancipatory training

49. The number of unemployed people living in India is around:
(A) 13.10 million (B) 9.02 million
(C) 7.98 million (D) 10.51 milion

50. Community development project was started on the basis of:
(A) Nalagarh Commission Report
(B) Royal Commission Report
(C) Ford Foundation Report
(D) GMF Enquiry Report

51. The population of India according to 2001 census is:
(A) 102.21 crores (B) 104.56 crores
(C) 107.56 crores (D) 102.70 crores

52. The sex ratio of India according to 2001 census is:
(A) 927 (B) 933
(C) 938 (D) 942

53. India's population projection for 2026 is:
(A) 1.21 billion (B) 1.27 billion
(C) 23.50 billion (D) 1.40 billion

54. The literacy rate of India according to 2001 census is:
(A) 75.85 (B) 54.16
(C) 62.67 (D) 65.38

55. The concept of multipurpose village level worker was evolved from:
(A) Etawah Pilot Project
(B) Community Development Project
(C) National Extension Service
(D) Gurgaon Project

56. The first bank to launch Kisan Credit Card:
(A) Punjab National Bank
(B) State Bank of India
(C) Bank of Baroda
(D) Allahabad Bank

57. Etawah district is in the state of:
(A) Madhya Pradesh
(B) Uttar Pradesh
(C) Haryana (D) Bihar

58. Nilokheri project was also called:
(A) Mazdoor Manzil (B) Kisan Manzil
(C) Mazdoor Madad (D) Kisan Madad

59. The term community development have originated in:
(A) Oxford (B) Cambridge
(C) Ohio (D) San Francisco

60. Firka development scheme was started by:
(A) Bihar government
(B) Bengal government
(C) Maharastra government
(D) Madras government

61. DRDA was established in:
(A) 1919 (B) 1980
(C) 1967 (D) 1947

62. The chairman of District Rural Development Agency is:
(A) Block Development Officer
(B) District Planning Officer
(C) District Panchayat President
(D) District Collector

63. The authority responsible for planning and implementation of rural development programme at state level is:
(A) Planning Commission
(B) State Development Committee
(C) Ministry of Rural Development
(D) Chief Minister

64. The chairman of State Development Committee is:
(A) Governor (B) Chief Minister
(C) State Agriculture Minister
(D) Minister of Rural Development

65. The Gandhian approach of rural development have its basic emphasis on:
(A) Village industries
(B) Development of inner man
(C) Local self government
(D) Production improvement

66. The annual outlay for the ministry of rural development for the year 2006-07 was:
(A) 36000 crores
(B) 31000 crores
(C) 21000 crores
(D) 14000 crores

67. India's position in the UNDP's Human Development Repost for 2006 was:
(A) 212 (B) 214
(C) 126 (D) 198

68. The National Commission of Women was set up is:
(A) 31stJanuary, 1991
(B) 31stJanuary, 1992
(C) 31stJanuary, 1993
(D) 31stJanuary, 1994

69. The chief proponent of the concept of "Empowerment"is:
(A) Alfred York (B) David Werner
(C) Paulo Fiere (D) Osgood

70. Which among the following is a correct equation:
(A) Motivation + Empowerment = Enabling
(B) Motivation + Enabling = Empowerment
(C) Enabling + Empowerment = Motivation
(D) Motivation + Empowerment + Enabling = 0

71. The transfer of authority from central government to state government implies:
(A) Decentralization (B) Delegation
(C) Devolution (D)Deconcentration

72. In India land reforms have been equated with total:
(A) Urban development
(B) Agrarian development
(C) Industrial development
(D) Educational development

73. The term land reforms should be used in broadest sense to serve as:
(A) Inscription (B) Prescription
(C) Description (D) Formulation

74. Ministry of Panchayat Raj was established on:
(A) 27thMay, 2004
(B) 15thAugust, 2002
(C) 2ndOctober, 2001
(D) 15thAugust, 2006

75. The backward class commission was headed by:
(A) B.N. Roy (B) Mahatma Gandhi
(C) Kaka Kalelkar (D) None of these

76. NIRD under Ministry of Rural Development is located at:
(A) New Delhi (B) Hyderabad
(C) Lucknow (D) Chennai

77. Journal of Rural Development is published from:
(A) MANAGE
(B) Ministry of Rural Development
(C) NIRD
(D) IARI

78. Which among the following is a rural development journal:
(A) Kurukshetra (B) Yojana
(C) Gram vikash (D) Gramin Soochna

79. The number of languages in which Yojana is published:
(A) 10 (B) 11 (C) 12 (D) 13

80. International Women's day is celebrated on:
(A) 8^{th}March (B) 8^{th}October
(C) 21^{st}November (D) 2^{nd}December

81. World Habitat day is celebrated on the first Monday of:
(A) October (B) December
(C) May (D) July

82. Tehsil Divas in Uttar Pradesh is organized on every month of:
(A) $1^{st}$and 2^{nd}Monday
(B) $1^{st}$and 3^{rd}Monday
(C) $1^{st}$and 3^{rd}Tuesday
(D) $2^{nd}$and 4rth Tuesday

83. The Lakhina experiment is associated with:
(A) District Administration
(B) Panchayat Raj Adminsitration
(C) Rural Development Administration
(D) Tribal Development Administration

84. Panchayati Raj system was adopted first in:
(A) Rajasthan and Madhya Pradesh
(B) Andhra Pradesh and West Bengal
(C) Rajasthan and Andhra Pradesh
(D) Tamilnadu and Rajasthan

85. The basic unit of Panchayati Raj institutions:
(A) Zila Parishad
(B) Grama Sabha
(C) Panchayat Samiti
(D) Block Panchayat

86. The 'Year of Grama Sabha' was observed in:
(A) 1999-2000 (B) 2001-2002
(C) 2004-2005 (D) 2006-2007

87. The total number of panchayats in the country:
(A) 2.31 lakhs (B) 2.23 lakhs
(C) 2.47 lakhs (D) 2.79 lakhs

88. The intermediate level in Panchayati Raj institutions:
(A) Zila Parishad (B) Grama Sabha
(C) Panchayat Samiti
(D) Block panchayat

89. The total number of panchayat at intermediate level:
(A) 6000 (B) 7000
(C) 8000 (D) 9000

90. The third tier of Panchayati Raj is:
(A) Zilla Parishad
(B) Grama Sabha
(C) Panchayat Samiti
(D) Block Panchayat

91. The responsibility of village panchayat include:
(A) Agricultural production
(B) Rural Industrial development
(C) Maternity and child welfare
(D) All of these

92. The need for constitutional sanction of village panchayats was first felt by:
(A) Balwantrai Mehta Committee
(B) Ashoka Mehta Committee
(C) GVK Rao Committee
(D) L M Singhvi Committee

93. Vikas Volunteer Vahini is sponsored by:
(A) DRDA (B) RRB's
(C) NGO's (D) NABARD

94. Rural infrastructural fund was set up by:
(A) RBI (B) NABARD
(C) IDRB (D) SBI

95. The Department of Rural Development under Government of India was started on:
(A) 1919 (B) 1928
(C) 1931 (D) 1947

96. People's plan was envisaged by:
(A) Jawaharlal Nehru (B) Visweswrayya
(C) M.N. Roy (D) Mahalanoibis

97. The report on All India Rural Credit Enquiry Survey was submitted on:
(A) 1947 (B) 1954
(C) 1960 (D) 1961

98. Community Development projects were inaugurated in India on:
(A) 26th January, 1952
(B) 2nd October, 1952
(C) 25th August, 1954
(D) 8th September, 1953

99. Grow More Food enquiry committee was constituted in:
(A) 1952 (B) 1953
(C) 1955 (D) 1956

100. Grow More Food enquiry committee was headed by:
(A) V.T. Krishnamachari
(B) M.S. Swaminathan
(C) M.K. Gandhi
(D) F.L. Brayne

101. Sarvodaya Scheme was started by:
(A) R.N. Ragore (B) Vinoba Bhave
(C) M.K. Gandhi (D) F.L. Brayne

102. Sarvodaya Scheme was started in:
(A) 1945 (B) 1946
(C) 1947 (D) 1948

103. Key Village Scheme was started in:
(A) 1949 (B) 1950
(C) 1951 (D) 1952

104. Rural Reconstruction Institute at Sundarban district was started by:
(A) R.N. Tagore (B) Daniel Hamilton
(C) M.K. Gandhi (D) F.L. Brayne

105. The structure of CD blocks were taken from:
(A) Etawah Project
(B) Firka Project
(C) GMF Campaign
(D) Nilokheri Experiment

106. The technical cooperation agreement of CDP was signed with:
(A) Africa (B) France
(C) USA (D) Netherland

107. CD is sponsored by:
(A) Government agencies
(B) Private agencies
(C) Voluntary agencies
(D) Both (A) and (B)

108. The community development programme was launched in India as:
(A) Development organisation
(B) Administrative setup
(C) As a means to an end
(D) Local self government

109. "A Guide to Community Development" was written by:
(A) M. Horton (B) D. Hamilton
(C) T. Moscher (D) Ensminger

110. The CDP was planned and conceived as peoples:
(A) Planned programme
(B) Local programme
(C) Self help programme
(D) Organize programme

111. The number of community development projects implemented in India:
(A) 54 (B) 55
(C) 56 (D) 57

112. The operational area of each community development project includes:
(A) 100 villages (B) 200 villages
(C) 300 villages (D) 400 villages

113. Under community development project each village level worker is appointed for:
(A) 1-5 villages (B) 5-10 villages
(C) 10-15 villages (D) 15-20 villages

114. "Extension Education in Community Development" is published by:
(A) Ministry of Planning
(B) Ministry of Food and Agriculture
(C) Ministry of Rural Development
(D) Ministry of Information and Broadcasting

115. National extension service was started in:
(A) 1952 (B) 1953
(C) 1954 (D) 1955

116. NES was considered basically as a:
(A) Method (B) Agency
(C) Project (D) Plan

117. Community Development Blocks was started in:
(A) 1952 (B) 1953
(C) 1954 (D) 1955

118. A community development project have an operational area of:
(A) 400-500 square miles
(B) 600-700 square miles
(C) 700-800 square miles
(D) 800-1000 square miles

119. The entire nation was covered under CD programme by the year:
(A) 1957 (B) 1958
(C) 1959 (D) 1960

120. The committee constituted to review CD and NES projects was headed by:
(A) Ashok Mehta
(B) Hanumantha Rao
(C) O.P. Dahama
(D) Balwant Ray Mehta

121. Panchayat Raj was introduced in India in:
(A) 1951 (B) 1952
(C) 1953 (D) 1954

122. Panchayat Raj institutions is formed based on:
(A) Ashok Mehta Committee
(B) Hanumantha Rao Committee
(C) Dantwala Committee
(D) Balwanthray Mehta committee

123. Panchayat Raj is a system of:
(A) Local government
(B) Local administration
(C) Local self-government
(D) None of these

124. The term Pánchayat Raj was christened by:
(A) M.K. Gandhi
(B) J.L. Nehru
(C) B.R. Mehta
(D) M.S. Mehta

125. The Panchayat Raj institutions are intended to strengthen:
(A)Structural planning
(B) Indicative planning
(C) Multilevel planning
(D) Functional planning

126. The origin of the word democracy is from:
(A) Roman (B) Sanskrit
(C) Greek (D) Persian

127. The word "democracy" means:
(A) Peoples participation
(B) Peoples rule
(C) Peoples government
(D) Peoples judgement

128. The process of dispersing decision making closer to the point of service or action:
(A) Decentralization
(B) Delegation
(C) Devotion (D) Deconcentration

129. The constitutional amendment act was passed in:
(A) 1996 (B) 2001
(C) 1992 (D) 1997

130. The election to panchayat is normally held after a period of:
(A) 3 years (B) 4 years
(C) 5 years (D) 6 years

131. The number of tiers for Panchayat in a state with population below 20 lakhs:
(A) Two (B) Three
(C) Four (D) Five

132. The proportion of seat in a panchayat to be filled from women through election:
(A) Two-third (B) One-third
(C) One-fifth (D) One-fourth

133. Which of the following states have no Panchayati Raj:
(A) Karnataka (B) Bihar
(C) Orissa (D) Meghalaya

134. The 73rd amendment act does not apply to the states of:
(A) Nagaland (B) Mizoram
(C) Meghalaya (D) All of these

135. The total number of district in India:
(A) 256 (B) 594
(C) 488 (D) 604

136. The responsibility of village panchayat include:
(A) Agricultural production
(B) Rural industrial development
(C) Maternity and child welfare
(D) All of these

137. The pioneer state in India to bring the whole of it under democratic decentralization:
(A) Uttar Pradesh (B) Bihar
(C) Rajasthan (D) Madhya Pradesh

138. The constitutional 74thamendment act relates to:
(A) Urban local bodies
(B) Election to Gram Sabha
(C) Free primary education
(D) Reservation in panchayat

139. Which of the following is a committee on block level planning:
(A) Hanumantha Rao Committee
(B) Dantewada Committee
(C) G.V.K. Rao Committee
(D) L.M. Singhvi Committee

140. Which of the following is a committee on district planning:
(A) Hanumantha Rao Committee
(B) Diwakar Committee
(C) Dantewada Committee
(D) Santhanam Committee

141. Ashok Mehta Committee recommended the establishment of:
(A) Mandal Panchayat
(B)Nagar Panchayat
(C) Panchayat Samiti
(D) Gram Panchayat

142. The need for constitutional sanction of village panchayat was first felt by:
(A) Balwantray Mehta Committee
(B) Ashok Mehta Committee
(C) G.V.K. Rao Committee
(D) L.M. Singhvi Committee

143. Ford foundation was invited to India on:
(A) 1958 (B) 1959
(C) 1960 (D) 1961

144. "India's food crisis and step to meet it" is a:
(A) Book on food situation in India
(B) Report of food foundation team
(C) Formula to solve food problems
(D) A slogan for campaign

145. The programme started based on the report of ford foundation:
(A) CDP (B) IADP
(C) IRDP (D) Etawah project

146. The IADP was launched in:
(A) 1960 kharif (B) 1960 rabi
(C) 1961 kharif (D) 1961 rabi

147. "Betting on the strong horse" is a slogan associated with:
(A) IRDP (B) HYVP
(C) IADP (D) SFMA

148. Linking farm plan with credit is a characteristic feature of:
(A) IRDP (B) IADP
(C) CDP (D) Etawah Project

149. CDP differs from IADP as the former have:
(A) Area based approach
(B) Decentralized administration
(C) Degree of investment
(D) Multi purpose approach

150. The extension approach used under IADP was:
(A) Proscriptive (B) Prescriptive
(C) Participatory (D) Commercial

151. The IAAP was launched in:
(A) 1962 (B) 1963
(C) 1964 (D) 1965

152. High Yielding Variety Programme was started in:
(A) 1962 (B) 1963
(C) 1964 (D) 1965

153. High yielding technologies are:
(A) Resource positive (B)Resource negative
(C) Resource neutral (D) None of these

154. Farm and Home unit of Akashvani was started in connection with:
(A) IAAP (B) HYVP
(C) IRDP (D) CDP

155. Radio Rural Forum was started in:
(A) 1963 (B) 1964
(C) 1965 (D) 1966

156. Broadcasting of rural programme in television started in:
(A) 1965 (B) 1966
(C) 1967 (D) 1968

157. Which of the following is not a key feature of T&V system:
(A) Professionalism
(B) Time bound work
(C) Single line of command
(D) Farming System Research

158. The base level extension functionary in T&V system:
(A) BDO (B) VLW
(C) AEO (D) SDAO

159. The first state in India to introduce Training and Visit system was:
(A) Punjab (B) Rajasthan
(C) Tamilnadu (D) Bihar

160. The ratio of extension agent to farmer in a typical T&V system is:
(A) 1-250 (B) 1-500
(C) 1-800 (D) 1-1500

161. T&V system of extension can be characterized as:
(A) Management system
(B) Supervisory system
(C) Integrated system
(D) Broad based system

162. Small Farmers Development Agency was set up in:
(A) 1969-70 (B) 1970-71
(C) 1971-72 (D) 1972-73

163. The basic approach followed by SFDA:
(A) Employment approach
(B) Integrated approach
(C) Welfare approach
(D) Target group approach

164. SFDA is meant for the farmers having land holding of:
(A) 1-2 hectare (B) 0-1 hectare
(C) 1-5 hectare (D) 2-5 hectare

165. SFDA-MFAL is associated with:
(A) Venkatpaiah committee
(B) Venkataraman committee
(C) Chelliah committee
(D) None of these

166. An agricultural labourer earns more than of his wage from agriculture:
(A) 33.33 per cent (B) 40.00 per cent
(C) 50.00 per cent (D) 66.66 per cent

167. A welfare programme for the rural poor:
(A) Krishi Vigyan Kendra
(B) Lab to Land
(C) National Demostration
(D) Minikit trials

168. The Ganga Kalyan Yojna is for:
(A) Consumption purpose of farmers
(B) Irrigation for small and marginal farmers
(C) Credit support scheme for rural women
(D) Credit support scheme for small farmers

169. Intensive Cattle Development Project was launched in:
(A) 1963-64 (B) 1964-65
(C) 1965-66 (D) 1966-67

170. The rural development programme launched on the basis of area development approach:
(A) DPAP (B) MWP
(C) RWP (D) IRDP

171. Desert Development Programme was launched on:
(A) 1977-78 (B) 1968-69
(C) 1971-72 (D) 1956-57

172. Tribal Area Development Programme was started in the year:
(A) 1970-71 (B) 1983-84
(C) 1984-85 (D) 1985-86

173. Integrated Wasteland Development Programme was launched on:
(A) 1989-90 (B) 1968-69
(C) 1971-72 (D) 1956-57

174. National Food Security Mission covers all except:
(A) Rice (B) Wheat
(C) Pulses (D) Sugarcane

175. Annual growth rate in agriculture sector expected during the eleventh plan:
(A) 8 per cent (B) 6 per cent
(C) 4 per cent (D) 3 per cent

176. Planning commission was setup in:
(A) 1949 (B) 1950
(C) 1951 (D) 1952

177. The first chairman of planning commission was:
(A) Dr. Rajendra Prasad
(B) Dr. S. Radhakrishnan
(C) Pt Jawaharlal Nehru
(D) Dr. B.R. Ambedkar

178. Planning commission is a:
(A) Statutory body (B) Advisory body
(C) Autonomous body (D) Elected body

179. Which one of the following is not an objective of India's economic planning:
(A) Population growth
(B) Industrial growth
(C) Self reliance (D) Poverty alleviation

180. The long term objective of development in India has been:
(A) High economic growth
(B) Economic self reliance
(C) Social justice (D) All of these

181. The National Development Council gets its administrative support from:
(A) Parliament
(B) Planning commission
(C) Finance commission
(D) Union cabinet

182. The Planning commission of India is a:
(A) Constitutional body
(B) Statutory body
(C) Non-statutory body
(D) Autonomous body

183. Who is not a member in the National Development council:
(A) Prime Minister
(B) President
(C) Chief Ministers
(D) Cabinet Ministers

184. Who was the first deputy chairman of planning commission of India?
(A) Jawaharlal Nehru
(B) Rajendra Prasad
(C) V.T. Krishnamachari
(D) C. Rajagopalachari

185. The main objective of National Development Council is:
(A) To promote common economic policies in all vital spears
(B) To mobilize the efforts and resources of the nation in support of plan
(C) To ensure balanced and rapid development of all part of the country
(D) To secure cooperation of states in the execution of the plan

186. The book entitled"Planned Economy for India" was the attempt of:
(A) Mahalanobis
(B) Viswesvarayya
(C) Jawaharlal Nehru
(D) M.N. Roy

187. The highest policy making body in matters relating to national planning is the:
(A) Planning commission
(B) Finance commission
(C) Interstate council
(D) National Development Council

188. The National Development Council was set up in:
(A) 1950 (B) 1951
(C) 1952 (D) 1953

189. Reviewing the implementation of national plans from time to time is the function of:
(A) Research Programmes Committee
(B) Union Council of Ministers
(C) Planning commission
(D) National Development Council

190. The Programme Evaluation Organization is an integral part of the:
(A) National Development Council
(B) Central Secretariat
(C) Department of Public Enterprises
(D) Planning Commission

191. The plan implementation and its impact evaluation is done by:
(A) NDC (B) Union cabinet
(C) PEO (D) Planning commission

192. Which of the following is not a function of planning commission in India?
(A) Formulation of a plan
(B) Execution of development plans
(C) Appraisal of plan progress
(D) Policy and administrative recommendation

193. The rolling plan concept in national planning was introduced by:
(A) Congress government
(B) UDF government
(C) Janata Party government
(D) NDA government

194. The credit for coining and applying the concept of indicative planning goes to:
(A) USSR (B) USA
(C) Britain (D) France

195. Economic planning is a subject included under the:
(A) Union List (B) State List
(C) Concurrent List (D) Unspecified List

196. In 1938 the National Planning Committee was set up under the chairmanship of:
(A) Jawaharlal Nehru
(B) Vallabhbhai Patel
(C) Moulana Abdul Kalam Azad
(D) Mohammad Ali Jinnah

197. Encyclopedia of social work in India is published by:
(A) NIRD (B) Planning commission
(C) ICSSR (D) UNESCO

198. The first five year plan was submitted to the parliament in:
(A)1950 (B) 1951
(C) 1952 (D) 1953

199. Harrod-Domar model was adopted in:
(A) First five year plan
(B) Second five year plan
(C) Third five year plan
(D) Fourth five year plan

200. Removal of poverty was the foremost objective of which five year plan:
(A) Fourth (B) Fifth
(C) Sixth (D) Seventh

201. Which decade of planning has been the best for agricultural growth:
(A) Fifties (B) Sixties
(C) Seventies (D) Eighties

202. Till which plan employment strategy has been growth linked:
(A) Third plan (B) Fourth plan
(C) Fifth plan (D) Sixth plan

203. Command Area Development Programme was launched during:
(A) Fifth five year plan
(B) Sixth five year plan
(C) Seventh five year plan
(D) Eight five year plan

204. The opening of Indian economy and signing of WTO happened during:
(A) Seventh five year plan
(B) Eight five year plan
(C) Ninth five year plan
(D) Tenth five year plan

205. Perspective plan is drawn out for a period of:
(A) One year (B) 2 to 3 years
(C) 5 to 10 years (D) 15 to 20 years

206. The term "Hindu rate of growth" was coined by:
(A) J.N. Bhagwathi
(B) K.N. Raj
(C) Raj Krishna
(D) Mahalanoibis

207. Atainment of economic self reliance and removal of poverty were the main objectives of:
(A) First five year plan
(B) Fourth five year plan
(C) Fifth five year plan
(D) Sixth five year plan

208. Which one of the following did the Mahalonobis model lay greater emphasis:
(A) Agriculture (B) Heavy industries
(C) Power generation
(D) Science and technology

209. India achieved self reliance in food production during:
(A) Fourth five year plan
(B) Fifth five year plan
(C) Sixth five year plan
(D) Seventh five year plan

210. Micro level planning is an example for:
(A) Imperative planning
(B) Five year planning
(C) Regional planning
(D) Target group planning

211. Planning in India as it is at present can be termed as:
(A) State planning
(B) Centralized planning
(C) Decentralized planning
(D) Indicative planning

212. Indicative planning started from:
(A) Eight five year plan
(B) Seventh five year plan
(C) Sixth five year plan
(D) Fifth five year plan

213. Five year plan is a:
(A) Perspective plan
(B) Medium term plan
(C) Adhoc plan (D) Contingent plan

214. Why was plan holiday necessary in the national planning:
(A) First Indo-Pakistan war
(B) Second Indo-Pakistan war
(C) Third Indo-Pakistan war
(D) Indo-China war

215. The highest priority of the first five year plan was given for:
(A) Agriculture and irrigation
(B) Medium and large industry
(C) Unemployment and poverty
(D) Communication and education

216. The goal of socialistic pattern of society was presented before the national in the:
(A) First five year plan
(B) Second five year plan
(C) Third five year plan
(D) Fourth five year plan

217. The concept of mixed economy in planning was introduced during:
(A) Second five year plan
(B) Fifth five year plan
(C) Seventh five year plan
(D) Ninth five year plan

218. A rolling plan is a plan for:
(A) One year (B) Two year
(C) Three year (D) Year to year basis

219. The idea of family planning as a concept of population control was widely popularized during:
(A) Fifth five year plan
(B) Sixth five year plan
(C) Seventh five year plan
(D) Eight five year plan

220. The current five year plan is:
(A) Tenth (B) Eleventh
(C) Twelfth (D) Thirteenth

221. The tenth five year plan ended on:
(A) 2007 (B) 2008
(C) 2009 (D) 2010

222. The eleventh five year plan ended on:
(A) 2009 (B) 2010
(C) 2011 (D) 2012

223. The GDP growth envisaged in the tenth plan is:
(A) 8.10 per cent
(B) 8.50 per cent
(C) 9.0 per cent
(D) 10.0 per cent

224. The percentage of reducing in poverty envisaged at the end of tenth five year plan is:
(A) 4 per cent (B) 5 per cent
(C) 6 per cent (D) 7 per cent

225. The approximate number of new employment opportunities envisaged in the tenth five year plan is:
(A) 50 million (B) 60 million
(C) 70 million (D) 80 million

226. The allocation of funds for rural development programme in the tenth plan is:
(A) 77474 crores (B) 82276 crores
(C) 85123 crores (D) 94000 crores

225. IRDP was introduced in India on:
(A) 1977-78 (B) 1978-79
(C) 1979-80 (D) 1980-81

226. The basic approach followed in IRDP is:
(A) Individual based (B) Group based
(C) Community based (D) Family based

227. IRDP was conceived essentially as a:
(A) Adult education programme
(B) Anti-poverty programme
(C) Anti-boarding programme
(D) Fisheries programme

228. The basic unit of development under IRDP is:
(A) District (B) Village
(C) CD blocks (D) Family

229. Which of the following is not a characteristic of IRDP:
(A) Unidimensional in nature
(B) Adopt a holistic approach
(C) Calls for higher coordination
(D) Poverty alleviation programme

230. The total credit disbursed in agricultural sector comes around:
(A) 1800 billion (B) 2000 billion
(C) 2200 billion (D) 2300 billion

231. The share of cooperatives in the total credit disbursed in agricultural sector is:
(A) 70 per cent (B) 60 per cent
(C) 50 per cent (D) 40 per cent

232. Short term credit to farmers with a limit of 3 lakhs is available at an interest of:
(A) 7 per cent (B) 8 per cent
(C) 9 per cent (D) 10 per cent

233. The total number of kisan credit distributed in India:
(A) 642 lakhs (B) 556 lakhs
(C) 612 lakhs (D) 621 lakhs

234. Total number of bank linked SHG in India:
(A) 1.25 lakhs (B) 2.25 lakhs
(C) 3.25 lakhs (D) 4.25 lakhs

235. The National Agricultural Insurance Scheme was launched on:
(A) 1999-2000 Rabi (B) 1999-2000 Kharif
(C) 2001-2001 Rabi (D) 2000-2001 Kharif

236. The recently launched weather insurance scheme of government of India:
(A) Ganga Beema (B) Varsha Beema
(C) Dhan Phuhar Beema
(D) Barsat Beema

237. The agricultural and Processed Food Products Export Development Authority (APEDA) came into existence in:
(A) 1986 (B) 1987
(C) 1988 (D) 1989

238. The GDP growth recorded in agriculture sector for the year 2006-07:
(A) 2.70 (B) 2.80
(C) 3.12 (D) 3.71

239. The special relief package for distress farmers recently announced does not cover the state of:
(A) Kerala (B) Tamilnadu

240. The IRDP beneficiaries should include SC/ ST at the rate of:
(A) 20 per cent (B) 25 per cent
(C) 30 per cent (D) 35 per cent

241. The SFDA has been merged into:
(A) MFAL (B) SFDA
(C) SISI (D) CAPART

242. The agency constituted for implementing IRDP:
(A) DRDA (B) SFDA
(C) NREP (D) TRYSEM

243. DRDA expands to:
(A) District Rural Development Authority
(B) District Rural Development Agency
(C) District Rural Development Act
(D) District Rural Development Association

244. DRDA is located at:
(A) Block level (B) District level
(C) Tehsil level (D) Village level

245. The meetings of District Development Committee is presided by:
(A) BDO (B) DDO
(C) CDO (D) District Collector

246. Minimum need programme was launched during:
(A) Third five year plan
(B) Fourth five year plan
(C) Fifth five year plan
(D) Seventh five year plan

247. TRYSEM was introduced in the year:
(A) 1978 (B) 1979
(C) 1979 (D) 1980

248. Training under TRYSEM is mainly given for:
(A) Young scientists (B)Rural Youth
(C) Young students (D) Young women

249. DWCRA was implemented in:
(A) 1982 (B) 1979
(C) 1978 (D) 1977

250. The sill requirement for the members of DWCRA was met through:
(A) SFDA (B) IRDP
(C) NREP (D) TRYSEM

251. The programme aimed at creating income generating assets through a mix of bank credit and government subsidies:
(A) SGSY (B) NREP
(C) SITRA (D) JGSY

252. The Swarnajayanti Gram Swarozgar Yojna (SGSY) was launched on:
(A) April, 1999 (B) October, 1999
(C) November, 1999 (D) March, 1999

253. SGSY was formed by integrating all of the below programmes except:
(A) SITRA (B) PMRY
(C) MWS (D) GKY

254. The beneficiaries under SGSY is:
(A) Projapathis (B) Rural youth
(C) Swarozgaris (D) Farm family

255. The share of centre and state in the funding of SGSY is:
(A) 75:25 (B) 60:40
(C) 50:50 (D) 25:75

256. The minimum net monthly earning envisaged for a beneficiary of SGSY is:
(A) 1500 (B) 2000
(C) 2500 (D) 3000

257. The largest self employment programme for rural poor:
(A) SGSY (B) NREP
(C) SJSRY (D) JGSY

258. SGSY adopts:
(A) Cluster approach
(B) Project approach
(C) Target group approach (D) All of these

259. The BPL families in a SHG formed under SGSY should be:
(A) 80 per cent (B) 83.33 per cent
(C) 90 per cent (D) 91.33 per cent

260. The number of people living under the below poverty line in India:
(A) 9.56 crores (B) 9.52 crores
(C) 10.72 crores (D) 11.21 crores

261. The reservation for women as beneficiary under SGSY is:
(A) 50.00 per cent (B) 60.00 per cent
(C) 40.00 per cent (D) 33.33 per cent

262. The reservation for SC/ST as beneficiary under SGSY is:
(A) 50.00 per cent (B) 60.00 per cent
(C) 40.00 per cent (D) 33.33 per cent

263. The maximum subsidy under SGSY for individuals
(A) 7000 (B) 7500
(C) 8000 (D) 9000

264. The maximum subsidy under SGSY for SHG's:
(A) 1.00 lakh (B) 1.25 lakhs
(C) 1.50 lakhs (D) 2.00 lakhs

265. Bank credit under SGSY is available on group corpus up to:
(A) 3 times (B) 4 times
(C) 5 times (D) 6 times

266. Swarna Jayannti Shahari Rozgar Yojna (SGSRY) was launched in the country from:
(A) 1st December, 1997
(B) 2nd October, 1998
(C) 14th November, 1998
(D) 2nd October, 1999

267. The programme exclusively for urban poor:
(A) SJSRY (B) PMRY
(C) NSAP (D) SGSY

268. The programme developed to cover one crore poorest of the poor:
(A) SFDA (B) TRYSEM
(C) NAEP (D) AAY

269. Which of the following programme have clubbed together to form SJSYR:
(A) NRY (B) UBSP
(C) PMIUPEP (D) All of these

270. Antyodaya Anna Yojna was launched in:
(A) 1999 (B) 2000
(C) 2001 (D) 2002

271. The amount of rice and wheat together supplied per family in a month under AAY:
(A) 20 kg (B) 25 kg
(C) 30 kg (D) 35 kg

272. The per kg rice and wheat under AAY is:
(A) Rs 2 and Rs 3 (B) Rs 3 and Rs 3
(C) Rs 2 and Rs 2 (D) Rs 3 and Rs 2

273. The number of children covered under the mid day meal scheme:
(A) 12 crores (B) 15 crores
(C) 20 crores (D) 21 crores

274. Mid Day Meal Scheme covers children studying at:
(A) Primary level
(B) Secondary level
(C) High School level
(D) Collegiate level

275. The Mid Day Meal Scheme was started in:
(A) 1995 (B) 2000
(C) 2005 (D) 2010

276. The nutrition content recommended in the mid Day Meal Scheme:
(A) 450 calories and 12 gram protein
(B) 500 calories and 30 gram protein
(C) 350 calories and 24 gram protein
(D) 300 calories and 12 gram protein

277. A scheme aimed at ensuring food security of senior citizens:
(A) SJSRY (B) PMRY
(C) NSAP (D)Annapurma Scheme

278. The Annapurna Scheme was started in:
(A) 1998 (B) 1999
(C) 2000 (D) 2001

279. The amount of food grains supplied to senior citizens not covered under National Old Age Pension Scheme under Annapurna Scheme:
(A) 5 kg/month (B) 10 kg/month
(C) 15 kg/month (D) 25 kg/month

280. The on-going National Social Assistance Programme (NSAP) provides benefits under:
(A) National Old Age Pension Scheme
(B) National Family Benefit Scheme
(C) National Maternity Benefit Scheme
(D) All of these

281. Day of elderly was observe on:
(A) December 1 (B) October 1
(C) September 1 (D) May 1

282. Life expectancy at birth in India according to the 2001-2005 estimates:
(A) 60 years (B) 62 years
(C) 63 years (D) 65 years

283. The scheme to provide safeguard against starvation during the period of natural calamity or during lean season:
(A) NREP
(B) Village Gramin Bank
(C) IRDP
(D) RLEGP

284. Which of the following is not a poverty removal programme:
(A) NREP (B) HYVP
(C) IRDP (D) RLEGP

285. Food for work was started in:
(A) 1977 (B) 1978
(C) 1979 (D) 1980

286. The food for work programme has been restructured as:
(A) IRDP (B) MFAL
(C) NREP (D) DPAP

287. NREP was started in:
(A) 1979 (B) 1980
(C) 1981 (D) 1982

288. NREP was launched to check:
(A) Unemployment (B) Population
(C) Brain drain (D) Drought

289. The programmes designed to create community assets on the one hand and providing wage employment on the other hand:
(A) NREP (B) IRDP
(C) JRY (D) DWCRA

290. Which of the following has been started to supplement NREP:
(A) DPAP (B) RLEGP
(C) IRDP (D) NATP

291. The National Rural Employment Programme dependent on:
(A) Rural employment
(B) Self employment
(C) Wage employment
(D) Revolving employment

292. The programme emphasizing on employment generation among the following is:
(A) CADP (B) JRY
(C) KADP (D) CDP

293. JRY started in:
(A) 1988 (B) 1989
(C) 1990 (D) 1991

294. JRY is combination of following programmes:
(A) FFW and NAEP
(B) PIREP and RLEGP
(C) SFDA and MFAL
(D) NREP and RLEGP

295. Jawahar Rozgar Yojana was started by:
(A) Jawaharlal Nehru
(B) Rajiv Gandhi
(C) Indira Gandhi
(D) Sanjay Gandhi

296. The Employment Assurance Scheme was launched on:
(A) 1992 (B) 1993
(C) 1994 (D) 1995

297. Jawahar Gram Samridhi Yojana (JGSY) was launched in:
(A) April, 1999 (B) October, 1999
(C) November, 1999 (D) March, 1999

298. Jawahar Gram Samridhi Yojana was formed modifying:
(A) IRDP (B) PMRY
(C) MWS (D) JRY

299. The programme aimed at creation of demand driven village infrastructure including durable assets at the village level to enable the rural poor to increase the opportunities for sustained employment:
(A) SGSY (B) PMRY
(C) JGSY (D) DWACRA

300. Jawahar Gram Samridhi Yojana is implemented at the level of:
(A) District panchayat
(B) Intermediate panchayat
(C) Village panchayat
(D) Nagar panchayat

301. The most recently started programme for providing employment to rural people is:
(A) JGSY (B) NREP
(C) JRY (D) SGRY

302. JGSY and EAS were merged to form:
(A) Sampoorna Grameen Rozgar Yojana
(B) National Social Assistance Programme
(C) Pradhan Mantri Gram Sadak Yojana
(D) SJSRY

303. Sampoorna Gramen Rozgar Yojana (SGRY) was started on:
(A) 1999 (B) 2000
(C) 2001 (D) 2002

304. A scheme exclusively implemented by the Panchayat Raj Institutions:
(A) SGRY (B) NSAP
(C) PMGSY (D) SJSRY

305. Under which scheme center supply food grains to state government free of cost every year:
(A) SGRY (B) NSAP
(C) PMGSY (D) SJSRY

306. The National Rural Employment Guarantee Act was passed in the parliament on:
(A) 2002 (B) 2003
(C) 2004 (D) 2005

307. The minimum wage prescribed for unskilled labours in India:
(A) 45 (B) 50
(C) 55 (D) 60

308. The minimum number of days of employment a year guaranteed under the NREG Act:
(A) 90 days (B) 100 days
(C) 120 days (D) 150 days

309. Jobs ubder NREG Act has to provided to registered job card holders within:
(A) 5 days (B) 15 days
(C) 20 days (D) 30 days

310. Prime Minister's Rozgar Yojana (PMRY) was started in:
(A) 1993 (B) 1994
(C) 1995 (D) 1996

311. The age group to get subsidy under PMRY:
(A) 18-35 (B) 20-40
(C) 15-35 (D) 25-50

312. The maximum amount of subsidy eligible per entrepreneur under PMRY:
(A) 10000 (B) 12500
(C) 15000 (D) 20000

313. The maximum income limit to get subsidy under PMRY:
(A) 75000 (B) 80000
(C) 90000 (D) 100000

314. The maximum finance one can get under PMRY for starting joint venture projects:
(A) 1 lakh (B) 2 lakhs
(C) 5 lakhs (D) 10 lakhs

315. The minimum level of education to make eligibility under PMRY:
(A) IV pass (B) VIII pass
(C) X pass (D) XII pass

316. Which of the following schemes does not come under Nehru Rozgar Yojana:
(A) TRYSEM (B) SUME
(C) SHASU (D) SUKE

317. Pradhan Mantry Gram Sadak Yojana was launched during:
(A) December, 2000 (B) August, 2002
(C) October, 2001 (D) August, 2003

318. The sharing of cost between center and state in the Pradhan Mantri Gram Sadak Yojana is:
(A) 100:0 (B) 75:25
(C) 60:40 (D) 50:50

319. The Pradhan Mantri Gramodaya Yojana is a programme related to:
(A) Rural sanitation
(B) Rural water supply
(C) Rural housing
(D) Primary education

320. The programme launched to meet the housing needs of the poor:
(A) IAY (B) JRY
(C) MWS (D) NREP

321. The programme launched to meet the housing needs of the poor in Uttar Pradesh:
(A) IAY (B) JRY
(C) MWS (D) KAY

322. IAY was launched as a sub scheme of:
(A) RLEGP (B) JRY
(C) MWS (D) NREP

323. IAY was de-linked from JRY and become an independent scheme by:
(A) 1994 (B) 1995
(C) 1996 (D) 1997

324. The ceiling on construction assistance under the IAY is:
(A) Rs. 15000/- per unit
(B) Rs. 25000/- per unit
(C) Rs. 35000/- per unit
(D) Rs. 45000/- per unit

325. Which of the following is the function of JRY:
(A) Land development and social forestry
(B) Soil and water conservation
(C) Flood protection and drainage
(D) All of these

326. The primary objective of JRY is:
(A) Creation of employment
(B) Creation of social assets
(C) Assistaing rural people
(D) All of these

327. A time bound plan for rural infrastructure development by the government of India:
(A) Bharat Nirman
(B) Bharat Parivartan
(C) Naya Bharat
(D) Hamara Bharat

328. Prime Minister's Rozgar Yojana strategy is embedded with:
(A) Self employment
(B) Wage distribution
(C) Credit disbursement
(D) Capital creation

329. The concept of Bharat Nirman covers all of the following areas except;
(A) Rural connectivity
(B) Rural housing
(C) Rural water supply
(D) Primary education

330. The number of un-electrified villages taken for electrification under Bharat Nirman:
(A) 0.50 lakhs (B) 1.25 lakhs
(C) 2.25 lakhs (D) 3.75 lakhs

331. The additional houses to be constructed for rural poor under Bharat Nirman:
(A) 60 lakhs (B) 62 lakhs
(C) 70 lakhs (D) 75 lakhs

332. The number of habitats not covered with pure drinking water in India:
(A) 55000 (B) 75000
(C) 100000 (D) 125000

333. All weather roads under Bharat Nirman connects habitats with population above:
(A) 5000 (B) 2000
(C) 1000 (D) 500

334. The targets aimed under Bharat Nirman is to be achieved by:
(A) 2008 (B) 2009
(C) 2010 (D) 2011

335. The decadal growth in population for the period 1991-2001 is:
(A) 13.86 per cent (B) 21.34 per cent
(C) 29.86 per cent (D) 38.54 per cent

336. The land availability per person as per 1991 census is:
(A) 0.42 hectares (B) 0.61 hectares
(C) 0.64 hectares (D) 0.75 hectares

337. The density of population of India as per 2001 census:
(A) 297 per km^2 (B) 324 per km^2
(C) 366 per km^2 (D) 378 per km^2

338. The proportion of our population in the working age group of 15-64 years:
(A) 62.9 per cent
(B) 67 per cent
(C) 60 per cent
(D) 55 per cent

339. Accelerated Rural Water Supply Programme was launched in:
(A) 1972-73 (B) 1973-74
(C) 1974-75 (D) 1975-76

340. National Drinking Water Mission was launched in:
(A) 1985 (B) 1986
(C) 1987 (D) 1988

341. Rajive Gandhi National Drinking Water Mission was launched on:
(A) 1991 (B) 1992
(C) 1993 (D) 1994

342. The recently launched Community-based rural drinking water supply programme:
(A) Swajaldhara (B) Jaldhara
(C) Neerdhara (D) Samritdhara

343. Farmers functional literacy programme was started in the year:
(A) 1968 (B) 1958
(C) 1978 (D) 1948

344. National adult education programme was launched in:
(A) 1968 (B) 1978
(C) 1988 (D) 1984

345. National Literacy Mission was launched in the year:
(A) 1990 (B) 1991
(C) 1992 (D) 1993

346. National Literacy Mission aims at imparting functional literacy to non-literates in the age-group of:
(A) 5-15 (B) 10-25
(C) 15-35 (D) 35-50

347. Operation black board aims to improve:
(A) Primary education
(B) Adult education
(C) Informal education
(D) Professional education

348. The constitutional amendment made to ensure free and compulsory education to all children:
(A) 86th Amendment (B) 74th Amendment
(C) 73rd Amendment (D) 72nd Amendment

349. Free and compulsory education to all children in the age group of 6-14 years is envisaged in:
(A) Sarva Shiksha Abhiyan
(B) Kasturba Gandhi Balika Vidyalaya
(C) National Literacy Mission
(D) National Adult Education Programme

350. All children complete eight years of elementary schooling under Sarva Shiksha Abhiyan by:
(A) 2008 (B) 2009
(C) 2010 (D) 2011

351. Sarva Siksha Abhiyan was started in:
(A) 1994 (B) 1995
(C) 1996 (D) 1997

352. National Service Scheme was started in:
(A) 1967 (B) 1968
(C) 1969 (D) 1970

353. Youth is classified to belong under the age group of:
(A) 5-15 (B) 10-25
(C) 15-35 (D) 35-50

354. Nehru Yuva Kendras was started in:
(A) 1962 (B) 1972
(C) 1982 (D) 1992

355. Nehru Yuva Kendra was started for:
(A) Working youth
(B) Non student youth
(C) Student youth
(D) School dropout

356. Kasturba Gandhi Balika Vidyalaya are meant for promoting education of:
(A) Girl children
(B) Senior citizen
(C) Rural youth
(D) Agricultural labourers

357. The organization under Department of Science and Technology established with the main goal of providing institutional support in protecting grassroots green innovations:
(A) MANAGE (B) NISTAT
(C) National Innovation Foundation
(D) National Science foundation

358. "Files to field programme" has been introduced in:
(A) Lucknow (B) Ahmedabad
(C) Ahmednagar (D) New Delhi

359. National Rural Health Mission was launched in:
(A) 2003 (B) 2004
(C) 2005 (D) 2006

360. The comprehensive housing scheme launched recently with a view to ensuring integrated provision of shelter, sanitation and drinking water:
(A) RLEGP
(B) JRY
(C) IAY
(D) Samagra Awas Yojana

361. The centrally sponsored Rural Sanitation Programme was launched in:
(A) 1985 (B) 1986
(C) 1987 (D) 1988

362. The most important agency responsible for distribution of agricultural credit is:
(A) Cooperative banks
(B) Commercial banks
(C) Money lenders
(D) NABARD

363. "Each for all and all for each" forms the basic principle of:
(A) Cooperative banks
(B) ICAR
(C) Planning Commission
(D) NABARD

364. National Agricultural Cooperative Marketing Federation of India Ltd has its headquarters at:
(A) Chennai (B) Lucknow
(C) New Delhi (D) Hyderabad

365. The first cooperative society was started in:
(A) 1904 (B) 1905
(C) 1906 (D) 1907

366. The minimum members required to register a co-operative society:
(A) 10 (B) 15
(C) 20 (D) 25

367. Nationalization of commercial banks happened in:
(A) 1967 (B) 1968
(C) 1969 (D) 1970

368. Lead bank scheme was introduced in:
(A) 1968 (B) 1969
(C) 1970 (D) 1971

369. Regional Rural Bank was set up in:
(A) 1972 (B) 1973
(C) 1974 (D) 1975

370. RRB were set up under the recommendation of:
(A) Nariman Committee
(B) Narasimhan committee
(C) Sen Committee
(D) CRAFICARD Committee

Chapter 11

ECONOMIC PRINCIPLES

1. The percentage share of agriculture in National economy is:
(A) 50 (B) 40 (C) 45 (D) 25

2. The first chairman of Agriculture Price Commission was:
(A) Sri. L.L. Jha
(B) Sri. M.L. Dantewala
(C) Sri Ashok Mehta
(D) Sri Dharam Narayan

3. Loan for marriage known as:
(A) Direct productive loan
(B) Unproductive loan
(C) Indirect productive loan
(D) None of these

4. Economy of a developing country is marked by:
(A) Predominance of agriculture
(B) Predominance of industry
(C) Predominance of transportation
(D) None of these

5. The chairman of S.F.D.A. is:
(A) Prime Minister
(B) Chief Minister
(C) Finance Minister
(D) District Magistrate

6. The owner of loan in state farming is:
(A) Farmer (B) Capitalist
(C) Government (D) None of these

7. A farmer who has land holding between 1 and 2 hectares will be called as:
(A) Marginal farmer (B) Small farmer
(C) Medium farmer (D) Large farmer

8. The Regional Rural Bank was established through:
(A) RRB Act 1974 (B) RRB Act 1975
(C) RRB Act 1976 (D) None of these

9. India's working population engaged in agriculture is about:
(A) 50-52% (B) 67-69%
(C) 75-80% (D) 90-95%

10. The population of agricultural goods manufactured with agricultural contents contributes to India's export by about:
(A) 30% (B) 50%
(C) 70% (D) 90%

11. The percentage share of agricultural export in total export of India is about:
(A) 10% (B) 18% (C) 25% (D) 30%

12. The growth of Indian economy can be sustained only if agriculture and allied activities grow at an annual growth rate of:
(A) 4% (B) 6% (C) 8% (D) 10%

13. The scheme of agriculture in the National income has been continuously:
(A) Increasing (B) Decreasing
(C) Constant (D) None of these

14. As compared to other developing countries, the share of agriculture in National income of India is:
(A) Much lower (B) Much higher
(C) Slightly higher (D) Slightly lower

15. The working population engaged in agriculture is very high only in:
(A) Developing countries
(B) Developed countries
(C) European countries (D) USA

16. The sugar industries in India depend for their raw materials on:
(A) Industrial product
(B) Mines and minerals
(C) Agriculture (D) Crafts

17. With the globalization of trade, the importance of the following has increased:
(A) Food processing
(B) Steel industries
(C) Machines and Tools
(D) Chemicals

18. The first five year plan in India was started on:
(A) Ist April 1958 (B) Ist April 1951
(C) Ist April 1952 (D) Ist April 1953

19. The chairman of the Planning Commission in India is:
(A) Minister of Planning
(B) Minister of Industries
(C) Prime Minister
(D) President of India

20. A marginal farmer owns cultivated land of:
(A) 1 to 2 hectares
(B) 2 to 3 hectares
(C) Below 1 hectare
(D) 3 hectares and above

21. A large farmer owns cultivated land of:
(A) 1 to 2 hectares (B) 2 to 3 hectares
(C) Below 1 hectare(D) Above 2 hectares

22. Which of the following organization governs international trade:
(A) U.N.O. (B) F.A.O.
(C) I.M.F. (D) W.T.O.

23. The fourth Five Year Plan gave first priority to:
(A) Industries (B) Transport
(C) Health and Education (D) Agriculture

24. The Agricultural Prices are recommended by:
(A) I.C.A.R. (B) C.S.I.R.
(C) C.A.C.P. (D) R.B.I.

25. The Agricultural Prices are recommended on the basis of:
(A) Average cost of production
(B) Minimum cost of production
(C) Bulk line cost
(D) Maximum cost of production

26. Which of the following organization refinance agriculture:
(A) I.D.B.I. (B) I.C.I.C.
(C) N.A.B.A.R.D. (D) S.B.I.

27. Regional Rural Banks provide credit to:
(A) Big farmers
(B) Weaker section of urban population
(C) Weaker section of rural population
(D) None of these

28. Loan for purchase of tractor is known as:
(A) Short term loan
(B) Medium term loan
(C) Long term loan (D) Cop plan

29. First priority to industry was given during:
(A) First Five Year Plan
(B) Second Five Year Plan
(C) Third Five Year Plan
(D) Fourth Five Year Plan

30. Agricultural Produce and Marketing Act was passed in:
(A) 1937 (B) 1935 (C) 1947 (D) 1949

31. The growth rate of food production in India after 1980 has been:
(A) Increasing (B) Decreasing
(C) Constant (D) Negative

32. Increase in Indian population is due to:
(A) High birth rate (B) Low birth rate
(C) High birth rate and low death rate
(D) Defective Family Planning

33. The export of Agricultural commodities since Nineties is:
(A) Increasing (B) Decreasing
(C) Stand still (D) Negative

34. I.R.D.P. Programme is meant for:
(A) Poverty alleviation
(B) Employment generation
(C) Disposal of agricultural produce
(D) Agro-processing

35. The Zamidari Abolition Act was passed in UP in:
(A) 1950 (B) 1952
(C) 1954 (D) 1956

36. Payment made to labourers for their services is called:
(A) Rent (B) Wages
(C) Interest (D) None of these

37. The agro-climatic Regional Planning approach was started in:
(A) Sixth Five Year Plan
(B) Seventh Five Year Plan
(C) Eighth Five Year Plan
(D) Ninth Five Year Plan

38. In India land reforms have been equated with total:
(A) Urban development
(B) Agrarian development
(C) Industrial development
(D) Educational development

39. The term land reforms should be used in broadest sense to serve as:
(A) Inscription (B) Prescription
(C) Description (D) Formulation

40. CAPART expands as:
(A) Council for Action and People Advancement in Rural Technology
(B) Council for Advancement of People Action and Rural Technology
(C) Commission for Action and People Advancement in Rural Technology
(D) Committee for Advancement of People Action and Rural Technology

41. CAPART is an autonomous wing of:
(A) Ministry of Agriculture and Cooperation
(B) Ministry of Rural Development
(C) ICAR
(D) Ministry of HRD

42. Khadi and Village Industries Commission was established on:
(A) 1957 (B) 1959
(C) 1976 (D) 1982

43. The International Institute of Rural Reconstruction is located at:
(A) Philipines (B) Hyderabad
(C) Bombay (D) New Delhi

44. A monthly newsletter published by Ministry of Rural Development:
(A) Grameen Bharat
(B) Grama Sabha
(C) Panchayat Sabdh
(D) Rural Panchayat

45. The Right to Information Act passed in the year:
(A) 2005 (B) 2006
(C) 2007 (D) 2008

46. Mission 2007 refers to:
(A) Medicinal Plants
(B) Micro irrigation
(C) Plantation sector
(D) Knowledge centres

47. The recent National Commission on Agriculture was headed by:
(A) M.S. Swaminathan
(B) Chidambaram
(C) Sharad Pawar (D) Panjab Singh

48. The millennium development goal of reducing world poverty by half is aimed by:
(A) 2020 (B) 2015
(C) 2010 (D) 2012

49. Zamidari system was introduced by:
(A) Mughals (B) British
(C) Marathas (D) All of these

50. The land revenue under Zamidari system was charged:
(A) Through village panchayat
(B) Through landlords
(C) Directly from the farmers
(D) All of these

51. 'Zamidari' system was abolished in India on:
(A) 1952 (B) 1954
(C) 1968 (D) 1960

52. In Mahalwari system land revenue was charged:
(A) Through landlords
(B) Directly from the farmers
(C) Through village panchayat
(D) All of these

53. The land revenue in Ryotwari system was charged:
(A) Through landlords
(B) Directly from the farmers
(C) Through village panchayat
(D) All of these

54. The department of agriculture, revenue and commerce was set up in:
(A) 1861 (B) 1871
(C) 1891 (D) 1899

55. The election of panchayat is normally held after a period of:
(A) 3 years (B) 4 years
(C) 5 years (D) 6 years

56. The number of tier for Panchayats in a state with population below 20 lakhs:
(A) Two (B) Three
(C) Four (D) One

57. The proportion of seat in a panchayat to be filled from women through election:
(A) Two-third (B) One-third
(C) One-fifth (D) One-fourth

58. Which of the following states have no Panchayati Raj:
(A) Karnataka (B) Bihar
(C) Orissa (D) Meghalaya

59. The 73rd amendment act does not apply to the stages of:
(A) Nagaland (B) Mizoram
(C) Meghalaya (D) All of these

60. The national level Consumer courts deals with the cases involving claims up to Rs.:
(A) 20 lakh (B) 40 lakh
(C) 60 lakh (D) 1 crore

61. While buying a product from a market which of the following logo assures you of quality?
(A) ISI or Agmark (B) LTTE
(C) KCF (D) BJP

62. Which day is celebrated as the National Consumer Day in India?
(A) 24th January (B) 24th December
(C) 24th October (D) 26th January

63. Why is 24th December celebrated as the National Consumer Day?
(A) It was the day when the Indian Parliament enacted the Consumer Protection Act 1986
(B) It was the day when the Indian Parliament celebrates Gandhi Jayanti
(C) On this day the first Consumer Court was set up in Delhi
(D) On this day the Supreme Court was setup

64. A chemist sold you a medicine of expiry date. Under which consumer right you can approach the Consumer court?
(A) Right to Informed
(B) Right to Information
(C) Right to safety
(D) Right to Equality

65. To which court you will approach if your claim is more than Rs 40 lakh and less than Rs one crore?
(A) State Consumer Court
(B) National Consumer Court
(C) District Consumer Court
(D) The Supreme Court

66. Which of the following is not a Consumer Right?
(A) Right to Informed
(B) Right to Information
(C) Right to safety
(D) Right to Equality

67. Find the odd one out:
(A) Right to Informed
(B) Right to Infcrmation
(C) Right to Education
(D) Right to Equality

68. Which of the following is not a Consumer Right?
(A) Right to Freedom of Religion
(B) Right to Information
(C) Right to safety
(D) Right to Equality

69. The planning commission was setup in:
(A) 1949 (B) 1950
(C) 1951 (D) 1952

70. The first chairman of planning commission:
(A) Dr. Rajendra Prasad
(B) Dr S Radhakrishnan
(C) Pt Jawaharlal Nehru
(D) Dr. B. R. Ambedkar

71. Planning commission is a:
(A) Statutory body
(B) Advisory body
(C) Autonomous body
(D) Elected body

72. Which one of the following is not an objective of India's economic planning:
(A) Population growth
(B) Industrial growth
(C) Self reliance
(D) Poverty alleviation

73. The long term objective of development in India has been:
(A) High economic growth
(B) Economic self reliance
(C) Social justice
(D) All of these

74. The National Development Council gets its administrative support from:
(A) Parliament
(B) Planning commission
(C) Finance commission
(D) Union cabinet

75. The planning commission of India is a:
(A) Constitutional body
(B) Statutory body
(C) Non-statutory body
(D) Autonomous body

76. Who is not member in the National Development Council:
(A) Prime minister (B) President
(C) Chief minister (D) Cabinet ministers

77. Who was the first deputy chairman of planning commission of India:
(A) Jawaharlal Nehru
(B) Rajendra Prasad
(C) V T Krishnamachari
(D) C Rajagopalachari

78. The main objective of National Development Council is:
(A) To promote common economic policies in all vital spears
(B) To mobilize the efforts and resources of the nation in support of plan
(C) To ensure balanced and rapid development of all parts of the country
(D) To secure cooperation of states in the execution of the plan

79. The book entitled 'Planned Economy for India' was the attempt of:
(A) Mahalanobis
(B) Viswesvarayya
(C) Jawaharlal Nehru
(D) M N Roy

80. The highest policy making body in matters relating to national planning is the:
(A) Planning commission
(B) Finance commission
(C) Interstate council
(D) National Development Council

81. The National Development Council was set up on:
(A) 1950 (B) 1951
(C) 1952 (D) 1953

82. Reviewing the implementation of national plans from time to time is the function of:
(A) Research Programmes committee
(B) Union Council of Ministers
(C) Planning commission
(D) National Development Council

83. The programme Evaluation Organization is an integral part of the:
(A) National Development Council
(B) Central Secretariat
(C) Department of Public Enterprises
(D) Planning Commission

84. The plan implementation and its impact evaluation is done by:
(A) NDC (B) Union Cabinet
(C) PEO (D) Planning commission

85. Which of the following is not a function of planning commission in India:
(A) Formulation of a plan
(B) Execution of development plans
(C) Appraisal of plan progress
(D) Policy and administrative recommendation

86. The rolling plan concept in national planning was introduced by:
(A) Congress Government
(B) UDF Government
(C) Janata Party Government
(D) NDA Government

87. The credit for coining and applying the concept of indicative planning goes to:
(A) USSR (B) USA
(C) Britain (D) France

88. Economic planning is a subject included under the:
(A) Union list
(B) State list
(C) Concurrent list
(D) Unspecified list

89. In 1938 the National Planning Committee was set up under the chairmanship of:
(A) Jawaharlal Nehru
(B) Vallabhbhai Patel
(C) Moulana Abdul Kalam Azad
(D) Mohammad Ali Jinnah

90. Encyclopedia of social work in India is published by:
(A) NIRD (B) Planning commission
(C) ICSSR (D) UNESCO

91. The first five year plan was submitted to the parliament in:
(A) 1950 (B) 1951
(C) 1952 (D) 1953

92. Harrod-Domar model was adopted in:
(A) First five year plan
(B) Second five year plan
(C) Fifth five year plan
(D) Sixth five year plan

93. Removal of poverty was the foremost objective of which five year plan:
(A) Fourth (B) Fifth
(C) Sixth (D) Seventh

94. Which decade of planning has been the best for agricultural growth:
(A) Fifties (B) Sixties
(C) Seventies (D) Eighties

95. Till which plan employment strategy has been growth linked:
(A) Third plan (B) Fourth plan
(C) Fifth plan (D) Sixth plan

96. Command Area Development Programme was launched during:
(A) Fifth five year plan
(B) Sixth five year plan
(C) Seventh five year plan
(D) Eight five year plan

97. The opening of Indian economy and signing of WTO happened during:
(A) Seventh five year plan
(B) Eight five year plan
(C) Ninth five year plan
(D) Tenth five year plan

98. Perspective plan is drawn out for a period of:
(A) One year (B) 2 to 3 years
(C) 5 years (D) 15 to 20 years

99. The term "Hindu rate of growth" was coined by:
(A) J N Bhargawathi (B) K N Raj
(C) Raj Krishna (D) Mahalanobis

100. Attainment of economic self reliance and removal of poverty were the main objectives of:
(A) First five year plan
(B) Fourth five year plan
(C) Fifth five year plan
(D) Sixth five year plan

101. Which one of the following did the Mahalonobis model lay greater emphasis:
(A) Agriculture
(B) Heavy industry
(C) Power generation
(D) Science and technology

102. India achieved self reliance in food production during:
(A) Fourth five year plan
(B) Fifth five year plan
(C) Sixth five year plan
(D) Seventh five year plan

103. Micro level planning is an example for:
(A) Imperative planning
(B) Five year planning
(C) Regional planning
(D) Target group planning

104. Planning in India as it is at present can be termed as:
(A) State planning
(B) Centralized planning
(C) Decentralized planning
(D) Indicative planning

105. Indicative planning started from:
(A) Eight five year plan
(B) Seventh five year plan
(C) Sixth five year plan
(D) Fifth five year plan

106. Five year plan is a:
(A) Perspective plan
(B) Medium term plan
(C) Adhoc plan (D) Contingent plan

107. Why was plan holiday necessary in the national planning:
(A) First Indo-Pakistan war
(B) Second Indo-Pakistan war
(C) Third Indo-Pakistan war
(D) Indo-China war

108. The highest priority of the first five year plan was given for:
(A) Agriculture and irrigation
(B) Medium and large industry
(C) Unemployment and poverty
(D) Communication and education

109. The goal of socialistic pattern of society was presented before the nation in the:
(A) First plan (B) Second plan
(C) Third plan (D) Fourth plan

110. The concept of mixed economy in planning was introduced during:
(A) Second plan (B) Fifth plan
(C) Seventh plan (D) Ninth plan

111. A rolling plan is a plan for:
(A) One year (B) Two year
(C) Three year (D) Year to year basis

112. The idea of family planning as a concept of population control was widely popularized during:
(A) Fifth plan (B) Sixth plan
(C) Seventh plan (D) Eight plan

113. The current five year plan is:
(A) Ninth plan (B) Eleventh plan
(C) Tenth plan (D) Twelfth plan

114. The eleventh five year plan ended on:
(A) 2011 (B) 2008
(C) 2009 (D) 2010

115. The GDP growth envisaged in the tenth plan is:
(A) 8.10 per cent (B) 8.50 per cent
(C) 9 per cent (D) 10 per cent

116. The percentage of reduction in poverty envisaged at the end of tenth plan is:
(A) 2 per cent (B) 5 per cent
(C) 10 per cent (D) 15 per cent

117. The approximate number of new employment opportunities envisaged in the tenth plan is:
(A) 50 million (B) 40 million
(C) 60 million (D) 30 million

118. The allocation of funds for rural development programme in the tenth plan is:
(A) 77,474 crores (B) 42,276 crores
(C) 85,123 crores (D) 94 million

119. The Panchayati Raj institutions are intended to strengthen by:
(A) Structural planning
(B) Indicative planning
(C) Multilevel planning
(D) Functional planning

120. The origin of the word 'Democracy' is derived from:
(A) Roman (B) Sanskrit
(C) Greek (D) Persian

121. The word 'Democracy' means:
(A) Peoples participation
(B) Peoples rules
(C) Peoples government
(D) People judgement

122. The term panchayati raj was christened by:
(A) M.K. Gandhi (B) J. L. Nehru
(C) B. R. Mehta (D) M. S. Mehta

123. Panchayati raj was introduced in India in:
(A) 1950 (B) 1952
(C) 1958 (D) 1967

124. Panchayati raj institution is formed based on:
(A) Ashok Mehta Committee
(B) Hanumantha Rai Committee
(C) Dantiwala Committee
(D) Balwant rai Mehta Committee

125. Panchayat raj is a system of:
(A) Local government
(B) Local administration
(C) Local self government
(D) None of these

126. Annual growth rate in agriculture sector expected during the eleventh plan:
(A) 8 per cent (B) 6 per cent
(C) 4 per cent (D) 3 per cent

127. The increase in the production of rice expected at the end of eleventh plan:
(A) 8 million tones (B) 10 million tones
(C) 12 million tones (D) 15 million tones

128. The rural development programmme launched on the basis of area development approach:
(A) DPAP (B) MWP
(C) RWP (D) IRDP

129. Small farmers Agri-business Consortium was set up in:
(A) 1992 (B) 1994
(C) 1971 (D) 1986

130. A farmer cooperative society among the following is:
(A) IFFCO (B) FCI
(C) ICAR (D) NIRD

131. A welfare programme for the rural poor:
(A) KVK
(B) Lab to Land
(C) National Demonstration
(D) Minikit Trial

132. The Ganga Kalyan Yojna is for:
(A) Consumption purpose of farmers
(B) Irrigation for small and marginal farmers
(C) Credit support scheme for rural women
(D) Credit support scheme for small farmers

133. The Food Corporation of India was set up in:
(A) 1960 (B) 1962
(C) 1965 (D) 1967

134. An Agricultural labourer earns more than of his wage from agriculture:
(A) 33.33 per cent (B) 40.00 per cent
(C) 50.00 per cent (D) 66.66 per cent

135. "India's food crises and steps to meet it" is a:
(A) Book on food situation in India
(B) Report of ford foundation team
(C) Formula to solve food problem
(D) A slogan for campaign

136. The third in chronological order among the following report is:
(A) Royal Commission Report
(B) Agricultural Administration Report
(C) GMF Enquiry Committee Report
(D) Nalagarh Committee Report

137. Bhoodan movement was initiated on:
(A) 1947 (B) 1948
(C) 1951 (D) 1963

138. Bhoodan movement was started by:
(A) Acharya Vinoba Bhave
(B) Mahatma Gandhi
(C) Jai Prakash Narayan
(D) Jawaharlal Nehru

139. Bhodan movement was started at:
(A) Santhinietan (B) Wardha
(C) Panchampalli (D) Pavnar

140. The land collected in bhoodan is distributed among:
(A) Marginal farmers
(B) Landlless peasants
(C) Medium farmers (D) Krishi pandits

141. The report on All India Rural Credit Enquiry Survey was submitted on:
(A) 1947 (B) 1954
(C) 1960 (D) 1961

142. Which of the following network connect agricultural produce, wholesale markets and rural markets:
(A) AFPINET (B)AGMARKNET
(C) HORTNET (D) VISTARNET

143. Which one is the land Information System Network linking all institutions involved in land and water management for agricultural productivity and production system:
(A) LISNET (B) ARINET
(C) AGRISNET (D) VISTARNET

144. Which one is the most effective method is identifying real rural needs:
(A) Election method
(B) Selection method
(C) Workshop method
(D) Socio-metric method

145. Which one is the purpose of social control:
(A) Social order (B) Stability
(C) Custom (D) None of these

146. Which one of the following is not an objective of India's economic planning:
(A) Population growth
(B) Industrial growth
(C) Self reliance
(D) Poverty alleviation

147. What is five year plan:
(A) Perspective plan
(B) Medium term plan
(C) Adhoc plan (D) Contingent plan

148. Which of the following is co-operation among family members:
(A) Non formal (B) Voluntary
(C) Formal (D) Informal

149. A tool for optimum utilization of the available resources in the maximization of the attainable income of the farmers is known as:
(A) Farm planning
(B) Home planning
(C) Programme planning
(D) Extension planning

150. Allocation of funds for different sectors of programme:
(A) Controlling (B) Budgeting
(C) Financing (D) Recruiting

151. Basically GATT is aimed at:
(A) Facilitating accords on the most favoured nation status to the trading partners
(B) Reciprocity and transparency in global trade
(C) Tariff reduction through trade negotiations
(D) All of above

152. By whom the Rural infrastructural fund set up:
(A) RBI (B) NABARD
(C) IDRB (D) SBI

153. Central co-operative banks and Cooperative societies were established under Sir Daniel Hemilton Scheme in:
(A) 1920 (B) 1922
(C) 1924 (D) 1926

154. First census of population in India was conducted in the year:
(A) 1761 (B) 1771
(C) 1781 (D) 1791

155. First Five year Plan of India adopted:
(A) Harrod Domar Model
(B) Lewis-Fei Model
(C) Maholonobis Model
(D) Keynesian Model

156. Food Grain Price Committee was appointed by the Government of India in:
(A) 1954 (B) 1958
(C) 1962 (D) 1964

157. Food grain production includes the production of:
(A) Cereals only (B) Pulses only
(C) Cereals and Pulses
(D) Cereals, pulses and Oilseeds

158. For which of following a cooperative society is an example:
(A) Formal group (B) Informal group
(C) Primary group (D) Territory group

159. For which of following logical framework technique is greatly used:
(A) Project planning (B) Input utilization
(C) Work scheduling (D)Policy formulation

160. For which of the following Kasturba Gandhi Balika Vidyalaya are meant for promoting education:
(A) Gild children
(B) Senior citizen
(C) Rural youth
(D) Agricultural labourers

161. Land Development Banks in India are owned by whom:
(A) Govt. of India
(B) State Governments
(C) Commercial banks
(D) Cooperative banks

162. Land holding of either 10 or more hectares has been classified as which of following in India:
(A) Semi-medium (B) Medium
(C) Semi large (D) Large

163. Land reforms have been equated with total:
(A) Urban development
(B) Agrarian development
(C) Industrial development
(D) Educational development

164. Lead bank scheme was introduced in the year:
(A) 1965 (B) 1969
(C) 1976 (D) 1964

165. Major commercial banks were nationalized in:
(A) 1965 (B) 1967
(C) 1969 (D) 1971

166. Mandi ordinance of agricultural production was enforced in:
(A) 1949 (B) 1950
(C) 1955 (D) 1959

167. Mention the type of correlation that operates between rurality and population density:
(A) Positive (B) Negative
(C) Accidental (D) Scattered

168. National demonstration was started in the year:
(A) 1975 (B) 1974 (C) 1963 (D) 1964

169. National demonstration is an example for:
(A) Firstline demonstration
(B) Frontline demonstration
(C) Futureline demonstration
(D) All of these

170. Planning at districts and block levels is generally considered as:
(A) Micro planning (B) Meso planning
(C) Macro planning (D) All of these

171. Planning in India as it is at present can be termed as which of the following:
(A) State planning
(B) Centralized planning
(C) Decentralized planning
(D) Indicative planning

172. Providing subsidy to farmers on introduction of farm machinery is an example for which of the following:
(A) Relative advantage
(B) Complexity
(C) Compatibility (D) Observability

173. The election to panchayat is normally held after a period of how many years:
(A) 3 years (B) 4 years
(C) 5 years (D) 6 years

174. The family in an agrarian society is:
(A) Patriarchal (B) Matriarchal
(C) Matrilocal (D) Nuclear

175. The Panchayati Raj institutions are intended to strengthen by which:
(A) Structural planning
(B) Indicative planning
(C) Multilevel planning
(D) Functional planning

176. The Planning Commission was set in India in accordance with the directive principles in:
(A) Article 38 (B) Article 39
(C) Article 42 (D) Article 51

177. The scheme of Regional Rural Banks was launched in India on:
(A) Ist Oct. 1975 (B) 15^{th}Aug. 1978
(C) 2^{nd}Oct. 1975 (D) 1^{st}Jan. 1970

178. The tool for optimum utilization of the available resources in the maximization of the attainable income of the farmers is:
(A) Farm planning
(B) Resource allocation
(C) Planned change
(D) Farm layout

179. The ——— year plan sought to maintain a clear cut distinction between the minimum support price and the procurement price:
(A) Second (B) Third
(C) Fifth (D) Sixth

180. Following the support price policy, the Government of India in 1984-85 raised the price of rice (paddy) by:
(A) 32.2% (B) 44.2%
(C) 36.8% (D) 26.7%

181. The Food Corporation of India was set up in:
(A) 1960 (B) 1962
(C) 1965 (D) 1967

182. Land holding of either 10 or more hectares have been classified as ——— in India:
(A) Semi-medium (B) Medium
(C) Semi-large (D) Large

183. Land reforms have been equated with total ——— development:
(A) Urban (B) Agrarian
(C) Industrial (D) Educational

184. Profits are defined as π = R-(FC + VC), where, FC = total fixed cost and VC = total variable cost and ———:
(A) R=Revenue (B) R = Reduction
(C) R = Regression (D) R = Co-efficient

185. According to Hung-Chao Tai land reforms aim at aiming just relationship among ——— population:
(A) Urban (B) Industrial
(C) Traders (D) Agricultural

186. The term "land reform" should be used in broadest sense to serve as ———— for action:
(A) Inscription (B) Prescription
(C) Description (D) Formulation

187. In Zamidari system was based on ———— to support British Imperialism in India:
(A) Exploitation (B) Elimination
(C) Regulation (D) Augmentation

188. The main objective of land reform measures undertaken after independence was removal of institutional and ———— obstacles:
(A) Emotional (B) Motivational
(C) Conventional (D) National

189. In Zamidari abolition, the elimination of intermediaries led to excessive burden of:
(A) Revenue (B) Reservation
(C) Supplementation (D) Compensation

190. The reform of tenancy laid down that the rent should not exceed ———— per cent of gross produce:
(A) 20-25 (B) 10-15
(C) 25-30 (D) 35-40

191. According to the Directorate of Marketing and Inspection ———— per cent may be taken as the average marketing margin for all surplus food grains in India:
(A) 15 (B) 20
(C) 25 (D) 30

192. NABARD Act was passed in————, 1979:
(A) January (B) February
(C) March (D) April

193. NABARD came into existence on July 1, ————:
(A) 1980 (B) 1981
(C) 1982 (D) 1983

194. According to the 1951-52 survey, the proportions of indebtness was about 60 per cent for all the families, of which ———— per cent were cultivators:
(A) 70 (B) 60 (C) 80 (D) 90

195. The average debt for an indebted cultivator's family during 1951-52 and 1961-62 rose by ———— per cent:
(A) 60 (B) 45
(C) 35 (D) 25

196. The spurt in Zero Marginal Productivity type of exercise was given by ———— in 1954:
(A) W. Arthur Lewis
(B) W. David Hopper
(C) W. Nelson Michael
(D) W. Benzene Shoemaker

197. Balwantrai Mehta Team stimulated an active consideration of———— through democratic bodies:
(A) Centralization (B) Mobilization
(C) Stabilization (D) Decentralization

198. Maharashtra Government appointed a committee in ———— under the chairmanship of Vasantrao Naik to examine the question of democratic decentralization in Maharashtra:
(A) 1961 (B) 1959
(C) 1957 (D) 1955

199. Maharatshtra Zila Parishads and Panchayat Samiti Act was inacted in:
(A) 1967 (B) 1965
(C) 1963 (D) 1961

200. In ———— village panchayats were established under the Bombay Village Panchayat Act, 1958:
(A) Madhya Pradesh (B) Gujarat
(C) Rajasthan (D) Maharashtra

201. According to 1981 census, ———— of working force was still engaged in the primary sector:
(A) 61 (B) 67 (C) 71 (D) 77

202. One of the significant models of economic development of a labour surplus economy has been advanced by:
(A) Hung-Chao Tai
(B) Huag P. Lewis
(C) Chan K. Heady
(D) Foe C. Gray

203. In an expanding industrial sector, technological development in industry tend to——— labour:
(A) Displace (B) Replace
(C) Engage (D) No change

204. In countries of——— labour problem has become all the more serious owing to demographic explosion:
(A) America (B) Europe
(C) Asia (D) Whole world

205. The problem of allocation of capital is essentially a farm management problem aiming at——— utilization of the available production resources:
(A) Maximizing (B) Optimizing
(C) Stabilizing (D) Minimizing

206. ——— production function is expressed as Y = m-arx, where Y = total production, x = quantity of variable factor, m = maximum output attained from fixed technical unit, r = ratio of addition of increments to production and a = maximum output added by the particular variable factor:
(A) Douglas (B) Bruce
(C) Spillman (D) Hillman

207. The constant elasticity of substitution function (CES) was used by J.W. Swan and R.M. Solow in:
(A) 1926 (B) 1936 (C) 1946 (D) 1956

208. Input-output production function is based on the general equilibrium model of ——— (1874):
(A) Leon Walras (B) Leon Michael
(C) Wilson Arrow (D David Solow

209. According to law of——— proportions, "If equal increments of one input are added, the inputs of other production services being held constant, beyond a certain point, the resulting increments of the product will decrease, that is, marginal product will diminish"
(A) Constant (B) Linear
(C) Variable (D) Inverse

210. ——— rightly remarked, "Let divisibility enter through door, the law of variable proportions rushes out through the window"
(A) M.M. Bober (B) C.A. Watson
(C) T. Sutovsky (D) R.M. Solow

211. When fertilizer input is increased from one to two units, total output increases from 103 to 174, that is, the output has increased only by——— per cent:
(A) 90 (B) 85 (C) 61.9 (D) 71.9

212. ——— was the first economist to give a clear formulation of price elasticity (Ep) which is the ratio of a relative change in quality of relative change in price (Ep = relative change in quantity or relative change in price):
(A) Alfred Marshall (B) A.M. Khusro
(C) E.J. Long (D) P. Dorner

213. Food Grain Price Commission was constituted in India by:
(A) 1951 (B) 1953
(C) 1947 (D) 1950

214. The Food Grain Price Commission was appointed in——— by the Govt. of India:
(A) 1954 (B) 1958
(C) 1970 (D) 1973

215. The Agricultural Price Commission (APC) was set up in ——— to evolve a balanced and integrated price structure:
(A) 1955 (B) 1965
(C) 1970 (D) 1973

216. Rural population is about——— of the world population and lives mainly in developing countries:
(A) One-third (B) One-fourth
(C) One-fifth (D) Two-third

217. In India——— of the population lives in villages:
(A) Half (B) One-third
(C) Three-fourth (D) Two-third

218. Efforts to revive panchayats in India were made in ——— by introducing Village Panchayat Act in several states:
(A) 1945 (B) 1935
(C) 1925 (D) 1915

219. The Village Panchayati Act came into existence in:
(A) 1956 (B) 1958
(C) 1960 (D) 1954

220. ———— concentrates on a rational structure with universal application in social institutions:
(A) Bureaucracy (B) Autocracy
(C) Anarchy (D) Monarchy

221. Formation of clique ——— an organization:
(A) Damages (B) Improves
(C) Stabilizes (D) None of these

222. Micro economics is concerned with:
(A) Analyzing the economic behaviour of the individual farm or business
(B) Analyzing the total economy of nation
(C) Only deals with scarcity of resources
(D) None of these

223. Macro economics deals with:
(A) Analyzing the economic behaviour of the individual farm or business
(B) Analyzing the total economy of nation
(C) Scarcity only (D) Net profit only

224. US economy is an example of:
(A) Mixed economy (B) Closed economy
(C) General economy
(D) Capitalistic economy

225. Jouzier definition of agricultural economics is related to:
(A) Prosperity to the enterprise
(B) Net profit (C) Humen behaviour
(D) Wealth activities

226. Capitalistic economy refers to:
(A) Private ownership of resources
(B) Public ownership of resources
(C) Public and private ownership of resources
(D) All of these

227. The law of diminishing returns states that:
(A) The output increases with increasing input levels continuously
(B) The output increases with input levels irregularily
(C) The output increases with increasing input levels upto certain level and than declines
(D) All of these

228. In Ist irrational stage of production function, marginal return is:
(A) Equal to average return
(B) Equal to total output
(C) Either lower or equal to average return
(D) All of these

229. In IIrd irrational stage of production function, the marginal return is always:
(A) Negative (B) Positive
(C) Equal to average (D) All of these

230. Agriculture differs with industry in:
(A) Nature plays an important role in agricultural production
(B) Land is the most important factor for agricultural production
(C) Elasticity of price and supply of agricultural products are lower than industrial products
(D) All of these

231. Which of the following function is related to production function analysis?
(A) Total physical product
(B) Average physical product
(C) Marginal physical product
(D) All of these

232. Principle of least-cost combination explains:
(A) Output is the function of two or more than two variable inputs
(B) Use of two variable inputs in such a way to get maximum profit
(C) Use of two variable inputs in such a way to get same level of output at comparatively lowest cost
(D) All of these

233. Law of equi-marginal return states that:
(A) Each unit of input is invested in different enterprises in such a way so that a farmer can get maximum marginal return
(B) Each unit of input is invested to get maximum return
(C) Each unit of input invested to get maximum average return
(D) None of these

234. Principal of equi-marginal return states that:
(A) Maximum satisfaction is obtained from a given level of expenditure on two goods when the last penny spent on either good gives the same amount of satisfaction
(B) Equal marginal return is obtained by substitution of one good by other
(C) Both (A) and (B)
(D) None of these

235. Factor-product relationship is also known as:
(A) Input-output relationship
(B) Output-output relationship
(C) Demand-supply relationship
(D) Output-Input relationship

236. Law of equi-marginal return also known as:
(A) Law of opportunity cost
(B) Alternative cost principle
(C) Both (A) and (B)
(D) Law of equi-marginal return

237. Law of variable proportions is also known as:
(A) Law of returns
(B) Law of factor-product relationship
(C) Law of input-output relationship
(D) All of these

238. Law of variable proportions deals with:
(A) Input-output relationship where the output is the function of input
(B) Cost of output and input
(C) Supply of output (D) All of these

239. Maximum net return from one variable input will occur when:
(A) Additional return from an additional unit of output from employing an additional unit of input is equal to the additional cost of that unit of input
(B) Value of marginal return is equal to cost of the input
(C) Marginal return x selling price of product = marginal input cost
(D) All of these

240. The production possibilities curve shows:
(A) Alternative combination of two outputs that may be produced with a given amount of resource during a period of time
(B) Combination of inputs that can produce equal level of production
(C) Combination of outputs that can be produced with same amount of input
(D) All of these

241. Indian agriculture is characterized by:
(A) Low productivity
(B) Subsistence farming
(C) Labour dominance
(D) All of these

242. The cost of production can be minimized by using the principle of:
(A) Law of return
(B) Law of opportunity cost
(C) Law of least-cost combination
(D) Cost principle

243. Who defined Economics is a wealth of science:
(A) Smith (B) Marshall
(C) Robins (D) Pigue

244. Who defined Economics as the study of human behaviour in the daily business of life?
(A) Smith (B) Marshall
(C) Robins (D) Pigue

245. Who defined economics as the study of human welfare?
(A) Smith (B) Marshall
(C) Robins (D) Pigue

246. Who defined economics as the study of choice making behaviour of human being?
(A) Smith (B) Marshall
(C) Robins (D) Pigue

247. Which of the following is a subject matter of economics?
(A) Consumption
(B) Production and distribution
(C) Public finance
(D) All of these

248. Which is not a factor of production?
(A) Land (B) Labour
(C) Capital (D) Crops

249. Principle of equi-marginal return is also known as:
(A) Product-product relationship
(B) Output-output relationship
(C) Substitution between output/products
(D) All of these

250. Law of diminishing marginal utility states that:
(A) Utility varies with time and space
(B) A commodity generates different utilities for different people and for the same person at different times
(C) Utility of additional unit of a commodity to any consumer commodity to already consumption increases
(D) All of these

251. Principle of least –cost combination is also known as:
(A) Substitution between inputs
(B) Factor-factor relationship
(C) Rate of technical substitution
(D) All of these

252. Which of the following is not a type of production function?
(A) Linear (B) Quadratic
(C) Square root (D) Rectangular

253. Maximum profit can be achieved in agribusiness when marginal return (MR) is equal to:
(A) Marginal cost (B) Average output
(C) Total output (D) Total cost

254. Production function illustrates the relationship between:
(A) Input levels and corresponding output
(B) Input levels and corresponding marginal input
(C) Demand and supply pattern
(D) Demand and competition

255. At optimum level of production or output, the marginal return should be:
(A) Positive (B) Negative
(C) Zero (D) Equal to average return

256. Which of the following function is related to an economic system?
(A) What goods and services are to be produced?
(B) How to best organize and combine recourse of production?
(C) Allocation or distribution of income to owners of resources
(D) All of these

257. At point of inflection in production function curve, MR is:
(A) >AR (B) <AR
(C) MR=AR (D) All of these

258. Marginal cost is equal to:

(A) $\dfrac{\text{Change in total cost}}{\text{Additional output}}$

(B) $\dfrac{\text{Change in total Revenue}}{\text{Additional output}}$

(C) $\dfrac{\text{Change in total Value Product}}{\text{Additional input}}$

(D) All of these

259. Marginal rate of substitution is equal to:

(A) $\frac{\text{Change in number of y units}}{\text{Change in number of x units}}$

(B) $\frac{\text{Change in price of x commodity}}{\text{Change in price of y commodity}}$

(C) $\frac{\text{Price of x commodity}}{\text{Price of y commodity}}$

(D) none of these

260. The main purpose of farmers is:
(A) High production
(B) Minimum cost of cultivation
(C) High living standard
(D) High net profit

261. If the increase or decrease one product increases or decreases the other product, the relationship between both products is known as:
(A) Complementary (B) Supplementary
(C) Competitive (D) Symbiotic

262. If increase or decrease in the production level of one product does not affect the production level of the other product, the relationship is called as:
(A) Complementary (B) Supplementary
(C) Competitive (D) Symbiotic

263. If increase in the production level of one product necessarily decreases the production level of the other product, the relationship is called as:
(A) Complementary (B) Supplementary
(C) Competitive (D) Symbiotic

264. Production possibility curve is also called as:
(A) Iso-revenue curve
(B)Opportunity curve
(C) Iso-return curve (D) All of these

265. Types of farming refers to:
(A) Types of enterprises are grown or taken on a farm
(B) Ownership of farm
(C) Both (A) and (B) (D) None of these

266. Types of farming refers to:
(A) Types of enterprises are grown or taken
(B) Ownership of farm and method or practices of production
(C) Both (A) and (B)
(D) None of these

267. Law of demand indicates:
(A) Inverse relationship between demand and price
(B) Linear relationship between demand and price
(C) Inverse relationship between market demand and marketable surplus
(D) Linear relationship between supply and demand

268. Price line or budget line indicates:
(A) A capacity of a consumer to purchase within a given budget
(B) A capacity of a consumer to provide marketable surplus
(C) The graph between demand and satisfaction
(D) None of these

269. Price elasticity of demand is always:
(A) Positive (B) Zero
(C) Negative (D) None of these

270. The demand of any commodity is affected by:
(A) Price of the commodity
(B) Taste and liking of the consumers
(C) Income of the consumers
(D) All of these

271. Which of the following item has the highest income elasticity?
(A) Bread (B) Butter
(C) Milk (D) Wine

272. Income elasticity is generally:
(A) Zero (B) Positive
(C) Negative (D) Infinite

273. The supply elasticity of perishable commodities is:
(A) Elastic (B) Inelastic
(C) Zero (D) Equal to one

274. The supply elasticity of bulky products is normally:
(A) Less elastic (B) High elastic
(C) Complete elastic (D) All of these

275. Supply of a commodity is governed by:
(A) Price of the commodity
(B) Price of production factors
(C) State of technology
(D) All of these

276. The price elasticity of demand for agricultural commodities is always:
(A) <1 (B) >1 (C) Zero (D) Infinite

277. The price elasticity of luxury items is always:
(A) <1 (B) >1 (C) Zero (D) Infinite

278. If the price elasticity is >1, the total revenue will be increased at:
(A) Decrease in price
(B) Increase in price
(C) No change in price
(D) Unknown

279. If the price elasticity is <1, the total revenue will be increased at:
(A) Decrease in price (B) Increase in price
(C) No change in price(D) Unknown

280. The classical function curve exhibits following production stage:
(A) Stage I – Irrational production stage
(B) Stage II- Rational production stage
(C) Stage III- Irrational production stage
(D) All of these

281. Marginal return is equal to:
(A) $\frac{\Delta Y}{\Delta X}$ (B) $\frac{Y}{X}$
(C) $\frac{Px}{Px}$ (D) $\frac{Px}{Py}$

282. In rational stage of production function, both marginal and average returns are:
(A) At decreasing rate
(B) At increasing rate
(C) Both are equal
(D) None of these

283. An isoquant line shows:
(A) Combination of two inputs that may produce a given level of output
(B) Combination of two inputs that may produce maximum level of output
(C) Different marginal returns
(D) None of these

284. An isocost line shows:
(A) Combination of two variable inputs that cost different amount
(B) Combination of two variable inputs that cost the same amount
(C) Combination of two variable inputs that produce the same level of production
(D) All of these

285. What is the problem of Indian agriculture?
(A) Low irrigation
(B) Low and erratic rainfall
(C) Indebtness of Indian farmers
(D) All of these

286. Which of the following principle is employed in production when resources are scare?
(A) Equi-marginal principle
(B) Comparative advantage
(C) Law of returns
(D) None of these

287. Production possibility curve indicates:
(A) Various combinations of the products y_1 and y_2 obtained with fixed level of resources combinations
(B) Maximum production with unlimited sources
(C) Marginal return with limited resources
(D) None of these

288. Operating ratio is equal to:

(A) $\frac{\text{Total expenses}}{\text{Gross income}}$ (B) $\frac{\text{Fixed expenses}}{\text{Gross income}}$

(C) $\frac{\text{Operating expenses}}{\text{Gross income}}$ (D) None of these

289. Gross ratio is equal to:

(A) $\frac{\text{Total expenses}}{\text{Gross income}}$ (B) $\frac{\text{Fixed expenses}}{\text{Gross income}}$

(C) $\frac{\text{Operating expenses}}{\text{Gross income}}$ (D) None of these

290. Fixed ratio is equal to:

(A) $\frac{\text{Total expenses}}{\text{Gross income}}$ (B) $\frac{\text{Fixed expenses}}{\text{Gross income}}$

(C) $\frac{\text{Operating expenses}}{\text{Gross income}}$ (D) None of these

291. Fixed cost is also known as:
(A) Over head cost (B) Prime cost
(C) Alternate cost (D) All of these

292. Variable cost is also known as:
(A) Overhead cost (B) Prime cost
(C) Alternate cost (D) All of these

293. Opportunity cost is also known as:
(A) Overhead cost (B) Prime cost
(C) Alternate cost (D) All of these

294. Indifference curve shows a:
(A) Relative willingness to consume one good rather than another
(B) Relationship of demand and supply
(C) Relationship between demand and price
(D) Relationship between supply and price

295. If price elasticity is -1, it indicates that:
(A) An equal change in demand and price
(B) Percentage change on quantity in smaller than the percentage price change
(C) Percentage quantity change is greater than the percentage price change
(D) None of these

296. If the price elasticity of demand is more than one, the elasticity is known as:
(A) Elastic (B) Inelastic
(C) Infinity (D) None of these

297. If the price elasticity of demand is less than one, the elasticity is known as:
(A) Elastic (B) Inelastic
(C) Infinity (D) None of these

298. Changes in supply elasticity occurs mainly during:
(A) Short period
(B) Long period
(C) Winter season
(D) Rainy or summer season

299. The elasticity of the agricultural commodity in very long run will be always:
(A) Elastic (B) Inelastic
(C) Both (A) and (B) (D) None of these

300. The supply slope is normally:
(A) Negative (B) Positive
(C) Both (A) and (B) (D) None of these

301. The luxury good has income elasticity:
(A) 1 (B) <1
(C) >1 (D) Between 0 to 1

302. Monopoly implies to:
(A) Single firm
(B) Special or unique product
(C) Firm is price maker
(D) All of these

303. Monopsony refers to:
(A) Single buyer
(B) Few buyers
(C) Many buyers
(D) No buyers

304. Market price of a commodity means:
(A) Balance between demand and supply of a commodity for a specific day
(B) It changes very frequently
(C) It increases or decreases and also remains nearly normal price line
(D) All of these

305. Normal price of a commodity means:
(A) It is almost fixed price of a commodity
(B) The consumption and production rates of a commodity remains same at its normal price
(C) Price plays a great role in normalizing the price of a commodity
(D) All of these

306. Agricultural commodities have greater price fluctuations mainly because of:
(A) Inelastic demand and supply
(B) Elastic demand and supply
(C) More control of government
(D) All of these

307. Which of the following is a government measure to deal with uncertainty in agriculture?
(A) Guaranteed agricultural prices
(B) Buffer stock schemes
(C) Crop insurance
(D) All of these

308. Duopoly refers to:
(A) Two either firms (sellers) or two buyers in the competition
(B) Few firms or buyers
(C) Many firms or buyers
(D) None of these

309. Uncertainity in agriculture refers to:
(A) Yield uncertainty
(B) Price uncertainty
(C) Tenurial uncertainty
(D) All of these

310. Uncertainty can be reduced mainly at farmers level by:
(A) Diversification (B) Flexibility
(C) Capital rationing
(D) All of these

311. What is the important measure to stabilize the agricultural prices and incomes?
(A) Buffer stock scheme
(B) Crop insurance scheme
(C) Control over agricultural production
(D) All of these

312. Instability in agriculture is caused by:
(A) Fluctuations in price
(B) Fluctuations in income
(C) Fluctuations in outcome
(D) All of these

313. Farm management should answer the following question:
(A) How to organize a farm
(B) How to operate it
(C) How to achieve the maximum net profit
(D) All of these

314. Farm management is:
(A) An art (B) A science
(C) A business (D) All of these

315. Farm layout means:
(A) Physical arrangement of farm in the form of plots, building, roads, channels etc.
(B) Planning of farm to maximize the profit
(C) Production of crops
(D) None of these

316. Farm layout is required because of:
(A) Newly acquired land
(B) Change in power and enterprises
(C) Change in technologies
(D) All of these

317. Farm profit is equal to:
(A) Gross revenue – cost
(B) Gross revenue – net returns
(C) Gross income – gross revenue
(D) All of these

318. Farm profit is affected by:
(A) Total produce
(B) Market and farm prices
(C) Input cost
(D) All of these

319. Farm budget refers to:
(A) A tentative estimation of expenses and returns of a farm
(B) A blueprint of production and cost
(C) Loan and its interest (D) All of these

320. Complete farm budget refers to:
(A) Farm budget for a entire farm as a single unit
(B) Farm budget separately for each enterprise
(C) It is a budget for a Gram Panchayat as a single unit
(D) It is a budget for whole area of the region

321. Partial farm budget refers to:
(A) Farm budget for a entire farm as a single unit
(B) Farm budget separately for each enterprise
(C) It is a budget for a Gram Panchayat as a single unit
(D) It is a budget for whole area of the region

322. Farm inventory refers to:
(A) List of items possessed by a farm on a specific date
(B) Land under cultivation
(C) Inventory of new farm
(D) All of these

323. Farm production records indicate:
(A) Financial position of farm
(B) Production in quantity
(C) Production in quality
(D) All of these

324. Which of the following is a farm production record?
(A) Livestock record
(B) Crop cultivation record
(C) Crop production record
(D) All of these

325. The motto of land reform in India is:
(A) Land to tiller
(B) Land to landlord
(C) Land to Talukedar (D) All of these

326. The land revenue was charged in cash first time during the reign of:
(A) Ashoka (B) Chandra Gupta I
(C) Tamur (D) Akbar

327. The land revenue during Kautilya era was:
(A) $1/6^{th}$ of total produce
(B) $1/5^{th}$ of total produce
(C) $1/4^{th}$ of total produce
(D) $1/3^{rd}$ of total produce

328. The land revenue during Sher Shah Suri reign was:
(A) $1/6^{th}$ of total produce
(B) $1/5^{th}$ of total produce
(C) $1/4^{th}$ of total produce
(D) $1/3^{rd}$ of total produce

329. The land revenue during Akbar regime was:
(A) $1/6^{th}$ of total produce
(B) $1/5^{th}$ of total produce
(C) $1/4^{th}$ of total produce
(D) $1/3^{rd}$ of total produce

330. Agricultural holding refers to:
(A) Total area of land which is held for cultivation as a single unit
(B) Total area of land used as a single unit
(C) Total area owned by a cultivator
(D) All of these

331. Operational holding refers to:
(A) Total area of land which is held for cultivation as a single unit
(B) Total area including fallow and barren used as a single unit
(C) Total area owned by a owner
(D) All of these

332. Fragmentation of holding refers to:
(A) It is a breaking up of a single holdings into a number of small pieces situated at a distance from each other
(B) Consolidating the small holdings
(C) Breaking of holdings into economic holdings
(D) None of these

333. An optimum holding refers to the size that:
(A) Should not be more than 3 times of economic holding
(B) Should be > 3 times of economic holding
(C) Should be equal to operational holding
(D) Should be <2 ha land

334. The cooperative movement was started first in:
(A) England (B) Germany
(C) USA (D) India

335. Which was the first cooperative society setup in the world?
(A) Rashdel Poineers Equitable Society
(B) Primary Cooperative Credit Society
(C) National Dairy Development Board
(D) Anand Milk Union Limited

336. Which of the following is a first cooperative credit society opened in the world?
(A) Reiffeisen Society, Germany
(B) Milk Cooperative Society, Denmark
(C) 4 H Club, USA (D) None of these

337. Which of the following is a first cooperative bank opened in the world?
(A) Cooperative Bank of China
(B) Bank of London
(C) Shultz Delite Bank of Germany
(D) IFCI Bank

338. Which of the non-institutional agency of that provides agricultural loan?
(A) Traders and money lenders
(B) Commercial banks
(C) Land Development Bank
(D) Regional Rural Banks

339. Which of the following credit agency that provides institutional loan to farmers?
(A) Cooperative Societies
(B) Commercial Banks
(C) Land Development Bank
(D) All of these

340. Which is the most important source of loan in India?
(A) Money lenders
(B) Land Development Banks
(C) Government (D) Cooperative Societies

341. Which of the following credit agency provides long term loan to farmers?
(A) State Land Development Bank
(B) Primary Land Development Bank
(C) Money lenders
(D) All of these

342. Primary Land Development Bank is established at:
(A) State level (B) District level
(C) Tehsil level (D) Village level

343. Central Land Development Bank is established at:
(A) State level (B) District level
(C) Tehsil level (D) Village level

344. Regional rural banks provide loan to:
(A) Small and marginal farmers
(B) Landless labourers
(C) Rural artisans
(D) All of these

345. Which type of loan is supplied agricultural service societies?
(A) Short term (B) Medium term
(C) Long term (D) All of these

346. National Agricultural Credit Council was constituted in year:
(A) 1952 (B) 1962
(C) 1969 (D) 1972

347. The main function of National Agricultural Credit Council is:
(A) To help in distribution of agricultural credit card estimation of total credit demand
(B) To provide agricultural loan to farmers
(C) To control over RBI activities
(D) To check the inflation rate

348. The main objective of Agricultural Finance Corporation is:
(A) To provide credit to agricultural schemes
(B) To encourage the role of commercial banks in agricultural development
(C) Both (A) and (B)
(D) None of these

349. Which type of loan is given by ARDC?
(A) Short term (B) Medium term
(C) Long term
(D) Medium and long term

350. Marketing involves the activities from:
(A) Production to end consumer
(B) Production only
(C) Production to middlemen
(D) Production to storage

351. Monopoly market refers to:
(A) Single producer or seller in market
(B) Single buyer in market
(C) More number of producers and single buyers
(D) None of these

352. Monopsony market refers to:
(A) Market having only one buyer or institution
(B) Buyer is free to decide the quantity to be purchased
(C) Buyer is free to change the price policy
(D) All of these

353. Duopsony refers to:
(A) Market having only two buyers
(B) Market having only two sellers
(C) Market having only two buyers and sellers each
(D) None of these

354. Oligopsony refers to:
(A) Few buyers in a market
(B) Only one buyer in a market
(C) Only one seller in a market
(D) None of these

355. Oligopoly refers to:
(A) More than sellers in market
(B) Both product homogeneity and differentiation may be found
(C) Product substitution is necessary
(D) All of these

356. Marketable surplus refers to:
(A) Actual amount that is marketed
(B) Excess amount than requirement for consumption
(C) Both (A) and (B) (D) None of these

357. Marketed surplus refers to:
(A) Actual amount that is marketed
(B) Excess amount than requirement for consumption
(C) Both (A) and (B) (D) None of these

358. Normally marketed surplus is higher than marketable surplus because of:
(A) To repay the loan
(B) Poor facilities of storage at farms
(C) Inability to estimate the total family consumption
(D) All of these

359. Which of the following is a market for agricultural commodities?
(A) Whole sale market
(B) Retail market
(C) Fairs
(D) All of these

360. Demand of a particular commodity can be estimated by:
(A) Yardstick approach
(B) Regression technique
(C) Simultaneous equation model
(D) All of these

361. Marketing middlemen refers to:
(A) Institutions or persons that work between the transfer process of a commodity from producers to consumers
(B) Institutions or persons that purchase the produce from producers
(C) Institutions or persons that sell the produce to consumers
(D) None of these

362. Kachcha commission agent is related to:
(A) Collection of agricultural produce
(B) Distribution of agricultural produce
(C) Both (A) and (B)
(D) None of these

363. Packka commission agent is related to:
(A) Collection of agricultural produce
(B) Distribution of agricultural produce
(C) Both (A) and (B)
(D) None of these

364. Market structure refers to:
(A) Organizational structure that affects the competition and price in market
(B) Behavioral pattern of buyers and sellers through which prices are affected
(C) It indicates about the role of commission agents in market only
(D) All of these

365. Market conduct refers to:
(A) Organizational structure that affects the competition and price of a commodity
(B) Behavioral pattern of market that affect the commodity price
(C) Involvement of brokers in marketing
(D) All of these

366. Which of the following is a important element of market structure?
(A) Degree of seller's concentration
(B) Degree of buyer's concentration
(C) Degree of product differentiation
(D) All of these

367. Which of the following is a important element of market conduct?
(A) Rules and regulations of production firms
(B) Product policy
(C) Product promotion policy
(D) All of these

368. Market clearing refers to:
(A) Equilibrium state of market price
(B) Stage of marketing at which brokers mediates in deciding the price of a commodity
(C) Change in price with increase supply
(D) None of these

369. Price discovery refers to:
(A) Process through which buyer and sellers decide a price of a commodity at a given place
(B) Stage of deciding equilibrium market price
(C) Search of market price
(D) All of these

370. Which of the following is a component of marketing cost?
(A) Labour (B) Transportation
(C) Packing and storage
(D) All of these

371. Price spread is also known as:
(A) Marketing cost
(B) Marketing margin
(C) Farm price (D) None of these

372. Which of the statement is correct for monopoly market?
(A) MC = MR (B) MC > AC
(C) MC < AC (D) None of these

373. New name of Indian Standard Institution is:
(A) Bureau of Indian Standards
(B) Institute of Indian Standards
(C) Standard Institute of India
(D) None of these

374. The Bureau of Indian Standard is situated at:
(A) Lucknow (B) New Delhi
(C) Hyderabad (D) Mumbai

375. The price which necessarily covers the variable cost and also the fixed price is:
(A) The long run price
(B) The short run price
(C) The market price
(D) The equilibrium price

376. Marginal Physical Product of a variable input is negative:
(A) Only in I stage (B) Only in II stage
(C) Only in III stage(D) In I and II stage

377. Average cost is equal to:
(A) Total cost + opportunity cost
(B) $\frac{\text{Total cost}}{\text{Output}}$
(C) $\frac{\text{Average variable cost}}{\text{Output}}$
(D) Cost B + Cost C

378. Fixed cost is equal to:

(A) $\frac{\text{Total fixed cost}}{\text{Output}}$ (B) $\frac{\text{Total cost}}{\text{Output}}$

(C) $\frac{\text{Total variable cost}}{\text{Output}}$

(D) None of these

379. Variable cost is equal to:

(A) $\frac{\text{Total cost}}{\text{Output}}$ (B) $\frac{\text{Total variable cost}}{\text{Unit of output}}$

(C) Cost A + Cost B + Cost C

(D) All of these

380. Marginal cost is equal to:

(A) $\frac{\text{Total cost}}{\text{Output}}$ (B) $\frac{\text{Increase in variable cost}}{\text{Increase in output}}$

(C) Cost C + Rent paid to leased rent

(D) $\frac{\text{Total variable cost}}{\text{Increase in output}}$

381. Demand is a:

(A) Desire and purchase of a commodity at a specified time

(B) Desire for a commodity but not able to purchase it

(C) Ability to purchase a commodity but not purchase it

(D) All of these

382. If the price elasticity of barley is -0.5, it means:

(A) Barley demand will be decreased by 0.5%, if the price increased by 1%

(B) Barley demand will be increased by 0.5% if the price is increased by 1%

(C) If the price of barley is increased by 0.5%, the demand will also be increased by 0.5%

(D) Price and demand both either increased or decreased by 0.5%

383. If the petrol price fall from Rs. 10 to 8 for 1 litre and as a result the demand of petrol is increased from 5 units to 9 units. What will be the price elasticity of demand?

(A) -1 (B) -2

(C) -3 (D) -4

384. Cross elasticity of demand refers to:

(A) $\frac{\text{\% change in quantity demanded of X}}{\text{\% change in income}}$

(B) $\frac{\text{\% change in quantity demanded of X}}{\text{\% change in price}}$

(C) $\frac{\text{\% change in income}}{\text{\% change in quantity demanded}}$

(D) None of these

385. Income elasticity of demand refers to:

(A) $\frac{\text{\% change in quantity demanded}}{\text{\% change in income}}$

(B) $\frac{\text{\% change in quantity demanded}}{\text{\% change in price}}$

(C) $\frac{\text{\% change in income}}{\text{\% change in quantity demanded}}$

(D) None of these

386. Price elasticity of supply is equal to:

(A) $\frac{\text{\% change in quantity supplied of a commodity A}}{\text{\% change in price of A}}$

(B) $\frac{\text{\% change in demand of commodity A}}{\text{\% change in price of A}}$

(C) $\frac{\text{\% change in income}}{\text{\% change in price}}$

(D) All of these

387. The production or supply of a product will be closed by a firm if:
(A) Production cost < cost of marginal product
(B) Production cost > cost of marginal product
(C) Variable cost > production cost
(D) All of these

388. Which of the following condition is required for perfect competition in the market?
(A) More number of firms in the market
(B) Homogeneous product
(C) Free entry or exit in the market
(D) All of these

389. What will happen if the market price of a commodity is higher than normal price?
(A) Demand will be reduced and supply will be increased in the market
(B) Demand will be increased
(C) Supply will be reduced
(D) None of these

390. What will happen if the market price of a commodity is lower from its normal price?
(A) Demand of a commodity will be increased
(B) Supply will be decreased
(C) Market price will increase towards the normal price line
(D) All of these

391. Income elasticity of inferior goods is:
(A) Positive (B) Negative
(C) Between zero to one
(D) Infinite

392. An income elasticity of necessary goods is:
(A) 1 (B) < 1
(C) >1 (D) Between zero to one

393. Which of the following is a feature of monopolistic competition?
(A) Many firms
(B) Firm is price maker
(C) Products/services differentiated through promotion, quality and design
(D) All of these

394. Cyclic fluctuations in prices of agricultural commodities means:
(A) Ups and downs in the prices after certain intervals
(B) Increase in prices after certain intervals
(C) Decrease in prices after certain intervals
(D) None of these

395. The fluctuations in agricultural prices due to operation of Cobweb theorem shows:
(A) Cyclic fluctuations
(B) Long term fluctuations
(C) Seasonal fluctuations
(D) All of these

396. Long term fluctuations in agricultural prices and incomes are caused mainly due to:
(A) Change in population
(B) Opening of new markets
(C) Development of new technology
(D) All of these

397. Monopoly is a perfect stage of:
(A) Imperfect competition
(B) Pure competition
(C) Monopolistic competition
(D) None of these

398. The condition of market where few buyers are present is called as:
(A) Monopsony (B) Oligopsony
(C) Oligopoly (D) Monopoly

399. Mechanisation of agriculture refers to:
(A) Use of power driven machines for various farm jobs
(B) Ploughing with tractor operated plough only
(C) Crushing of sugarcane with crusher only
(D) All of these

400. Mechanisation of agriculture is required for:
(A) Better agricultural yield
(B) Multiple cropping
(C) Generation of employment
(D) All of these

401. How much farm power is available per hectare?
(A) 0.5 KW (B) 1.0 KW
(C) 1.5 KW (D) 2.0 KW

402. The wage is not paid to:
(A) Permanent labour
(B) Contract labour
(C) Family labour (D) Casual labour

403. Level of farm mechanization is affected by:
(A) Size of holding
(B) Labour requirement
(C) Intensification of crop production
(D) All of these

404. Farm planning refers to:
(A) Reorganization and/or alteration of present farm organization and operation in order to get maximum return from limited resources
(B) Activities for crop production on farm
(C) Physical arrangement of farm resources
(D) All of these

405. The basic objective of farm planning is:
(A) To guide a farmer to move in the right direction
(B) To indicate the destination
(C) Proper use and allocation of input
(D) All of these

406. Farm price means to:
(A) The price that farmer receives for his produce
(B) Price prevailing in the market
(C) Price declared by Govt.
(D) All of these

406. Farm plan refers to:
(A) Systematic pre-thinking of future activities related to farm business
(B) It is a tentative future plan for farm business
(C) It is related to maximize the farm profit
(D) All of these

407. Farm production efficiency can be measured with following overall standard?
(A) Input per unit of output
(B) Total production index
(C) Output per hectare
(D) All of these

408. What is the objective of keeping farm records?
(A) To know annual performance of farm
(B) To know the present worth
(C) To compare the performance over a period of years
(C) All of these

409. The main objective of land reform in India is:
(A) Abolition of intermediates
(B) Confirment of ownership right to tenants
(C) Imposition of ceiling on agricultural land holdings
(D) All of these

410. The maximum limit is land holding under best soil is:
(A) 10-18 acres (B) > 27 acres
(C) > 34 acres (D) None of these

411. The maximum limit of land holding if the soil is poor is:
(A) 20-22 acres (B) > 37 acres
(C) > 54 acres (D) None of these

412. What is the main provision of land consolidation act?
(A) Fixation of minimum standard area for regulating transfers
(B) Prohibition of fragmentation below standard area
(C) Filling of objectives, appeals and penalties
(D) All of these

413. UP consolidation of Holding Act was passed in years :
(A) 1935 (B) 1943
(C) 1953 (D) 1958

414. UP consolidation of Holding (Amendment) Act was passed in years:
(A) 1938 (B) 1943
(C) 1953 (D) 1958

415. Royal commission on agriculture was appointed in year:
(A) 1919 (B) 1921
(C) 1927 (D) 1932

416. Farmers can get long term loan from:
(A) Money lenders
(B) Land Development Banks
(C) Government
(D) All of these

417. Long term loan is given mainly for:
(A) Purchase of seeds, fertilizers and weedicide
(B) Purchase of agricultural implements
(C) Purchase of land and its development
(D) All of these

418. Farmers can get short term loan from:
(A) Money lenders
(B) Land Development Banks
(C) Government
(D) All of these

419. Short term loan is given mainly for:
(A) Purchase of seeds, fertilizers and weedicide
(B) Purchase of agricultural implements
(C) Purchase of land and its development
(D) All of these

420. What is the main principle of good system of credit to agriculture?
(A) Repayment capacity
(B) Risk bearing capacity
(C) Estimation of returns to credit
(D) All of these

421. The lead Bank Scheme was started in 1969 on the recommendation of:
(A) Nariman Committee
(B) All India Rural Credit Survey Committee
(C) Committee on Cooperative Credit
(D) Expert group of RBI on credit scheme

422. Abolition of bounded labour was recommended by:
(A) Dhebar Commission
(B) National Agriculture Commission
(C) Planning Commission
(D) Labour Commission

423. The bounded labour practiced was abolished in India by the notification made on:
(A) January 26, 1975
(B) August 15, 1975
(C) October 2, 1975
(D) October 25, 1975

424. Agricultural marketing involves the operations:
(A) Movement of food and raw material from field to final consumer
(B) Handling of produce at farm
(C) Processing, grading and packing
(D) All of these

425. The main objective of agricultural marketing is:
(A) To make available agricultural produce to consumer at optimum prices
(B) To help farmers to get maximum profit from their produce
(C) To maintain optimum market prices of agricultural commodities
(D) All of these

426. Functional approach of agricultural marketing deals with:
(A) Different functions
(B) Different institutions
(C) Different commodities
(D) None of these

427. Which of the following is a facilitating function of agricultural marketing:
(A) Standardization
(B) Financing
(C) Risk bearing (D) All of these

428. Which of the following is physical function of agricultural marketing?
(A) Buyers and sellers
(B) Transportation
(C) Processing and storage
(D) All of these

429. Which of the following is known as a middlemen?
(A) Whole sellers (B) Retailers
(C) Commission agents (D) All of these

430. Specialised market refers to:
(A) Sell-buy of specialized kind of a commodity
(B) Sell-buy of different kind of a commodity
(C) Communication, transport and storage play an important role
(D) All of these

431. Terminal market are found at:
(A) Large cities and sea ports
(B) Village level
(C) Town levels (D) District level

432. Local market generally found at:
(A) Large cities and sea ports
(B) Village level
(C) Town levels (D) District level

433. Which of the marketing system was prevalent in the ancient time?
(A) Barter system
(B) Modern system
(C) European system
(D) None of these

434. Computer based market refers:
(A) Prices are decided on internet or phone
(B) Exchange of commodities are based on computer information
(C) Payment can also be made through computer
(D) All of these

435. Complete competitive market should have:
(A) Large number of sellers and buyers
(B) No product differentiation
(C) Dynamic factors of production
(D) All of these

436. The market that is organized under some rules and regulations made by trade body federation etc. is known as:
(A) Regulated market
(B) Organized market
(C) Economic market
(D) Unorganized market

437. The market that is under control of government rules and regulations is called :
(A) Regulated market
(B) Organized market
(C) Economic market
(D) Unorganized market

438. Agricultural products are characterized as:
(A) Bulky raw material (B) Perishable
(C) Essential for life (D) All of these

439. Agricultural production is characterized as:
(A) Annual variability
(B) Seasonal variation
(C) Varying costs of production
(D) All of these

440. Which of the following statement is correct?
(A) Marketable surplus > marketed surplus
(B) Marketable surplus < Marketed surplus
(C) Marketable surplus = marketed surplus
(D) All of these

441. Marketable surplus is affected by:
(A) Prices of agricultural produce
(B) Support prices
(C) Availability of resources and its prices
(D) All of these

442. Marketed surplus is affected by:
(A) Family size (B) Production level
(C) Sale prices (D) All of these

443. Which of the following government agency is responsible for marketing and distribution of agricultural produce in the country?
(A) FCI
(B) State Trading Corporation Ltd
(C) Agricultural Price Commission
(D) All of these

444. Which of the following government agency is responsible for marketing and distribution of tobacco produce in the country?
(A) FCI
(B) State Trading Corporation Ltd
(C) Tobacco Board
(D) Agricultural Price Commission

445. Which of the following government agency is responsible for international trading of agricultural produce in the country?
(A) FCI
(B) State Trading Corporation Ltd
(C) Agricultural Price Commission
(D) All of these

446. Crop area production and productivity data are available with:
(A) Directorate of Economics and Statistics
(B) Forward Market Commission
(C) Central Storage Corporation
(D) All of these

447. Marketing margin can be defined as:
(A) Part of total cost that goes to marketing firms
(B) It is a difference between producer's selling cost and consumer's purchasing cost
(C) It is the produce of total services of marketing firms
(D) All of these

448. Absolute margin equal to:
(A) Sale price – Purchase price
(B) Sale rice + Purchase price
(C) Sale price – Total marketing cost
(D) None of these

449. Net margin is equal to:
(A) Absolute margin – absolute marketing expenditure
(B) Sale price – Purchase price
(C) Sale price – total cost
(D) Absolute margin/sale price x 100

450. Marketing margin can be calculated by:
(A) Concurrent margin method
(B) Time lag method
(C) Both (A) and (B) (D) None of these

451. Which of the following is a first regulated market opened in India?
(A) Karanjia cotton market
(B) Brar cotton market
(C) Cashew market of Kerala
(D) None of these

452. The new name of Agricultural Price Commission is:
(A) Commission on Agricultural Cost and Prices (CACP)
(B) Commission on agricultural Prices
(C) Agricultural Prices and Cost Commission
(D) None of these

453. The CACP came into existence in year:
(A) 1982 (B) 1985
(C) 1987 (D) 1990

454. Which of the following basis is adopted in determination of prices of agricultural commodities in our country?
(A) Production cost (B) Parity cost
(C) Prevalent market price
(D) All of these

455. Which of the following aspect is related to macro economics?
(A) Agricultural employment
(B) Instability of agriculture
(C) Tenurial system
(D) All of these

456. The principle of least-cost combination of two factors indicates that profit can be maximized at:

(A) $\frac{\Delta X2}{\Delta X2} = \frac{Px1}{Px2}$ (B) $\frac{\Delta X2}{\Delta X1} = \frac{Px2}{Px1}$

(C) $\frac{\Delta X1}{X2} = \frac{Px1}{Px2}$ (D) None of these

457. In the production possibility curve, the maximum profit will be at:

(A) $\frac{\Delta Y2}{\Delta Y1} = \frac{-Py1}{Py2}$

(B) $Py2, \Delta y2 = Py1, \Delta y1$

(C) Both (A) and (B) (D) None of these

458. Which of the following is representing Cobb-Douglas production function?

(A) $Q = KL^{\alpha}C^{\beta}$ (B) $Q = L^{\alpha}C^{\beta}M^{\chi}$

(C) $Q = KC^{\alpha}M^{\beta}$ (D) All of these

459. Elasticity of production is equal to:

(A) $\frac{\Delta Y1}{\Delta X} \times \frac{X}{Y}$ (B) $\frac{MP}{AP}$

(C) $\frac{\text{Present change in output}}{\text{Present change in input}}$

(D) All of these

460. If MP, AP and EP represent marginal product, average product and elasticity of production, which of the following inference is correct?

(A) When MP = AP, Ep=1

(B) When MP>AP, Ep>1

(C) When MP<AP, Ep<1

(D) All of these

461. The least cost combination of inputs will be at:

(A) $\frac{\Delta X1}{\Delta X2} = \frac{px2}{PX1}$ (B) $\frac{\Delta X1}{\Delta X2} > \frac{px2}{px1}$

(C) $\frac{\Delta px1}{\Delta px2} < \frac{py2}{py1}$ (D) None of these

462. The size of holding that gives full employment to the farm family and reasonable living standard is known as:

(A) Family holding
(B) Optimum holding
(C) Economic holding
(D) Non-economic holding

463. Operating holding refers to:

(A) Land area owned by a farmers
(B) Cropped area cultivated by a farmer
(C) Net land area cultivated by a farmer
(D) Land area leased in by a farmer

464. The contribution of livestock to gross farm income in mixed farming is:

(A) At least 10% (B) At least 30%
(C) At least 40 % (D) More than 50%

465. Under diversified farming, the income from the single enterprise should not be:

(A) < 50% of total farm income
(B) > 50% of total farm income
(C) > 33% of total farm income
(D) None of these

466. Under specialized farming, the income from the single enterprise should not be:

(A) < 50% of total farm income
(B) > 50% of total farm income
(C) > 75% of total farm income
(D) None of these

467. Ranching refers to:

(A) Growing of crops only
(B) Rearing of livestock only
(C) Growing of crops and rearing of livestock
(D) All of these

468. Price elasticity of demand is equal to:

(A) $\frac{\text{Percentage change in quantity demanded}}{\text{Percentage change in price}}$

(B) $\frac{\text{Percentage change in supply of quantity}}{\text{Percentage change in price}}$

(C) $\frac{\text{Percentage change in quantity demanded}}{\text{Percentage change in quantity supplied}}$

(D) All of these

469. The income elasticity is generally:
(A) 1% increase income increase only 0.2% demand
(B) 1% increase income decreases 0.2 % demand
(C) Both A and B (D) None of these

470. International trade is more difficult in arrange than inter regional trade within the country because:
(A) Factors of production are less mobile
(B) Each country uses a different currency
(C) Government interferes with foreign trade for political reasons
(D) All of these

471. Risk and uncertainty differs in:
(A) Risk has known probability but uncertainty has no probability
(B) Risk is temporary by uncertainty is sure
(C) Risk is more fatal than uncertainty in agriculture
(D) None of these

472. The difference of average price and minimum support price of a particular commodity is paid to each farmer if the average price is lower than minimum support price, under the scheme of:
(A) Crop Insurance
(B) Forward price
(C) Deficiency payment
(D) All of these

473. Labour efficiency can be increased by:
(A) Improving farm layout
(B) Farm work simplification
(C) Training of labourers
(D) All of these

474. Farm management refers to:
(A) Managing the farm
(B) The art of managing the farm
(C) The art of managing the farm successfully as measured by the test of profitability
(D) All of these

475. Farm building is important because of:
(A) It saves feed, both in quantity and quality
(B) It increases safety in farm work
(C) It helps to increase production
(D) All of these

476. Which of the following state is most mechanization in India?
(A) UP (B) Punjab
(C) Haryana (D) Kerala

477. Which of the following is a partial standard of measuring farm production efficiency?
(A) Crop yield index
(B) Production per unit livestock
(C) Yield per hectare of the most important crop
(D) All of these

478. Farm survey refers to:
(A) An enquiry in the composition activities and living conditions of farmers and farm
(B) Collection of information related expenses and profit of the farm
(C) Visit of farm (D) All of these

479. Which of the following step (s) are involved in farm survey?
(A) Planning of survey and collection of data
(B) Analysis of data
(C) Presentation and interpretation of the findings
(D) All of these

480. Which type of sampling method is used for collection of data if the population heterogeneous?
(A) Random sampling
(B) Stratified sampling
(C) Purposive sampling
(D Sequential sampling

481. Sequential sampling method is widely used in industries for:
(A) Quantity control
(B) Quality control
(C) Efficiency measures
(D) None of these

482. Cluster sampling method is used for:
(A) Quality control
(B) Geological surveys
(C) Efficiency measures
(D) All of these

483. The system of accounting refers to:
(A) Posting of entry or transactions in books of account
(B) Entry of items in register
(C) Listing of production records
(D) Physical verification of records

484. Toy holding refers to:
(A) Very small size of holding, generally in sq. m and sq. yards
(B) Very large size of holding generally > 10 acre of land
(C) Both (A) and (B) (D) None of these

485. Who proposed the concept of agricultural banks?
(A) Sir Willian Wedderburn and Justice Ranade
(B) Dr, Dadgil
(C) H. Dupernix
(D) Henry Wolff

486. Which of the following credit agency provides short and medium credit to farmers?
(A) State Cooperative Banks
(B) District Cooperative Bank
(C) Cooperative Credit Society
(D) All of these

487. Which of the following is not accepted as mortgage against long term loan by land development bank?
(A) Land (B) House
(C) Waterlogged soil (D) All of these

488. Regional Rural banks were established on:
(A) 26 September, 1975
(B) 19 July, 1969
(C) Ist July, 1965
(D) 2 October, 1969

489. Market refers to:
(A) A place where producers and sellers are assembled for selling and purchasing of goods
(B) Sellers and producers have free but competitive relationship
(C) Prices of commodities remain at normal prices
(D) All of these

490. Market process refers to:
(A) Activities through which a commodity is reached from producer to final consumer
(B) Grading and storage of a commodity
(C) Middlemen through which a commodity is either sold or purchase
(D) None of these

491. Which of the following approach useful and simple to study the agricultural marketing?
(A) Functional approach
(B) Institutional approach
(C) Commodity approach
(D) Behavioral approach

492. What is difference between Dalal and Commission agent?
(A) Commission agents have more rights than Dalal
(B) Dalal does not have physical control of produce
(C) Commission agents can sell the produce and charge the commission
(D) All of these

493. Which of the following statement is correct?
(A) All regulated markets should be organized markets
(B) All organized markets should be regulated markets
(C) All regulated markets may not be organized markets
(D) All unorganized markets may not be regulated markets

494. Which of the following middlemen accept the ownership of produce during the marketing process?
(A) Itinerant traders
(B) Village traders
(C) FCI (D) All of these

495. Which of the following agency is responsible for creating awareness about quantity control of agricultural products?
(A) Export Inspection Standard Council of India
(B) Indian Standard Institute
(C) Indian Society for Quality control
(D) All of these

496. Which of the following is related to market development?
(A) Advertisement (B) Quality control
(C) Packing (D) All of these

497. Which of the following is a trade practices of marketing?
(A) Personal selling
(B) Advertisement based sale
(C) Sales by inspection
(D) All of these

498. Percentage marketing margin is equal to:
(A) Sale price – purchase price
(B) Absolute margin/sale price x 100
(C) Sale price – total marketing cost
(D) None of these

499. Agricultural Production Mandi Act was passed in UP in year:
(A) 1962 (B) 1964
(C) 1966 (D) 1968

500. Which of the following is a feature of a regulated market?
(A) Proper marketing system
(B) Proper communication of marketing information
(C) Availability of optimum grading, storage and packing facilities
(D) All of these

501. The farm-business income is:
(A) Farm net income
(B) The farm expenditure
(C) The family labour income
(D) The family labour income + unpaid charge of family labour + interest on owned capital

502. AGMARK is a symbol of:
(A) Productivity (B) Total production
(C) Quality
(D) Quality and purity of product

503. Demand and supply of agricultural goods is:
(A) Price determining force
(B) Price elasticity curve
(C) Demand oriented force
(D) None of these

504. The marketing function which creates time utility is:
(A) Cold storage (B) Storage
(C) Marketing (D) None of these

505. The minimum support is announced by the government to support the:
(A) Retailers (B) Whole sellers
(C) Producer (Farmers)
(D) Local leaders

506. Presence of only two sellers of a commodity is a market is called:
(A) Monopoly (B) Deuopoly
(C) Both of these (D) None of these

507. The cost of production of the crop can be minimized by using economic principle:
(A) Law of diminishing return
(B) Law of opportunity cost
(C) Cost principles
(D) Principles of least cost combination

508. Princile of equi-marginal returns is applied when:
(A) The resources are unlimited
(B) The resource price are high
(C) The prices of resources are varying
(D) The resources are limited

509. Do the total cost and the variable cost increase with increase in the output?
(A) Never (B) Yes
(C) Partially (D) With decrease in output

510. When a new variety of crop is introduced:
(A) Partial budgeting is required
(B) Full budgeting is a must
(C) No budgeting is needed
(D) Only planning is needed

511. The objective of preparation of calendar of operations is:
(A) To prepare a cropping scheme
(B) To calculate the cost of production
(C) To assess the need of human and bullock power
(D) To adjust farm enterprises

512. The farm plan would be based on the availability of efficiency of various factors of production:
(A) True (B) False
(C) Partially true
(D Incomplete statement

513. Profit is maximised when:
(A) Marginal cost < Marginal return
(B) Marginal cost = Marginal return
(C) Marginal cost > Marginal return
(D) None of these

514. Which is not a system of farming:
(A) State farming (B) Collective farming
(C) Cooperative farming
(D) Dry farming

515. Price elasticity of supply of food products is generally:
(A) Infinite (B) Less than one
(C) More than one (D) Zero

516. Specilised farming is a example of:
(A) Type of farming (B) System of farming
(C) Capitalistic farming (D) None of these

517. One of the important fixed cost on a farm is:
(A) Seed cost (B) Land rent
(C) Plant protection cost
(D) Fertilizer cost

518. Regulated market ensures:
(A) Procurement price
(B) Support price
(C) Remunerative price
(D) Fair price

519. Farm budget is a tool of farm management analysis:
(A) True (B) False
(C) Never include cost
(D) It is a futile exercise

520. Subsistence farming is characterized by:
(A) Enough surplus or sale
(B) Distress sale
(C) Can hire large number of labourers
(D) The resource structure is rich

521. Mixed farming is when:
(A) Crop production is combined with livestock enterprise and 50% - 50% income is from both
(B) Entire farm land is sown under mixed cops
(C) Only milch cattle are raised on the farm
(D) Income from milch animal is 95%

522. Extension workers should be well versed with economic principles:
(A) Never (B) Partially
(C) Without it they cannot give effective advice from business point of view
(D) This is not necessary

523. When marginal product is more than marginal cost, further investment on production would be:
(A) Profitable (B) Harmful
(C) Equal (D) No effect

524. The record of farm produce is entered in:
(A) Logbook (B) Inventory book
(C) Farm stock register
(D) Cash book

525. Price theory is a part of:
(A) Micro economics (B) Macro economics
(C) Farm management
(D) Marketing

526. Farming in India in accordance with the Nagpur Resolution should be through:
(A) Cooperative Joint farming
(B) Cooperative Better farming
(C) Cooperative Tenant farming
(D) Cooperative Collective farming

527. When marginal product is declining but remains more than zero, the total product will/will be:
(A) Zero
(B) Increase with increasing rate
(C) Increase with increasing rate
(D) Initially decreasing than increasing

528. Which one of the following is not include in the factors of production:
(A) Enterprise
(B) Organization
(C) Land
(D) Time

529. When a farm is managed by Government officials it is called a:
(A) Peasant farm
(B) Capitalistic farm
(C) Collective farm
(D) State farm

530. When the depreciation of an asset is measured by divide the first cost by its estimated life, it is called:
(A) Diminishing value method
(B) Sum of the years digit method
(C) Straight line method
(D) Revaluation methods

531. When $\dfrac{\Delta x1}{\Delta x2} = \dfrac{\Delta x2}{\Delta x1}$ — then we can :
(A) Employ any combination of x1 and x2
(B) Do nothing
(C) Go on substituting x1 and x2
(D) Employ x2 only

532. Level of optimum production is available in:
(A) I stage of production
(B) II stage of production
(C) III stage of production
(D) None of these

533. Which one of the following kind of budgeting would you advise to a new farm planner:
(A) Enterprise budgeting
(B) Complete budgeting
(C) Partial budgeting
(D) None of these

534. Which of the following is the most effective way to overcome the defects of agricultural marketing?
(A) Cooperative marketing
(B) Forward marketing
(C) Regulated marketing
(D) Monopoly

535. A farm plan can be prepared by using:
(A) Partial budgeting
(B) Complete budgeting
(C) Input-output model
(D) Break even point

536. When marginal product in increasing at increasing rate then:
(A) Total product increases at a decreasing rate
(B) Total product increases at a increasing rate
(C) Average product decreases
(D) None of these

537. Farming system approach ensures the best use of:
(A) Land (B) Labour
(C) Capital
(D) Land + Labour + Capital + Management

538. The price which necessarily covers the variable cost and also the fixed cost is:
(A) The long run price
(B) The short run price
(C) The market price
(D) The equilibrium price

539. Read the following statements and mark your answer according to code given below:

I. For variable input; APP = MPP in the Ist stage

II. For variable input; APP = MPP at the end of Ist stage

III. For variable input; APP = MPP at the start of Ist stage

IV. For variable input; APP = MPP in the IInd stage

Answer code

(A) I and III are correct
(B) II and III are correct
(C) I and IV are correct
(D) II and IV are correct

540. MPP (Marginal Physic Product) of a variable input is negative:
(A) Only in I stage (B) Only in II stage
(C) Only in III stage (D) In II and III stage

541. Match list I with list II and select the correct answer using the code given below:

	List I (Farm Income)	List II (Measure)
a.	Family labour income	1. G.I. – cost A1
b.	Farm business income (owner farmer)	2. G.I. – cost A2
c.	Net income	3. G.I. – cost B
d.	Farm business income (tenant farmer)	4. G.I. – cost C

Code:

	a	b	c	d
(A)	2	1	3	4
(B)	1	2	4	3
(C)	1	2	3	4
(D)	3	1	4	2

542. Marketing surplus is:
(A) Total production – Total family requirement
(B) Total consumption – Marginal consumption
(C) Total cost – marginal cost
(D) Total produce – Total cost

543. Line passing through the least cost points in the isoquants map is called:
(A) Isoline (B) Ridge line
(C) Expansion path (D) Isoquant curve

544. The demand of the commodity is different mainly by its:
(A) Quality (B) Price
(C) Storage (D) Supply

545. Formula which indicates the least-cost combination point is:
(A) MR = MC (B) $\Delta Y/\Delta x1$
(C) MRSY1, Y2
(D) MRSx1 x2 = PX2/PX1

546. The chief function of money is:
(A) Reserve base of credit expansion
(B) Providing liquidity
(C) Medium of exchange (D) None of these

547. Support prices of agricultural commodities are worked out by:
(A) Finance ministry
(B) Agricultural Price Commission
(C) Food and Agricultural ministry
(D) None of these

548. According to the law of Diminishing Return the optimum profit will be at the point where:
(A) MP = AP (B) MP > AP
(C) AP > MP (D) MC = MP

549. The best level of output is obtained at point where:
(A) MR < MC (B) MR = MC
(C) MR > MC (D) MR = AC

550. Most prevalent "system of farming" in India is:
(A) Cooperative farming
(B) Collective farming
(C) Peasant farming (D) State farming

551. When the depreciation of an asset is measured by dividing the first cost by its estimated life, it is called:
(A) Diminishing value method
(B) Sum of years digit method
(C) Straight line method
(D) Revaluation method

552. The commission for agricultural cost and prices is located in:
(A) Lucknow (B) Hyderabad
(C) New Delhi (D) Faridabad

553. The record of farm produce is entered in:
(A) Log book (B) Cash book
(C) Farm stock register
(D) Inventory book

554. The marginal cost –curve intersects the average cost-curve when average cost is:
(A) Maximum (B) Minimum
(C) Rising (D) Falling

555. The overhead cost is also known as:
(A) Variable cost (B) Fixed cost
(C) Total cost (D) Marginal cost

556. The subsidiary service of a marketing system is:
(A) Welfare (B) Assembling
(C) Dispersion (D) Processing

557. According to law of diminishing return, the optimum [profit will be at the point where:
(A) MP = AP (B) MP > AP
(C) AP > MP (D) MC = MR

558. The vertical demand curve for a commodity shows that the demand is:
(A) Perfectly inelastic (B) Highly elastic
(C) Unit elastic (D) Perfectly elastic

559. The demand of agricultural product is mainly dependent on:
(A) Increase in agricultural holding
(B) Increase in income
(C) Increase in population
(D Increase in cost

560. Type of farming is based on:
(A) Law of comparative advantage
(B) Law of equi-marginal return
(C) Law of diminishing return
(D) Law of substitution

561. The ultimate aim of farm management is to:
(A) Increase gross income
(B) Optimize factor utilization
(C) Reduce total cost (D) None of these

562. Which one of the following is a gift of nature?
(A) Land (B) Labour
(C) Capital (D) None of these

563. The efficiency of marketing channel is determined mainly on the basis of:
(A) Price spread
(B) Difference between retail and wholesale prices
(C) Regulated market
(D) Producer's share in consumer's rupees

564. Which one is not the function of a production manager?
(A) Planning (B) Organisation
(C) Marketing (D) Monitoring

565. Which system of farming does not involve farmer?
(A) Collective farming (B) Dry farming
(C) State farming
(D) Diversified farming

566. Livestock insurance is included to:
(A) Fixed cost (B) Special cost
(C) Variable cost (D) Marginal cost

567. Farm management is a:
(A) Intra-farm science
(B) Inter-farm science
(C) Inter-marginal farm science
(D) International farm science

568. Which one is not a tool of farm management?
(A) Farm planning (B) Farm budgeting
(C) Production and cost functions
(D) Farm stock keeping

569. One of the important fixed cost on a farm is:
(A) Seed cost (B) Land cost
(C) Irrigation cost
(D) Fertilizer cost

570. When marginal product is at the maximum, the marginal cost is at the:
(A) Higher point
(B) Lowest point
(C) Parallel
(D) No change

571. If the marginal product is more than marginal cost than further investment in production is:
(A) Profitable
(B) Harmful
(C) No effect
(D) Equal

572. If the substitution ratio is less than the price ratio, than the cost:
(A) Remains the same
(B) Increases
(C) Decreases
(D) None of these

573. Principle of equi-marginal return is applied when:
(A) The resources are unlimited
(B) The price of resources is high
(C) The prices of resources vary
(D) The resources are limited

574. Which of the following is also referred to as "Opportunity cost"?
(A) Money cost
(B) Alternate cost
(C) Social cost
(D) Fixed cost

575. The law of diminishing return refers to an eventual fall in:
(A) Marginal product of the variable factor
(B) Average income of the farm
(C) Total production of all factors
(D) Total earning of the farm

576. Which one of the following is not correctly matched?

Type of market	Basis of classification
(A) Unorganised market	Organization
(B) Regulated market	Control
(C) Perfect market	Competition
(D) Terminal market	Area

577. Which of the following cost is subtracted from gross income (GI) to get the value of family labour income?
(A) Cost A1 (B) Cost A2
(C) Cost B (D) Cost C

578. Marketng surplus is:

(A) Total production	Total family consumption
(B) Total consumption	Marginal consumption
(C) Total cost	Marginal cost
(D) Total production	Total cost

579. Read the following statements and mark your answer according to code given below:
I. For variable input, APP = MPP in the I stage
II. For variable input, APP = MPP at the end of I stage
III. For variable input, APP = MPP at the start of II stage
IV. For variable input, APP = MPP in the II stage

Answer code
(A) I and III are correct
(B) II and III are correct
(C) I and IV are correct
(D) II and IV are correct

580. The concept of farming is best explained by:
(A) Secretarial planning approach
(B) Mixed farming approach
(C) Holistic farming approach
(D) Diversified farming approach

581. When MPP is zero, TPP is at its maximum, because:
(A) Further addition to MPP is negative
(B) Further addition to MPP is positive
(C) MC = MR
(D) Elasticity of productive is more than one

582. When marginal product increases at increasing rate, then:
(A) Total product increases at a decreasing rate
(B) Total product increases at a increasing rate
(C) Average product decreases
(D) None of these

583. When increase or decrease in the level of one product does not affect the production level of the other product, it is known as:
(A) Supplementary relationship
(B) Complementary relationship
(C) Competitive relationship
(D) None f these

584. According to A. Marshall the " law of diminishing return" is applicable to:
(A) Industry (B) Agriculture
(C) Service sector (D) All of these

585. Why people prefers to receive payments in money?
(i) Because a person holding money can easily exchange it for any commodity or service that he/she might want
(ii) Because money raises the status of a person
(A) (i) is true but (ii) is false
(B) (i) is false but (ii) is true
(C) Both (i) and (ii) are true
(D) Both (i) and (ii) are false

586. What is barter system?
(i) A system under which goods are directly exchanged without the use of money
(ii) A system under which different currencies are exchanged
(iii) A system under which banks work as mediator between buyer and seller
(A) Only (i) is true (B) (iii) is true
(C) Both (i) and (ii) are true
(D) Both (i) and (ii) are false

587. What is double coincidence of wants?
(A) A situation under which seller is ready to sell the product
(B) A situation under which buyer is ready to buy the product
(C) A situation under which both buyers and sellers are ready to exchange the product
(D) A situation under which neither the buyer nor the seller is ready to exchange the product

588. Which of the following is most important medium of exchange?
(A) Money (B) Gold
(C) Silver (D) Copper

589. Which of the following are part of medium forms of money?
(i) Paper notes and coins
(ii) Demand deposits
(ii) ATM card
(A) Only (i) (B) Only (i) and (ii)
(C) All of these (D) None of these

587. Which of the following is not a feature of money?
(A) It includes currency – paper notes and coins and demand deposits
(B) It is without any use of its own
(C) It is authorized by the government of the country
(D) Gold has high liquidity so it is also included in the money

588. The modern currency is without any use of its own but still it is accepted as a medium of exchange because:
(A) The currency is authorized by the World Bank
(B) The currency is authorized by the government of the country
(C) The currency is authorized by RBI
(D) There is no other thing which can be used as a medium of exchange

589. Which of the following issues currency notes in India?
(A) RBI (B) SBI
(C) ICICI (D) President

590. Which of the following is not true with reference to Indian currency?
(A) It is issued by the Reserve Bank of India
(B) As per Indian law, no other individual or organization is allowed to issue currency
(C) No individual in India can legally refuse a payment made in rupees
(D) All the currency notes are signed by the President of the country

591. The rupees is widely accepted as a medium of exchange in India which of the following is/are appropriate reason for the same?
(i) It is issued by the RBI on the behalf of the central government
(ii) The law legalizes the use of rupee as a medium of payment
(iii) It is the only currency available in India
(A) Only (i) and (iii) (B) Only (ii) and (iii)
(C) Only (i) and (ii) (D) All of these

592. Why the deposits in the bank accounts are known as demand deposits?
(i) Because these can be withdrawn on demand
(ii) Because these creates demand of currency in the market
(iii) Because the banks pay an interest rate on the deposits
(A) Only (i) (B) Only (i) and (ii)
(C) Only (i) and (iii) (D) All of these

593. Which of the following facility is provided by the demand deposits
(i) Credit card (ii) Cheque
(A) Only (i)
(B) Only (ii)
(C) Only (i) and (ii)
(D) All of these

594. What is a cheque?
(A) It is a paper instructing the bank to pay money
(B) It is a paper instructing the bank to deposit money
(C) It is a paper issued by bank against a loan
(D) It is a rate of interest

595. For which purpose the bank's use the major portion of their deposits?
(A) To pay the depositors who come to withdraw money
(B) They invest in share market
(C) To extend loans
(D) They deposit the excess money to RBI

596. Which of the following work as a mediator between those who have surplus funds and those who are in need of these funds?
(A) RBI (B) Commercial banks
(C) Landlords (D) Government of India

597. What is the major source of revenue for the commercial banks?
(A) The difference between what is charged from borrower and what is paid to depositors
(B) Revenue from the share market
(C) Fixed deposits
(D) None of these

598. Read the given statement and choose the right word from the given below:
An agreement in which the lender supplies the borrower with money, goods or services in return for the promise of future payment.
(A) Credit (B) Demand deposit
(C) Currency (D) Draft

599. In rural areas, the main demand for the credit is:
(A) For industralisation
(B) For crop production
(C) For marriages
(D) For education

600. ....is an asset that the borrower owns and uses this as a guarantee to the lender until the loan is repaid:
(A) Will (B) Collatoral
(C) Fixed deposit (D) Education

601. Interest rate, collateral and documentation requirement, and the mode of repayment together comprise. What is called the:
(A) Terms of credit
(B) Procedure of credit
(C) Method of credit (D)Terms of trade

602. Which of the following is a formal source of credit?
(A) Money-lender (B) Relatives
(C) Trader
(D) Commercial Banks

603. Which of the following is a informal source of credit?
(i) Money-lender (ii) Relatives
(iii) Trader (iv) Commercial Banks
(A) Only (i) and (ii)
(B) Only (ii), (iii) and (iv)
(C) Only (iii) and (iv) (D) All of these

604. Which of the following agency supervises the formal sources of credit?
(A) RBI (B) SBI
(C) CBI (D) Police

605. Which of the following is the main source of credit for the rural households in India:
(A) Commercial Banks
(B) Co-operative societies
(C) Land-lords
(D) Money lenders

606. Which of the following charges higher rate of interest from the borrower?
(A) Informal sources of credit
(B) Formal sources of credit
(C) Both formal as well as informal sources of credit
(D) Government of India

607. Which of the following is the main source of credit for the rich house holds?
(A) Informal (B) Formal
(C) Both formal and informal
(D) Neither formal nor informal

608. With reference to source of credit find the odd one out:
(A) Money lenders (B) Traders
(C) Friends (D) Commercial banks

609. Poor households are still dependent on formal sources of credit. Which of the following is/are reasons for the same?
(i) Banks are not present everywhere in rural India
(ii) Banks charges higher rate of interest than other sources
(iii) Bank loans require proper documents and collateral but poor don't have these
(A) Only (i) (B) Only (ii)
(C) Only (i) and (iii) (D) Only (ii) and (iii)

610. Which of the following promotes for house hold to take loan from money lender?
(i) In most of the cases there is no need for any collatoral
(ii) They charges low rate of interest
(iii) The borrower can, if necessary approach the money lender even without repaying their earlier loans
(A) Only (i) (B) Only (ii)
(C) Only (i) and (iii) (D) Only (i) and (ii)

611. What is SHG?
(A) Self Help Groups
(B) Self Handicraft Groups
(C) Social Help Groups
(D) Society Help Groups

612. A typical Self Help Group has how members:
(A) 100-150 (B) 150-200
(C) 15-20 (D) 5-6

613. People prefer to take loan from SHG because:
(A) No rate of interest is charges
(B) Rate of interest is less than what the information sources charges
(C) Rate of interest is less than what the formal sources charges
(D) Most of members are women

614. Which of the following is correctly matched:
(i) Banks : Formal source of credit
(ii) SHG : Informal source of credit
(iii) Land : Collatoral
(A) Only (i) (B) Only (ii)
(C) Only (iii) (D) Only (i) and (ii)

615. Who takes most of the decisions regarding in the SHG's?
(A) Banks (B) Members
(C) NGO's (D) RBI

616. What is MNC?
(A) Multi Nation Corporations
(B) Multinational Corporations
(C) Moving National Corporations
(D) Mid National Cooperatives

617. A company that owns or controls production in more than one nation. Choose the correct word:
(A) MNC (B) MVC
(C) KMC (D) VMC

618. Which of the following is not a feature of MNC's?
(i) A MNC owns or controls production in more than one nation
(ii) The goods and services are produced globally
(iii) Production is organized in a complex way
(iv) Most of the MNCs are investing in developed countries
(A) Only (i) and (ii) (B) Only (ii) and (iii)
(C) Only (iii) and (iv) (D) Only (iv)

619. Which of the following country provides cheap manufacturing location?
(A) China (B) Mexico
(C) USA (D) UK

620. Which of the following is not an important locational factor for a MNC?
(A) Cheap manufacturing location
(B) Availability of cheap labour
(C) Market
(D) Type of government i.e., democratic or non democratic

621. MNC are investing in India which of the following is/are appropriate reason for the same?
(i) Availability of cheap skilled as well as unskilled labour
(ii) Market
(iii) Good infrastructure
(iv) Efficient Government
(A) Only (i) and (ii) (B) Only (ii) and (iii)
(C) Only (iii) and (iv)(D) All of these

622. What is investment?
(A) Money that is spent to buy assets such as land, building, machines and other equipment
(B) Money that is deposited in banks
(C) Money that is spent on imports
(D) Profit of the businessman

623. Investment made by MNC is called:
(A) Internal investment
(B) External investment
(C) Foreign investment
(D) Multi investment

624. Which of the following methods is/are used by MNC's to spread/increase their production?
(A) They buy up local companies
(B) They place orders to the local producers
(C) They produce jointly with local producers
(iv) They sell their products at cheap rates
(A) Only (i) and (ii)
(B) Only (i), (ii) and (iii)
(C) Only (ii) and (iii)
(D) Only (iii) and (iv)

625. What is globalisation? Select the most appropriate answer:
(A) It is the process of rapid integration or interconnection between countries
(B) It is increase in foreign trade
(C) It is the process under which developing countries are using modern technology
(D) It is the process of movement of people from one region to another

626. Which of the following is not a feature of foreign trade?
(i) It connects different markets of the world
(ii) Choice of goods in the market rises
(iii) Prices of similar goods in the two markets tend to become equal
(iv) Developing countries gain more that the developed countries
(A) Only (i) and (ii) (B) Only (iv)
(C) Only (iii) and (iv) (D) All of these

627. .... is the process of rapid integration or interconnection between countries:
(A) Privatisation (B) Liberalisation
(C) Globalisation (D) None of these

628. Which of the following has been the major factor that has stimulated the globalization process:
(A) Technology (B) Ship industry
(C) Democracy (D) Socialism

629. Which of the following is a trade barrier?
(i) Technology (ii) Tax on imports
(iii) Tax on exports
(A) Only (i)
(B) Only (i) and (ii)
(C) Only (ii) and (iii)
(D) Only (iii)

630. Why the countries use trade barriers?
(i) Countries put up trade barriers to increase or decrease foreign trade
(ii) Countries put up trade barriers to decide what kind of goods and how much of each, should come into the country
(iii) To create social and political equality
(A) Only (i) (B) Only (i) and (ii)
(C) Only (iii) (D) Only (i) and (ii)

631. The Indian government, after independence has put trade barriers to foreign trade and foreign investments:
(A) To protect the producers within the country from foreign competition
(B) To promote Swadeshi Movement
(C) Because foreign goods were very expensive
(D) Because foreign goods were not liked by the Indian people

632. When was India's foreign trade liberalized?
(A) 1960s (B) 1970s
(C) 1980s (D) 1990s

633. Removing barriers or restrictions setup by the government is known as:
(A) Privatisation (B) Globalisation
(C) Liberalisation (D) Industrialisation

634. Which of the following is/are result of liberalization?
(i) Businesses are allowed to make decisions freely
(ii) Government removes unwanted restrictions from trade
(iii) Government provides political and fundamental rights to all
(iv) Poverty is removed
(A) Only (i) and (ii)
(B) Only (i), (ii) and (iii)
(C) Only (iii) and (iv) (D) Only (iv)

635. Indian government wants to check the free flow of cheap Chinese electronic goods into Indian market. Which of the following method can government use?
(A) Trade barrier
(B) Complaint to WTO
(C) Government can increase export tax
(D) Government can complaint to World Bank

636. What is WTO?
(A) World Tower Organisation
(B) World Trade Organisation
(C) World Tax Organisation
(D) World traders Organisation

637. What is the basic aim of WTO?
(A) To check the free flow of goods and services across the world
(B) To bridge gap between rich and poor countries
(C) To liberalise international trade
(D) To support MNC

638. How many countries are the members of WTO?
(A) 95 (B) 120 (C) 149 (D) 200

639. Which of the following organization/s liberalises international trade?
(i) WTO (ii) MNC
(iii) Commercial bank
(A) Only (i)
(B) Only (i) and (ii)
(C) Only (ii) and (iii)
(D) Only (iii)

640. Which of the following has benefitted the most from the globalization?
(A) Small producers
(B) Farmers
(C) Poor
(D) Consumers

641. Which of the following is/are Indian MNC's?
(i) Samsung (ii) Nokia
(iii) Tata motors (iv) Asian Paints
(A) Only (i) and (ii) (B) Only (ii) ad (iii)
(C) Only (iii) and (iv) (D) Only (i) and (iv)

642. Which of the following has benefitted the most from the process of globalisation?
(A) IT (B) Agro based industries
(C) Textile industries
(D) Iron and steel industry

643. Which of the following have suffered the most because of the globalisation?
(A) IT
(B) Small manufectures of vegetable oil, capacitors, plastic etc.
(C) MNC's (D) Automobile industries

644. Which of the following has benefitted the most from the globalization?
(A) People with education, skill and wealth
(B) People living below poverty line
(C) Farmers (D) Local traders

645. Which of the following method can be used by the government for a fair globalization?
(A) It can impose trade barriers
(ii) It can negotiate at the WTO for fairer rules
(iii) It can align with other developing countries
(iv) It can close its market for foreign trade
(A) Only (i) and (ii) (B) Only (ii) and (iv)
(C) Only (i), (ii) and (iii) (D) All of these

646. Who is a consumer?
(A) Anyone who purchase goods and services from the market and pays for these
(B) Anyone who sells goods and services in the market
(C) Anyone who manufactures goods
(D) Anyone who invests in the market

647. Anyone who produce goods and services and sell these in the market is known as:
(A) Producer (B) Consumer
(C) Buyer (D) Entrepreneur

648. Under which of the following condition the market do not work in a fair manner?
(i) When producers are few and powerful whereas consumers purchase in small amounts and are scattered
(ii) When consumers are few and producers are in a large number
(A) Only (i) and (ii) (B) Only (i)
(C) Only (iii) (D) Only (ii)

649. Which of the following is an appropriate reason for the Consumer Movement?
(A) It arose out dissatisfaction of the consumers as many unfair practices were being indulged in by the sellers
(B) It arose because of spread of democracy
(C) It arose because of cheap goods and services
(D) It arose because of MNC's

650. Which of the following factor's gave birth to the Consumer Movement in India?
(i) Food shortage
(ii) Hoarding
(iii) Black marketing and adulteration
(iv) Government of India
(A) Only (i) and (ii)
(B) Only (i), (ii) and (iii)
(C) Only (iii) (D) Only (i) and (ii)

651. The Consumer Movement in an organized form was started in:
(A) 1950s (B) 1960s
(C) 1970s (D) 1980s

652. Why is the year 1986 known as a turning point in the history of Consumer Movement?
(A) COPRA was enacted
(B) CHOPRA was enacted
(C) Policy of liberalization was followed in this year
(D) WTO was established

653. What is COPRA?
(A) Consumer Preservation Act
(B) Consumer Protection Act
(C) Consumer Protection Activities
(D) None of these

654. "We do find bad quality products in the market". Which of the following is an appropriate reason for the same?
(i) Production is being done by MNC's
(ii) Consumer Movement is still very weak
(iii) Consumer are not aware of their rights and duties
(iv) Prevailing laws to safe guard the consumers are very weak
(A) Only (i)
(B) Only (ii), (iii) and (iv)
(C) Only (iii) and (iv)
(D) Only (i) and (ii)

655. What is MRP?
(A) Minimum Retail Price
(B) Maximum Retail Price
(C) Marginal Retail Price
(D) Marginal Revenue Price

656. If a manufacture has sold you a pressure cooker with improper safety value, which of the following right can help you to pull up the manufacturer?
(A) Rights to safety
(B) Right to Choice
(C) Right to Consumer Education
(D) Right to Representation

657. Under which right the producer is suppose to provide details about the products, its ingredient, price hatch number, date of manufacturing, etc?
(A) Right to Informed
(B) Right to Information
(C) Right to Choose
(D) Right to Represent

658. What is RTI?
(A) Right to Informed
(B) Right to Information
(C) Right to Choose
(D) Right to Represent

659. Which of the following right you will use in case you want to have information about the functioning of a government department?
(A) Right to Informed
(B) Right to Information
(C) Right to Choose
(D) Right to Seek Redressal

660. You have joined an education institution and had paid the fee for the whole year but you found that the institution is not providing what it promised. Now you want to change the institution but the management refused to refund your fee. Which of your right is being violated?
(A) Right to Informed
(B) Right to Information
(C) Right to Choose
(D) Right to Represent

661. Suppose you want to buy toothpaste, and the shop owner says that he/she can sell the toothpaste only if you buy a toothbrush. Which of your right is being violated by the shopkeeper?
(A) Right to Informed
(B) Right to Information
(C) Right to Choose
(D) Right to Seek redressal

662. Which of the following guides consumers regarding how to file cases in the Consumer Court?
(A) PRO
(B) Consumer Forums
(C) Police
(D) A lawyer

663. Under which of the following right a consumer can represent his/her own case by himself?
(A) Right to Informed
(B) Right to Information
(C) Right to Choose
(D) Right to Represent

664. Under which of the following rights the government gives advertisements in the newspapers about the consumer's rights and print slogans like "Jago Grahak Jago":
(A) Right to Informed
(B) Right to Information
(C) Right to Education
(D) Right to Represent

665. Under COPRA a .... tier judicial machinery has been setup to provide justice to the exploited consumers:
(A) 2 (B) 3
(C) 4 (D) 5

666. The district level Consumer courts deals with the cases involving claims up to Rs.:
(A) 20 lakh
(B) 40 lakh
(C) 60 lakh
(D) 80 lakh

667. The state level Consumer courts deals with the cases involving claims up to Rs.:
(A) 20 lakh - 1 crore
(B) 40 lakh- 2 crore
(C) 60 lakh - 3 crore
(D) 80 lakh - 4 core

Chapter 12

SOCIOLOGY AND SOCIAL WORKS

1. Sociology is the study of:
   (A) Individual (B) Society
   (C) Group (D) Social behaviour
2. Father of sociology is:
   (A) August Comte (B) Herbert Spenser
   (C) Emile Durkhein (D) Max Weber
3. Basic concept of sociology is:
   (A) Family (B) Community
   (C) Village (D) Society
4. Who is known as father of sociology?
   (A) Max Waber (B) J.B. Chitamber
   (C) H. Spencher (D) August Compte
5. Scope of sociology is situated by:
   (A) Education (B) Trade
   (C) Community (D) Individual
6. Extension can be considered to have developed from:
   (A) Sociology (B) Psychology
   (C) Education (D) Philosophy
7. Basic unit of society is:
   (A) Neighbourhood (B) Group
   (C) Family (D) Community
8. What was SPA?
   (A) It was a Seven Party Alliance
   (B) It was a Six Party Alliance
   (C) It was a Socialist Party Alliance
   (D) It was a Separate Party Alliance
9. Which of the following is/are true with reference to pressure groups?
   1. They attempt to influence government policies.
   2. They aim to directly control or share political power.
   (A) 1 is true but 2 is false
   (B) 1 is false but 2 is true
   (C) Both 1 and 2 are true
   (D) Both 1 and 2 are false
10. Read the statement and select the correct answer:
    These are a loose organization and their decision making, is more informal and flexible.
    (A) Political parties
    (B) Pressure groups
    (C) Movements
    (D) Sectional interest groups
11. Which of the following seek to promote the interest of a particular section or group of society?
    (A) Movement (B) Political party
    (C) Sectional Interest Group
    (D) Public Interest Group
12. Society can be considered as:
    (A) Abstract (B) Concrete
    (C) Both (A) and (B) (D) None of these

13. Which of the following is/are true with reference to movement?
(i) They depend on spontaneous mass participation
(ii) The decision making is more informal and flexible
(iii) They can raise any specific or general issue
(iv) Their ultimate goal is to hold political power
(A) Only(i), (ii), (iii) (B) Only (i), (ii), (iv)
(C) Only (i), (ii) (D) Only (iii), (iv)

14. Which of the following is a single issue movement?
(A) Narmada Bachao Andolan
(B) Women's Movement
(C) Save Earth Movement
(D) Trade Union

15. Which among the following is the special feature that distinguishes a pressure group from a political party?
(A) Parties take political stances, while pressure groups do not bother about political issues
(B) Pressure groups are confined to a few people, while political parties involve large number of people
(C) Pressure groups do not seek to get into power while political parties do
(D) Pressure groups do not seek to mobilize people, while political parties do

16. Consider the following statements about the pressure groups and political parties:
(i) Pressure group are organized expression of the interests and view of specific social sections
(ii) Pressure groups take positions on political issues
(iii) All pressure groups are political parties
Which of the statements give above are correct?
(A) i, ii and iii (B) i and ii
(C) ii and iii (D) i and iii

17. Which among the four meanings is closest to the term 'Overurbanisation'?
(A) Urbanisation at the cost of rural development
(B) Urbanisation without the essential infrastructure
(C) Rapid urbanization
(D) Emergence of megalopolis

18. Which of the following promote collective good?
(i) Political party
(ii) Public Interest Group
(iii) Sectional Interest Group
(A) Only (i) (B) Only (ii)
(C) Only (iii) (D) Only (i) and (ii)

19. Cities in primary phase of urbanization display:
(A) Heterogenetic transformation
(B) Generative transformation
(C) Parasitic transformation
(D) Orthogenetic transformation

20. Who is Partisan?
(A) A person who is strongly committed to a party
(B) A person who is strongly committed to public interest group
(C) A person who is strongly committed a religion
(D) A person who is political leader

21. How the criteria of selecting the candidate differ in USA and India?
(A) In USA members and supporters of a party chose its candidates whereas in India top party leaders choose the candidates for contesting election
(B) In India members and supporters of a party choose its candidates whereas in USA top party leaders choose the candidates for contesting election
(C) In USA the President of the country chose the candidates who can context election whereas in India the top party leaders chose the candidates for contesting election
(D) None of the mentioned above

22. Consider the following two statements:
Assertion (A): Members of urban society and held together by organic solidarity
Reason (R): Individualism is the characteristic of urban social structure
Select the correct answer:
(A) Both A and R are true and R is the correct explanation of A
(B) Both A and R are true by R is not the correct explanation of A
(C) A is true, R is false
(D) A is false, T is true

23. Why India has evolved a multiparty system? Choose the most appropriate answer:
(A) Because India has a multireligion culture
(B) Because India is divided into states
(C) Because India is a poor country
(D) Because of social and geographical diversity

24. Which of the following are components of a political party?
(i) The leader
(ii) The active members
(iii) The followers
(iv) The electoral symbol
(A) Only (i) and (ii)
(B) Only (i), (ii) and (iii)
(C) Only (iv) (D) Only (iii) and (iv)

25. A party that secure atleast 6% of the total votes in an election to the Legislative Assembly of a State and wins atleast two seats is registered as:
(A) National Party
(B) States Party
(C) International Party
(D) None of these

26. The rural areas surrounded by the city s known as:
(A) Suburban
(B) Hinterland
(C) Urban village
(D) Slum

27. Democratic government is a legimate government. Which of the following is true with reference to this?
(A) It accommodates social diversities
(B) It is people's own government elected by the people
(C) It leads to peaceful harmoniuous life
(D) None of these

28. Which of the following is not true with the reference to democracy?
(A) Democratic governments take decision very fast
(B) Democracy ensures that decision making is based on norms and procedures
(C) Democratic government is a legitimate government
(D) Democratic government is an accountable government

29. Culture treats all human products and learned abilities:
(A) Equally (B) Differently
(C) Insignificant (D) Great

30. Which of the following challenge is faced by every democracy in one form or another:
(A) Challenge of expansion
(B) Challenge of deepening of democracy
(C) Foundational challenge
(D) None of these

31. Studies on political and social inequalities in democracy show that:
(A) Democracy and development go together
(B) Inequalities exist in democracies
(C) Development under a democracy is not much different from that of dictatorships
(D) Dictatorship is better than democracy

32. In the context of assessing democracy, which among the following is the odd one out. Democracies need to ensure:
(A) Free and fair elections
(B) Dignity of the individual
(C) Everyone has freedom to what they choose
(D) Equal treatment before law

33. In India, the bigger cities have higher rate of population growth than the smaller urban centres. This is primarily because of:
(A) Natural rate of growth
(B) Better medical facilities
(C) Rural urban-migration
(D) Social mobility of city dwellers

34. Which of the following is true with reference to Indian democracy:
(i) Democracy in India has strengthened the claims of the disadvantaged and discriminated castes for equal status and equal opportunity
(ii) Democracy in India has created economic equality
(A) Both (i) and (ii) are true
(B) Both (i) and (ii) are false
(C) (i) is true but (ii) is false
(D) (i) is false but (ii) is true

35. On which of the following issues most of the democratic governments have failed?
(i) Economic Equatibility
(ii) Removal of poverty (iii) Accountability
(A) (i) and (ii)
(B) Only (i)
(C) Only (ii)
(D) Only (i) and (iii)

36. What is challenge?
(i) A challenge is a difficulty that carries within it an opportunity for progress
(ii) A challenge is a difficulty that can not be overcome
(A) (i) is true but (ii) is false
(B) (i) is false but (ii) is true
(C) Both (i) and (ii) are true
(D) Both (i) and (ii) are false

37. Urban areas formed by the fusion of towns through spreading industry and housing is known as:
(A) Metropolis
(B) A megalopolis
(C) A conurbation
(D) Suburbia

38. This involves bringing down the existing non-democratic regime, keeping military away from controlling government and establishing a sovereign and functional state:
(A) Challenge of expansion
(B) Challenge of deepening of democracy
(C) Foundational challenge
(D) None of these

39. When the balance between the material and non-material culture is disrupted by rapid social change, the material culture changes rapidly and the non-material culture is left behind. This situation is called:
(A) Left culture (B) Far-culture
(C) Cultural lag (D) Taken culture

40. This involves applying the basic principle of democratic government across all the regions, different social groups and various institutions:
(A) Challenge of expansion
(B) Challenge of deepening of democracy
(C) Foundational challenge
(D) None of these

41. A megalopolis is:
(A) The primary city created by the merging of cities
(B) The city with a population of tem million
(C) A sprawling greater urbanized area created by the merging of two or more metropolitan areas
(D) A succession of cities along a highway

42. Ensuring greater power to local governments, expansion of federal principle to all the units of the federation, inclusion of women and minority groups is covered under:
(A) Challenge of expansion
(B) Challenge of deepening of democracy
(C) Foundational challenge
(D) None of these

43. The group tries to prohibit the actions that it considered undesirable. In sociology this process is called:
(A) Social change (B) Social development
(C) Social balance (D) Social control

44. Which of the following is true with reference to democracy?
(i) More and more countries are adopting democratic type of government
(ii) This shows that people are totally satisfied with the working of democracies
(A) Both (i) and (ii) are true
(B) Both (i) and (ii) are false
(C) (i) is true but (ii) is false
(D) (i) is false but (ii) is true

45. Which of the following challenge is/are not being faced by the Indian democracy:
(i) Foundational challenge
(ii) Challenge of expansion
(iii) Challenge of deepening of democracy
(A) Only (i)
(B) Only (ii)
(C) Only (iii)
(D) All of these

46. "Some norms within a culture are explicit, out in the open, every one in the culture is able to formulate the rule". This may be defined as:
(A) Explicit culture
(B) Explicit norms
(C) Implicit culture
(D) Implicit norm

47. All the suggestions or proposals about overcoming various challenges to democracy are called:
(A) Democratic reforms
(B) Constitutional reforms
(C) Social reforms (D) Economic reforms

48. Organised Social Life emerges from:
(A) Social organization
(B) Social interaction
(C) Social culture
(D) Social group

49. Most of the Indian Political parties lack internal democracy now select the appropriate challenge:
(i) Foundational challenge
(ii) Challenge of expansion
(iii) Challenge of deepening of democracy
(A) Only (i) (B) Only (ii)
(C) Only (iii) (D) All of these

50. Consider the following two statements:
Assertion (A): Urban communities and rural communities are exclusive of each other
Reason (R): Urban communities and rural communities differ in their characteristics
Select the correct answer:
(A) Both A and R are true and R is the correct explanation of A
(B) Both A and R are true by R is not the correct explanation of A
(C) A is true, R is false
(D) A is false, T is true

51. Who among the following should carry political reforms?
(i) Political activities (ii)Political parties
(iii) Movements
(iv) Politically conscious citizens
(A) Only (i) and (ii) (B) Only (ii) and (iii)
(C) Only (iii) and (iv)(D) All of these

52. Explicit norms are learned through:
(A) Formal means
(B) Informal means
(C) Inbuilt in human
(D) Human beings learn it automatically

53. What is RTI?
(A) Right to Intention Act
(B) Right to Information Act
(C) Right to Transparency Act
(D) None of these

54. Which of the following law empowers the people to find out what is happening in government and act as a watch dog of democracy:
(A) RTO (B) TRX
(C) RTI (D) ART

55. Which of the following is/are most appropriate for democratic reforms?
(i) By legal ways
(ii) By improving the quality of political participation by ordinary citizens
(iii) Democratic reforms should be carried out mainly by political activities, parties, movements and citizens
(A) Only (i) is true
(B) Only (ii) is true
(C) Both (i) and (ii) are true
(D) Both (ii) and (iii) are true

56. According to Census of India, 2001, what is the population percentage of scheduled caste?
(A) 16.2 (B) 26.3
(C) 36.2 (D) 46.2

57. Many states have banned people who have more than two children from contesting panchayat election. Which of the following is true with reference to this:
(i) This has resulted in denial of democratic opportunity to many poor and women
(ii) This has strengthened Indian democracy
(A) (i) is true but (ii) is false
(B) (i) is false but (ii) is true
(C) Both (i) and (ii) are true
(D) Both (i) and (ii) are false

58. Role:
(A) Is a set of expectation
(B) Cannot be defined without referring to another
(C) Both (A) and (B)
(D) None of these

59. For Sumner, the groups with which the individual identifies himself are his:
(A) Living group (B) Group identity
(C) In-group (D) Belonging group

60. What is the major cause of sexual division of labour in many families?
(A) Cender division
(B) Economic division
(C) Political division
(D) Cultural division

61. What is a secular state?
(A) Allows freedom to practice any religion
(B) There is no official religion
(C) Prohibits discrimination on religious grounds

62. Which of the following consists of concepts (ideas, notions, doctrine, interests) and a structure?
(A) Customs (B) Folkways
(C) Culture (D) Institutions

63. What is a major cause behind failure of the Parliament to reserve at least one-third seats for women representation?
(A) Lack of consensus among political parties
(B) Refusal by President to give consent to the Bill
(C) Rejection of the Bill in Rajya Sabha
(D) None of these

64. —— is official religion for the India state:
(A) There is no official religion
(B) Hinduism
(C) Sikhism (D) Christianity

65. In the Indian context, urban is generally described as a:
(A) Function of population size and density
(B) Permanent settlement of cities
(C) Community enjoying modern facilities with low density of population
(D) Class of people characterized by higher education

66. What does caste hierarchy mean?
(A) A shift from rural areas to urban areas
(B) Shift from one occupation to another
(C) A ladder like formation in which all caste groups are placed from the highest to the lowest
(D) None of these

67. Non-fraternal polyandry is practiced among:
(A) Khasis of Garo Hills
(B) Nayars of Malabar Hills
(C) Todas of Nilgiri Hills
(D) Hos of rajmahal Hills

68. In which of the following are one third of the seats reserved for women in India?
(A) Lok Sabha (B) State Assemblies
(C) Government jobs (D)Panchayats

69. The focus of human relation approach is the study of modern industrial set-up is:
(A) Worker's morale
(B) Profit making
(C) Management's benefit
(D) Solution of union grievances
(D) All of these

70. A desire to have impact or control over others is:
(A) Need for achievement
(B) Need for power (C) Need for security
(D) Need for self-actualization

71. What is fundamental rights?
(A) The situation when a particular community tries to promote its own interest at the cost of other communities
(B) The right to freedom of religion, our constitution provides to all citizens freedom to profess, practice and propagate any religion, or not to follow any
(C) The state to intervene in the matters of religion in order to ensure equality within religious communities
(D) The various kinds of harassment, exploitation and violence against women

72. Ecological process of urban growth is:
(A) Invasion (B) Migration
(C) Natural growth (D) All of these

73. Name any reformers who advocated and worked to eastablish a society in which caste inequalities are absent:
(A) Jotiba Phule
(B) Dr. B.R. Ambedkar
(C) Periyar Ramaswami
(D) All of these

74. A town which serves the hinterland and return is served by the hinterland is called:
(A) Central place (B) Growth centre
(C) Main centre (D) Service centre

75. A person, who says that religion is the principal basis of community:
(A) Communalist
(B) Feminist
(C) Secularist
(D) Patriachal

76. Arousal during the time of sleep is:
(A) Zero (B) Low
(C) Medium (D) High

77. Most people are more energetic and alert at the high point of their:
(A) Jet lag
(B) Circadian rhythms
(C) Yerks Dodson law
(D) Shift Work

78. The term serial monogamy' refers to a type of monogamy where remarriage is:
(A) Not permitted after marriage has been terminated by death or divorce
(B) Permitted after the first marriage has been terminated by death or divorce
(C) Permitted but with husband's younger brother or wife's younger sister
(D) All of these

79. What is feminist movement?
(A) The movements which aimed at equality of men and women in all spheres of life
(B) The movements which related to women only
(C) The movements which related to man only
(D) None of these

80. Intrinsic motivation is closely related which of the hierarchical level of Maslow:
(A) Lower needs
(B) Middle needs
(C) Higher needs
(D) All of these

81. Urban areas formed by the fusion of towns through spreading industry and housing is called:
(A) In built area (B) Suburb
(C) Conurbation (D) Spreading

82. How to improve the women participation in politics:
(A) To make it legally binding to have a fair proportion of women in the elected bodies
(B) By reserving some seats in the Lok Sabha and the State Assemblies for women
(C) Political parties should also give due representation to women members
(D) By raising the literacy rate
(E) All of these

83. Poluandry is practiced by:
(A) Hos, Todas, Nambudris
(B) Todas, Kotas, Khasas
(C) Kotas, Gonds, Moplas
(D) Nadan, Ongis, Jarwas

84. A desire to meet some internalized standard of excellence is called:
(A) Need for achievement
(B) Need for power
(C) Need for security
(D) Need for self-actualization

85. What is communal politics?
(A) Organisation of people into social groups for the purpose of marriage, work and diet
(B) Using different regions in politics
(C) Both (A) and (B)
(D) None of these

86. An approach to study Urban Sociology is:
(A) Ecological (B) Community
(C) Associational (D) All of these

87. What is communalism?
(A) The situation when a particular community tries to promote its own interest at the cost of other communities
(B) The right to freedom of religion, our constitution provides to all citizens freedom to profess, practice and propagate any religion, or not to follow any
(C) The state to intervene in the matters of religion in order to ensure equality within religious communities
(D) The various kinds of harassment, exploitation and violence against women

88. Physical contact is considered as an example for:
(A) Primary motive
(B) Secondary motive
(C) Stimulus motive (D) All of these

89. Most of the activities we think of work are:
(A) Intrinsic reward
(B) Extrinsic reward
(C) Both (A) and (B) (D) None of these

90. What is equality:
(A) The situation when a particular community tries to promote its own interest at the cost of other communities
(B) The right to freedom of religion, our constitution provides to all citizens freedom to profess, practice and propagate any religion, or not to follow any
(C) The state to intervene in the matters of religion in order to ensure equality within religious communities
(D) The various kinds of harassment, exploitation and violence against women

91. Which of the following statements about India as a secular state is incorrect?
(A) Allows freedom to practice any religion
(B) There is no official religion
(C) Prohibits discrimination on religious grounds
(D) It reserves seats for religious minorities

92. What is domestic violence?
(A) The situation when a particular community tries to promote its own interest at the cost of other communities
(B) The right to freedom of religion, our constitution provides to all citizens freedom to profess, practice and propagate any religion, or not to follow any
(C) The state to intervene in the matters of religion in order to ensure equality within religious communities
(D) The various kinds of harassment, exploitation and violence against women

93. A common form of religion among tribes widely separated by time and space is what the sociologists call:
(A) Monotheism (B) Polytheism
(C) Totemism (D) Shintoism

94. Homogeneous societies exist in:
(A) India and USA
(B) India and Swedan
(C) Germany and Swedan
(D) Swedan and USA

95. The criminal tribe of Gurgaon is:
(A) Gowari (B) Bawaria
(C) Danteswari (D) Mewati

96. Rural and urban centres have co-existed in India, except for a brief interlude during:
(A) Indus valley
(B) Medieval period
(C) Rig-vedic period
(D) Gupta period

97. Maslow called the less powerful but humanly important actualization needs as:
(A) Basic needs (B) Growth needs
(C) Meta needs (D) All of these

98. Factors responsible for the origin of social difference:
(A) Birth (B) Choice
(C) Economic status (D) Religion
(E) All of these

99. The largest slum of Asia is:
(A) Dharavi (B) Kamatipura
(C) Basti (D) Ahata

100. On the basis of blood relationships a family may be:
(A) Conjugal or consanguineous
(B) Matrilineal or patrilineal
(C) Nuclear or extended
(D) Matrilocal or patrilocal

101. Reference groups are:
(A) Marine engineers
(B) Airforce pilot
(C) Nuclear scientists
(D) All of these

102. What are the overlapping social differences?
(A) When one social difference overlaps another difference
(B) When one social issue cross-cuts another social issue
(C) Either (A) or (B)
(D) Neither (A) nor (B)

103. Andamanese are:
(A) Negrito (B) Caucasoid
(C) Mongoloid (D) Alpian

104. What are cross-cutting social differences?
(A) When one social difference overlaps another difference
(B) When one social issue cross-cuts another social issue
(C) Either (A) or (B)
(D) Neither (A) nor (B)

105. The distribution of IQ in a population approximates to a:
(A) Bell curve
(B) Vertical curve
(C) Horizontal curve
(D) All of these

106. ——— is a system of government in which the power is divided between a central authority and various constituent units of the country:
(A) Federalism (B) Unitary
(C) Dectatorship (D) Democracy

107. Which of the following terms means providing security and stability?
(A) Stabilization (B) Synbolisation
(C) Mooring (D) Cherish

108. "Feminist movement" are aimed at:
(A) Liberty (B) Equality
(C) Participation (D) Power

109. Marriage of a woman to her husband's brother is known as:
(A) Sororate
(B) Polyandry
(C) Levirate
(D) Cross-cousin marriage

110. Which of the following is not a feature of federalism?
(A) There are two or more levels of government
(B) Different tiers of government govern the same citizen
(C) Sources of revenue for each level of government are clearly specified
(D) The central government can order the state government

111. The need that occupies the bottom most position in the hierarchy of Maslow:
(A) Psychological need
(B) Biological need
(C) Physiological need
(D) Physical need

112. Maslow believed that higher needs are expressed only when the physiological needs are:
(A) Satisfied (B) Dissatisfied
(C) Saturated (D) Stabled

113. ———— is a coming together federation:
(A) USA (B) India
(C) Pakistan (D) Sri Lanka

114. Ecology refers to the relationship between organisms and their:
(A) Environment (B) River system
(C) Purity of air (D) Fauna

115. In holding together federation the central government tends to be:
(A) More powerful than the state
(B) Weak
(c) Less powerful than the state
(D) A dictator

116. Air, food , water, sleep, sex etc cover the:
(A) Physiological need
(B) Biological need
(C) Psychological need
(D) Physical need

117. Indian Constitution provides for a ——— tier of government:
(A) Two (B) Three
(C) Four (D) Five

118. After 'Industrial Revolution' divorce rate have:
(A) Increased
(B) Decreased
(C) Remained the same
(D) Not taken place

119. ——— includes subjects of national importance such as defence, foreign affairs, banking, etc:
(A) State List
(B) Union List
(C) Concurrent List
(D) Residency List

120. Which of the following is not a subject of the Union List?
(A) Foreign affairs (B) Currency
(C) Banking (D) Law and order

121. Marriage is a:
(A) Folkway (B) More
(C) Social institution (D) Social norm

122. In local self government institutions atleast one third of all positions are reserved for:
(A) Men (B) Women
(C) Children (D) Scheduled caste

123. Successful economic development in complex societies require not only a balance between industrialization and agricultural development but also a balanced growth of;
(A) Distributive system
(B) Urbanization
(C) Foreign trade (D) Mordernization

124. Which of the following orders best represents the development of industrial society?
(A) Manorial system, guild system, domestic system and industrial society
(B) Guild system, manorial system, domestic system and industrial society
(C) Domestic system, guild system, manorial system and Industrial society
(D) Manorial system, domestic system, guild system and Industrial society

125. Disparity in income levels of different economic entities are measured on:
(A) Supply curve (B) Demand curve
(C) Cost curve (D) Lorenz curve

126. Which one of the following religious is not theorcentric?
(A) Hinduism (B) Christianity
(C) Buddhism (D) Islam

127. It is a council consisting of several ward members, often called the panch and a president or a sarpanch is:
(A) Gram Panchayat
(B) Panchayat Samiti
(C) Zila Parishad (D) Mayor

128. An essential condition for a social group is:
(A) Absence of stratification
(B) Social relation
(C) Mutual assistance
(D) Unintrupted organization

129. A few gram panchyats are grouped together to form:
(A) Gram Panchayat
(B) Panchayat Samiti
(C) Zila Parishad (D) Mayor

130. Primary condition for the formation of a social group requiring is voluntary membership is:
(A) Lack of stratification
(B) A common interest or objective
(C) Lack of any form of control
(D) Personal contact

131. A collection of people crossing one road at a signed point is a:
(A) Quasi group
(B) Social class
(C) Status group
(D) Secondary group

132. An elected Chairperson of the Municipal Corporation:
(A) Gram Panchayat
(B) Panchayat Samiti
(C) Zila Parishad
(D) Mayor

133. Which one of the following does not constitute an element of social structure?
(A) Status and Role
(B) Groups and Sub-groups
(C) Institutions
(D) Sub-culture

134. The rural local government is known as?
(A) Gram Sabha (B) Zila Parishad
(C) Panchayati Raj (D) Gram Panchayat

135. Teenage group is first and foremost an example of a:
(A) Peer group (B) Play group
(C) Football group (D) Secondary group

136. A point of similarity between human and animal society is that both:
(A) Want self-preservation
(B) Wish to perpetuate
(C) Have tendency to live in groups
(D) All of these

137. Functions of Gram Sabha is:
(A) It is the decision making body of the entire village
(B) The village panchayat works under the supervision of the Gram Sabha
(C) It approves the annual budget of the Gram Panchayat
(D) All of these

138. The formal structure of any organization includes:
(A) Personal roles and relations
(B) Cliques
(C) Delegation of authority
(D) Personal interests

139. What type of groups are united only be one purpose?
(A) Primary
(B) marginal
(C) Spatial
(D) Functional

140. The element of interaction is lacking in an:
(A) Integration (B) Aggregation
(C) Association (D) Assimilation

141. The change from the occupational role to the retirement role is an example of situation involving a degree of:
A) Role detachment
(B) Role discontinuity
(C) Preferential role
(D) Primary role

142. Important difficulties of the local government in India are:
(A) Most states have not transferred significant powers to the local governments
(B) There is a shortage of resources
(C) Elections are not held regularly
(D) The Gram Sabhas are not held regularly
(E) All of these

143. Secondary group is:
(A) City (B) Political party
(C) Labour union (D) All of these

144. The following four scholars contributes to structure and function relating to concepts and theories:
1. Robert K. Merton
2. Auguste Comte
3. T. Parsons 4. B. Mallinowski

Which one of the following shows the correct chronological order in which their contributions were made
(A) 1, 2, 3, 4 (B) 2, 1, 4, 3
(C) 2, 4, 3, 1 (D) 2, 3, 4, 1

145. Specific features of federalism:
(A) The power is divided between a central authority and its various constituent units
(B) Different tiers of the government govern the same citizens
(C) The fundamental provisions of the government cannot be unilaterally changed by one level of government
(D) It has a dual objective, i.e., to safeguard and promote the unity of the country, and also to accommodate the regional diversity
(E) All of these

146. The precedence of one role over another is called:
(A) Role primary (B) Role order
(C) Preferential role
(D) Primary role

147. When two people interact with each other, they form:
(A) Dyadic group (B) A society
(C) An institution (D) An association

148. Who coordinates the activities of all the Block Samitis in the whole district?
(A) Gram Sabha
(B) Panchayati Raj
(C) Zila Parishad
(D) None of these

149. State of relative isolation of the occupants of role that results from the tendency of persons occupying a given role is called:
(A) Role aloofness
(B) Role isolation
(C) Role insulation
(D) Isolation

150. 'Roles do not exist by themselves in isolation from one another, each role has its complimentary or associated role or roles' This statement may well define the concept of:
(A) Role-set
(B) Role congruence
(C) Multiple role
(D) Statuses

151. Specific features of the third tier government is:
(A) Locals have better knowledge
(B) Direct participation of the people
(C) Foundation of democracy
(D) Reduction of burden of the Central Government
(E) All of these

152. Which of the following country witnessed an extraordinary popular movement to restore democracy:
(A) Pakistan (B) India
(C) Nepal (D) Bolvia

153. Consider the following statements:
The dangers of ethnocentrism are very grave in India effects of bringing unity among people of India are defeated because ethnocentrism:

1. Involves a double moral standard – one for in-group and another for out-group
2. Distorts self and originally of every individual
3. Makes individuals to look at the culture of out-groups with hatred an even contempt

Which of these statements are correct?
(A) 1, 2, 3 (B) 1, 2
(C) 2, 3 (D) 1, 3

154. Process of socialization starts:
(A) After birth of the child
(B) Before child's birth
(C) After getting maturity
(D) In youth only and not earlier than that

155. Match list I with list II and select the correct answer:
List I (Concepts of functional theory)
a. Functional Imperative
b. Net balance of functional consequence
c. Universal functionalismd. Positivistic organicism
List II (Scholar associated with concepts)
1. R. K. Merton 2. B. Mallinowski
3. Auguste Comte 4. T. Parsons
5. Marion J Levy
Codes:

	a	b	c	d
(A)	4	3	1	2
(B)	1	2	3	4
(C)	4	1	2	4
(D)	5	3	1	4

156. A society in which cultural groups are not forced into assimilation, but are able to coexist with one another in mutual respect and with a fair share of the national resources is described as;
(A) A melting pot (B) Democratic
(C) Pluralistic (D) A mosaic

157. Which one the following is the best criterion of contrast between a rural and urban community:
(A) Geographical location
(B) Specific age of population
(C) Specific density of population
(D) Diversification of occupation

158. Which of the following king was killed under mysterious mass care of the royal family in 2001?
(A) King Gyanendra (B) King Birendra
(C) King Koirala (D) King Puspa Kamal

159. Which of the following statements is more correct:
(A) Sociology is a natural science
(B) Sociology is a social science
(C) Sociology is an applied science
(D) Sociology is a normative science

160. The relation existing between a type writer and a desk can be called:
(A) Material (B) Cultural
(C) Physical (D) Structural

161. "The detachment of the performer from the role he/she is playing/performing' may be categorized as:
(A) Role detachment (B) Aloofness
(C) Role –distance (D)Distant role

162. Origin of society was due to:
(A) Force (B) God's will
(C) Evolution (D) Individual

163. Social function may be defined as:
(A) A social event to which many important people are invited
(B) An activity or consequence associated with a structure
(C) A responsibility which is assigned to a person or group of persons
(D) Any event which is considered important by the members of a society

164. Which of the following characteristics distinguishes man from the other animals?
(A) Ability to stand erect
(B) Ability to adapt to the environment
(C) Ability to make tools
(D) Ability to live in group

165. What does society exclude:
(A) Differences (B) Interdependence
(C) Reciprocity (D) Time boundness

166. The concept of 'Sanskritization' is closely related to the concept of:
(A) Reference group (B) Primary group
(C) Interest group (D) Tertiary group

167. A social situation where two or more conflicting groups resolve not to fight further to adjust with each other leaving the causes of conflict unresolved and agreeing to get certain benefits on mutual basis is conceptualized as the process of:
(A) Cooperation (B) Adjustment
(C) Integration (D) Accommodation

168. If society has to depend on external force in social control it shows its:
(A) Weakness rather than strength
(B) Strength rather than weakness
(C) Unity rather than separateness
(D) Separationess rather than unity

169. The unity and stability of Indian society apart from its hydraulic nature depends upon:
(A) Religion and culture
(B) Caste and religion
(C) Caste and clan
(D) Practice of untouchability

170. Institutions are characterized by:
(A) Social relationship
(B) Approval by culture
(C) Competition (D) Both (A) and (B)

171. Youth is a:
(A) Group (B) Congregation
(C) Age group (D) Social category

172. The national culture which recognize the importance of family:
(A) Roman (B) Indian
(C) American (D) China

173. A tight cluster of houses surrounded by the field of the villagers is:
(A) Scattered settlement
(B) Linear settlement
(C) Nucleated settlement
(D) Circular settlement

174. Ceremonies that mark a critical transition in life of an individual from one phase of life cycle to another are called:
(A) Role taking (B) Rites of passage
(C) Status crystallization
(D) Status-set

175. The first social group in modern life that importance self sufficiency:
(A) Neighborhood (B) Family
(C) Community (D) Village

176. Competition can be seen at which of the following five levels?
(A) Economic, cultural, society, political, racial
(B) Economic, normal, cultural, political, ethnic
(C) Economic, educational, religious, cultural, political
(D) Economic, educational, social, religious, political

177. Value judgement is more in:
(A) Folkways (B) Rituals
(C) Mores (D) Taboos

178. Lack of congruity between expectations associated with social roles taken on consecutively by an individual is called:
(A) Role detachment
(B) Role discontinuity
(C) Distant role (D) Consequence

179. Behaviour of an individual in a particular role providing a pattern or model upon which another individual bases his behaviour in performing the same role is called:
(A) Role model (B) Role pattern
(C) Role actor (D) Model

180. The feature which distinguish youth culture is that:
(A) It is a culture of leisure than work
(B) Social relationships are organized around the peer groups, not families
(C) Youth groups are particularly interested in style
(D) All of these

181. Scattering of homestead settlement are not found in:
(A) Himalayan foothills
(B) Highlands of Gujarat
(C) Satpura range of Maharastra
(D) Plains of Uttar Pradesh

182. In the class system, the status is:
(A) Ascribed (B) Achieved
(C) Hereditary (D) Acquired

183. An informal group of individuals meeting regularly to indulge in common activities which may be outside the framework of law is called:
(A) Syndicate (B) Gang
(C) Occult (D) Clique

184. Socialisation is a matter of:
(A) Learning (B) Biological inheritance
(C) Socialising (D) Division of labour

185. Monomorphic opinion leadership is prevalent in:
(A) Developed societies
(B) Developing societies
(C) Under developed societies
(D) Democratic societies

186. An essential element for the existence of a society is:
(A) High level of education
(B) Belief in common political ideology
(C) Sociability among the members
(D) Share in private property

187. The most important reason for which man needs society is that he:
(A) Wants to live in the world
(B) Needs it both for the sake of life as well as good life
(C) Wants to become an ideal citizen
(D) Feels that society is a super power

188. A type of marriage known as Levirate is the one that takes place between man and a:
(A) Sister and his deceased wife
(B) Widow of his deceased brother
(C) Woman of inferior caste
(D) Woman of superior caste

189. Family is included under:
(A) Primary group (B) Secondary group
(C) Pertaining group (D) None of these

190. Which among the following is not the element of the social system:
(A) Belief (B) Sentiment
(C) Norms (D) Slaves

191. Maturation is primarily:
(A) A social process
(B) Analogous to socialization
(C) Responsible for accelerating process of socialization
(D) A physical process

192. The term 'situs' is used to denote differentiation in society evaluated in terms of:
(A) Superiority or inferiority
(B) Rich and poor
(C) Educated and uneducated
(D) Healthy and sick

193. "Reflexive role taking":
(A) Is taking the role of another by viewing one self from the point of view of the other
(B) Allows a person to become object in himself as other's see him
(C) Concept is similar to the Cooley's concept of the looking-glass self
(D) All of these

194. When, due to the attraction of new opportunities, people migrate to the town it is called:
(A) Pull factor of migration
(B) Push factor of migration
(C) Both (A) and (B) (D) None of these

195. Barter is an exchange of:
(A) Service for service
(B) Goods for service
(C) Goods for goods
(D) All of these

196. Muslim marriage the amount to be paid to the bride by the bride-groom at the time of divorce is known as:
(A) Talaq (B) Mehar
(C) Khula (D) Dahej

197. Consider the following pairs are correctly contrasted:
1. Primogenitures : Ultimogeniture
2. Exogamy : Endogamy
3. Family of orientation : Family of procreation
4. Residually joint family : potentially joint family

Select the correct answer from below:
(A) 1, 2, 3 (B) 1, 2
(C) 1,3, 4 (D) 1, 2, 3, 4

198. Which one of the following is not the basic element of a community:
(A) Relationship with environment
(B) Legal binding
(C) Relationship with territory
(D) Psychological feelings

199. The groom does not leave his house hold in:
(A) Patrilocal residence
(B) Matrilocal residence
(C) Avunculocal residence
(D) Neo-local residence

200. The main objective of Hindu marriage is:
(A) Procreation
(B) Sexual gratification
(C) Dharma
(D) Group formation

201. Society is a web of social relationship according to:
(A) Maclver
(B) Gidding
(C) Ginsberg
(D) Cooley

202. Society is not a make, but a growth. This view is related to:
(A) Evolutionary theory
(B) Divine origin theory
(C) Matriarchal theory
(D) Patriarchal theory

203. Which type of residence does not help the formation of extended family?
(A) Virilocality (B) Uxorilocality
(C) Avunculocality (D) Neolocality

204. In the early colonial period in the urban areas the structure of economic opportunities of people in the urban areas was;
(A) Minimised
(B) Enlarged
(C) Remained the same
(D) Constantly in flux

205. An in-group is not characterized by:
(A) Jealousy and competition
(B) Intimate relationship
(C) Sense of achievement
(D) Element of sympathy

206. In hindi, the only classificatory kinship terminology is:
(A) Samaadhin
(B) Damaad
(C) Bahu
(D) Chacha

207. Married couples go to live with a brother of the groom's mother in:
(A) Patrilocal residence
(B) Uxorilocal residence
(C) Avunculocal residence
(D) Neo-local residence

208. Which among the following is a reference group:
(A) Doctors
(B) Nuclear scientists
(C) Marine engineers
(D) All of these

209. Marriage is considered to be necessary because it:
(A) Is only way to acquire paternal property
(B) Is the only way of sexual gratification
(C) Provides societal sanction
(D) Is regularized and provides socially sanctioned sex gratification

210. The practice of levirate is found in:
(A) Matrilineal societies
(B) Patrilineal societies
(C) Bi-lineal societies
(D) All of these

211. The criticism against sociology having its own-subject matter is:
(A) Sociology is merely a mixture of varied study disciplines having social content
(B) The subject matter of sociology is disintegrated into various other social sciences
(C) Subject matter of sociology has been derived from so many social sciences
(D) All of these

212. Usefulness of language is that it help in:
(A) Increase of knowledge
(B) Making social contacts easy
(C) Preservation of cultures
(D) All of these

213. Marriage within the caste is called:
(A) Endogamy (B) Exogamy
(C) Sagotra (D) Hypergamy

214. Social stratification implies:
(A) Social justice (B) Social inequality
(C) Social injustice (D) Social equality

215. The rule whereby the married couple establish their own residence is called:
(A) Avunculocal residence
(B) Bilocal residence
(C) Patrilocal residence
(D) Neolocal residence

216. Rural community react towards a stranger by:
(A) Hostility
(B) Hospitality
(C) Either hospitality or hostility
(D) Indifference

217. A large number of reference groups are a typical characteristic of:
(A) Large, complex societies
(B) Small, simple societies
(C) Tribal societies
(D) All of these

218. The simplest definition of urban social structure has been given in terms of:
(A) Demographic (B) Industry
(C) Education (D) Professions

219. The caste system is a:
(A) Social institution
(B) Religious institution
(C) Economic institution
(D) Political institution

220. Which one of the following defines family which is created as a result of marriage and the off springs consequently born?
(A) Family of orientation
(B) Family of procreation
(C) Extended Family
(D) Joint Family

221. The concept of cultural lag explains that there is faster rate of change in the:
(A) Economic sector of society in comparison to non-economic sectors
(B) Political domain of society in comparison to economic domain
(C) Technological sector of society in comparison to cultural sectors
(D) Religious activities of society in comparison to cultural activities

222. An important limitations of our social life in modern times is:
(A) Process of standardization
(B) Growth of agriculture
(C) Growth of science and technology
(D) Growth of industry

223. Indicíduality in sociological sense means:
(A) Depicting strength of character
(B) Physical detachment of two units
(C) Depicting strength of character as an individual
(D) Physical attachment of two units

224. The cause of social standardization is:
(A) Advancement of mass media
(B) Growth of means of communication
(C) Growth of means of transportation
(D) All of these

225. In which of the following systems flexibility of mobility exists:
(A) Estate system (B) Caste system
(C) Class system (D) None of these

226. Nuclear family in its volarized form consists of:
(A) Parents of the same children and children of the same parents
(B) Parents of the different children and children of the same parents
(C) Parents of the same children and children of the different parents
(D) None of these

227. Preferential marriage with the father's sister's daughter occurs among:
(A) Garos of Assam
(B) Gonds of MP
(C) Bhutias of UP
(D) Kadar of South India

228. 'Makka parage marriage' is:
(A) Is performed when an okka does not have son to continue the line
(B) Is one in which man does not leave his natal okka to become member of wife's okka
(C) Is 12 pieces of gold not transferred
(D) All of these

229. If maternal uncle comes first among all male relatives, then this kinship usage is called:
(A) Maternity (B) Avunculate
(C) Mamalate (D) Maternealate
(E) All of these

230. In double descent system, one inherits property from:
(A) Father's matrilineal relatives and mother's patrilineal relatives
(B) Father's patrilineal relatives and mother's matrilineal relatives
(C) Father's patrilineal and matrilineal relatives
(D) Mother's matrilineal and patrilineal

231. What does culture normally reflects in modern society:
(A) Group culture
(B) National culture
(C) Religious culture
(D) Geographical culture

232. The system in which a woman is shared by the ahole group is called:
(A) Group marriage (B) Levirate
(C) Polyandry (D) Endogamy

233. When individual stands up against the group, the process is called:
(A) Conflict (B) Competition
(C) Alienation (D) Unsocialization

234. Mark out the factors responsible for cohesion of rural society:
(A) Common traits
(B) Common objectives
(C) Similar experiences
(D) All of these

235. To marry a woman of a higher caste is called:
(A) Hypogamy (B) Hipergamy
(C) Castegamy (D) Sororate

236. What differentiates state from nation:
(A) Territory (B) Government
(C) Sovereignty (D) Political unity

237. The practice of being mate, actual or potential, to one's husband's brother is called:
(A) Family marriage (B) Levirate
(C) Moeity (D) Continuity

238. Sharda Act (1929) is:
(A) Sati Prohibition Act
(B) Widow Remarriage Act
(C) Age of Consent Act
(D) Child Marriage Restraint act

239. Which of the following is the basis of Gadhian state:
(A) Non-violence
(B) Force
(C) Satyagraha
(D) Confederation

240. When individual is ex-communicated or membership of the group is prohibited for him/her, the process is called:
(A) Isolation
(B) Cooperation
(C) Tolerance
(D) Discrimination

241. Which one of the following does not explain why domestic violence is so common place?
(A) Emotional instability within the family
(B) Good deal of violence in family is even tolerated and approves
(C) Social approval of wife battering
(D) Violence within the family reflects broader patters violent behaviour is society

242. Rampur village study conducted by Oscar Lewis was about:
(A) Caste system in Indian village
(B) Jajmani system
(C) Redistribution in village economy
(D) Agricultural production

243. The social norms:
(A) Give cohesion to a society
(B) Lead society to disintegration
(C) Make society rigid
(D) Destroy the freedom of the individuals

244. Consider the following two statements:
Assertion (A): Joint families are gradually disappearing from the rural social scene
Reason (R): Due recognition is not given to individual achievement in joint families
Select the correct answer:
(A) Both A and R are true and R is the correct explanation of A
(B) Both A and R are true by R is not the correct explanation of A
(C) A is true, R is false
(D) A is false, T is true

245. Conformity to norms depends on:
(A) Effectiveness of sanctions
(B) Ideal character of the norms
(C) Cohesiveness of the group
(D) Sense of cooperation in the group

246. Lower-class culture illustrates:
(A) Sub-culture (B) Counter culture
(C) Dominant culture
(D) Contra culture

247. Society is the total social heritage of folkways, mores and institutions, of habits, sentiments and ideals. This is which view of society?
(A) Structural (B) Cultural
(C) Functional (D) None of these

248. Conformity to norms in industrial society in comparison to agrarian society is:
(A) Less (B) More
(C) Equal (D) Exceptional

249. The individual united with the group is the process:
(A) Integration (B) Socialization
(C) Cooperation (D) Accommodation

250. To process of learning that takes place in a group situation is called:
(A) Socialization
(B) Culturisation
(C) Routinisation
(D) Acculturation

251. Cognitive culture consists of the which aspects of a cultural system?
(A) Mental (B) Social
(C) Economical (D) Religious

252. Role conflict is a major phenomenon of:
(A) Primitive society
(B) Agrarian society
(C) Tribal society
(D) Industrial society

253. Consider the following two statements:
Assertion (A): Today there is strong emphasis on economic growth accompanied by social justice
Reason (R): The earliest emphasis on economic growth has led to wider disparities between rich and poor, rural and urban people in India
Select the correct answer:
(A) Both A and R are true and R is the correct explanation of A
(B) Both A and R are true by R is not the correct explanation of A
(C) A is true, R is false
(D) A is false, T is true

254. Which of the following cannot be called an institution:
(A) Family (B) Church
(C) Political party (D) Student union

255. 'Potlach' is an instance of:
(A) Status competition
(B) A system of setting debts
(C) A system of capital accumulation
(D) A system of ceremonial exchange of goods

256. Consider the following characteristics:
1. Economic relations based on barter and exchange
2. Manufacturing of capital goods
3. Absence of specialization based on technical ability
4. Absence of regular market

Select the correct answer from below:
(A) 1, 2, 3 (B) 1, 2
(C) 1, 3, 4 (D) 1, 2, 3, 4

257. Which of the following is not a characteristic of a social group:
(A) Reciprocal relations'
(B) Common interests
(C) Definite we feelings
(D) Similar ethnic background

258. What cause guilt:
(A) Depression
(B) Violation of morale
(C) Repressed depression
(D) Eroticism

259. Relationship between individual and culture is that in which:
(A) Individual is simultaneously product, carrier and creator of the culture
(B) Society and culture accounts for similarities of personality among the individuals but not individuality of personality
(C) Total complex of culture determines the behvaiour of the individuals
(D) Neither individual creates culture nor does culture create the individual

260. From which point of view, society is defined as a complex of groups in reciprocal relationship, interacting upon one another and helping each person to fulfil his wishes in association with his fellows?
(A) Societal
(B) Cultural
(C) Functional
(D) Structural

261. Which of the following methods cannot be used to reach an agreement in a primary group:
(A) Authority
(B) Compromise
(C) Voting
(D) Coercion

262. Match list I with list II and select the correct answer:

List I (Books)
a. History of Trade Unionism
b. Labour problems in Indian Industry
c. The new men of power
d. Union democracy

List II (Authors)
1. V.V. Giri
2. S.M. Lepset
3. C.W. Mills
4. C.W, Punekar
5. Sydney and Beatrice Webb

Codes:

	a	b	c	d
(A)	4	3	1	2
(B)	1	2	3	4
(C)	5	1	3	2
(D)	5	3	1	4

263. In secondary group, membership is:
(A) Voluntary
(B) Compulsory
(C) Spontaneous
(D) Abrupt

264. "Looking glass self". A self development concept was developed by:
(A) Cooley (B) G.M. Mead
(C) S. Freud (D) Mac Iver

265. Consider the following two statements:

Assertion (A): In Amerian culture, the term 'uncle' is used to refer to four kind of relatives

Reason (R): For American, the distinction between different kind of uncles is relatively unimportant

Select the correct answer:

(A) Both A and R are true and R is the correct explanation of A
(B) Both A and R are true by R is not the correct explanation of A
(C) A is true, R is false
(D) A is false, T is true

266. Which of the following is not a feature of joint family:

(A) Joint property
(B) Common religious
(C) Common occupation
(D) Small size

267. Consider the following two statements:

Assertion (A): In an industrial concern, when the workers do not get their proper wages, they are likely to protest and go on strikes after getting themselves organised

Reason (R): The trade unions are formed on the nasis of common interests of the workers and their organized associational efforts in the form of movements

Select the correct answer:

(A) Both A and R are true and R is the correct explanation of A
(B) Both A and R are true by R is not the correct explanation of A
(C) A is true, R is false
(D) A is false, T is true

268. Match list I with list II and select the correct answer:

List I (Categories)

a. Agriculture b. Industrial
c. Pastoral d. Hunting

List II (Stages of evolution)

1. First 2. Second
3. Third 4. Fourth

Codes:

a	b	c	d	
(A)	4	3	1	2
(B)	1	2	3	4
(C)	2	4	3	1
(D)	5	3	1	4

269. Which of the following is not a characteristic of caste:

(A) Social and religious hierarchy
(B) Restrictions in social intercourse
(C) Freedom of marriage
(D) Endogamy

270. When norms of the social system are more modern, opinion leadership is more of:

(A) Polymorphic type
(B) Neomorphic type
(C) Multimorphic type
(D) Onomorphic type

271. Match list I with list II and select the correct answer:

List I

a. Endogamy
b. Patriarchial family
c. Karta
d. Anuloma

List II

1. Head of the Family
2. Marriage outside the caste
3. Marriage inside the group
4. Male is the head of the family
5. Marriage of upper caste of man with a lower caste woman

Codes:

	a	b	c	d
(A)	4	3	1	2
(B)	1	2	3	4
(C)	3	4	1	5
(D)	5	3	1	4

272. Caste system is harmful because:

(A) It denies social mobility
(B) It hinders national unity
(C) It prohibits inter-caste marriage
(D) It separates social and political life

273. Sociology is concerned mainly with the study of:
(A) Ethical improvement of society
(B) Structure, function changes in society
(C) Evolution of high social consciousness
(D) Social and cultural reformation

274. The principle according to which the tasks of each worker are simplifies and integrated with those of other workers is:
(A) Babbage principle
(B) Peters principle
(C) Parkinson's law
(D) Taylorism

275. Sociogram is best described as a sociological:
(A) Concept (B) Method
(C) Theory (D) Technique

276. The specialization of work tasks by which different occupations are combination within a production system is called:
(A) Production system
(B) Production relations
(C) Mode of population
(D) Division of labour

277. "Individuality arises only if the commonly recedes" was said by:
(A) Max Weber (B) Karl Marx
(C) Emile Durkhei (D) Aldous Huxley

278. A polygraph machine is used to determine:
(A) Physiological changes
(B) Why a person is lying
(C) Which emotion a person is expressing
(D) Whether a person has criminal tendencies

279. The memories impression left on the brain are called:
(A) Memory images (B) Memory traces
(C) Memory maps (D) Memory potentials

280. The process by which the individual learns to conform to the norms of the group is called:
(A) Social stratification (B) Socialization
(C) Social interaction (D) Social dynamics

281. 'Culture is the handiwork of man and the medium through which he achieves has/her ends'. Who hold this?
(A) G.H. Mead (B) F. Boas
(C) Kroeber (D) Malinowski

282. First scientific step towards growth of language is believed to be:
(A) Voice of sages (B) Imitation
(C) Symbol (D) Emotions

283. A family of orientation is one:
(A) In which the individual is born
(B) Establishes by marriage
(C) Consists of husband and wife only
(D) Husband, wife and their children

284. Consider the following two statements:
Assertion (A): Industrial revolution brought about decline motivation and sense of personal responsibility
Reason (R): It replaced independent proprietorship with dependence on employer
Select the correct answer:
(A) Both A and R are true and R is the correct explanation of A
(B) Both A and R are true by R is not the correct explanation of A
(C) A is true, R is false
(D) A is false, T is true

285. Which of the following are the salient features of a market economy?
1. Existence of black money and white money
2. Existence of agriculture and industry
3. Existence of commercial agriculture with subsidiary farming
4. Modern industry and commercial agriculture co-existing with subsistence farming and traditional handicrafts

Select the correct answer from below:
(A) 1, 2, 3 (B) 1, 2
(C) 2, 3, 4 (D) 1, 2, 3, 4

286. The most important element in the institution of joint family is:
(A) Living in the same house
(B) Sharing of common kitchen
(C) Sharing of sense of belongingness
(D) Pursueing the same occupation

287. Consider the following two statements:
Assertion (A): One's affinal relations may also be one's consanguineal relations
Reason (R): Generaligical network in simple societies does not exhaust kinship network
Select the correct answer:
(A) Both A and R are true and R is the correct explanation of A
(B) Both A and R are true by R is not the correct explanation of A
(C) A is true, R is false
(D) A is false, T is true

288. For development of modern sapitalism, it is necessary to have:
(A) Formal and value rationalities
(B) Formal and instrumental rationalities
(C) Instrumental and value rationalities
(D) Instrumental rationality

289. Thanophobia is:
(A) Fear of fire (B) Fear of death
(C) Fear of water (D) Fear of animals

290. Incompatibility between two or more different roles that an individual is expected to play is termed:
(A) Competition (B) Role set
(C) Role conflict (D) Role strain

291. Which one of the following is not as characteristic feature of the production systems of pre-literate societies?
(A) Absence of organized labour market
(B) Cooperativeness
(C) Non existence of monetary incentives and rewards
(D) Put ownership of the mean of production

292. 'Social Mobility' in the villages is less because:
(A) Life is static there
(B) People are orthodox
(C) Geographical mobility is absent there and there is no sizeable division of labour because of lack of industries, etc
(D) Government does not help them economically

293. The total output of goods and services in a national economy is called:
(A) Gross national product
(B) Total income
(C) Net income
(D) Gross domestic product

294. The form of agronomy, in which land is broken with a digging stick or not with consequent limitation upon the amount of food that can be produced is called:
(A) Agriculture (B) Horticulture
(C) Pastoral (D) Food gathering

295. Socialism is not connected with:
(A) Private property should be abolished
(B) Means of production and distribution must be in the hands of the community
(C) Defects of social organization arise from private ownership and the desire of profit
(D) Laissez faire

296. Capitalism has:
(A) Stimulated progress
(B) Widened the gap between the employer and employee
(C) Brought a civilization without culture
(D) All of these

297. Projection is:
(A) Accusing another person for your weaknesses
(B) Finding a safe victim for your aggression
(C) Distorting truth to find excuse
(D) Day dreaming and escaping into a make believing world

298. Sensation and knowledge is:
(A) Attention (B) Remembering
(C) Recognition (D) Perception

299. The practice of a man varying more than one women is termed:
(A) Polygamy (B) Polyandry
(C) Hypergamy (D) Polygamy

300. The industrial system of production is distinguished from other systems on the basis of availability of:
(A) Skilled labour
(B) Means of production
(C) Capital
(D) Inanimate source of energy

301. Which of the following, according to Perlman, accounts for the rise of trade unions?
(A) Need for identity
(B) Job scarcity
(C) Exploitation
(D) Need for solidarity

302. The sociogram is most useful in representing a persons:
(A) Objective class position
(B) Opinion about public issues
(C) Personality structure
(D) Informal relations with a structure

303. The custom or practice of identifying a father through one of his children is known as:
(A) Couvade (B) Teknonyiny
(C) Linkage (D) patrimony

304. In which of the following sections in India the joint family is mostly absent?
(A) Urban business class
(B) Professional class
(C) Landless wage-earners
(D) Rural land owning class

305. Materilineality in India today is confined to:
(A) Kerala only
(B) South and Bihar
(C) South and Assam
(D) South and Gujarat and parts of MP

306. Consider the following two statements:
Assertion (A): In all societies avoidance of one kind or another is observed between certain relatives
Reason (R): The basic purpose of such avoidance is to prevent such sexual intimacy as would amount to incost
Select the correct answer:
(A) Both A and R are true and R is the correct explanation of A
(B) Both A and R are true by R is not the correct explanation of A
(C) A is true, R is false
(D) A is false, T is true

307. Socializing persons into basic values and practices of society is the characteristic function of:
(A) Family institution
(B) Economic institution
(C) Educational institution
(D) Political institution

308. Assimilation:
(A) Bring about cultural diffusion
(B) Is a historical process
(C) Has culture contact
(D) All of these

309. Intuitive, association or personal thinking is called:
(A) Inductive thinking
(B) Deductive thinking
(C) Illogical thinking
(D) Logical thinking

310. Social change refers to change in:
(A) Value system (B) Dress
(C) Food habits
(D) Habitation patterns

311. The word 'consanguineal' literally means:
(A) Living together
(B) Kith and kin
(C) Of one totemic group
(D) Of one blood group

312. Which of the following is a primary group:
(A) Family (B) Students of sociology
(C) Social workers (D) All of these

313. Secondary kin is:
(A) Sister's husband
(B) Brother's wife
(C) Wife's brother (sala)
(D) All of these

314. The change in social value is really:
(A) Change in soft point of culture
(B) Change in hard point of culture
(C) Social taboo
(D) Ethnocentrism

315. Which sociology emphasizes that sociology should be emancipated from the physical sciences and develop its own thing on the basis of social data?
(A) Proto Sociology
(B) Historical Sociology
(C) Scientific Sociology
(D) Industrial Sociology

316. Autophobia is:
(A) Fear of heights (B) Fear of men
(C) Fear of water (D) Fear of being alone

317. An attitude arrived at without sufficient exploration of the facts is:
(A) Prejudice (B) Stereotype
(C) Public opinion (D) Perception

318. With increase of productive forces:
(A) Man's mastery over natural increases
(B) Man becomes more conscious of nature
(C) Man turns into guardian of nature
(D) Man becomes enslaved to nature

319. Which of the following relations, in its primary from is not the incest taboo?
(A) Father and daughter
(B) Mother and son
(C) Brother and sister
(D) Uncle and niece

320. Cultural relatively refers to:
(A) Different in standards of measurement in culture
(B) Cultural lag
(C) Time gap between cultural development
(D) None of these

321. A person who actually initiates the action in the community is:
(A) Professional leader
(B) Assured leader
(C) Operational leader
(D) Popular leader

322. Which of the following can be considered as a force of production?
(A) Tractor (B) Steam engine
(C) Wind mill (D) All of these

323. The term role-set has been coined to refer of the:
(A) Total complex of other social positions with which any particular social position is characteristically connected
(B) Total number of occupants of the correlative social positions when a particular occupant of a particular position is taken as a point of reference
(C) Both (A) and (B)
(D) None of these

324. The unification of individual's Socio-psychological behaviour is reflected in:
(A) The way individual behaves
(B) His personality
(C) His social interaction
(D) His culture

325. Material forces of production are:
(A) Continuously expanding
(B) More or less stagnant
(C) Moving towards scarcity
(D) Potentially distractive

326. Which of the following is an informal method of social control:
(A) Belief (B) Coercion
(C) Law (D) Education

327. The 'production' is influenced by:
(A) Social structure (B) Technology
(C) Culture (D) All of these

328. Concept of social evaluation was first given by:
(A) Augste Comte
(B) Herbert Spenscer
(C) Emile Durkheim (D) Lester C Ward

329. Transformation of charismatic leadership into institutionalized leadership is:
(A) Routinisation (B) Institutionalisation
(C) Categorisation (D) Politicicisation

330. A society does not include:
(A) Differences (B) Interdependence
(C) Reciprocity (D) Time boundness

331. Group of lineal kinsman is:
(A) Grand father-grand mother, father-mother, son-daughter and her husband
(B) Mother's mother, mother's father, son – daughter, uncles
(C) Grand father-grand mother, father-mother, son-daughter
(D) Grand father, maternal grand father, maternal uncle, paternal uncle

332. When the trade and money become developed activities, an institution becomes
(A) More varied (B) Less varied
(C) More dynamic (D) Less dynamic

333. Consider the following group:
1. A political party of ones choice
2. Ones religion
3. Ones enemy
4. Ones favourite teacher
Among these groups, reference group (s) would includes
(A) 1, 2, 3, 4 (B) 1, 3, 4
(C) 2, 3, 4 (D) 2, 4

334. Social change means a change in:
(A) Occupation
(B) Administration
(C) Standards of living
(D) Social structure

335. Specific ways of looking at and understanding are:
(A) Perspective (B) Theory
(C) Model (D) School

336. The idea of sociometry was first put forward by:
(A) Fisher
(B) Osgood
(C) Mereno
(D) William Stephenson

337. Factor which led to the weakening of working class power and unionism is:
(A) Capital (B) Technology
(C) Voluntary organization
(D) Secondary organization

338. The process by which culture traits spread from one culturally distinct group to another is known as:
(A) Evolution (B) Assimilation
(C) Accommodation
(D) Diffusion

339. Two or more individual who mutually choose each other in sociometry:
(A) Fraternal twins (B) Identical wins
(C) Semantic twins (D) Clique

340. Which one of the following sequences occur in cultural change?
(A) Diffusion, Assimilation, Acculturation
(B) Acculturation, Contraculturation, Assimilation
(C) Contraculturation, Assimilation, Diffusion
(D) Diffusion, Acculturation, Assimilation

341. Social vacuum theory was proposed by:
(A) Sorokin (B) Mc Gregor
(C) Curt Lewin (D) Gilford Pinchot

342. A grouping together of workers to assert/demand their rights is called:
(A) Consciousness (B) Association
(C) Unionism (D) Assertiveness

343. Training given to a person for acquiring membership in a non-membership group is called:
(A) Conformity (B) Deviance
(C) Re-socialisation
(D) Anticipatory socialization

344. 'Khulah' among Muslims means:
(A) When wife demands divorce
(B) When husband demands divorce
(C) When husband has to pay Mehar
(D) When wife has to pay Mehar

345. Anomie means:
(A) Anomaly (B) Anonymous
(C) Lack of rules (D) Normative

346. Rural economy in India goes back to the:
(A) Indus valley civilization
(B) Vedic civilization
(C) Harsha Vardhana
(D) Maurya period

347. Materialization must start from:
(A) Man (B) Individual
(C) Group (D) Society

348. Cultural relativism means that:
(A) There are no universal cultural norms that we must follow
(B) One's view of social institutions is coloured by his cultural perspective
(C) The function and meaning of a that are relative to its cultural setting
(D) One believes that one's own culture is superior than others

349. Conditional stimulus is otherwise known as:
(A) Neutral stimulus
(B) Specific stimulus
(C) General stimulus
(D) Applied stimulus

350. Malinowski was of the opinion that which basic need impelled man to action?
(A) Cultural (B) Social
(C) Biological (D) Psychological

351. The panchayat president in a village is an example for:
(A) Professional leader (B) Lay leader
(C) Action leader (D) None of these

352. A town which serves its hinterland and in return is served by the hinterland, is called a:
(A) Service Centre (B) Growth Centre
(C) Growth Pole (D) Central Place

353. Psychology studies the behaviour of:
(A) Human (B) Animal
(C) Human and animal
(D) None of these

354. The values and moral standards of the child is mostly influenced by:
(A) Teachers (B) Friends
(C) Parents (D) None of these

355. The total sum of the means which provide objects required for human sustenance and comfort is called:
(A) Factory
(B) Relation of production
(C) Technology (D) Guild

356. A kinship group is called a broad range or narrow range one on the basis of:
(A) Number of persons it includes
(B) Number of lineal kins
(C) Closeness of relations
(D) Types of the group

357. Among Hindu the group initiating marriage is:
(A) Boy's group
(B) Girl's group
(C) Both take initiatives
(D) None of these

358. Main objective of 'Nikah' among Muslims is:
(A) Religious sanctification of marriage
(B) Procreation and legalization of children
(C) Finding groom for daughter/sister
(D) To purify the bride and groom

359. Psychology was first defined as:
(A) Science of soul
(B) Science of mind
(C) Science of consciousness
(D) Science of unconsciousness

360. The legal right over the use of his/her invention is called:
(A) Law of return (B) Social obligation
(C) Law of patents (D) Social security

361. The object of a motivated and directed sequence of behaviour is called:
(A) Drive (B) Need
(C) Response (D) Goal

362. When a special role is given to the father's sister, it is called:
(A) Avunculate (B) Amitate
(C) Potential relation (D) Patriarchal

363. Psychology originated from:
(A) Physiology (B) Philosophy
(C) Biology (D) Sociology

364. A person or group which has a great amount of wealth is called:
(A) Affluent (B) Capitalist
(C) Owner (D) Entrepreneur

365. The right to give a maiden in marriage is held in orderly manner by the group:
(A) Grand father, father, mother, brothers, kinsman
(B) Father, mother, grand father, kinsman, brother
(C) Father, mother, grand father, kinsman, brother
(D) Father, grand father, mother, brother kinsman

366. Discovery of new knowledge about the behaviour is based on:
(A) Experimentation (B) Observation
(C) Both (A) and (B) (D) None of these

367. The term Marx used as a substitute for what he calls 'relations of production' is:
(A) Social ensemble
(B) Class relations
(C) Social order
(D) Class order

368. The school of thought in psychology is known as:
(A) University of Psychology
(B) College of Psychology
(C) Institute of Psychology
(D) School of Psychology

369. The two forms of preferential marriage were:
(A) Generalized and restricted
(B) Levirate and sororate
(C) Exogamy and endogamy
(D) Polygyny and polyandry

370. Which one of the following is not an ecological process of urban growth?
(A) Centralisation (B) Decentralisation
(C) Invasion (D) Commutation

371. Which one of the following cities may be called a megalopolis?
(A) Hyderabad (B) Chennai
(C) Mumbai (D) Ahmedabad

372. The process by which socio-cultural identity of one group is merged into that of another because of its accepting the latter's culture is called:
(A) Assimilation (B) Integration
(C) Diffusion (D) Cultural contact

373. Hindu Marriage Act of 1955 permits divorce on the grounds of:
(A) Insanity, leprosy, adultery
(B) Economic incompetence, non – religiousness
(C) Insanity and illiteracy
(D) Gambling, alcoholism, insanity

374. The early school of psychology was named:
(A) Functionalism (B) Behaviourism
(C) Gestatism (D) Structuralism

375. Beti Vyavhar, refers to:
(A) Relationship between daughter's family of orientation with that of family of procreation
(B) Respect towards daughter's husband and father's sister's husband
(C) Relationship in which social groups can intermarry
(D) Due consideration towards daughter's education, health, food and nutrition

376. Introspection was the main method used in:
(A) Functtionalism (B) Behaviorism
(C) Gestatism (D) Structuralism

377. Advertisement is used to create:
(A) Awareness among consumers
(B) Consumer's acceptance
(C) Consumer's demand
(D) All of these

378. The German word "Gestalt" means:
(A) Structure (B) Element
(C) Configuration (D) Combination

379. The Hindu Adoption and Maintenance Act, 1956 is also applicable to:
(A) Muslims, Christians, Jains
(B) Buddhists, Jains, Sikhs
(C) Buddhists, Sikhs, Chritians
(D) Buddhists, Jains, Muslims, Sikhs

380. Psychoanalysis was founded in Australia by:
(A) Adler (B) Karl Jung
(C) Sigmund Freud (D) Anna Freud

381. Mechanisms within an organisms that activate behaviour or direct it toward some goal is:
(A) Drive (B) Need
(C) Motivation (D) Incentive value

382. The key idea of psychoanalyis is:
(A) Conscious motivation
(B) Subconscious motivation
(C) Unconscious motivation
(D) None of these

383. An expert on the problem of ageing is called:
(A) Gerontologists (B) Psychologist
(C) Biologist (D) All of these

384. Levirate is practiced by:
(A) Jats of UP (B) Todas of Nilgiri
(C) Santhals of Bihar (D) Gonds of MP

385. Most of the sound we hear in everyday life are the result of:
(A) Simple wave (B) Complex wave
(C) Primary wave (D) Secondary wave

386. Examples of hypergamy were found among:
(A) Kulin Brahmans of Bengal
(B) Chamars of Agra
(C) Ahirs of Mathura
(D) Khans of Aligarh

387. Classical experience by a person is called:
(A) Pavlov (B) Bolls
(C) Kamin (D) Menzel

388. Innate motives based on biological needs that must be met for survival is:
(A) Primary motives
(B) Stimulus motives
(C) Secondary motives
(D) Learned motives

389. Storage of information from past experience closely related to learning:
(A) Information (B) Motivation
(C) Memory (D) Cognition

390. Which of the following may become a reference group for a given person?
(A) The group to which the person belongs or does not belong
(B) Only that group to which the person does not belong
(C) Only that group to which the person belongs
(D) Only that group of which the person is a member only on the basis of achievement

391. 'Consciousness of its own and because of its only consciousness, realizes its unity and enters into conflict'. This statement of Marx comes under:
(A) Social conflict (B) Dialectic
(C) Social ensemble (D) Alienation

392. The conditional stimulus is Pavlov's experiment was:
(A) Light (B) Bell
(C) Food (D) Saliva

393. The response that reliably follow an unconditional stimulus as:
(A) Conditional response
(B) Unconditional response
(C) Both (A) and (B)
(D) None of these

394. Four types of Hindu Marriage or of eight types bring purification to ancestors on the sides of father and mother are:
(A) Brahma, Gadharva, Arsha, Daiva
(B) Daiva, Arsha, Brahma, Gandharva
(C) Prajapatya, Daiva, Asura, Brahma
(D) Brahma, Daiva, Arsha, Prajapatya

395. Visual sensory channel holds information up to:
(A) 4 seconds (B) 3 seconds
(C) 2 seconds (D) 1 second

396. Mate selection by trials in which young man has to prove his prowess before he can claim the hand of any girl is practice among:
(A) Nadars (B) Kadars
(C) Bhils (D) Gonds

397. The rules for word order insentences is known as:
(A) Grammer (B) Syntax
(C) Symbols (D) Phonemes

398. Thinking in cognitive psychology is equated with:
(A) Nervous system (B) Brain
(C) Computer (D) Programming

399. Innate need for stimulation and information to nervous system is called:
(A) Primary motives
(B) Stimulus motives
(C) Secondary motives
(D) Learned motives

400. Short term memory holds the information received from the sensory register up to:
(A) 40 seconds (B) 35 seconds
(C) 30 seconds (D) 25 seconds

401. The degree to which incoming information is processed so that it can be tied to or integrated with existing memories:
(A) Maintenance (B) Elaboration
(C) Cognition (D) Information

402. A computer is equated in cognitive psychology with:
(A) Brain (B) Nervous system
(C) Spinal code (D) Synapse

403. Ultimogeniture refers to a system of inheritance by:
(A) Eldest son (B) Eldest daughter
(C) Youngest son (D) All the sons

404. Cognition and dissonance respectively means:
(A) Thoughts and clashing
(B) Clashing and thoughts
(C) Thoughts and balancing
(D) All of these

405. Beginning with a very large request and then backing down to a smaller one is:
(A) The door in the face approach
(B) The foot in the door approach
(C) Both (A) and (B)
(D) None of these

406. Forgetting caused by the prior learning of other materials is called:
(A) Retroactive interference
(B) Proactive interference
(C) Reasoning interference
(D) None of these

407. The members of an agrarian society exhibits:
(A) Strong in-group feelings
(B) Strong rivalries and conflict
(C) Secondary relationships
(D) Impersonal relationships

408. "Society in action" is otherwise known as:
(A) Social control
(B) Social differentiation
(C) Social interaction
(D) Social welfare

409. A person obtaining a score of 100 in the IQ sale is considered:
(A) Average intelligence
(B) High intelligence
(C) Low intelligence (D) None of these

410. Money is a:
(A) Medium of exchange
(B) Standard of value
(C) Means of payment (D) All of these

411. Two types of definition of society are:
(A) Structural and international
(B) Functional and structural
(C) Evolutionary and diffusive
(D) Structural and evolutionary

412. IQ score are not very dependent until the age of:
(A) Six (B) Seven
(C) Eight (D) Nine

413. Khasi tribes living in the hills of Meghalaya are:
(A) Matrilineal (B) Patrilineal
(C) Bi-lineal (D) Cognatic

414. Achievement is classified under:
(A) Primary motive
(B) Secondary motive
(C) Stimulus motive
(D) All of these

415. The principle or set of principles by which ego's consanguineal relatives are determined is technically called:
(A) Rules of descent
(B) Rules of kinship
(C) Rules of grouping
(D) Rules of lineage

416. The close similarity in IQs would be observed for:
(A) Parents and children
(B) Identical twins reared apart
(C) Fraternal twins reared together
(D) Siblings reared together

417. The key sense involved in maintaining balance:
(A) Kinesthesisense
(B) Vestibular sense
(C) Transduction
(D) Receptors

418. Colour blindness is due to the efficiency of:
(A) Rod cells
(B) Cone cells
(C) Both (A) and (B)
(D) None of these

419. The position of unequal distribution of land in village generally results in:
(A) Injustice (B) Social revolution
(C) Caste conflict
(D) Illiteracy and malnutrition

420. One seldom witnesses psychological conflict in the villages because:
(A) There are no cinemas, or other recreation centers which create psychological crisis
(B) Villagers lead a contended life
(C) Villagers live in the lap of nature
(D) Villagers live a contended life

421. Unlike rural community, urban community is:
(A) More static
(B) Less familiar
(C) More familiar
(D) More traditional

422. Crime is usually:
(A) Similar in the rural and urban areas
(B) Higher in the rural areas than in the urban areas
(C) Higher in the big cities than in the rural areas
(D) Lower in the metropolitan cities than in the small towns

423. Many motivated activities are thought to begin with a:
(A) Drive (B) Need
(C) Response (D) Goal

424. Transitional step between crude barter and exchange with 'true money' is known as:
(A) Money barter
(B) Reciprocity
(C) Gift
(D) Consumption

425. Society is:
(A) Natural (B) Deliberate
(C) Practical (D) Functional

426. Drives activates a:
(A) Drive (B) Need
(C) Response (D) Goal

427. The tribe having low amount of bride price is:
(A) Nayar (B) Meena
(C) Rengma Naga (D) Kuki

428. Motives based on learned psychological needs is known as:
(A) Learned motives
(B) Stimulus motives
(C) Secondary motives
(D) Both (A) and (B)

429. A border line person has the IQ of:
(A) 70 to 84 (B) 55 to 69
(C) 50 to 55 (D) 45 to 55

430. Those individuals most likely to develop heart diseases have labeled:
(A) Type A personality
(B) Type B personality
(C) Both (A) and (B)
(D) None of these

431. Thirst is classified under:
(A) Primary motive (B) Secondary motive
(C) Stimulus motive (D) All of these

432. Quality of work is tied more to:
(A) Extrinsic factors (B) Intrinsic factors
(C) Both (A) and (B) (D) None of these

433. The reaction we experience when our sensory channels are stimulated:
(A) Feeling (B) Sensation
(C) Perception (D) Emotion

434. According to Brihaspati, offspring from:
(A) Five maternal generations and seven paternal generations are sapinda and they cannot inter-marry
(B) Seven maternal generations and five paternal generations are sapinda and they cannot inter-marry
(C) Three maternal and five paternal generations cannot inter-marry
(D) Three paternal and five maternal generations are spainda and they cannot inter marry

435. Exploration, manipulation and curiosity provide evidence for the existence of:
(A) Stimulus drives (B) Social motives
(C) Sensory deprivation
(D) Curiosity drives

436. The libido which flows from the life instincts are called:
(A) Thanatos (B) Eros
(C) Psyche (D) Conscience

437. Consider the following statements about cultural relativism:
1. Each culture is holistic
2. Culture complex is the smallest unit of culture
3. Culture is shared by the members of a given group
4. Physical and biological features of people are determined by culture
Which of these statements are correct?
(A) 1, 2, 3 (B) 1, 3, 4
(C) 2, 3 (D) 2, 4

438. The smallest meaningful units of speech is called:
(A) Phonemes (B) Morphemes
(C) Heuristics (D) Programming

439. Need cause a psychological thought or feeling called:
(A) Drive (B) Need
(C) Response (D) Goal

440. Biological needs are essential because they maintain:
(A) Thermostat (B) Homeostasis
(C) Incentive vale (D) Need

441. The tribe having very high bride price is:
(A) Hos (B) Munda
(C) Kharia (D) Nayars

442. The word Androgyny literally means:
(A) Man-man (B) Woman-woman
(C) Man-woman (D) None of these

443. The hereditary aspect of a person's emotional nature is denoted by:
(A) Temperament (B) Character
(C) Traits (D) Personality

444. A very direct way to learn about personality is to engage a person in:
(A) Conversation (B) Observation
(C) Projective test (D) None of these

445. A tendency in generalize a favourable or unfavourable impression to unrelated details of personality is:
(A) Halo effect
(B) Behavioural assessment
(C) Aggressiveness
(D) All of these

446. The cause of aggressive and destructive urge is:
(A) Eros (B) Psyche
(C) Libido (D) Death instinct

447. Discharging pent up feeling, usually of hostility, on object less dangerous than those which initially aroused the emotions;
(A) Sympathism
(B) Sublimation
(C) Compensation
(D) Displacement

448. Social learning theories emphasize:
(A) Identification and imitation
(B) Imitation and expectancy
(C) Identification and expectancy
(D) Feeding and expectancy

449. Match list I with list II and select the correct answer:
List I (Industry)
a. Textile b. Petroleum Industry
c. Banking Industry d. Dot com industry
List II (Characteristic)
1. Capital intensive 2. Labour intensive
3. Flexible labour 4. Services
Codes:

	a	b	c	d
(A)	2	1	4	3
(B)	1	2	3	4
(C)	5	1	3	2
(D)	5	3	1	4

450. The most widely used statistics in psychology is:
(A) Correlation coefficient
(B) Frequency distribution
(C) Arithmatic mean
(D) None of these

451. Stress may occur at:
(A) Physical level (B) Psychological level
(C) Biological level (D) Both (B) and (C)

452. Modern society is sometimes called:
(A) Organisational society
(B) Associational society
(C) Group society (D) Cultural society

453. The term buyers market and seller market refers to:
(A) Perfect competition
(B) Monopolistic market
(C) Availability or non-availability of goods
(D) Price rise

454. An ideal economic system:
(A) Must be a function of social system
(B) Basic aim should be profit
(C) Must be correlation between different social and economic system
(D) Must be independent of social system

455. The Nutritional Approach to poverty tries to measure:
(A) The health of the urban population
(B) Poverty on the basis of minimum good requirements
(C) Relative poverty of the poor as compared to the rich
(D) Poverty on the basis of the income level of the rural population

456. In order to maintain good economic condition of the people not owning any piece of land the government should:
(A) Settle the village people in town
(B) Provide land to them in some other village
(C) Settle then near river banks
(D) Undertaken the programme of cottage industries in the villages

457. Preventing painful or dangerous thoughts from entering consciousness is known as:
(A) Reaction formation (B) Regression
(C) Repression (D) Interjection

458. Whatever experienced by a person is called:
(A) Feeling (B) Sensation
(C) Perception (D) Emotion

459. Increasing feeling of worth by identifying self with person or institution of illustrious standing is known as:
(A) Interjection (B) Compromise
(C) Identification (D) Displacement

460. "Hitler saw the liquidation of the Jews not as reprehensible but as a noble crusade" is an example of:
(A) Projection
(B) Repression
(C) Regression
(D) Rationalization

461. "Sour grape" mechanism and "Sweet lemon" mechanism are the types of:
(A) Sublimation
(B) Intellectualization
(C) Rationalization
(D) None of these

462. "The fox who, unable to reach clusters of luscious grapes, decided that they were sour and not worth having anyway" is an example of:
(A) Sour grapes mechanism
(B) Sweet lemon mechanism
(C) Intellectualization (D) None of these

463. "Every dark cloud have a silver lining" and "Everything happens for the best" are the examples of:
(A) Regression
(B) Sweet lemon mechanism
(C) Repression (D) Sublimation

464. The audible range of sound waves for human beings is between:
(A) 20 and 30000 Hertz
(B) 20 and 2000 Hertz
(C) 20 and 20000 Hertz
(D) 10 and 30000 Hertz

465. Which of the following is not classified as a psychosomatic disorder:
(A) Hypertension (B) Colitis
(C) Eczema (D) Thymus

466. Match list I with list II and select the correct answer:
List I (Theory of trade unions)
a. Economism b. Incorporation
c. Oligarchy d. Trusteeship
List II (Scholar)
1. Lenin 2. Michels
3. Trotsky 4. Weber
5. Gandhi
Codes:

	a	b	c	d
(A)	4	3	1	2
(B)	1	3	2	5
(C)	5	1	3	2
(D)	5	3	1	4

467. Most outward symptoms of stress disappear during the stage of:
(A) Alarm reaction
(B) Stage of resistance
(C) Stage of exhaustion
(D) None of these

468. The term sociology was coined in:
(A) 1798 (B) 1815
(C) 1839 (D) 1957

469. Social motives are important components of:
(A) Emotion (B) Behaviour
(C) Personality (D) None of these

470. The authoritaton personality tends to be prejudiced against all out-groups, a quality referred to as:
(A) Discrimination (B) Scapegoating
(C) Ethnocentric (D) None of these

471. Individuals will work hard on various tasks to the degree that they expect such effort will yield positive outcome:
(A) Expectancy theory
(B) Goal setting theory
(C) Locus of control (D) Valence theory

472. Motivation resulted out from the intention to reach certain goals:
(A) Goal setting theory
(B) Locus of control
(C) Valence theory (D) None of these

473. The period of transition from childhood to adulterated is known as:
(A) Young age (B) Adolescence
(C) Middle age (D) None of these

474. Psephology is a word used to describe the study of:
(A) Criminal (B) Sexual relations
(C) Elections (D) Adult delinquency

475. "A motor cycle should have two wheels and an engine and handle bars" forms an example for:
(A) Relational concepts
(B) Disjunctive concept
(C) Conjunctive concept (D) None of these

476. Leviate Marriage is:
(A) Marriage between cross cousin
(B) Marriage with the daughter of the sister
(C) Marriage with brother's daughter
(D) Marriage with deceased husbands brother

477. Consider the following two statements:
Assertion (A): The greater the urbanization, the larger the number of secondary groups
Reason (R): The greater the social division of labour, the larger is the scope of individualism
Select the correct answer:
(A) Both A and R are true and R is the correct explanation of A
(B) Both A and R are true by R is not the correct explanation of A
(C) A is true, R is false
(D) A is false, T is true

478. Value is a belief that something is:
(A) Moral
(B) Very productive in society
(C) Good and desirable (D) Cultural

479. A test that gives the same score every time when it is administrated is:
(A) Valid (B) Reliable
(C) Objective (D) Credible

480. The principle which explains the significant ritual and social role of mother's brother in a patrilineal society is known as:
(A) Consanguinity
(B) Affinity
(C) Complementary afflictions
(D) Joking relationship

481. Which one of the following is not a function of the family in industrial society?
(A) To act as a production unit
(B) To transmit social values and culture to the children
(C) To act as a consumption unit
(D) To impart education to children

482. The push theories of motivation is generally attached to:
(A) Cognitive (B) Drive
(C) Incentive (D) None of these

483. Attitudes are expressed in:
(A) Belief component
(B) Emotional component
(C) Action component (D) All of these

484. Match list I with list II and select the correct answer:
List I
a. Max Muller b. E.B. Tylor
c. Emile Durkheim d. Auguste Comte
List II
1. Animism 2. Totemism
3. Tetishisin 4. Naturisim
Codes:

	a	b	c	d
(A)	4	3	1	2
(B)	4	2	3	1
(C)	5	1	3	2
(D)	5	3	1	4

485. 'Gilded age' stands for:
(A) Age of gild-relation
(B) Age of gold
(C) Age of gladiators
(D) Age of gladness and happiness

486. The thinking type that operates through the right brain:
(A) Divergent thinking
(B) Creative thinking
(C) Horizontal thinking (D) All of these

487. A marriage in which a woman of upper caste marriages a man of the lower caste is known as:
(A) Exogamy (B) Hypogamy
(C) Hypergamy (D) Endogamy

488. The sequence of stages in motivation is called as:
(A) Motivational cycle
(B) Emotional stages
(C) Physiological emotional cycle
(D) None of these

489. The book "I'm OK You're OK" is written by:
(A) Thomas and Amy Harris
(B) Eric Berne
(C) Thomas Pele (D) Paul Stephenson

490. Motives is an individual is:
(A) Permanent and rigid
(B) Modifiable only
(C) Changeable (D) Formed only once

491. When those who have risen above the lowest poverty levels indulge in civic unrest against the privileged, it is an expression of;
(A) Relative deprivation
(B) Social inequality
(C) Social mobility (D) Factionalism

492. The movement of People from central cities of communities in the surrounding areas is known as:
(A) Sub urbanization
(B) Under urbanization
(C) Over-urbanisation (D) Urbnaization

493. In which one of the following types of cities do we find the elite class at the centre of the city?
(A) Pre-industrial (B) Modern
(C) Historical (D) Commercial

494. Which one of the following terms is used for spread of population and industry beyond the borders of urban centres?
(A) Urban sprawl (B) Urbanisation
(C) Urban centre (D) Suburbanisation

495. Consider the following two statements:
Assertion (A): As less developed countries begin to industrialize their cities also expand
Reason (R): Industrialization is a necessary condition of urban growth
Select the correct answer:
(A) Both A and R are true and R is the correct explanation of A
(B) Both A and R are true by R is not the correct explanation of A
(C) A is true, R is false
(D) A is false, T is true

496. Dayabhaga is generally followed in:
(A) Rajasthan (B) West Bengal
(C) Punjab (D) Maharastra

497. A group of relatives which includes all those consanguineous relatives to whom an individual feels some sense of obligation is called:
(A) Kindred (B) Coagulate
(C) Agnate (D) Uterine

498. The incentive in youth club is:
(A) Recognition (B) Aspiration
(C) Motivation (D) None of these

499. The concept of urban recycling refers to:
(A) Pollution control mechanisms
(B) Relocation of industries in the city
(C) Renovation of old buildings and construction of new units
(D) Relocation of the city at another site

500. Which among the following is more likely to lead towards a more liberal attitude towards parenthood without marriage?
(A) Immorality
(B) Surplus of women and economic difficulty
(C) Prostitution (D) Polygyny

501. The philosophy of history as a distinct branch of speculation is a creation of:
(A) Thirteenth Century
(B) Eighteenth century
(C) Nineteenth Century
(D) Twentieth Century

502. The element that does not comes under the cognitive behaviour process is:
(A) Receiving (B) Comprehending
(C) Synthesis (D) Analysing

503. The relationship between the individual and culture is one in which:
(A) Society and culture account for similarities of personality among individuals but not the individuality of personality
(B) The total complex of culture determine the behaviour of the individual
(C) The individual is simultaneously the product, the carrier and creator of culture
(D) Neither the individual creates culture nor does the culture create the individual

504. What is psychologists interest in motivation study:
(A) Actions and overt behaviour
(B) Factors that cause behaviour
(C) Achievement and success
(D) Repression

505. Which one the following systems correctly represents the marriage of a widower with the sister of his wife?
(A) Levirate (B) Sororate
(C) Polygyny (D) Monogamy

506. Which one of the following tribes is matrilineal?
(A) Chakesang (B) Jaintia
(C) Mizo (D) Kuki

507. Receiving, valueing, responding etc. are dimensions of:
(A) Cognitive domain
(B) Affective domain
(C) Psychomotive domain
(D) All of these

508. A role play situation in which respondent play his own role:
(A) Socio-drama (B) Psycho-drama
(C) Arto-drama (D) None of these

509. Stress arouses:
(A) Emotion (B) Hostility
(C) Fear (D) Anxiety

510. "In the game of baseball, a strike is either a swing and a miss or a pitch down the middle or a foul ball" forms an example for:
(A) Relational concepts
(B) Disjunctive concept
(C) Conjunctive concept
(D) None of these

511. A method in which people are asked to respond to questions by expressing their personal evaluation:
(A) Attitude scale
(B) Interview
(C) Self report method
(D) None of these

512. Thinking directed towards discovery of a single established correct answer is:
(A) Divergent thinking
(B) Convergent thinking
(C) Creative thinking
(D) All of these

513. Cultural relativism means:
(A) Function and meaning of a trait are relative to its cultural setting
(B) Culture is relatively rare
(C) Culture evolution is relative
(D) Cultural diffusion is relative

514. Among the simple societies, agriculture is the:
(A) First level of development
(B) Second level of development
(C) Third level of development
(D) Fourth level of development

515. Ability to attain a desired goal is referred to:
(A) Skill (B) Attitude
(C) Competency (D) Motivation

516. The propaganda technique that omit important information that might support an unfavourable view:
(A) Card staking
(B) Snobbery
(C) Straw person
(D) None of these

517. The ego state in transactional analysis that rest on the taught concept of life:
(A) Parent (B) Child
(C) Adult (D) None of these

518. The earliest Indian religious and philosophical thinking concerning society is found in:
(A) Vedas
(B) Puranas
(C) Mahabharatha
(D) Ramayana

519. Relations of production primarily consist of:
(A) Social relationships arising out of the production process
(B) Differential requirements of classes in society
(C) Ideal material needs of humans in history
(D) Individual motives for acquisition in society

520. Which of the following explanations stand for acculturation?
(A) The coming into contact of two cultures as a result of which one is influenced by the other
(B) Spread of cultural items and features from the major centres of civilization to smaller regions
(C) Institutions and cultural patterns which are universally found in all human societies
(D) An individual absorbing the cultural traits of other groups

521. Word sociology is made up of two words:
(A) Societies and Logy
(B) Societia and Logistia
(C) Socious and Logos
(D) Socia and Logos

522. The two words Socious and Logos were from which language?
(A) Latin and Greek
(B) Arabic and Persian
(C) German and French
(D) Spanish and Greek

523. According to Auguste Comte's opinion, sociology is very near to:
(A) Anthropology
(B) Social Anthropology
(C) Biology (D) Physics

524. Which among the following is the youngest social science?
(A) Sociology (B) Political Science
(C) Economics (D) Statistics

525. 'Kibbutz' system of family life is present in:
(A) Israel
(B) Japan
(C) India
(D) America

526. Who among the following has distinguished between lions and foxes as types of elites?
(A) Pareto (B) Shils
(C) Weber (D) Mannhiem

527. The method employed for measuring socio-economic status of the people where maximum importance attached to the occupation of a person is:
(A) Differential scale
(B) Edward method
(C) Scale analysis (D) Role distance

528. Value of Sociological Theory lies in:
(A) Number of questions raised
(B) Describing society
(C) Describing thinkers
(D) Descending behavioural patterns

529. Achievement motivation is characterized by:
(A) Sense of excellence
(B) Sense of unique achievement
(C) Sense of long term involvement
(D) All of these

530. The first and last stage in general reasoning are:
(A) Conception and Judgement
(B) Conception and Perception
(C) Perception and Judgement
(D) Perception and Conception

531. Realism holds that:
(A) Group is as real as the person
(B) Group is not real
(C) Person is real
(D) Society is real

532. Orientalists were scholars who studied the cultures of:
(A) India and China
(B) Greek and Roman
(C) Egypt and Mesopotamia
(D) Far east countries

533. Marx studies society in:
(A) Holistic way
(B) Conceptual way
(C) Methodological way
(D) Factual way

534. The element of social action according to Weber is:
(A) Actor (B) Situation
(C) Ends (D) All of these

535. Relation of production constitute relationships between:
(A) Things and things
(B) People and things
(C) People and people
(D) Factory and worker

536. The feeling of ethnocentrism by a group in a multilateral society results in:
(A) Strengthening of loyality to all cultures
(B) Weakening of loyality to all cultures
(C) Cultural harmony
(D) Cultural conflict

537. Motion of function has mainly been used by:
(A) Biological science (B) Social science
(C) Physical science (D) Literature

538. Which of the following is an example of Sanskritisation?
(A) Monogamy among present Nayars
(B) Brahma form of marriage among Namboodiris
(C) Sapta padi among Kshatriyas
(D) Sapta padi among the peasant castes of KA

539. Consider the following statements about cultural relativism:
1. It is the tendency to evaluate other cultures in terms of our own culture
2. Each society is evaluated in terms of its own circumstances and values
3. The contemporary society is evaluated in relation to its past
4. Actions that are moral in some societies might be immoral in others

Which of these statements are correct?
(A) 1, 2 (B) 1, 4
(C) 2, 3 (D) 2, 4

540. Absorption of culture is similar to:
(A) Integration (B) Assimilation
(C) Adjustment (D) Accommodation

541. The private property is:
(A) Water (B) Factory where one works
(C) Books in the neighborhood library
(D) Copyright of a writer

542. Consider the following two statements:
Assertion (A): Not all sub cultures are counter cultures
Reason (R): The term subculture does not imply that a group is in conflict with the dominant culture
Select the correct answer:
(A) Both A and R are true and R is the correct explanation of A
(B) Both A and R are true by R is not the correct explanation of A
(C) A is true, R is false
(D) A is false, T is true

543. Village settlement is found to exist in:
(A) Paleolithic age (B) Mesolithic age
(C) Neolithic age (D) Early stone age

544. In the settlement pattern of vil[illegible]s in India, most important factor is:
(A) Kinship (B) Caste
(C) Religion (D) Occupation

545. For field observation, on which method should sociologists rely?
(A) Interview (B) Questionnaire
(C) Primary Source (D) Sample method

546. Unsound reasoning technique that are used to change how people think:
(A) Fallacies (B) Paralogic
(C) Euphemism (D) Equivocation

547. Indology means:
(A) Detailed description about India
(B) A branch of Archaeology
(C) Study of India and its culture
(D) Study of Indian census

548. What is the primary methodology of functionalists?
(A) Conjecture and guess
(B) Secondary source
(C) Field work (D) Census data

549. For Spencer, society is a/an:
(A) Super organism
(B) Organic whole
(C) Organicism
(D) Collective whole

550. Factors influencing social organization of the village is:
(A) Caste system
(B) Its closeness to the towns
(C) Existence of good means of communication and transport
(D) All of these

551. The history of functional analysis may be traced in:
(A) Durkheims causal functional analysis
(B) Malinowski's study of culture
(C) Comte's positivist analysis
(D) Marx's dialectcal analysis

552. Ethnocentrism is dysfunctional to:
(A) Intra-group relations
(B) Peer group relations
(C) Inter-group relations
(D) Community relations

553. The questionnaire method of data collection is applicable if there is a/an:
(A) Overpopulation (B) Big area
(C) Hetergeneous area (D) All of these

554. The villagers always stick to their general behaviouristic patterns because they:
(A) Are very small in number
(B) Are more cultured
(C) Are well socialized
(D) Follow tradition, custom and folkways

555. The evolutionary theories, cultural parallels are explained through:
(A) Diffusion
(B) Psychic unity of mankind
(C) Similarity of environment
(D) Similarity of thought and space

556. Crime is absent in the villages but it is rampant in the cities because:
(A) Villagers have no time to do so
(B) Villages are small and people can be caught easily
(C) The villagers are controlled by their families
(D) The villagers are fatalists and are god fearing people and they shudder at the idea of committing crimes

557. A sociological theory is a set of ideas which provides an explanation for:
(A) Continental society
(B) Humans society
(C) Human behaviour
(D) Global society

558. In an aggregation, the individuals collected lack:
(A) Socialisation
(B) Mutual responsiveness
(C) Desire to respond to external stimulus
(D) Social norms

559. According to functionalist school of thought, human nature s just like:
(A) Natural gas (B) Water
(C) Mirror (D) Wax

560. The essential characteristic of the rural society is:
(A) Individualism (B) Parochialism
(C) Heterogeneity
(D) Face-to-face relation

561. The idea of 'Global Village' implies:
(A) One government administers all countries
(B) Socio-cultural differences are obliterated
(C) Network of communications binds peoples of the world
(D) Agriculture become the most important and global occupation

562. Term correctly defining belief and value systems used by Robert Redfield in the analysis of folk-urban continuum is:
(A) Local tradition (B) Folk tradition
(C) Little tradition (D) Great tradition

563. For Weber, all human action is directed by:
(A) Individual (B) Rationality
(C) Meaning and motive
(D) Society

564. Which term is equivalent to group marriage?
(A) Polygyny (B) Exogamy
(C) Cenogamy (D) Endogamy

565. Group Mind theory believes that society:
(A) Is above the individual
(B) Comes along with the individuals
(C) And individuals have no relations
(D) Is below the individual

566. Features of market exchange is:
(A) Buyers approach the sellers
(B) Sellers wish to exchange their goods because they need the money
(C) Fluctuation in price depends on supply
(D) All of these

567. If a person physically assaults another, it is an example of which action?
(A) Affective action (B) Brave action
(C) Rational action (D) Fearful action

568. Functionalism in Social Anthropology arose as a critique of:
(A) Positivism and Organicism
(B) Evolutionism and Diffusionism
(C) Individualism and Positivism
(D) Evolutionism and Particularism

569. Which of the following statements about culture are correct?
1. Each culture is holistic
2. Culture complex is the smallest unit of culture
3. Culture is shared in common by the members of a given group
4. Physical and biological features of people are determined by culture
(A) 1, 2, 3 (B) 2, 3, 4
(C) 1, 3 (D) 2, 4

570. Marriage among Nayars of Kerala is called:
(A) Shadia vivaha (B) Sambandham
(C) Troille (D) Karanvam

571. A social theory which starts from a conception of society and postulates ways in which society influences the individual is a:
(A) Macroscopic theory
(B) Katascopic theory
(C) Microscopic theory
(D) Productive theory

572. Which of the following are likely to happen in a hitherto prominent culture if there is reaction against acculturation?
1. It will combine the old one the new in a new mosaic
2. It will condemn the cultural degeneration
3. It will defend its heritage
4. It will express its resentment through aggression or phantasy

Select the correct answer
(A) 1, 2, 3 (B) 1, 2, 4
(C) 1, 3, 4 (D) 2, 3, 4

573. The tendency of a person to reject the culture of his own group is known as:
(A) Egocentrism (B) Jingoism
(C) Zealotism (D) Xenocentrism

574. Production and exchange of goods and service for the satisfaction of wants is called:
(A) Productive action
(B) Exchange
(C) Economic action
(D) Rational action

575. Self sufficiency of Indian villages in recent times is no longer there due to:
(A) Education
(B) Proximity to the urban areas
(C) Modern means of communication
(D) Economic growth

576. Cultural relativism:
(A) In a recognition in which each culture is judged in terms of its own standards
(B) Is an argument that differed parts of culture are inter-related
(C) Means any culture can be evaluated in absolute terms
(D) Means that different cultures are related to each other in terms of some universal elements

577. Characteristic of little community defined by Robert Redfield is:
(A) Smallness (B) distinctiveness
(C) Self-sufficiency (D) All of these

578. Sociologically 'Culture' denotes:
(A) Refinement in a person
(B) Learned behaviour of man
(C) Growth of micro-organisms in a laboratory
(D) Cultivation of aesthetics

579. The monetary authority in India is the:
(A) State Bank of India
(B) Reserve Bank of India
(C) Planning Commission
(D) Ministry of Finance

580. Match list I with list II and select the correct answer from the codes given below
List I
a. Westernization and Sanskritization
b. Universalization and Parchialization
c. Great tradition and little tradition
d. Universalism and Particularism
List II
1. McKim Marriott 2. Robert Redfield
3. M.N. Srinivas 4. Talcott Parsons
5. R.K. Merton
Codes:

	a	b	c	d
(A)	2	1	4	3
(B)	3	1	2	4
(C)	3	4	2	1
(D)	5	3	1	4

581. 'Ekapatrnivrata' stands for:
(A) Ideal of having only one wife
(B) Ideal of having sex with only one person
(C) Ideal of providing with need to one's wife
(D) Ideal of having only one husband

582. Which one of the following terms correctly defines beliefs and value systems used by Robert Redfield in the analysis of folk-urban continuum?
(A) Great tradition
(B) Little tradition
(C) Universal tradition
(D) Local tradition

583. The process of borrowing cultural items from other cultures is known as:
(A) Acculturation (B) Socialisation
(C) Enculturation (D) Acquisition

584. Criteria used by Sorokin in distinguishing a rural society from an urban society is:
(A) Mobility
(B) Heterogeneity
(C) System of interaction
(D) All of these

585. The principle that a culture must be understood and judged on its own terms without reference to the values of another culture is known as:
(A) Cultural specificity
(B) Cultural pluralism
(C) Cultural hegemony
(D) Cultural relaticism

586. Which one of the following is known as human capital?
(A) Returns on industry
(B) Returns on savings
(C) Returns on education
(D) Returns on business

587. A subculture that adheres to a set of belief and values, radically objects the society's dominant culture and prescribes as alternative one is:
(A) Idioculture
(B) Subculture
(C) Counterculture
(D) Real culture

588. The marriage of a widower with the sister of his dead wife is correctly represented by:
(A) Sororate system
(B) Levirate system
(C) Monogamy system

589. Match list I with list II and select the correct answer:
List I (concepts)
a. Culture contact b.Cultural universal
c.Cultural relativismd. Cultural pluralism

List II (meanings)
1. Adopting new cultural pattern
2. Striking uniformity among cultures
3. No universal standards can be used to evaluate a culture
4. Cultures co-exist without evaluating each other

Codes:

	a	b	c	d
(A)	2	1	4	3
(B)	1	2	3	4
(C)	1	3	2	1
(D)	5	3	1	4

590. Which of the following statements correctly defines the terms 'ethnocentrism'?
(A) Judging other cultures with the help of cultural 'ideal types'
(B) Judge other species by comparing them with human species
(C) Looking at other cultures with a sense of inferiority
(D) Judging other cultures by comparison with one's own

591. 'Disperate polyandry' stands for:
(A) When husbands of women are brothers
(B) When husbands of women are related in some way
(C) When husbands are not related to each other
(D) When husbands are engaged in one type of economic activity

592. A basic distinction between culture and civilization is:
(A) Culture is what we have achieved and civilization is what we aspire to achieve
(B) Culture is what comes out of intelligence and civilization is not
(C) Culture is secondary and civilization is not
(D) Culture is what we are and civilization is what we have

593. The cities are not marked by:
(A) Dense settlement (B) Advance culture
(C) Permanence
(D) Occupational specialization

594. Which one of the following pairs is not correctly matched?
(A) Dewar – Bhabhi relationship : Joking relationship
(B) Marriage with mother's brother : Experimental marriage
(C) Cross – cousin : Preferential form of marriage
(D) Marriage with father's brother : Incest taboo

595. The Hindu Widow Remarriage Act of 1956 legalised the remarriage of Hindus of:
(A) Bengal (B) High caste
(C) All castes (D) North India

596. Consider the following two statements:
Assertion (A): The culture of the dominant class controls the dominant meanings
Reason (R): The structure of relations between classes perpetuates itself by reproducing its characteristics in the domain of culure
Select the correct answer:
(A) Both A and R are true and R is the correct explanation of A
(B) Both A and R are true by R is not the correct explanation of A
(C) A is true, R is false
(D) A is false, T is true

597. Everything is right somewhat but not everywhere. This refers to:
(A) Cultural-universals
(B) Cultural-relativity
(C) Counter-culture
(D) Cultural-plurasim

598. When cultural traits spread from one society to another the process is called:
(A) Diffusion (B) Parallelism
(C) Evolution (D) Pluralism

599. Which one of the following is not part of little tradition?
(A) Folk Tales (B) Legends
(C) Ramayana (D) Assimilation

600. The concept of cultural lag explains that there is faster rate of change in the:
(A) Economic sector of society in comparison to non-economic sectors
(B) Political domain of society in comparison to economic domain
(C) Technological sectors of society in comparison to cultural sectors
(D) Religious activities of society in comparison to cultural activities

601. When cultural norms and values are transmitted from generation to generation, the process is called:
(A) Cultural reproduction
(B) Acculturation
(C) Cultural diffusion
(D) Cultural stimulation

602. According to the theory of cultural lag:
(A) Both material and non-material culture undergo changes simulataneously
(B) Changes in non-material culture are faster
(C) Material culture changes faster than non-material culture
(D) Sometimes material culture and at other times non-material culture changes faster

603. Consider the following two statements:
Assertion (A): New technology is received more quickly than values
Reason (R): Change in non-material culture always lag behind changes is material culture
Select the correct answer:
(A) Both A and R are true and R is the correct explanation of A
(B) Both A and R are true by R is not the correct explanation of A
(C) A is true, R is false
(D) A is false, T is true

604. Consider the following two statements:
Assertion (A): It is witnessed that even the rational persons put some food and eatables on the burial pit of dead persons
Reason (R): Modes of action persist more obstinately than the modes of thinking
Select the correct answer:
(A) Both A and R are true and R is the correct explanation of A
(B) Both A and R arc true by R is not the correct explanation of A
(C) A is true, R is false
(D) A is false, T is true

605. The narratives passdown in oral transmission is termed as:
(A) Legend (B) Folkways
(C) Folklore (D) None of these

606. What does acculturation signify?
(A) Growing up and internalizing the values and norms of one's society
(B) Spread of a cultural element from the centre of origin to other societies
(C) Interchange of elements of culture between societies
(D) Incorporation of an element of culture from another culture, while retaining its identity

607. The statement that Indian culture is superior to western culture is an example of:
(A) Cultural relativity
(B) Ethnocentrism
(C) Fundamentalism
(D) Xenophobia

608. Which of the following concepts correctly refers to the process by which cultural traits are transmitted from one society to another?
(A) Acculturation (B Universalisation
(C) Diffusion (D) Assimilation

609. Which one of the following terms is used for the argument that there is no superior international or universal morality, that the moral and ethical rules of all cultures deserve equal respect:
(A) Cultural parallelism
(B) Ethnocentrism
(C) Cultural relativism
(D) Cultural determinism

610. Match list I with list II and select the correct answer:

List I

a. Cultural absolutes
b. Cultural universals
c. Cultural alternatives
d. Cultural specialities

List II

1. The forms of behaviour recognized by a society as valid but which cut across class, occupational or sex lines
2. Those aspects of behaviour that characterize groups within the larger social whole
3. Those norms which are fixed and do not differ from culture to culture
4. Those forms of behaviour that are expected of any normal member of a society

Codes:

	a	b	c	d
(A)	4	3	1	2
(B)	1	2	3	4
(C)	1	3	2	1
(D)	5	3	1	4

611. Consider the following statement
Culture is
1. learned
2. Transmitted from one generation to another
3. Non-material 4. Dynamic

Which of these statements are correct?

(A) 1, 2, 3 (B) 2, 3, 4
(C) 1, 2, 4 (D) 3, 4

612. The tendency to see one's own ideas beliefs and practices superior to that of others is called:
(A) Xenophobia
(B) Self–orientation
(C) Egocentrism
(D) Ethnocentrism

613. Which one of the following terms is used to explain the state of cultural system in which the cultural elements change in unequal rates disrupting the harmony between material and non-material aspects?
(A) Cultural disintegration
(B) Cultural disorganization
(C) Cultural lag
(D) Cultural ambivalence

614. The Rules of Residence where a couple upon being married, go to live with the brother of the groom's mother are known as:
(A) Patrilocal (B) Matrilocal
(C) Avunculocal (D) Neolocal

615. Consider the following two statements:
Assertion (A): Man is the product of the cultural milieu in which he is born
Reason (R): Culture determines every aspect of human personality
Select the correct answer:
(A) Both A and R are true and R is the correct explanation of A
(B) Both A and R are true by R is not the correct explanation of A
(C) A is true, R is false
(D) A is false, T is true

616. Who among the following inherits property among Khasis of Meghalaya?
(A) Eldest son
(B) Youngest son
(C) Youngest daughter
(D) Oldest daughter

617. In a country, a separatist group has succeeded in seceding from the political union. This has raised the hopes of other centrifugal forces in the country. The possibility of the other centrifugal forces getting active would be the product of the separatist group:
(A) Manifest function
(B) Latent function
(C) Latent dysfunction
(D) Manifest dysfunction

618. When a whole way of life is in the process of change under influence of another culture, we call it?
(A) Acculturation (B) Diffusion
(C) Assimilation (D) Socialization

619. In which of the following books CH-Cooley introduced the concept of 'Primary Group' characterized by intimate, face-to-face association?
(A) Social Structure (B) Human Groups
(C) The Nature of Human Groups
(D) Social Organisation

620. Match list I with list II and select the correct answer:
List I
a. Ethnocentrism b. Cultural lag
c. Cultural Universals d. Cultural traits
List II
1. Items of a culture, belief values and typical ways of doing things
2. Division of labour, the incest taboo, marriage, the family, rite de passage and ideology
3. The tendency to judge other cultures in terms of one's own customs and values
4. A situation that develops when new patterns of behaviour conflict with traditional values
5. The reaction people may have when encountering cultural traditions different from their own

Codes:

	a	b	c	d
(A)	4	3	1	2
(B)	1	2	3	4
(C)	3	4	2	1
(D)	5	3	1	4

621. Which of the following are characteristics of the social structure of the society?
1. A society has a structure of inter-related roles
2. Society is a social type of group with a comprehensive social system
3. A society tends to survive for a long period
4. A society is economically self-sufficient
Choose the correct answer using the codes given below:
(A) 1, 2, 3 (B) 1, 2, 4
(C) 1, 3, 4 (D) 2, 3, 4

622. The process by which the group-values are internationalized by the individual is called:
(A) Assimilation (B) Integration
(C) Socialisation (D) Accommodation

623. Which of the following are example of an organization?
(A) Prison (B) University
(C) Family (D) Club
Choose the correct answer using the codes given below:
(A) 1, 2, 4
(B) 1, 2, 3
(C) 1, 3, 4
(D) 2, 3, 4

624. Which one of the following statements is correct?
(A) Social structure and social organization are two concepts – denoting the same phenomena
(B) Both social structure and social organization have no empirical referents
(C) An organization function to accomplish a goal, while a structure does not
(D) Social structure has no relation with function

625. The only ritual in the life of Hindu female is:
(A) Educational initiation
(B) Marriage
(C) Birth of a son
(D) When she conceives

626. Which of the following pairs is not correctly matched?
(A) Social Group – Orderly relations
(B) Aggregate – Sense of belonging
(C) Primary Group – Emotional attachment
(D) Reference Group – Non membership

627. When cultural norms and values are transmitted for generation to generation, the process is called:
(A) Cultural reproduction
(B) Acculturation
(C) Cultural diffusion
(D) Cultural stimulation

628. Which one of the following is an appropriate example of crowd?
(A) An aggregate of people at a bus stand
(B) An aggregate of people at a market place
(C) An aggregate of people in a procession
(D) An aggregate of people assembled when a house is set on fire

629. Which one of the following statements is correct?
(A) Primary groups can never arise within secondary groups
(B) Primary groups may arise within secondary groups
(C) Secondary groups are less important than primary groups
(D) Primary groups are less important than secondary groups

630. Any small group marked by continuous close interaction of a highly personal and emotionally supportive nature is called:
(A) Reference Group
(B) Primary Group
(C) Interest Group
(D) Secondary Groups

631. The essence of secondary group is its:
(A) Consciousness of kind
(B) Degree of performance
(C) Face-to-face relationship
(D) Casualness of relationship

632. The characteristic setting within which intimate, spontaneous, person-to-person interaction takes place is called:
(A) A co-operative group
(B) A primary group
(C) A model group
(D) A reference group

633. When two cultural groups come in contact with each other and consequently, in due course of time, one group borrows so heavily from the other that it loses its own identity and gets itself finally absorbed in the other group. This process is known as:
(A) Enculturation (B) Socialisation
(C) Integration (D) Assimilation

634. The thesis that the more economically developed a society becomes, the more it resembles other industrialized countries has been conceptualized as:
(A) Globalisation (B) Convergence
(C) Westernization (D) Diffusion

635. Which one of the following is not ecological process of urban growth?
(A) Centralisation (B) Decentralisation
(C) Invasion (D) Commutation

636. The view that the tribal people should be allowed to retain their special characteristic while becoming a part of the mainstream is known as:
(A) Cultural relativism (B) Acculturation
(C) Assimilation (D) Isolationism

637. A city is distinguished from a village by:
(A) Preponderance of secondary occupation
(B) Occupational immobility
(C) High birth rate
(D) Informal social control

638. Which one of the following terms correctly defines beliefs and value systems used by Robert Redfield in the analysis of Folk-Urban continuum?
(A) Great Tradition
(B) Little Tradition
(C) Universal Tradition
(D) Local Tradition

639. An institution is best described by a/an:
(A) Prescribed set of moral principles
(B) Established form or conditions of procedure
(C) Socially established organization
(D) Socially established system of values

640. If there are strong tendencies both to approach and to avoid the same goal, it is:
(A) Double approach conflict
(B) Double avoidant conflict
(C) Approach avoidant conflict
(D) None of these

641. Social distance is often associated with a scaling technique developed by:
(A) Thurstone (B) Likert
(C) Moreno (D) Bogardus

642. When a collectivity of individuals is emotionally aroused and is engaged in, or is ready to engage in, violent action it is designed as:
(A) A crowd (B) A gang
(C) A mob (D) A group

643. Socialisation takes place:
(A) During childhood
(B) During young age
(C) During the whole life
(D) At different intervals

644. All members of society occupy a number of social positions called:
(A) Role (B) Prestige
(C) Esteem (D) Status

645. The group in which a person adopts as his standard for faishoning his behaviour is called a:
(A) Primary group (B) Reference group
(C) Religious group (D) Secondary group

646. Consider the following two statements:
Assertion (A): All groups are social relationships, but not all social relationships are groups
Reason (R): A group involves some degree of cooperation among its members for the attainment of common goals
Select the correct answer:
(A) Both A and R are true and R is the correct explanation of A
(B) Both A and R are true by R is not the correct explanation of A
(C) A is true, R is false
(D) A is false, T is true

647. Consider the following two statements:
Assertion (A): Small size and face-to-face interaction are conditions for the formulation of primary groups
Reason (R): Social interaction is more frequent and internate when the number of participants is small
Select the correct answer:
(A) Both A and R are true and R is the correct explanation of A
(B) Both A and R are true by R is not the correct explanation of..
(C) A is true, R is false
(D) A is false, T is true

648. Match list I with list II and select the correct answer:
List I (Groups)
a. In-Group b. Peer-Group
c. Primary-Group d. Secondary-Group
List II (Characteristics of Group)
1. Face-to-face relations, smallness, emotional attachment
2. Largeness, impersonal relations, contractual conditions
3. Similar status and roles of group members
4. Relations of peace order and subjectively attached feeling
5. Relation of mutual adjustment, cooperation and integration
Codes:

	a	b	c	d
(A)	4	3	1	2
(B)	1	2	3	4
(C)	4	3	1	2
(D)	5	3	1	4

649. A group become most conscious of its unity:
(A) When its members have common interests
(B) When it is threatened by another group
(C) When its members live in relative isolation
(D) When it is at the breaking point

650. Who among the following maintains that better group solidarity results with removal of misunderstanding through conflicts?
(A) Sumner (B) Dahrendorf
(C) Weber (D) Simmel

651. Minorities giving up their ancestral customs and adopting the culture of the majority is known as:
(A) Integration (B) Assimilation
(C) Accommodation (D) Socialisation

652. Which one of the following explanations is the best for the term 'Gemeinschaft'?
(A) It is a society with rigid social structure
(B) It is a small society in which with good relations
(C) It is a society in which most relationships are either personal or traditional
(D) It is a society which has caste/class differences

653. Which of the following are basic characteristics of social groups?
1. Physical presence of two or more persons
2. Interaction/with more persons
3. Definite pattern of relationship
4. Consciousness of members for belonging to the group
Choose the correct answer using the codes given below:
(A) 1, 2, 3 (B) 1, 2, 4
(C) 1, 3, 4 (D) 2, 3, 4

654. First stage of human society was that of:
(A) Agriculture (B) Pastoralist
(C) Hunting and food gathering
(D) Cottage industries

655. Culture:
(A) Traits can be acquired through socialization and habits
(B) Is something collective
(C) Has always links with the past
(D) All of these

656. Which of the following are characteristics associated with primary group?
1. Small size 2. Self interest
3. Similairty of background
4. Intensity of shared interests
Choose the correct answer using the codes given below:
(A) 1, 2, 3 (B) 1, 2, 4
(C) 1, 3, 4 (D) 2, 3, 4

657. A group which does not allow a person to join similar other groups at one and the same time is called:
(A) Disjustive group(B) In group
(C) Closed group (D) Congregate group

658. Which one of the following has elements of inherent co-operation and conflict?
(A) Association (B) Competition
(C) Adjustment (D) Assimilation

659. Which one of the following is important to differentiate between the social group and the social category?
(A) Size of the groups
(B) Reciprocal communication
(C) Geographical location
(D) Formal rules and regulation

660. Social relationship involves:
(A) Cooperation (B) Conflict
(C) Cooperation and conflict
(D) Face-to-face contact

661. When an immigrant group loses its previous cultural identity and adopts that of the host group it is called:
(A) Assimilation (B) Integration
(C) Acculturation (D) Culturation

662. Which among the following is based on direct cooperation:
(A) Family (B) Parliament
(C) Government (D) General assembly

663 A definite we feeling is the characteristic feature of :
(A) Racial group
(B) Primary group
(C) Nationality group
(D) Inherited group

664. Match list I with list II and select the correct answer:

List I (Concepts)

a. Primary Group b. Secondary Group
c. Reference Group d. In-Group

List II (Examples)

1. We, Indians
2. Undivided Hindu family
3. MTNL Labour Union
4. American students to an immigrant student
5. Sick patients

Codes:

	a	b	c	d
(A)	2	3	4	1
(B)	1	2	3	4
(C)	4	3	1	2
(D)	5	3	1	4

665. Norms are imposed through which means of social control?
(A) Formal and legal
(B) Formal and informal
(C) Cultural (D) Informal and legal

666. A group to which we compare ourselves when we judge our own status:
(A) Kin group (B) Peer group
(C) Secondary group
(D) Reference group

667. Which one of the following is not an illustration of integration?
(A) The merging of Princely status into Indian union after independence in 1947
(B) Introduction of three language formula in education
(C) Providing reservations to SC/ST communities
(D) Enacting various legislations for various religious groups

668. The primary factor in reference group behaviour is:
(A) Jealosy (B) Relative deprivation
(C) Out-going (D) Membership

669. Silent Trade is a form of:
(A) Barter (B) Money barter
(C) Market exchange
(D) Gift exchange

670. Sociology have divided society into hierarchical stages on the basis of:
(A) Technology, instruments, means of living
(B) Technology, use of fire, use of wheel
(C) Means of living
(D) Economy, means of living

671. Trait is:
(A) Social trend at present
(B) Social trend of the past
(C) Smallest unit of culture
(D) Smallest unit of personality

672. Match list I with list II and select the correct answer:

List I

a. Domestic Group
b. Local Descent Group
c. Lineage d. Clan

List II

1. Khandan 2. Pariwar
3. Kul 4. Gotra
5. Pravara

Codes:

	a	b	c	d
(A)	4	3	1	2
(B)	2	1	3	4
(C)	5	1	3	2

673. The hunting and food gathering stage is further divided into simple and developed stages on the basis of use of:
(A) Tools and animals hunted
(B) Stone and no use of stone
(C) Fire and no use of fire
(D) Wheel and no use of wheel

674. Sex, age and caste are example of:
(A) Achieved status
(B) Ascribed status
(C) Pre-set status (D) Status image

675. Society exclude:
(A) Reciprocity (B) Differences
(C) Interdependence (D) Adherence to time

676. A feature of society is:
(A) Interdependence
(B) Cooperation and conflict
(C) Completion (D) All of these

677. Sumner called his model of society as:
(A) Developmental model
(B) Progressive model
(C) Consensus model
(D) Conflict model

678. Necessary condition for a population to be defined as a society is:
(A) Common culture
(B) Shared interaction
(C) Common territory
(D) All of these

679. Culture has two essential qualities:
(A) Learned and shared
(B) Transmitted and shared
(C) Learned and forgotten
(D) Shared and communicated

630. Peer group is a group whose members share similar:
(A) Play ground
(B) Circumstances
(C) Values
(D) Study circle and books

681. Peer groups are often of a similar:
(A) Age (B) Height
(C) Race and colour (D) Culture

682. The process by which an individual learns the culture of their society is called:
(A) Socialisation (B) Internalization
(C) Sanskritisation (D) Modernization

683. Every culture contains a large number of guidelines which direct conduct in a particular situation is called:
(A) Norms (B) Culture
(C) Folkways (D) Mores

684. A norm is a:
(A) Specific guide to action
(B) Culture of society
(C) Guideline for socialization
(D) Guideline for social interaction

685. The descendants of a common male ancestor are called:
(A) Agnates
(B) Cognates
(C) Lineal kin
(D) Collateral kin

686. A group of people who come together to pursue a mutual goal and who behave according to established norms is called:
(A) Casual group
(B) Casual crowd
(C) Conventional crowd
(D) Organisational crowd

687. Norms are enforced by:
(A) Positive sanction
(B) Negative sanction
(C) Order
(D) Positive ad negative sanction

688. A social group is a:
(A) System of social interaction
(B) Gathering of people
(C) Organisation of people for purpose
(D) Association of people

689. Among the following which does not indicate status:
(A) Bharat Ratna
(B) Labourer
(C) Title knighthood
(D) Use of prefix 'Pandit'

690. Society is defined as a group of individuals:
(A) Who share common culture
(B) With common blood group
(C) Each of whom interacts with each of the other members some what frequently
(D) Who share common territory

691. Social position:
(A) May be occupied more than one person
(B) Is occupied by only one person
(C) Same person occupies many social positions
(D) All of these

692. Who among the following does not enjoy esteem:
(A) Dentist
(B) Police officer
(C) Brilliant student
(D) Corrupt official

693. A social position has which of the following parts?
(A) Obligations and rights
(B) Rights and conflicts
(C) Cooperation and conflict
(D) Rights and statues

694. The prestige of social position may have in:
(A) Status (B) In time
(C) Never (D) Society

695. Personality is expressed through:
(A) Habits, tendencies and thinking
(B) Habits ad behaviour
(C) Thinking
(D) Tendencies and thinking

696. To prepare one or future roles is:
(A) Futursation
(B) Prediction
(C) Anticipatory socialization
(D) Internalisation

697. Groups from which the individual extracts his behaviour and cultural norms are called:
(A) Cultural groups
(B) Homogeneous groups
(C) Reference groups
(D) Ought-to-be groups

698. Against any individual or group, discrimination refers to:
(A) Negative action
(B) Negative attitude
(C) Negative labeling
(D) Negative report

699. Match list I with list II and select the correct answer:
List I (Concepts)
a. Affinal kin b. Descent
c. Agnates d. Lineage
List II (Explanation)
1. Male ancestor
2. Relatives of three generations
3. Wife 4. Mother
5. Ambilateral
Codes:

	a	b	c	d
(A)	4	3	1	2
(B)	3	5	1	2
(C)	5	1	3	2
(D)	5	3	1	4

700. Behaviour:
(A) Is to a large extent cultural
(B) And culture are same
(C) Is not determined by culture
(D) Is dependent on some great man's behaviour

701. Non-material culture may consist of:
(A) Words people use, the ideas, customs and belives they hold and the habits they follow
(B) Non-manufactured items
(C) Manufactured items which people use in cultural way
(D) Culture which teaches a particular behaviour for particular situation

702. A cluster of related traits is called:
(A) Culture norms
(B) Culture complex
(C) Cultural collectivity
(D) Cultural uniformity

703. An institution is a:
(A) Defined way of behaviour
(B) Cluster of traits
(C) Series of complexes centering upon an important activity
(D) Institutionalised relations

704. Which of the following are the salient features of a market economy?
1. Institutionalized norms
2. Ubiquitous in all societies
3. System of exchange of goods and services
4. Specialized economic functions

Select the correct answer from below:
(A) 1, 2, 3 (B) 1, 2
(C) 1, 3, 4 (D) 1, 2, 3, 4

705. The acquisition of foreign culture by the subject people is called:
(A) Assimilation
(B) Acculturation
(C) Colonisation
(D) Subjectisation

706. A matriarchal is wherein:
(A) Marriage relations are permanent
(B) Descent is reckoned through the mother
(C) Authority of the family with father
(D) Al children succeed to the property

707. Associations:
(A) Have fixed and written rules
(B) And formal organizations are qite similar
(C) Are voluntarily formed
(D) All of these

708. Which of the following is not a characteristic of secondary group:
(A) Formal structure
(B) Large size
(C) Compulsory membership
(D) Indirect relations

709. Roles and statuses are often most:
(A) Conflicting (B) Unconnected
(C) Complementary (D) Contradictory

710. An act is considered legitimate or morally acceptable when it is in harmony with:
(A) Norms
(B) Accepted values
(C) Folkways
(D) Mores

711. Match list I with list II and select the correct answer:

List I (Name of Family)
a. Gharawad b. Okka
c. Nok d. Illam

List II (Tribes/Castes)
1. Coorgs 2. Nayars
3. Garos 4. Namboodiris
5. World Revolution and family pattern

Codes:

	a	b	c	d
(A)	4	3	1	2
(B)	2	1	3	4
(C)	5	1	3	2
(D)	5	3	1	4

712. Adaptation and assimilation:
(A) Is the process that brings about cultural diffusion
(B) Is a historical process
(C) Both (A) and (B)
(D) None of these

713. Competition between two or more goal is the characteristic feature of;
(A) Approach avoidant conflict
(B) Double avoidant conflict
(C) Double approach conflict
(D) None of these

714. The idea of folk-urban continuum was developed on the basis of studies conducted in:
(A) Mexico (B) Canada
(C) India (D) China

715. Society is an abstract organization which prevails where:
(A) Ever there is existence of social relationship
(B) There is group
(C) There is social stratification
(D) There is law and order

716. First-hand study of a small group is called:
(A) Ethnomethodology
(B) Ethnography
(C) Biography
(D) Sampling

717. The institutional basis of social reconstruction according to Comte is:
(A) Family (B) Morality
(C) Religion (D) All of these

718. Providing emotional support to the individual at the time of individual or social crisis is called:
(A) Emotional function
(B) Buttress function
(C) Calm function (D) Societal function

719. Ego's wife is lies:
(A) Primary conanguineal kin
(B) Primary affinal kin
(C) Secondary kin
(D) Tertiary kin

720. An aggregate of onlookers viewing a common event where the crowd has only a momentary existence and no internal organization is called:
(A) Casual crowd (B) Quasi crowd
(C) Onlooking group (D) Quasi group

721. No social relationship can exist without:
(A) Face to face contact
(B) Physical awareness of the presence
(C) Likeness of interest
(D) Carry common life activities

722. Identify the quasi-group from among the following:
(A) Crowd (B) Mob
(C) Trade union (D) State

723. What makes a society:
(A) People
(B) Place of their residence
(C) Reciprocity (D) Time boundness

724. Which among the following statements is correct?
(A) No community can be self sufficient these days
(B) Every community is sele sufficient these days
(C) The term community denotes almost uniformly and permanently shared lives of people over a definite range
(D) All of these

725. The social arrangement in which a group of people plans the activities which others are obliged to participate in is called:
(A) Informal organization
(B) Formal organization
(C) Association (D) Social group

726. Match list I with list II and select the correct answer:
List I (Term)
a. Avoidance b. Joint family
c. Gotra d. Sororate
List II (Concept)
1. Marriage 2. Kinship group
3. Exogamy 4. Kinship usage
5. Cult

Codes:

	a	b	c	d
(A)	4	2	3	1
(B)	3	5	1	2
(C)	5	1	3	2
(D)	5	3	1	4

727. Values:
(A) Guide a person's behaviour
(B) Guide a person's judgement
(C) Are an important part of every culture
(D) All of these

728. Norms is a:
(A) Rule
(B) Standard for action
(C) Pattern for action (D) All of these

729. Which of the following refers to norms that are looked on by the members of society as not being extremely important and may be violated without any severe punishment?
(A) Mores (B) Values
(C) Norms (D) Folkways

730. It is a centrist party in its ideological orientation, the party espouses secularism and welfare of weaker sections and minorities. Identify the party:
(A) Bhartiya Janta Party
(B) Indian National Congress
(C) Bahujan Samaj Party
(D) Communist Party of India Marxist

731. The norms that are looked on by the members of a society as being extremely important and violation of which results in severe punishment are:
(A) Norms (B) Values
(C) Mores (D) Folkways

732. Non-conformity to social norms is called:
(A) Social change (B) Deviance
(C) Social conflict (D) Struggle

733. A form of social interaction wherein two or more persons work together to gain a common end is called:
(A) Association (B) Cooperation
(C) Organisation (D) Accommodation

734. The party seeks to represent and secure power for the Bahujan Samaj which includes the Dalits, Adivasis, OBC etc:
(A) INC (B) BSP
(C) BJP (D) CPI(M)

735. Accommodation is essentially a process of:
(A) Adjustment
(B) Cooperation
(C) Group formation
(D) Competition

736. Two types of accommodation are:
(A) Acclimatisation and Naturalisation
(B) Assimilation and Acculturation
(C) Association and Organisation
(D) Assimilation and Naturalisation

737. "Sociology is a scheme for scientific social development". This statement is given by:
(A) Hegal
(B) Maciver
(C) C. Duncan Mitchell
(D) Auguste Comte

738. Acclimatisation refers to:
(A) Coming to term with new climate, type of soil, weather, air
(B) Coming to term with new political situation
(C) Coming to term with new economic situation
(D) Coming to term with new educational and learning situation

739. The party enjoys strong support in West Bengal. Supports socialism, seculism and democracy. Identify the party:
(A) INC (B) BSP
(C) BJP (D) CPI(M)

740. Naturalisation refers to:
(A) Coming to term with new political situation
(B) Coming to term with new economic situation
(C) Coming to term with new social milieu such as folkway, mores, etc
(D) Coming to term with new educational and learning situation

741. "Process of uniting of formerly separate groups into one group with obliteration of separate group differences and identification" is called:
(A) Competition (B) Assimilation
(C) Integration (D) Acclimatisation

742. Why is democracy better form of government than others:
(i) It improves the quality of decision making
(ii) It allows the citizens to do whatever they like
(A) (i) is true but (ii) is false
(B) (i) is false but (ii) is true
(C) Both (i) and (ii) are true
(D) Both (i) and (ii) are false

743. Which of the following is part of little tradition?
(A) Folk-tales (B) Legends
(C) Marriage songs (D) All of these

744. "Interaction without social contact" is:
(A) Integration
(B) Assimilation
(C) Competition
(D) Association

745. The term 'Diffusion' as used by anthropologists refers to the spread of element of:
(A) Customs (B) Way of life
(C) Culture (D) Norms

746. Consider the following two statements:

Assertion (A): Cities are centre of social heterogeneity and impersonal relations

Reason (R): Cities are generally populous

Select the correct answer:

(A) Both A and R are true and R is the correct explanation of A

(B) Both A and R are true by R is not the correct explanation of A

(C) A is true, R is false

(D) A is false, T is true

747. Levirate means:

(A) Man marrying his wife's sister while the wife is alive

(B) Man marrying his wife's sister after the death of his wife

(C) Woman marrying her husband's brother while the husband is still alive

(D) Woman marrying her husband's brother after the death of her husband

748. Folkways:

(A) Result from the frequent repetition of acts

(B) Are habits in the individual

(C) Are customs in the group

(D) All of these

749. Most of the social differences are based on:

(A) Birth (B) Choice

(C) Caste (D) Religion

750. Normative culture is that sub-division of non-material culture that consists of:

(A) Norms

(B) Values

(C) Mores

(D) Standard way

751. Who coined the term "Sanskritization":

(A) M.N. Srinivas

(B) A.K. Dahama

(C) Auguste Comte

(D) Weber

752. Which of the following is true with reference to democracy:

(i) Democracy allows us to correct mistakes

(ii) It can address all the socio-economic and political problems

(A) (i) is true but (ii) is false

(B) (i) is false but (ii) is true

(C) Both (i) and (ii) are true

(D) Both (i) and (ii) are false

753. The sector of economy that deals with raw materials including agriculture, fishing, mining and forestry is called:

(A) Primary (B) Agriculture

(C) Basic (D) Main

754. Primary civilisation of India is divided into:

(A) Little tradition and great tradition

(B) Big tradition and small tradition

(C) Educated tradition and uneducated tradition

(D) Rural tradition and caste tradition

755. Which of the following is not basic outcome of democracy?

(A) It produces a government that is accountable to the citizens and responsive to the needs and expectations of the citizens

(B) It creates economic equality

(C) It leads to better economic growth rate

(D) It reduces inequality in income

756. Which of the following statements regarding social differences is not correct?

(A) The social differences are mostly based on accident of birth

(B) Some of the social differences are based on our choices

(C) Every social difference does not lead to social division

(D) Social differences are not found in homogeneous societies

757. Little tradition refers to:

(A) Folk or unlettered peasants

(B) Urban culture

(C) Caste system

(D) Village jajmani system

758. Feeling as others feel refers to:
(A) Imitation (B) Identification
(C) Suggestion (D) Sympathy

759. To findout the degree of relationship between two existing traits we go for :
(A) Correlation study (B) Factor analysis
(C) Analysis of variance (D) None of these

760. Two aspects of social control are:
(A) Pressure oriented and value oriented
(B) Pressure oriented and norms oriented
(C) Culture oriented and customs oriented
(D) Religious oriented and education oriented

761. The organized for social behaviour and their repetition are called:
(A) Culture (B) Value
(C) Customs (D) Norms

762. Where there is life, there is society:. This statement is given by:
(A) Auguste Comte (B) Aristotle
(C) Confucius (D) Maciver and Page

763. Arrangement of person indicates:
(A) Culture (B) Group
(C) Position (D) Folkways

764. The expectations between a focal position and a number of different counter positions make up:
(A) Multiple role (B) Plural role
(C) Role set (D) Role relation

765. Which of the following is/are normally missing in a non democratic country?
(i) Transparency (ii) Elections
(iii) Equality among citizens
(A) (i) and (ii)
(B) Only (i)
(C) Only (ii)
(D) Only (i) and (iii)

766. Social processes are those which throughout the life are:
(A) Continuous
(B) Halted
(C) Broken
(D) Not achieved

767. Consider the following two statements:
Assertion (A): There is a proliferation of voluntary associations in cities
Reason (R): The hold of collective consciousness declines with an increase in organic solidarity
Select the correct answer:
(A) Both A and R are true and R is the correct explanation of A
(B) Both A and R are true by R is not the correct explanation of A
(C) A is true, R is false
(D) A is false, T is true

768. A person who participates in a group without ever becoming a member ever becoming a member of that group is called:
(A) Marginal man
(B) Distance man
(C) Distant man
(D) Symbolic man

769. Comte's intellectual efforts aimed at:
(A) Knowledge, capability to predict, control
(B) Belief, knowledge, control
(C) Analysis, objectivity, control
(D) Belief, control, assumption

770. In his/her study of human behaviour, sociologists:
(A) Determine which values are of greatest importance
(B) Treat values of data which he studies
(C) Attempts to justify the values of his society if he/she is good at sociology
(D) Leaves values completely along since they have no place in human behaviour

771. Face to face association and cooperation is more possible in the:
(A) Transitory group
(B) Ad-hoc group
(C) Primary group
(D) Basic group

772. "A person who is overtly concerned with vocational security, excessively methodical, more concerned with rules than organizational goal and unimaginative in solving problems" is a/an:
(A) Bureaucratic personality
(B) Idle personality
(C) Organisational personality
(D) Puritan personality

773. The Jharkhand movement is organized around:
(A) Illiteracy (B) Job deprivation
(C) Isolation (D) All of these

774. The regular movement of pastoral nomads with their livestock in search of fresh pastorage is called:
(A) Migration (B) Exodus
(C) Transhumance (D) Movement

775. "Study of positive Philosophy" is a book written by:
(A) Aristotle (B) Hippocrates
(C) Alter (D) Auguste Comte

776. When two or more persons meet it is:
(A) Social organization (B) Social group
(C) Social systems (D) Society

777. Three major views of social structure are:
(A) Structural-functionalist, structuralist, Marxist
(B) Interactionist, Marxist, functionalist
(C) Symbolic, Marxist, structuralist
(D) Structuralist, functionalist, interactionalist

778. The system of production for exchange supports the system of:
(A) Feudal economics
(B) Socialist economics
(C) Capitalist economics
(D) Post-industrial economics

779. The Indian type of family is characterized by:
(A) Polyandrous (B) Conjugal
(C) Matrilineal (D) Consanguinous

780. A barren wife or a sterile husband is an example of:
(A) Role strains (B) Role other
(C) No role (D) Role lacking

781. Providing cooperative attitudes, faiths, hope and charity is the characteristic function of:
(A) Religious institution
(B) Economic institution
(C) Political institution
(D) Educational institution

782. Folkways that deals with moral aspects are:
(A) Taboos (B) Laws
(C) Mores (D) Norms

783. A good illustration of social group is:
(A) Family
(B) A crowd in a street
(C) People in a circus tent
(D) A market

784. Which of the following is not related to the community:
(A) Group of people (B) Particular name
(C) Definite locality (D) Temporary

785. Mark out the factor that does not affect density of population:
(A) Climate (B) Surface of land
(C) Fertility of soil (D) Pollution

786. A social group is represented by:
(A) A golf club (B) Students of physics
(C) Members of trade union congress
(D) Family

787. An aggregate which lacks structure or organization and whose members may be unaware or less aware of the existence of the grouping has been called:
(A) Community
(B) Social group
(C) Primary group
(D) Institution

788. The term "Rurabinate" meaning rural and urban people was coined by:
(A) Donald Tuttle (B) M.K. Gandhi
(C) J.F. Davidson (D) Paul Leagans

789. What is one party system?
(i) A country where only one party is allowed to control and run the government
(ii) A country where only one communist party is allowed to work
(A) (i) is true but (ii) is false
(B) (i) is false but (ii) is true
(C) Both (i) and (ii) are false
(D) Both (i) and (ii) are true

790. To achieve consensus, primary group follows method of:
(A) Enumeration (B) Integration
(C) Compromise (D) All of these

791. The class of objects having more than one feature in common:
(A) Relational concepts
(B) Disjunctive concept
(C) Conjunctive concept
(D) None of these

792. Social organization means:
(A) Planned activities
(B) Organized activities
(C) Calculated activities
(D) Defined activities

793. Musicians are a/an:
(A) Aggregate (B) Group
(C) Category (D) Class

794. An example of institution is:
(A) Family (B) Marriage
(C) Kinship (D) All of these

795. Which of the following country has one party system?
(A) India (B) China
(C) USA (D) UK

796. If several parties compete for power, and more than two parties have a reasonable chance of coming to power either own strength or alliance with others, we call it a..........:
(A) Biparty system
(B) Multi-party system
(C) One-party system
(D) None of these

797. Status groups are not stratified according to:
(A) Type of dwelling
(B) Food habits
(C) Recreational activities
(D) Wealth acquisition

798. Nazism is a classical example of:
(A) Social stratification
(B) Social disintegration
(C) Ethnocentrism
(D) Cultural conflicts

799. Informal relationship do exist in the best form in which of the following institutions:
(A) Family (B) Economic
(C) Political (D) Religious

800. A stable society is a prerequisite for:
(A) Bureaucracy (B) Culture
(C) An integrated personality
(D) Family

801. Lack of relationship with other culture is called:
(A) Isolation (B) Insularity
(C) Separation (D) Polarity

802. An individual migrating from a village to a city is a example of:
(A) Changing group (B) Migration
(C) Leaving group (D) Exodus

803. The most short lived or social groups is:
(A) Crowd (B) Congregation
(C) Association (D) Organisation

804. Which of the following country has two party system?
(i) United States of America
(ii) United kingdom
(iii) India
(iv) China
(A) Only (i) and (ii) (B) Only (ii) and (iii)
(C) Only (iii) and (iv) (D) All of these

805. A situation in which large number of people attempt to obtain or achieve the same thing at the same time is called:
(A) Boom (B) Acquisition
(C) Fad (D) Fashion

806. 'Cultural is essentially a set of rules' according to:
(A) Malinowski (B) Redcliffe Brown
(C) Parsons (D) Merton

807. Marriage due to mutual affection is called:
(A) Gandharva (B) Rakshasa
(C) Daiva (D) Asura

808. Cultural pluralism may represent:
(A) Cultural diversity
(B) Incomplete assimilation
(C) Incomplete acceleration – acculturation
(D) Unity in diversity

809. Parliament is an example of:
(A) Delegate group
(B) Representative group
(C) Expressive group (D) Political group

810. Which of the following country has adopted multi party system?
(A) India (B) China
(C) United States (D) United Kingdom

811. Separate from other sets of related roles which are also arranged in hierarchies'. It is:
(A) Status set
(B) Stratification
(C) Situses set
(D) Role-relation

812. The form of marriage under which girl is given in marriage to a bridegroom with cloths and ornaments is called:
(A) Brahma (B) Asura
(C) Paisacha (D) Rakshasa

813. Tribal culture is attached to:
(A) Land, nature and rituals
(B) Falk tradition
(C) Their economy and customs
(D) Religious only

814. When several parties in a multiparty system join hands for contrasting election. Choose the right word:
(A) Partisan (B) Multiparty system
(C) Alliance (D) Defection

815. 'Related occupational specialties arranged hierarchically parallel to and 697. Populat culture refers to the culture of:
(A) Political class
(B) Culture of masses or of subordinate classes
(C) Culture of rich
(D) Culture of educated

816. The area in which similar cultural traits are found is called:
(A) Cultural state
(B) Cultural region
(C) Cultural area
(D) Cultural territory

817. Which of the following are true for multiparty system?
(i) It leads to political instability
(ii) It allows a variety of interests and opinions to enjoy political representation
(A) (i) is true (ii) is false
(B) (i) is false (ii) is true
(C) Both (i) and (ii) are true
(D) Both (i) and (ii) are false

818. In a primary relationship, which of the following is involved in the interaction?
(A) Partial person (B) Total person
(C) Total society (D) A group in society

819. Man, for the first time, became tool maker during the:
(A) Plaeolithic age (B) Mesolithic age
(C) Chalcolithic age (D) Meolithic age

820. A subjective evaluation of personality, particularly with regard to a person's desirable or undesirable qualities is called:
(A) Personality (B) Character
(C) Temperament (D) Self concept

821. All illiterate form a/an:
(A) Community (B) Category
(C) Aggregate (D) Social group

822. A correlation coefficient of zero indicates:
(A) Nonexistent relationship
(B) Existent relationship
(C) Neutral relationship
(D) None of these

823. The phenomenon of 'Cultural Relativism' is based on the assumption that:
(A) All cultures are equal and respectable
(B) Culture is acquired through learning
(C) Culture is genetically transmitted
(D) Culture determines make up of human being

824. Ethnic groups are also called:
(A) Deprived groups
(B) Waifed groups
(C) Separate groups
(D) Primordial collectivities

825. Which among the following holds true in case of values:
(A) Values are inherited
(B) Values are learned
(C) Values are never flexible
(D) All of these

826. The stage of demographic transition cycle wherein low mortality forms the main reason for population explosion:
(A) 1st stage
(B) 2nd stage
(C) 3rd stage
(D) 4th stage

827. An example of subculture is:
(A) American culture
(B) Indian culture
(C) Nepali culture
(D) Andamani culture

828. In sociology, 'Society' refers to the:
(A) Members of a specific in-group
(B) Patterns of the norms of interaction
(C) Congregation of people
(D) People with laws and customs

829. Informal rules of behaviour based on habits and tradition:
(A) Mores (B) Traditions
(C) Values (D) Folkways

830. The movement of people outside a society is called:
(A) Immigration (B) Emigration
(C) Migration (D) All of these

831. In India language has:
(A) Helped in promoting national unity
(B) Stood in the way of national integration
(C) Created no problem at any level
(D) Played healthy role in national life

832. A system of relationship between cells is:
(A) Group (B) Society
(C) Organism (D) Aggregation

833. Which of the following is covered in the equalitarian socialization process?
(A) Playground mates (B) Friends
(C) Classmates (D) All of these

834. Main criticism against social contact theory about the origin of state is that it believes that man came:
(A) Along with society
(B) After the society
(C) Before the society
(D) Long after society had established itself

835. Which of the following is the process by which the individual learns to conform to the norms of the group?
(A) Conformity (B) Socialisation
(C) Integration (D) Assimilation

836. The members of modern social system are:
(A) More emphatic
(B) More cosmopolite
(C) Less emphatic
(D) Both (A) and (B)

837. Take the odd man out:
(A) Competition (B) Accomodation
(C) Assimilation (D) Cooperation

838. Setlement pattern based on economic consideration:
(A) Cross road (B) Circular
(C) Nuclear (D) Line

839. 'Man is a social animal.' This is said by:
(A) Aristotle (B) Freud
(C) Maclver (D) Rousseau

840. "Table manners' falls under the category of:
(A) Mores (B) Folkways
(C) Customs (D) Values

841. Socialisation is different from socialism because the latter is:
(A) Concerned with the present structure of society
(B) Theory of future structure of society
(C) Way of social living and behaving
(D) Not concerned with social structure

842. Slips of letters may point to:
(A) Preconscious motives
(B) Conscious motives
(C) Unconscious motives
(D) All of these

843. The sociology of law involves the study of how social conditions influence:
(A) Enforcement of law
(B) Making of law
(C) Making and enforcement of law
(D) Quick deliverance of justice

844. When cultural traits spread from one society to another, the process is called:
(A) Diffusion (B) Parallelism
(C) Evolution (D) Pluralism

845. The rule that one marry within one's own caste is called:
(A) Hypogamy (B) Hypergamy
(C) Endogamy (D) Exogamy

846. Society helps to:
(A) Undermine social tensions
(B) Ignore social tensions
(C) Increase social tensions
(D) Regularise social tensions

847. A codifies system of mores is generally called:
(A) Norms (B) Laws
(C) Customs (D) Values

848. Social institutions exist or continue on account of:
(A) Common origin (B) Common history
(C) Likeness (D) Differences

849. In society difference grow due to:
(A) Isolation (B) Non-socialisation
(C) Specialisation (D) Socialisation

850. The word family have its root in:
(A) Greek (B) Roman
(C) Latin (D) British

851. Class struggle is most intensive in:
(A) Capitalistic society
(B) Primitive society
(C) Feudal society (D) Socialistic society

852. Individual defines society because society frustrates individual's attempt for the sake of:
(A) Interests of weaker sections of society
(B) Elite interests (C) Collective interests
(D) Promoting group interests

853. The concept of gemeinschaft and gasellschaft was developed by:
(A) E.M. Rogers (B) F. Tonnies
(C) Max Webber (D) Keynes

854. The gap between the extremes of social pyramid is the highest in:
(A) Rural society (B) Urban society
(C) Peri-urban society
(D) Peri-rural society

855. An important advantage of socialization is that it makes one:
(A) Realise the importance of wealth
(B) Member of political system
(C) Leave one's isolation
(D) Independent group of life

856. What is homogeneous society:
(A) When one social difference overlaps another difference
(B) When one social issue cross-cuts another social issue
(C) A society that has similar kinds of people, especially where there are no significant ethnic differences
(D) None of these

857. A person with mental ability for below average is termed as:
(A) Borderline (B) Average
(C) Mentally retarded (D) Dull normal

858. In society differences are always:
(A) Environmental
(B) Adopted
(C) Inborn
(D) Both adopted and natural

859. Society is a system of:
(A) Usages and procedures
(B) Groupings and divisions
(C) Rigid rules
(D) Unbreakable customs

860. Everything is right somewhat but not everywhere. This refers to:
(A) Cultural universals
(B) Cultural relativity
(C) Counter-culture
(D) Cultural pluralism

861. The type of correlation that operates between rurality and population density:
(A) Positive
(B) Negative
(C) Accdental
(D) Scattered

862. Ethnocentrism is dysfunctional of:
(A) Intra-group relations
(B) Peer-group relations
(C) Inter-group relations
(D) Community relations

863. Wide area of specialization is the characteristic feature of:
(A) Rural society (B) Urban society
(C) Peri-urban society
(D) Peri-rural society

864. When cultural norms and values are transmitted from generation to generation, the process is called:
(A) Cultural reproduction
(B) Acculturation
(C) Cultural diffusion
(D) Cultural stimulation

865. Dreams have been described as the royal road to the:
(A) Conscious (B) Subconscious
(C) Unconscious (D) None of these

866. A basic distinction between culture and civilization is that:
(A) Culture is what we have achieved and civilization is what we aspire to achieve
(B) Culture is what comes out of intelligence, and civilization is not
(C) Culture is secondary and civilization is primary
(D) Culture is what we are and civilization is what we have

867. Which among the following is not considered under basic rural institution:
(A) Zila Parishad (B) Gram panchayat
(C) Village school (D) Cooperative society

868. The minimal prerequisite of a social relationship is:
(A) Physical contact
(B) Face to face contact
(C) Mutual awareness (D) Harmony

869. Sumner referred to his model of society as a:
(A) Developmental model
(B) Conflict model
(C) Consensus model (D) Progressive model

870. Efforts against fragmentation is an example of:
(A) Socialization (B) Social movement
(C) Acculturation (D) Disorganization

871. The major reason for polyandry is:
(A) Woman's high sexual demand
(B) Population explosion
(C) Scarcity of woman
(D) Particular type of economy

872. What percentage of seats in local bodies are reserved for women candidates in India?
(A) Two-third (B) One-third
(C) One-fourth (D) None of these

873. Polyandry may be classified into two groups:
(A) Levirate and sororate
(B) Adelphic and Non-adelphic
(C) Hypergamy and hypogamy
(D) Cross cousin and parallel cousin

874. Totalitarian society give way to which type of institution:
(A) Catalystic (B) Voluntary
(C) Prescriptive (D) Lazerfaire

875. Number of tertiary kins are:
(A) One hundred fifty one
(B) One hundred forty one
(C) One hundred thirty one
(D) One hundred twenty one

876. In which system of marriage can one woman has many husbands with all of whole she cohabits in turn and it is not necessary that these husbands he brothers?
(A) Group marriage
(B) Non-fraternal polyandry
(C) Adelphic polyandry
(D) Joint marriage

877. What is the term used to describe the use of religion as a tool for dividing people?
(A) Diversity (B) Communalism
(C) Communism (D) Regionalism

878. In the context of democracies which of the following idea is correct-democracies have successfully eliminated:
(A) Conflicts among people
(B) Economic inequalities among people
(C) Differences of opinion about how marginalized sections are to be treated
(D) The idea of political inequality

879. Which of the following is true with respect to democracy?
(i) People prefer democracy over dictatorship
(ii) Democratic countries are free of corruption
(iii) Democratic government is a legitimate government
(A) (i) and (ii) (B) Only (i)
(C) Only (ii) (D) Only (i) and (iii)

880. Nearly universal norm for marriage in all societies is:
(A) Lineage endogamy (B) Clan endogamy
(C) Clan exogamy
(D) Parallel-cousin marriage

881. Co-operation is a:
(A) Conjunctive process
(B) Disjunctive process
(C) Dissociative process
(D) Disintegrative process

882. Wars and violence are evidence for:
(A) Life instinct (B) Libido
(C) Death instinct (D) None of these

883. 'ILA' in muslim marriage is:
(A) Muslim woman gives something as a compensation to the husband at the time of divorce
(B) Husband takes a vow to abtain from sexual intercourse for at least four months and if it is complete, it means divorce
(C) Muslim woman ask for divorce
(D) Cousin marriage

884. An arrangement in which a male can have sexual relationship with any female member without social restrictions is called:
(A) Endogamy (B) Promiscuity
(C) Maitree (D) Co-habition

885. Institutinalization is a means of:
(A) Socialization (B) Social control
(C) Accommodation (D) Interaction

886. Polygamy is a term covering:
(A) Leviate and sororate
(B) Endogamy and Exogamy
(C) Polygamy and polyandry
(D) Cross cousin marriage and parallel cousin marriage

887. Bringing prior experience to bear on problem solving is:
(A) Created images (B) Stored images
(C) Mentally rotate (D) Synesthesia

888. Sharda Act of 1929:
(A) Permitted widow remarriage among Hindus
(B) Abolished child marriage
(C) Permitted woman to earn ideas
(D) Permitted lower castes to marry in higher caste

889. Modernization must start from:
(A) Man (B) Individual
(C) Group (D) Society

890. Relatlives born out through marriage ties are called:
(A) Affinal kin (B) Consanguineal kin
(C) Secondary kin (D) Cognatics

891. Competition means:
(A) Ways of emerging as superior
(B) Accommodation
(C) Assimilation
(D) Fixation of status in society

892. Descriptive term of kinship:
(A) Describes speaker's exact relation
(B) Refers to family ties
(C) Refers to kin
(D) Refers to age

893. Which among the following technique enhances creative thinking:
(A) Brainstorming (B) Role play
(C) Simulation games (D) Case study

894. Cross-cousin marriage is prevalent among:
(A) Gond of MP (B) Meenas of Rajasthan
(C) Bhil of MP (D) Todas of Tamil nadu

895. Which is not an institution:
(A) Marriage (B) Theatre
(C) Inheritance (D) Property

896. Gifts, dowries and property were offered on the eve of marriage by:
(A) Red Indians of America
(B) All the tribes
(C) African tribes
(D) No tribes at all

897. Marrying one's mother's brother's daughter is:
(A) A from of incest prohibited in all civilized societies
(B) A kind of cross-cousin marriage
(C) Necessary in matrilineal society
(D) A common custom among all tribes

898. A form of polygyny where co-wives are sisters is called:
(A) Adelphic (B) Sistoral
(C) Sororal (D) Matriarchal

899. The ego is guided by:
(A) Pleasure principle (B) Life instinct
(C) Reality principle (D) Death instinct

900. Hindus consider marriage as obligatory because:
(A) Is is important for harmonious relations between sexes
(B) Sexual gratification outside marriage is sin
(C) Birth of a son is necessary for the discharge of obligation to the ancestors
(D) Marriage is precondition for entering into Grihastha Ashrama

901. The relationship based on the blood ties is called:
(A) Blood kinship (B) Parallel kinship
(C) Paternal kinship
(D) Consanguineous kinship

902. Under which 'ism' the norms be described:
(A) Idealism (B) Realism
(C) Utopianism (D) None of these

903. What differentiates an agricultural society from an industrial society:
(A) The environment
(B) Mode of occupation
(C) Social mobility
(D) All of these

904. If offsprings inherit the father's name, they will be described as:
(A) Name taking (B) Patrilineal
(C) Patrinomy (D) Patriarchal

905. Foundation of the Indian culture is:
(A) Railway system (B) Spiritualism
(C) Scriptures (D) English language

906. A type of polygamy in which several men and several women form a marriage relationship with one another is called:
(A) Group alliance
(B) Parallel marriage
(C) Cenogamy (D) Promisquity

907. Culture pertains to:
(A) Social heritage (B) Social etiquette
(C) Religious and moral attitudes
(D) None of these

908. Classifactory term of kinship refers to:
(A) Kin (B) Relationship
(C) More relationship rather than to kin
(D) Family ties

909. Which among the following has no mass, occupies no space, and is invisible:
(A) Memory (B) Learning
(C) Intelligence (D) None of these

910. The Bombay prevention of Hindu Bigamous Marriage Act 1947 enforced:
(A) Monogamy in the Bombay state
(B) Monogamy in India
(C) Monogamy among Parsis of Bombay
(D) Polygamy in the Bombay state

911. 'Uncle' is an example of:
(A) Classificatory term (B) Discriptive term
(C) Expected behaviour
(D) Role relation

912. Relationship of production is:
(A) Not a human relationship at all
(B) Not merely the ownership of means of production
(C) Not a cooperative relationship between individuals
(D) Essentially an exploitative relationship between producers

913. In Hindi 'sala' is an example of:
(A) Classifactory term
(B) Discriptive term
(C) Marriage alliance
(D) Expectations

914. Which of the following is a belief system:
(A) Magic (B) Totemism
(C) Fetishism (D) All of these

915. The practice to marry outside the village is prevalent in:
(A) Tribes of Chhota Nagpur
(B) Tribes of Nilgiri
(C) Tribes of Malabar
(D) Tribes of Andaman

916. Meaning of words and concepts and the rules for using them in language is contained in:
(A) Episodic memory
(B) Short term memory
(C) Long term memory
(D) Somantic memory

917. When husband and wife live alternatively at each other's residence, marriage system is called:
(A) Changing (B) Alternative
(C) Complementary (D) Contradictory

918. Relationships of production:
(A) Can generate changes in forces of production
(B) Can be at conflict with forces of production
(C) Have essentially no relationship or forces of production
(D) Can also dominate and generate changes in the forces of production

919. If a village made up of several clans, each located in a fairly definite part of the village, each of the clans is technically called:
(A) Clan continuity (B) Clan integral
(C) Composite clan (D) Clan continuity

920. Forgetting of past events together with the inability to foam new memories:
(A) Transient Global Amnesia
(B) Dream Amnesia
(C) Defense Amnesia
(D) Childhood Amnesia

921. The inability to encode and store new information in memory is called:
(A) Dream Amnesia
(B) Defensive Amnesia
(C) Childhood Amnesia
(D) Anterograde Amnesia

922. Classification between primary, secondary and tertiary kin is given by:
(A) Murdock
(B) Morgan
(C) Westermarck
(D) Parsons

923. Man's behaviour in society is determined by two forces namely:
(A) Formal and informal
(B) Natural and unnatural
(C) Physical and social
(D) Psychological and philosophical

924. Number of primary kins are:
(A) Eight (B) Seven
(C) Six (D) Five

925. A crowd is distinguished by:
(A) Transitory nature
(B) Continuous nature
(C) Permanent nature
(D) Stable nature

926. Polyandry and transient marriage relations were more common in primitive times than monogamy or polygamy. This view is followed from:
(A) Metronymic theory
(B) Matriarchal theory
(C) Patronymic theory
(D) Patriarchal theory

927. Two types of marriage among Muslims are:
(A) Polyandry and polygyny
(B) Beena and Muta
(C) Beena and Meena
(D) Monogamy and polygamy

928. Acculturation, social integration and assimilation might be described as:
(A) Steps in social movement
(B) Steps inassimilation process
(C) Forms of accomodation
(D) Anti-competitive behaviour

929. A specialized foam of direct observation is called:
(A) Observation
(B) Behavioural assessment
(C) Rating scale (D) Situational testing

930. For motivation, need and incentives respectively provide the:
(A) Pull and Push (B) Push and Pull
(C) Pull and Pull (D) Push and Push

931. Curiosity is an example for:
(A) Primary motive
(B) Secondary motive
(C) Stimulus motive (D) All of these

932. The basic speech sounds are called:
(A) Phonemes (B) Morphemes
(C) Heuristics (D) Programming

933. The sexual energy that underlines the biologically based urges is:
(A) Eros (B) Thanotos
(C) Psyche (D) Libido

934. A peaceful coexistence of distinct ethnic groups in society is called:
(A) Integration (B) Assimilation
(C) Amalgamation (D) Cultural Pluralism

935. When several sibs or clans combine to constitute a still wider grouping it is called:
(A) Clans grouping (B) Clan community
(C) Phratry (D) Joint family

936. Under which mode of production is labour power bought and sold?
(A) Asiatic (B) Ancient
(C) Feudal (D) Capitalist

937. The concept of primary groups was introduced by:
(A) Cooley (B) Darkheim
(C) Ginsberg (D) Sorokin

938. Most employer-employee relationships involve social interaction of the type:
(A) Cooperation (B) Conflict
(C) Exchange (D) Competition

939. Which of the following does not sociology consider as an example of social action:
(A) A man writing a letter to his friend
(B) A wife reading her husband's letter
(C) Father admonishing his son for misbehaviour
(D) Teacher, correcting the scripts of his student

940. The essential function of the family is:
(A) Satisfaction of sex needs
(B) Procreation and rearing of children
(C) Socialization of children
(D) Transmission of culture

941. All the property is distributed in all the sons and females have no property right. This system is called:
(A) Mitakshara (B) Matridual
(C) Vishakra (D) Pratikata

942. The term 'group-dynamics' refers to:
(A) Formation of large groups
(B) Adjustive changes in small groups
(C) Disintegration of a small group
(D) Interdependence of several groups

943. Which is not an organization:
(A) College (B) Factory
(C) Prison (D) None of these

944. Parallel cousin is also called:
(A) Ortho-cousin (B) Orio-cousin
(C) Couvade (D) Neo-cousin

945. Aggregates of fully established and accepted rules, procedures and norms:
(A) Value (B) Institution
(C) Folkways (D) Community

946. 'Okka parage marriage' is:
(A) Is preferred when on okka does not have son to continue the line
(B) Is 12 pieces of gold transferred from the man's okka to wife's okka
(C) Is one in which husband acquires membership of his wife okka
(D) All of these

947. When child bearing is deliberately avoided, the family is called:
(A) Non-conception family
(B) Confined family
(C) Arrested family
(D) Trying family

948. 'Preferential marriage' is a marriage in which ego's spouse is someone who is:
(A) Already related to ego
(B) His/her husband/wife's brother
(C) His/her husband/wife's sister
(D) Primary kin of ego

949. Number of secondary kins are:
(A) Thirty one (B) Thirty two
(C) Thirty three (D) Thirty four

950. Meaning of words and concepts and the rules for using them in language is contained in:
(A) Episodic memory
(B) Short term memory
(C) Long term memory
(D) Somantic memory

951. Status and rank characterize the members of:
(A) Tribe (B) Caste
(C) Society (D) Clan

952. A social mechanism for preventing certain kinship bonds from falling into neglect in a matrilineal society is:
(A) Amitate (B) Control mechanism
(C) Kin cherishing(D) Kin-contact

953. Creative thinking type with novel and unusual suggestions:
(A) Fluency (B) Flexibility
(C) Originality (D) Divergence

954. Which of the following can be appropriately called a mode of production?
(A) Feudal (B) Tribal
(C) Agricultural (D) Pastoral

955. The first stage in creative thinking among the following is:
(A) Incubation (B) Preparation
(C) Orientation (D) Verification

956. Consider the following two statements:
Assertion (A): Reference group formation is functional to the individual and society
Reason (R): Reference group formation leads to anticipatory socialisation
Select the correct answer:
(A) Both A and R are true and R is the correct explanation of A
(B) Both A and R are true by R is not the correct explanation of A
(C) A is true, R is false
(D) A is false, T is true

957. The term 'amnesia' means:
(A) Forgetting (B) Motivated forgetting
(C) Loss of memory (D) Disease of memory

958. The dog's mouth, after training watered whenever the bell was sounded:
(A) Conditional response
(B) Conditional stimulus
(C) Classical conditioning
(D) Unconditional response

959. 'Agricultural Foundamentalism' was given by:
(A) Davis (B) Leagens
(C) Fisher (D) Lehman

960. Folkways have:
(A) Moral validity
(B) Conventional validity
(C) Legal validity (D) All of these

961. IQ is calculated based on the formulae:
(A) Mental age + chronological age/100
(B) 100/Mental age x chronological age
(C) Mental age / chronological age x 100
(D) Chronological age/Mental age x 100

962. The nature of human is:
(A) Not immutable but historical
(B) An indeterminate unconsciousness
(C) Determined by psychological traits
(D) A manifestation of ecological factors

963. When more than one wife is shared by several brothers, it is called:
(A) Polyandry (B) Polygyny
(C) Polygynandry (D) Polygamy

964. Patriarchal and matriarchal theories about the origin of society believe that society emerged as a result of:
(A) Social contract
(B) Expansion of family
(C) Evolution of several factors combined together
(D) Use of brute force

965. Levirate marriage is also called:
(A) Nantra (B) Mantra
(C) Tandra (D) Kantra

966. Bringing prior experience to bear on problem solving is:
(A) Created images (B) Stored images
(C) Mentally rotate (D) Synesthesia

967. The Hindu Marriage Act of 1955 prescribes the minimum age of marriage is:
(A) 18 years for boys and 15 years for girls
(B) 21 years for boys and 18 years for girls
(C) 25 years for boys and 21 years for girls
(D) 18 years for boys and 16 years for girls

968. A family according to the rule of lineage may be:
(A) Patrilocal and matrilocal
(B) Patrilineal and matrilineal
(C) Patriarchal and matriacrchal
(D) Polygamous and polygynous

969. Hindu Marriagc Act of 1955 is:
(A) Bars marriage within five generations on the agnatic side
(B) Bars marriage within three generations on mother's side
(C) Permits marriage between cross cousins where it is customary
(D) All of these

970. Class is based on:
(A) Co-operation
(B) Competition
(C) Individual capability
(D) Economic level

971. Ritual kinship refers to:
(A) Legally defined marriage relationship
(B) Kins related to one another through blood
(C) Kinship on the basis of functional social relationship
(D) Persons related to each other through rituals

972. Functional fixedness prevents:
(A) Learning
(B) Memory
(C) Insight
(D) Image

973. The greater the degree of elaboration given to an item the greater the extent to which it is:
(A) Forgotten (B) Retained
(C) Remembered (D) Saved

974. 'Domestic group, consists of:
(A) Man and wife
(B) Man and his wife and their unmarried children
(C) Man, wife and their married sons and their children
(D) Man, wife and their married sons and daughters

975. Neighbours in an urban society are mostly engaged in:
(A) Unrelated occupation
(B) Related occupation
(C) Specially separated
(D) None of these

976. In scientific explanation, attention is concentrated upon the relation between:
(A) A particular fact and a universal law
(B) Observation and fact
(C) Happenings and observation
(D) Observation and particular custom

977. 'Bow and arrow ceremony' is prevalent among:
(A) Todas (B) Bhutias
(C) Jorwas (D) Kadars

978. If sibs are sub-divided into groups with known historical ancestors, they are called:
(A) Lineages (B) Ancestory
(C) Phratry (D) Moeity

979. The number of standardized inkblots in the Rorschach Inkblot test is:
(A) Five (B) Seven
(C) Nine (D) Ten

980. Outgoing, friendly and talkative special characteristics of:
(A) Introvert (B) Extrovert
(C) Melancholic (D) Sanguine

981. After the death of the head of the family, property is given only to eldest son. But on the basis of Karma, he has to secure all the family members economically. This system is called:
(A) Dayabhaga (B) Ultimogeniture
(C) Primogeneture (D) Patriarchal

982. A 10 year old child with a mental age of 12 has an IQ of:
(A) 100 (B) 120
(C) 12 (D) 10

983. The group which matrilineal is:
(A) Kuki, Nayar, Meena, Kadar
(B) Nayar, Khasi, Meena, Apatani
(C) Nayar, bant, Khasi, Apatani
(D) Nayar, Khasi, Bant, Garo

984. Number of secondary kins are:
(A) Thirty one (B) Thirty two
(C) Thirty three (D) Thirty four

985. Creative thinking type with novel and unusual suggestions:
(A) Fluency (B) Flexibility
(C) Originality (D) Divergence

986. A 10 year old child with a mental age of 10 has an IQ of:
(A) 100 (B) 120
(C) 12 (D) 10

987. When more than one wife is shared by several brothers, it is called:
(A) Polyandry (B) Polygyny
(C) Polygynandry (D) Polygamy

988. Under Islamic law is:
(A) Mother, wife and daughter are three female heirs
(B) Daughter is a primary heir like the son. Her share fluctuates, depending upon the number of surviving children
(C) Daughter's daughter, however does not stand on a par with the son's daughter
(D) All of these

989. A man who compares the social significance of tribal rituals in two preliterate societies is most likely a:
(A) Sociologist
(B) Archaeologist
(C) Social Anthropologist
(D) Ethnologist

990. Sub-nuclear family consists of:
(A) A husband only (B) A wife only
(C) Husband and wife only
(D) Husband, wife and son only

991. Which of the following is held to be the correct theory of the origin of society:
(A) Theory of divine origin
(B) Evolutionary theory
(C) Historical theory
(D) Genetics theory

992. The most basic function of a family includes:
(A) Procreation and economic security
(B) Procreation, love and affection
(C) Procreation and role play
(D) Procreation and socialization

993. Marriage by purchase is called:
(A) Arsha marriage
(B) Asur marriage
(C) Rakshas marriage
(D) Paishach marriage

994. When the couple lives at the house of groom's father's sister, it is called:
(A) Patrilocal (B) Duolocal
(C) Amitolocal (D) Susolocal

995. Patriarchal theory about the origin of society believes that:
(A) Society became possible as a result of explanation of family headed by father
(B) Society became possible as a result of expansion of family headed by mother
(C) Society resulted because of use of force
(D) Society was given to man by God

996. 'Stem family' stands for:
(A) Small extended family
(B) Joint family
(C) Nuclear family
(D) Family of corporate business houses

997. The first stage in creative thinking among the following is:
(A) Incubation (B) Preparation
(C) Orientation (D) Verification

998. IQ is calculated based on the formulae:
(A) Mental age + chronological age/100
(B) 100/Mental age x chronological age
(C) Mental age / chronological age x 100
(D) Chronological age/Mental age x 100

999. Children produced by inter-caste marriages were called:
(A) Vratya (B) Mratya
(C) Shankar (D) Mahav

1000. Muslims Bants first celebrate their marriage in:
(A) Hindu way and then Islamic way
(B) Islamic way and then Hindu way
(C) Christian way and then Islamic way
(D) Nayars way and then Islamic way

1001. Eunomia stands for:
(A) A state of societal health or well-being
(B) A state of normlessness
(C) Equal status (D) Equal wealth

1002. Co-operation among family members is:
(A) Nonformal (B) Voluntary
(C) Formal (D) Informal

1003. Female headed families are frequent among:
(A) Agricultural communities
(B) Industrial communities
(C) Semi-nomadic pastoral communities
(D) Post industrial communities

1004. Which among the following is built on the premise that "the best way to learn how a person reacts to certain types of situations is to stimulate those situations":
(A) Rating scale (B) Observations
(C) Situational tests
(D) Behavioral assessment

1005. The process of imitation may be perceptual or:
(A) Ideational (B) Spontaneous
(C) Deliberate (D) Conscious

1006. System of 'gerentocracy' in tribals separated:
(A) Old from young
(B) Un-married youth from much married, polygynous elders
(C) Much powerful; old with obedient youth
(D) Economically rich and active youth with non-active and poor old

1007. Modern trend in the developmental activity of the village economy is that the villages should:
(A) Become smaller in number
(B) Have all the amenities of the towns
(C) Be wiped out
(D) Become bigger in size

1008. Marriage alliance between three or more persons is called:
(A) Polyandry (B) Trigamy
(C) Polygamy (D) Exogamy

1009. A system of interrelated is an engaged in problem solving to attain a common goal is:
(A) Social institution
(B) Social stratum
(C) Social action (D) Social system

1010 'Filiocentric' is the term used for:
(A) The family where care of children is important
(B) Child centered family
(C) The family where it provides education
(D) The family having more than six children as a member

1011. When cohabition with any ,male is exclusive over a period of time although that woman is still regarded as married to someone, this condition may be described as:
(A) Secondary marriage
(B) Exclusive marriage
(C) Primary marriage
(D) Extended marriage

1012. Which among the following technique enhances creative thinking:
(A) Brainstorming (B) Role play
(C) Simulation games (D) Case study

1013. Mythical ancestors are common in:
(A) Clans and phratries
(B) Clans and joint family
(C) Clans and lineage
(D) Phraties and lineage

1014. The basic foam of social stratification is:
(A) Primary and secondary group
(B) Rich and poor
(C) Male and female group
(D) Class and caste

1015. Families in which both husband and wife have careers are called:
(A) Achieved family
(B) Career family
(C) Goal oriented family
(D) Dual career family

1016. Rural and urban communities combine into one by:
(A) Neighbourhood (B) Country
(C) Community (D) Region

1017. The social relationship deriving from blood ties, real and supposed, and marriages are collectively called:
(A) Family (B) Kinship
(C) Lineage (D) Joint family

1018. Institution require specific association to sustain them is of importance for:
(A) Determining size of association
(B) Institutions of religious only
(C) Complex societies only
(D) Understanding the basic distinction between institutions and other norms

1019. Functional fixedness prevents:
(A) Learning (B) Memory
(C) Insight (D) Image

1020. A system under which a woman is permitted to marry a person of the lower caste but vice-versa is not allowed is called:
(A) Hypergamy (B) Hypogamy
(C) Promiscuity (D) Polygamy

1021. The family in which one is born is:
(A) Family of procreation
(B) Family of orientation
(C) Achieved family
(D) Ascribed family

1022. A cooperative society is an example for:
(A) Formative group (B) Informal group
(C) Primary group (D) Territory group

1023. Task performance and social integration are essential to the family for:
(A) Social interaction (B) Members
(C) Well being (D) Acceptance

1024. Which among the following has no mass, occupies no space, and is invisible:
(A) Memory
(B) Learning
(C) Intelligence
(D) None of these

1025. Hobhouse's theory of social development describes the growth of:
(A) Property (B) Knowledge
(C) Technology (D) Wealth

1026 'Lubra' stand for:
(A) Aboriginal women or wife
(B) Educated Women
(C) Woman who is training another woman/girl for house keeping
(D) Mentally retarded women

1027. The cause for much pain in divorce proceedings is:
(A) Loss of money
(B) Separation between wife and husband
(C) Custody of the children
(D) Allegations at each other

1028. A group does not necessitate:
(A) Common territory
(B) Common ancestry
(C) Common interests
(D) Common food habits

1029. The institution of family gradually developed into:
(A) Political institution
(B) Kinship organization
(C) Economic organization
(D) Economic frontier and caste

1030. The term 'stridhanam' means:
(A) Woman's purity
(B) Woman's husband
(C) Woman's wealth
(D) Woman's attachment towards custom and tradition

1031. The importance of Panchayat Raj is:
(A) It helps the people to directly participate in decision making
(B) It helps in the decentralization of power
(C) It reduces the burden of the central government
(D) All of these

1032. All the panchayat samitis or mandals in a district together constitute the:
(A) Gram Panchayat
(B) Panchayat Samiti
(C) Zila Parishad
(D) Mayor

1033. Education, forests, trade unions come under:
(A) State List
(B) Union List
(C) Concurrent List
(D) Residency List

1034. A phratry is a grouping of clans which are related by traditions of:
(A) Joint family (B) Common descent
(C) Ritual (D) Family tradition

1035. Which of the following is considered the most objective measure of personality:
(A) Rating scales
(B) Personality questionnaires
(C) Projective test
(D) Thematic Apperception Test

1036. The tribes/caste having high standard and rate of marriage stability is:
(A) Nayars (B) Ho
(C) Muria (D) Kharia

1037. ———— is a holding together federation:
(A) USA (B) India
(C) Pakistan (D) Sri Lanka

1038. The process by which the group-values are internalized by the individual is called:
(A) Assimilation (B) Integration
(C) Socialisation (D) Accommodation

1039. A social relationship in society involves:
(A) Presence of institutions
(B) Reciprocal awareness
(C) Complication of social and political problems
(D) Consciousness of the existence of other individuals

1040. Levirate means:
(A) A man marrying his wife's sister while the wife is still alive
(B) A man marrying his wife's sister after the death of his wife
(C) A woman marrying her husband's brother while the husband is still alive
(D) A woman marrying her husband's brother after the death of her husband

1041. Two types of casual explanation are:
(A) Scientific and Historical
(B) Positive and negative
(C) Positivist and Phenomenological
(D) Scientific and Sociological

1042. A group of people organized for a particular purpose is called:
(A) Community (B) Association
(C) Society (D) Institution

1043. An example of role-set is:
(A) A teacher's relationship with has pupils, colleagues, the principal and parents
(B) A teacher as husband, father, sportman and social activist
(C) A teacher becoming later a head of the department, a principal and a registrar
(D) A student becoming a husband, a father and a grand father over a period of time

1044 'Zarduga' system of family life is found in:
(A) Japan (B) Yugoslavia
(C) Itely (D) Canada

1045. Which one of the following is not a reason for social difference due to accident by Birth?
(A) Male-Female (B) Tall-Short
(C) Rich-Poor (D) Fair-Dark

1046. Which theory posits the view that society is the family writ large?
(A) Force theory
(B) Evalutionary theory
(C) Matriarchal theory
(D) Patriarchal theory

1047. Society defined as a complex of groups in reciprocal relationships, interacting upon one another and helping each other to fulfill his/her wishes in association with his fallows. This is according to:
(A) Functional view
(B) Interactional view
(C) Structural view
(D) Cultural view

1048. Two types of culture are:
(A) Dominant and non-dominant
(B) Positive and negative
(C) Self and others (D) Our and their

1049. Fuctionalism regards society as a system, which means that:
(A) The terms, society and system are interchangeable
(B) The parts of society are discrete and isolated
(C) Society is the only example of a category called system
(D) Society is divided into a number of parts which are inter-connected

1050. The Rules of Residence where a couple upon being married, goes to live with the brother of the groom's mother are called:
(A) Patrilocal (B) Matrilocal
(C) Avunculocal (D) Neolocal

1051. Which of the following reflects that India is a secular state?
(A) No official religion
(B) Freedom of religion
(C) No religious discrimination
(D) All of these

1052. Which of the following is the element of difference between community and society:
(A) Definite locality
(B) A group of persons
(C) Likeness of interest
(D) Sentiment of oneness

1053. Which one of the following is not a reason of difference out of choice?
(A) Following a particular religion
(B) Following a particular profession
(C) Living in a particular colony
(D) Fair-Dark

1054. The ordered social relationships that grow out of the values, norms, statuses and role, patterning various activities to fulfil society's needs are called:
(A) Associations (B) Organisations
(C) Institutions (D) Clubs

1055. The social order is maintained largely by:
(A) Law (B) Socialisation
(C) Division of labour (D) State

1056. The term society in sociology is used to refer the:
(A) Persons living in an area
(B) Persons professing the same religion
(C) System of social relationships
(D) Organized relations between individuals

1057. Mnemonic techniques mostly rely on:
(A) Thinking (B) Association
(C) Learning (D) None of these

1058. How many items can your short term memory handle:
(A) About three (B) Exactly five
(C) About seven (D) Exactly ten

1059. An organized predisposition of an individual to think, feel and behave towards a referent or cognitive object is otherwise called:
(A) Individual attitude
(B) Individual vales
(C) Individual personality
(D) Individual response

1060. A type of marriage known as Levirate is the one that takes place between a man and a:
(A) Sister of his deceased wife
(B) Widow of his deceased brother
(C) Woman of an inferior caste
(D) Woman of a superior caste

1061. Among the following, which country suffered disintegration due to political fights on the basis of religious and ethnic identities?
(A) Belgium
(B) India
(C) Yugoslavia
(D) Netherland

1062. Which of the following statements is true:
(A) Society means likeness
(B) Society implies differences
(C) Society means mutual aid
(D) Both (A) and (B)

1063. Shyness, social withdrawal, and tendency not to talk much are the characteristics of:
(A) Introvert
(B) Extrovert
(C) Melancholic
(D) Sanguine

1064. A perfect market is:
(A) Where prices are determined by law of supply and demand
(B) Where buyers and sellers meet face to face
(C) Where there is no monopolistic competition
(D) Where goods are not in short supply

1065. Absorption of culture is:
(A) Abolishment of culture
(B) Conflict in cultures
(C) Merger of one culture into other culture
(D) Superiority of one culture over other

1066. In dealing with social division, which one of the following statements is not true about democracy?
(A) Due to political competition in a democracy, social divisions get reflected in politics
(B) In a democracy, it is possible for communities to voice their grievances in a peaceful manner
(C) Democracy is the best way to accomplish social diversity
(D) Democracy always leads to disintegration of society on the basis of social divisions

1067. A society that values man more and gives them power over woman:
(A) Feminist society
(B) Heterogenous society
(C) Patriachal society
(D) Communist society

1068. Feature of ceremonial exchange is that it:
(A) Involves sentiments
(B) Is a social exchange
(C) Does not involve immediate return
(D) All of these

1069. An example of integration is:
(A) Introducing the three languages formula in education
(B) Merging of princely status into Indian Union after independence
(C) Providing reservations to SC/ST communities
(D) All of these

1070. Modelling is a valuable therapeutic tool in the case of:
(A) Learning (B) Conditioning
(C) Phobias (D) None of these

1071. Property can be tangible or intangible. The intangible property is:
(A) Factory unit
(B) Truck plying on roads
(C) Copyright of an author
(D) Liquor shop

1072. Thematic Apperception Test was developed by:
(A) Cattel (B) Malvin Lee
(C) Philip Zimbardo
(D) Henry Murray

1073. Maturation is a process over which individual has:
(A) No control at all
(B) Only remote control
(C) Direct control
(D) Positive and effective control

1074. A man or a woman who believes in equal rights and opportunities for woman and man:
(A) Communist (B) Socialist
(C) Feminist (D) Communalist

1075. The easiest type of conflict to resolve is:
(A) Approach-avoidance
(B) Avoidance-avoidance
(C) Ambivalence
(D) Approach-approach

1076. The number of sketches in the original Thematic Apperception Test is:
(A) Twenty (B) Twenty five
(C) Fifteen (D) Thirty

1077. Example of direct exchange is:
(A) Redistribution
(B) Barter exchange
(C) Gift exchange
(D) Reciprocity

1078. The literacy rate among women in India is about ———, whereas it is ——— for men is:
(A) 54%, 76% (B) 50%, 80%
(C) 76%, 54% (D) 100%, 80%

1079. Sex ratio in India is:
(A) 880 (B) 927
(C) 300 (D) 400

1080. The study of the nature and acquisition of knowledge:
(A) Gerontology (B) Psychology
(C) Biology (D) Epistemology

1081. Presents freely given to kin, friends and neighbors without any explicit statement is called:
(A) Reciprocity
(B) Generalized reciprocity
(C) Mutual exchange (D) Kula system

1082. Pronatalism refers to:
(A) View that a person should not have as many children as possible
(B) View that only the child can provide sustenance and salvation to the family
(C) Strong positive value a society places on children
(D) View that vitality of the family depends upon the strong physique of the child

1083. ———— provides that equal wages should be paid to equal work:
(A) The Equal Wages Act
(B) Right to Equality
(C) Right to Wages
(D) Right to Information

1084. A type of displaced aggression among the following is:
(A) Escape (B) Scapegoating
(C) Ambivalence (D) All of these

1085. The main goal of 'Food for work' programme is:
(A) Giving relief to unemployed
(B) Equal distribution of food grains
(C) Cheap labour force
(D) Scarcity of food to be overcome

1086. The connection between nerve cells is called:
(A) Synapses (B) Dendrites
(C) Neurons (D) Axon

1087. In what way are goods divided among members of the society?
(A) Accumulation (B) Reciprocity
(C) Distribution (D) Dissemination

1088. The first state farm started during:
(A) The Gupta period
(B) The Mauryan period
(C) The Mughal period
(D) The British period

1089. Information is passed from one neuron to another by chemicals known as:
(A) Nerve impulses
(B) Neurotransfer
(C) Neurotransmitters
(D) None of these

1090. The Relative Deprivation Approach attempts to:
(A) Measure poverty of those below the poverty line
(B) Determineous privileges of the rich
(C) Assess the minimum food requirements
(D) Assess the deprivation of a section of population as compared to the others

1091. Difference between culture and civilization is:
(A) Culture is all slow process whereas civilization is a quick process
(B) Culture is internal but civilization is external
(C) Culture consists of ideals, values mental and emotional groups of life
(D) All of these

1092. Which among the following refers to perception of the world around us:
(A) Perception (B) Cognition
(C) Emotion (D) Sensation

1093. The type of movement an infant makes when it is started as:
(A) Grasping reflex (B) Rooting reflex
(C) Moro reflex (D)Adaptive reflex

1094. A visible and controversial development of the economy, is the domination of industries and markets by fewer and fewer corporations. This situation is called
(A) Monopoly (B) Oligopoly
(C) Competition (D) Merger

1095. What is the term used for a movement that seeks to secure equality of status for women?
(A) Racial movement
(B Feminist movement
(C) Civil Rights movement
(D) Equality movement

1096. The desires that are frequently revealed in dreams, reveries, jokes ad slip of the tongue, as well as under the influence of alcohol or drugs:
(A) Suppression (B) Repression
(C) Interjected (D) None of these

1097. The principles a government uses to determine its income and expenditures and the relationship between them is called:
(A) Fiscal policy (B) Budget
(C) Economic policy (D) Trade policy

1098. Insects warming around an outdoor light is an example of:
(A) Congregation (B) Group
(C) Aggregation (D) Gathering

1099. The descendants of a common male ancestor are called:
(A) Agnates (B) Cognates
(C) Lineal kin (D) Collateral kin

1100. The rural economy is contributed by:
(A) Cottage industry
(B) Animal husbandry
(C) Money lending (D) All of these

1101. What is the major cause of declining Girl Child ratio in India?
(A) Parents preference for sons
(B) Sex selected abortions
(C) Both (A) and (B) (D) None of these

1102. Naturalistic observation only provides description of:
(A) Experience (B) Feeling
(C) Behaviour (D) None of these

1103. A subfield of study called sociology of economic development has been developed after:
(A) Industrial Revolution
(B) First world war
(C) Second world war
(D) Great depression of 1930s

1104. The nervous system that get activated in resting and quite states:
(A) Somatic nervous system
(B) Parasympathic nervous system
(C) Sympat etic nervous system
(D) None of these

1105. Sociality is a quality an socialization is a:
(A) Virtue (B) Process
(C) Biological inheritance
(D) None of these

1106. A system where goods are exchanged without any meeting of the two parties to the transactions. This is:
(A) Perceived trade (B) Inner trade
(C) Silent trade (D) Concealed trade

1107. The appropriate number of nerve cells in human brain:
(A) 75 billion (B) 100 billion
(C) 150 billion (D) 200 billion

1108. Which of the following is not a feature of secularization:
(A) Rationality (B) Differentiation
(C) Scientific attitude
(D) Ceremonialism

1109. People will only be eradicated with the removal of inequality in general which requires the transformation of the economic structures in which it is embedded. Who holds this?
(A) Ralph Milliband (B) Halt
(C) Goode (D) Wilkinson

1110. The rapid withdrawal of a person's hand from a hot radiator is:
(A) Input (B) Reflex
(C) Emotion (D) Sensation

1111. Which of the following is disjunctive social process:
(A) Competition
(B) Cooperation
(C) Assimilation
(D) Integration

1112. Which of the following factors lead to the growth of cities:
(A) Industrialization
(B) Educational facilities
(C) Fertility of land
(D) Housing facilities

1113. The controversy between sociologists and psychologists is on:
(A) Nature-nurture argument
(B) Nature-nature argument
(C) Nurture-nurture argument
(D) All of these

1114. A characteristic feature of a rural community is:
(A) Formal relationship
(B) Rational will
(C) Community sentiment
(D) A vast number of association

1115. The number of seats reserved for women in Panchayati raj bodies is what part of total seats?
(A) One-forth (B) One-sixth
(C) One-third (D) Half

1116. Mob behaviour is:
(A) Irrational and suggestive
(B) Rational and suggestive
(C) Irrational but not suggestive
(D) Rational but nt suggestive

1117. Which is the highest tier of the Panchayati Raj is?
(A) Zila Parishad (B) Panchayati Raj
(C) Gram Sabha (D) None of these

1118. The specific achievements of the local self-government in India:
(A) It has made the country more united and stronger
(B) It has also made the administration easier
(C) Both (A) and (B)
(D) None of these

1119. Every adult of the village who is 18 years age constitute:
(A) Gram Sabha
(B) Zila Parishad
(C) Panchayat Smiti
(D) Gram Panchayat

1120. Which type state in India:
(A) Democratic (B) Secular
(C) Welfare (D) All of these

1121. Most important condition for growth of the early cities is:
(A) Surplus food production
(B) Increase in population
(C) Discoveries of new trades
(D) Technological innovations

1122. Colonial urban economy of India is characterized by:
(A) Destruction of traditional handicrafts
(B) Growth of new classes of the bourgeoisie
(C) Beginning and development of modern industries
(D) All of these

1123. Which of the following is the result of capitalism:
(A) Competition (B) Cooperation
(C) Integration (D) Democracy

1124. Which of the following is not a common reaction of frustation:
(A) Ambivalence
(B) Aggression
(C) Displaced aggression
(D) Persistence

1125. Cities and towns came into existence due to:
(A) Movement of population
(B) Growth of agriculture
(C) Industrialisation
(D) Development of commerce

1126. The family in agrarian society is:
(A) Patriarchal (B) Matriarchal
(C) Matrilocal (D) Nuclear

1127. Social stratification means classification of society:
(A) Into castes
(B) Based on economic conditions
(C) Biological characteristics
(D) All of these

1128. What determines group life:
(A) We feeling
(B) Feeling of unity
(C) Common purpose
(D) All of these

1129. The essential character of primary group is:
(A) Lack of identification
(B) Affection and love
(C) Identification of interest
(D) Continuous contact

1130. The standard of living in the villages is mostly of the same pattern whereas it varies in the towns because:
(A) There is no industries in the villages
(B) There is no social disintegration on the villages
(C) From agriculture the villagers obtain the same type of economic satisfaction and face the same problems
(D) Villages are dependent upon land and the same profession creates similar standard of living

1131. What are advantages of linguistic states?
(A) It has made the country more united and stronger
(B) It has also made administration easier
(C) Both (A) and (B)
(D) None of these

1132. Who has explained Sociology as the science with the most methods and the finest results?
(A) Henry Poincare
(B) Henry Maine
(C) Auguste Comte
(D) Herbert Spencer

1133. Much of 'Mutual sympathy' is observed in villages and much of callousness in the cities because:
(A) In small communities, greater sympathy can be observed
(B) Geographical conditions in villages lead to this situation
(C) Small incomes in villages create much sympathy amongst them
(D) Agricultural work leads to sympathy

1134. The portion of community which is marked off from the rest by social status is known as:
(A) Tribe (B) Class
(C) Race (D) Caste

1135. Intimate, spontaneous and face to face interaction takes place in a:
(A) Cooperative group
(B) Primary group
(C) Model group (D) Reference group

1136. Which among the following is a positive aspect of bureaucracy:
(A) Red tapism
(B) Mechanical adherence to rules
(C) Objective orientation
(D) Indifference to individual feelings

1137. Which of the following is not a characteristic of conflict:
(A) Conflict is a conscious action
(B) Conflict is an impersonal activity
(C) Conflict lacks continuity
(D) Conflict is universal

1138. Factor contributing to the growth of traditional societies is:
(A) Perpetuation of tribal and caste affiliation
(B) Perpetuation of kinship and clanship
(C) Continuation of agriculture as the dominant activities
(D) All of these

1139. The process of borrowing cultural items from another culture is known as:
(A) Acculturation (B) Socialization
(C) Unculturation (D) Acquisition

1140. Term used to refer to an ordinary fortified, town where inland trade was an important activity of people is:
(A) Rajdhani (B) Matha
(C) Kheta (D) Nagara

1141. Social psychology is that part of the general psychology which has a particular relevance to:
(A) Abnormal psychology
(B) Human interaction
(C) Social phenomena
(D) Group behaviour

1142. The process through which on ethic group loses its distinctive characteristic and become indistinguishable from any other social segment is called:
(A) Assimilation (B) Diffusion
(C) Multiculturalism (D) Pluralism

1143. The standard of living in the villages is mostly of the same pattern, whereas it varies in the towns because:
(A) There is no social disintegration in the villages
(B) There is no industries in the villages
(C) Villages is no dependent upon land and the same profession creates similar standards of living
(D) From agriculture the villagers obtain the same type of economic satisfaction facing the same problems

1144. Socialization is a process of converting a biological organization into:
(A) Human being
(B) Super human being
(C) Modern man (D) A civilized person

1145. The people shifting from rural to urban areas seem to adjust to city life more smoothly if they maintain:
(A) Caste hierarchy (B) Kin ties
(C) Bank balance (D) Huge library

1146. Residues are:
(A) Changing
(B) Relatively permanent
(C) Non-existent (D) Can be moulded

1147. Which one is not the factor of caste system:
(A) Purity and pollution
(B) Hierarchical order
(C) Hereditary occupation
(D) Common ancestor worship

1148 'Goa' is an example of:
(A) Religious city (B) Industrial city
(C) Port city (D) Commercial city

1149. Caste is a class gained by:
(A) Status (B) Birth
(C) Power (D) Social relationship

1150. A conjunctive social process among the following:
(A) Conflict (B) Competition
(C) Assimilation (D) All of these

1151. Which among the following is an organization:
(A) Hospital (B) Court
(C) Army (D) All of these

1152. Criteria for a place to be a town or urban area is:
(A) Minimum of 5000 persons
(B) At least 75% of the working occupations are non-agricultural
(C) Density of not less than 400 persons per kilometer
(D) All of these

1153. Which of the following is not a form of conflict:
(A) War (B) Litigation
(C) Feud (D) Examination

1154. Interaction between members or groups with different culture is called:
(A) Touch of culture
(B) Cultural diffusion
(C) Culture contact
(D) Acculturation

1155. The recent trend in the urban population in India is a:
(A) Steady increase
(B) State of stagnation
(C) Both (A) and (B)
(D) None of these

1156. Socialization is a process involving:
(A) Setting up the social norms
(B) Gradual changing of organism
(C) Declaring everything to society
(D) Training to adopt to society

1157. Mark out the essential elements of organization:
(A) Goals
(B) Norms
(C) Sanctions
(D) All of these

1158. Spatial feature of urbanization in India has been:
(A) Localised in nature
(B) Balanced in nature
(C) Both of these
(D) None of these

1159. Besides ———, there are other languages recognized by our constitution:
(A) English, 21 (B) Hindi, 22
(C) Hindi, 21 (D) Punjabi, 18

1160. A residence are characterized by structural decay, overcrowding, lack of recreational space and neglect of neighbourhood facilities is called:
(A) Middle town (B) Worker's residents
(C) Slums (D) Zone in transition

1161. The response that is learned in classical conditioning is:
(A Conditional response
(B) Unconditional response
(C) Unconditional stimulus
(D) Conditioned response

1162. Putting coded information into memory is:
(A) Retrieval (B) Encoding
(C) Storage (D) None of these

1163. Tendency of man to consider his own culture as of having high value and superior to all others:
(A) Cultural value (B) Assimilation
(C) Ethnocentrism (D) Accumulation

1164. Man's behaviour in society is determined mainly by the two forces namely:
(A) Formal and informal
(B) Natural and unnatural
(C) Physical and social
(D) Psychological ad philosophical

1165. The collective efforts to bring a new social order is referred as:
(A) Social evolution
(B) Social conflict
(C) Social movement
(D) Social planning

1166. The type of social interaction that emerges most significantly during emergencies:
(A) Coercion (B) Exchange
(C) Conflict (D) Cooperation

1167. The ending of craft production resulted in a shift of power from workers to:
(A) Capitalist (B) Trade union
(C) Government (D) Management

1168 Consider the following two statements:
Assertion (A): Culture is abstract
Reason (R): Culture is manifested in behaviour and in the results of behaviour, but is neither the behaviour itself nor the tangible results

Select the correct answer:
(A) Both A and R are true and R is the correct explanation of A
(B) Both A and R are true by R is not the correct explanation of A
(C) A is true, R is false
(D) A is false, T is true

1169. Property for power, and unlimited individual acquisition of wealth was a product is:
(A) Feudalism (B) Bureaucracy
(C) Capitalism (D) Aristocracy

1170. In the process of urbanization, the concept of 'Community Living Land' became:
(A) Goal oriented
(B) Casual
(C) Informalised
(D) Personalised

1171. An example of community is:
(A) College
(B) Village
(C) Neighbourhood
(D) None of these

1172. Pareto has described how many classes of residues:
(A) Two (B) Ten
(C) Six (D) Five

1173. Urban slums are the marks of:
(A) Industrial society
(B) Capitalist society
(C) A society in transition towards industrialism
(D) Post-industrial society

1174. Permanent settlement of 1793 created in bengal, and later in other parts of Indian provided;
(A) Tenants the right of ownership
(B) Definite proprietary rights for the zamidaris
(C) Abolition of zamidari system
(D) Abolition of share-cropping system

1175. Which one is an institution:
(A) Family (B) State
(C) Party system (D) All of these

1176. Geographical demarcation is a character of:
(A) Society (B) Community
(C) Group (D) None of these

1177. The 'bourgeoise' type of society was expounded by:
(A) Max weber (B) Karl Marx
(C) Hegel (D) T. H. Green

1178. What was the basic aim of Nepal's movement of 2008?
(A) To restore democracy
(B) To restore military rule
(C) To restore old dynasty
(D) To reappoint king Birendra

1179. What are the advantages of local governments?
(A) Constitutional status for local government has helped to deepen democracy in our country
(B) It has also increased women's representation and voice in our country
(C) This allows people to directly participate in decision making
(D) Local people have better ideas and knowledge about the local problems
(E) All of these

1180. The main difference between caste and class is that:
(A) Caste operates in a classless society
(B) Caste is always closed, and class is always open
(C) The status given to an individual is ascribed in the case of caste, whereas in class it is achieved
(D) The individual suffers most in the case of caste, whereas in class he does not

1181. Which of the following steps is not related to decentralization?
(A) Regular elections to local bodies
(B) 1/3 of all positions are reserved for women
(C) Some seats are reserved for SC and ST
(D) Formation of a coalition government

1182. Which among the following economic orders mainly depends on the geo-physical environment?
(A) Primitive economic order
(B) Agrarian economic order
(C) Both (A) and (B)
(D) None of these

1183. Which of the following distinguish the animal society from human society:
(A) Speech (B) Physical strength
(C) Mode of living (D) Culture

1184. Which one of the following traits of metropolis is explained by its being called a racial and cultural melting pot?
(A) Heterogeneity (B) Mobility
(C) Density (D) Size

1185. Pattern of settlement in rural areas is:
(A) Nucleated (B) Linear
(C) Scattered (D) Circular

1186. The form of a agronomy in which land is broken with a plough making possible large-scale food production is called:
(A) Agriculture (B) Horticulture
(C) Earth digging (D) Plantation

1187. Village community is constituted by:
(A) Joint family
(B) Sentimental attachment
(C) Agriculture (D) All of these

1188. The situation where every response is followed by a reward is termed as:
(A) Continuous reinforcement
(B) Fixed reinforcement
(C) Ratio reinforcement
(D) Controlled reinforcement

1189. Which of the following are units of social structure:
(A) Institution (B) Group
(C) Association (D) All of these

1190. The industrial system of production is distinguished from other systems on the basis of availability of:
(A) Skilled labour
(B) Means of production
(C) Capital
(D) Inanimate source of energy

1191. A government formed by the coming together of atleast two political parties is called the:
(A) Union government
(B) Combination government
(C) Coalition government
(D) Mixed government

1192. Which one of the following statements is correct regarding language policy of the Indian government?
(A) Hindi is standard or our national language
(B) Our Constitution did not give the status of national language to any one language
(C) English is identified as our national language
(D) Every state has its own national language

1193. When power is taken away from Central and State governments and given to local government, is it called:
(A) Distribution (B) Centralisation
(C) Reorganisation
(D) Decentralisation

1194. Which among the following is not an association:
(A) Trade union (B) Status
(C) Tennis club (D) All of these

1195. The haphazard of out growth of cities, often maxing land-use in unplanned ways, is most precisely known as:
(A) Urban sprawl (B) Strip cities
(C) Megalopolis (D) Urban periphery

1196. A few gram panchayats are grouped together to form what is usually called a:
(A) Panchyat Samiti
(B) Panchayat Raj
(C) Local government
(D) State government

1197. The key element in cognitive interpretation of instrumental conditioning is:
(A) Anticipation (B) Expectation
(C) Imitation (D) Interpretation

1198. Which one of the following factors best identifies the essential differences between capitalist and socialist economies?
(A) Means of production
(B) Ownership of means of production
(C) Complexity of the industrial system
(D) Work organization

1199. The process through which information coming from sense is "transformed", reduced, elaborated and used" in:
(A) Recognition (B) Recall
(C) Cognition (D) Memory

1200. The branch of psychology that deals with cognitive process is called:
(A) Developmental psychology
(B) Clinical psychology
(C) Experimental psychology
(D) Cognitive psychology

1201. Assimilation is essentially a:
(A) Social process
(B) Political process
(C) Historical process
(D) Psychological process

1202. Societies everywhere, in the course of economic production divide the human resources available to them in terms of:
1,Age 2.Sex
3.Physical ability
4.Formal; qualifications
Select the correct answer from below:
(A) 1 (B) 1, 2
(C) 1, 2, 3 (D) 1, 2, 3, 4

1203. Municipalities are set up in ———— whereas Municipal Corporations are in ————:
(A) Big towns, small towns
(B) Small towns, big towns
(C) Villages, towns
(D) Towns, villages

1204. The conflict between capitalists and workers is an example of:
(A) Personal conflict (B) Class conflict
(C) Racial conflict (D) Political conflict

1205. Socialization is a process which:
(A) Ceases at maturity
(B) Continues through life
(C) Influences society
(D) None of these

1206. "People who live close to one another who interest with one another frequently, and who feel they have some common traits or values they share with one another," constitute:
(A) Social group (B) Social institution
(C) Association (D) Community

1207. A formally organized group having limited objectives that are specifically spelled out and well understood is known as:
(A) Community (B) Association
(C) Primary group (D) Society

1208. Rural and urban centres share some common facets of life. They show:
(A) Antagonism (B) Contradiction
(C) Cooperation (D) Interdependence

1209. The conflict between capitalists and workers is an example of:
(A) Personal conflict
(B) Class conflict
(C) Racial conflict
(D) Political conflict

1210. Which of the following are factors of production in the economic system, according to Pearsons and Smelsar?
1,Land 2.Capital
3.Labour 4.Enterpreneurship
5.Specialisation
Select the correct answer from below:
(A) 1, 2, 3 (B) 1, 2, 5
(C) 1, 2, 3, 4 (D) 1, 2, 3, 4, 5

1211. ——— is a council consisting of several ward members, often called panch, and a president or sarpanch:
(A) Municipal Corporation
(B) Gram Sabha
(C) Gram panchyat
(D) Municipality

1212. Placing blame for difficulties upon others or attributing one's own unethical desires to others is known as:
(A) Repression (B) Projection
(C) Regression (D) Identification

1213. Which one of the following economies tend to spread consumerism?
(A) Barter (B) Market
(C) Mixed (D) Balanced

1214. Both Municipalities and Municipal Corporations are:
(A) Controlled by elected bodies consisting of people's representatives
(B) Controlled by the Union Government
(C) Controlled by the State Government
(D) Independent bodies

1215. Sensory input from the environment that informs us about something that is happening there:
(A) Information (B) Cognition
(C) Recognition (D) Encoding

1216. Which one of the following is the smallest unit of the culture:
(A) Cultural trait
(B) Cultural complex
(C) Cultural pattern (D) Cultural areas

1217. 'Patlach' is a tribal ceremony in which property is:
(A) Acquired (B) Displayed
(C) Destroyed (D) Renunciated

1218. Who said this "Status is the rank-order position assigned by a group to a role or to a set of roles":
(A) Lundberg (B) Maclver
(C) Kingsley Davis (D) Ogburn

1219. The tribe as a group is closet to:
(A) Class (B) Association
(C) Caste (D) Commonity

1220. What is the focus of human relations approach in the study of modern industrial set up?
(A) Worker's morale (B) Profit-making
(C) Management benefit
(D) Solution of union grievances

1221. The subjects that do not fall in any of the three lists comes under:
(A) State List (B) Union List
(C) Concurrent List (D) Residency List

1222. Both the Union and state governments can make laws on the subjects mentioned in the:
(A) State List (B) Union List
(C) Concurrent List (D) Residency List

1223. The process wherein a person achieves his goals and gratifies his needs in a substitutive fashion:
(A) Repression (B) Fantasy
(C) Regression (D) Identification

1224. Which one of the following factors was responsible for rapid industrialization of Japan as compared to China?
(A) Natural resources
(B) Population resources
(C) Better economic resources
(D) Sate policy

1225. Latest learning is an example of:
(A) Cognitive learning
(B) Insight learning
(C) Imitation
(D) All of these

1226. During medieval period, the artisans used to carry on their trade through:
(A) Commercial groups
(B) Markets
(C) Communities
(D) Guilds

1227. An association is characterized by:
(A) Norms (B) Customs
(C) Folkways (D) Usage

1228. Daniel Larner has measured rigorously inter-relation between non-economic variables and:
(A) Urbanisation (B) Industrialization
(C) Rural-urbanization
(D) Westernization

1229. Behaviours that can be learned only with greater difficulty:
(A) Prepared behaviour
(B) Unprepared behaviiour
(C) Contra prepared behaviour
(D) All of these

1230. A formally organized group having limited objectives that are specifically spelled out and well understood is known as:
(A) Community (B) Association
(C) Primary group (D) Society

1231. In common sense, the term 'society' is equivalent to:
(A) Boundaries (B) Organisation
(C) Group (D) Imbalance

1232. Match list I with list II and select the correct answer:
List I (Factors of production)
a. Land b. Labour
c. Capital d. Organisation
List II (Shares of income)
1. Profit 2. Interest
3. Wages 4. Rent

Codes:

	a	b	c	d
(A)	4	3	1	2
(B)	1	2	3	4
(C)	4	3	2	1
(D)	5	3	1	4

1233. 'Alienation' is most likely to occur if the person is a:
(A) Bank clerk
(B) Weaver
(C) Landless labourer
(D) Assembly line worker

1234. Match list I with list II and select the correct answer:
List I (Type of productive system)
a. Factory system b. Guild system
c. Domestic system
d. Putting-out system
List II (Occupational categories found therein)
1. Artisans 2. Operatives
3. Apprentices 4. Merchant enterpreneurs

Codes:

	a	b	c	d
(A)	4	3	1	2
(B)	1	2	3	4
(C)	2	3	1	4
(D)	5	3	1	4

1235. When we attempt to remove or surmount obstacles to satisfy our needs, it is called:
(A) Compromise behaviour
(B) Withdrawal behavious
(C) Attack behaviour
(D) All of these

1236. Consider the following two statements:
Assertion (A): Commercial revolution was the main reason for the development of medieval cities
Reason (R): Colonization and conquests opened up the East to the Western market

Select the correct answer:
(A) Both A and R are true and R is the correct explanation of A
(B) Both A and R are true by R is not the correct explanation of A
(C) A is true, R is false
(D) A is false, T is true

1237. Which one of the following is not an economic institution in the strict sociological sense?
(A) Contract (B) Market
(C) Division of labour (D) Factory

1238. 'Price mechanism' means:
(A) Price manipulation by producers
(B) Price manipulation by consumers
(C) Spontaneous play of market forces which determines prices
(D) Regulated market

1239. Norms without moral tones are called:
(A) Myths (B) Folkways
(C) Mores (D) Fads

1240. Frustation arouses:
(A) Emotion (B) Hostility
(C) Fear (D) Anxiety

1241. Among some communities the term used for mother and her sister is the same. This indicates the prevalence of:
(A) Leveret (B) Polygyny
(C) Sorprate (D) Polyandry

1242. At calm and relaxed stage, which among the following nervous system operates:
(A) Sympathetic system
(B) Parasympathetic system
(C) Peripheral system
(D) None of these

1243. When a person befiriends many people, he has higher need for:
(A) Achievement
(B) Power
(C) Affiliation
(D) Recognition

1244. Match list I with list II and select the correct answer:

List I

a. Meet my hunsbands

b. Why get married and ruin a perfect relationship

c. This was my first wife, Savitri was my 3rd. They are remarried now

d. My one and the only husband

List II

1. Monogamy
2. Polyandry
3. Consensual Union
4. Serial Monogamy

Codes:

	a	b	c	d
(A)	4	3	1	2
(B)	1	2	3	4
(C)	2	3	4	1
(D)	5	3	1	4

1245. Consider the following two statements:

Assertion (A): Family is the primary social unit

Reason (R): Family gives protection to its members and supports them financially

Select the correct answer:

(A) Both A and R are true and R is the correct explanation of A

(B) Both A and R are true by R is not the correct explanation of A

(C) A is true, R is false.

(D) A is false, T is true

1246. Consider the following two statements:

Assertion (A): Drug abuse has been on the increase among youths in recent times

Reason (R): The control of the family on its children is declining

Select the correct answer:

(A) Both A and R are true and R is the correct explanation of A

(B) Both A and R are true by R is not the correct explanation of A

(C) A is true, R is false

(D) A is false, T is true

1247. The ability to learn a particular kind of skill required for a specific situation is:

(A) Intelligence (B) Aptitude

(C) Attitude (D) None of these

1248. Which of the following is the strongest need:

(A) Hunger (B) Thirst

(C) Safety (D) Sex

1249. A separation between the "biological father" and "social father" is found in a family known as:

(A) Bilocal (B) Conjugal

(C) Polyandrous (D) Patriarchal

1250. According to Muslim law MEHR is:

(A) Te bride price paid at the time of marriage

(B) The panality paid by the husband for ill treating the wife

(C) The agreed payment made by the husband at the time of divorce

(D) The amount paid to the priest at the time of marriage

1251. Which one of the following is correct for the definition of 'sororate'?

(A) Custom of marrying the sister of the deceased wife

(B) Practice of marrying the brother of the increased husband

(C) Custom of marrying the sister's daughter

(D) Practice of marrying the mother's brother's daughter

1252. Which one of the following pairs is not an example of consanguineous kins?

(A) Brother-sister

(B) Father-son

(C) Husband-wife

(D) Mother-daughter

1253. When a slum is inhabited largely or exclusively by persons of a single ethnic or racial group, it may be called:

(A) Concentration (B) Locality

(C) Domination (D) Ghetto

1254. Which of the following best defines the term kinship?
(A) The social recognition and expression of generalogical relationship
(B) A socially recognized relationship
(C) The recognition of culturally conditioned relationship
(D) The grouping of relatives

1255. The behaviour that reflects the inner state is called:
(A) Covert behaviour
(B) Overt behaviour
(C) Exposed behaviour
(D) Concealed behaviour

1256. Legislative provision for intercaste marriage in India was provided by:
(A) The Hindu Marriage Act 1955
(B) The Special Marriage act, 1872
(C) The Hindu Code Bill, 1956
(D) The Gains of learning Act, 1930

1257. When a person tries to convince others, we can assume that he has:
(A) Need for affiliation
(B) Need for power
(C) Need for recognition
(D) Need for actualization

1258. A cross-cousin marriage means a marriage between:
(A) Uterine-kins
(B) Affinal kins
(C) Children of the sibling relation of the same sex
(D) Children of the siblings of opposite sex

1259. The industrial Revolution in England took place through:
(A) Wool industry
(B) Mining industry
(C) Cotton industry
(D) Lumbering industry

1260. Joking relationship take place between:
(A) Husband and wife
(B) Mother and daughter
(C) Father and son
(D) Brother-in-law and sister-in-law

1261. The latent disposition to strive for a good aim is called:
(A) Achievement motivation
(B) Aspiration (C) Desire
(D) Ambition

1262. Avoidance is practiced in the society to:
(A) Show respect towards elders
(B) Show displeasue to the rurals
(C) Prevent sexual contact between certain relatives
(D) Raise the status of women

1263. The concept closely associated with gesselschaft is:
(A) Primary relationship
(B) Secondary relationship
(C) Traditional relationship
(D) Kinship based relationship

1264. The rule of residence by which the married couple go to live with a brother of the groom's mother is called:
(A) Matrilocal (B) Patrilocal
(C) Avuneulocal (D) Neolocal

1265. Consider the following two statements:
Assertion (A): Polyandry is much restricted in distribution in our contemporary society
Reason (R): The practice of polyandry depends upon a number of socio-demographic factors
Select the correct answer:
(A) Both A and R are true and R is the correct explanation of A
(B) Both A and R are true by R is not the correct explanation of A
(C) A is true, R is false
(D) A is false, T is true

1266. Consider the following two statements:
Assertion (A): Monogamy is the most common form of marriage in the contemporary society
Reason (R): Status of woman get enhanced through mongamy

Select the correct answer:
(A) Both A and R are true and R is the correct explanation of A
(B) Both A and R are true by R is not the correct explanation of A
(C) A is truc, R is falsc
(D) A is false, T is true

1267. Match list I with list II and select the correct answer:
List I
a. I.P. Desai b. A.M. Shah
c. K.M. Kapadua d. M.S. Gore
List II
1. Household dimensions of family in India
2. Marriage and family in India
3. Urbanisation and family
4. Some aspects of family in Mahwa
5. World Revolution and family pattern

Codes:

	a	b	c	d
(A)	4	3	1	2
(B)	4	1	2	3
(C)	5	1	3	2
(D)	5	3	1	4

1268. During which stage most children learn to accept rules and standards of morality:
(A) Neonate stage (B) Early childhood
(C) Late childhood (D) Adulthood

1269. The term virilocal residence refers to the establishment of their residence by a newly married couple:
(A) With the bride's parents
(B) With the groom's parents
(C) Independently
(D) With the mother's brothers of the bridge

1270. Who has to pay 'Dowery' to whom?
(A) Wife to husband
(B) Husband to wife
(C) Bride's father to groom
(D) Groom's father to bride

1271. Consider the following two statements:
Assertion (A): If we want to understand society and social life, it is important to ignore family
Reason (R): It is unlikely that any society has ever existed without some social arrangement that could be termed family
Select the correct answer:
(A) Both A and R are true and R is the correct explanation of A
(B) Both A and R are true by R is not the correct explanation of A
(C) A is true, R is false
(D) A is false, T is true

1272. Which of the following is most closely related to happiness:
(A) Wealth (B) Religion
(C) Age (D) Social relationship

1273. Consider the following two statements:
Assertion (A): Kin terms are 'role terms'
Reason (R): Kin terms designate a biological as well as social relationship
Select the correct answer:
(A) Both A and R are true and R is the correct explanation of A
(B) Both A and R are true by R is not the correct explanation of A
(C) A is true, R is false
(D) A is false, T is true

1274. Colonists forcing natives to become slaves is an example of social interaction known as:
(A) Conflict (B) Coercion
(C) Exchange (D) Competition

1275. For a person, an adopted son is:
(A) An affinal kin
(B) A consanguineous kin
(C) Neither an affinal nor a consanguineous kin
(D) An affinal as well as a consanguineous kin

1276. Match list I with list II and select the correct answer:

List I (Forms of Marriage)

a. Marriage with mother's brother

b. Fraternal polyandry

c. Both fraternal and non-fraternal polyandry

d. Levirate

List II (Tribes)

1. Todas 2. Tiketans

3. Coorgs

4. Deshastha Brahmans of Kernataka

Codes:

	a	b	c	d
(A)	4	3	1	2
(B)	4	1	2	3
(C)	5	1	3	2
(D)	5	3	1	4

1277. Consider the following two statements:

Assertion (A): Socially sanctioned sex gratification is a basis for marriage

Reason (R): Marriage regulates sexual relations b/w individuals

Select the correct answer:

(A) Both A and R are true and R is the correct explanation of A

(B) Both A and R are true by R is not the correct explanation of A

(C) A is true, R is false

(D) A is false R is true

1278. In which of the following communities is the marriage with one's father's sisters daughter permitted?

(A) Coorgs and Marathas of northern Maharastra

(B) Marathas of southern Maharastra and okkaligas of Mysore (Karnataka)

(C) Marathas of entire Maharastra

(D) The Coorgs

1279. Social stratification of a consequence of:

(A) Differentiation and evaluation

(B) Differentiation and classification

(C) Classification and satisfaction

(D) Classification and tabulation

1280. A group based on kinship is called:

(A) Voluntary group

(B) Involuntary group

(C) Delegate group (D) Associated group

1281. 'Dudhlautana' as a form of cross-cousin marriage is prevalent among the:

(A) Gonds of Madhya Pradesh

(B) Santhals of Orissa

(C) Mudas of chhota Nagpur

(D) Bhils of Rajasthan

1282. Marriage between single pairs, without involving the rights of exclusive cohabitation with each other is referred to as:

(A) Synadasuian family

(B) Conjugal family

(C) Patriarchial family

(D) Promiscurous family

1283. A 'Tarwa' is composed of a woman:

(A) Her sons

(B) Her husband, her sons and son's daughters and sons

(C) Her daughters, husbands and daughters

(D) Her son's wives and daughter

1284. The advantage of polyandry is:

(A) Divorce is not easy

(B) Division of property is not easy

(C) Members feel security

(D) All of these

1285. Tracing out descent in male line for one purpose and in female line for other is called:

(A) Double male descent

(B) Double female descent

(C) Double unilineal descent

(D) Double unilineal descent

1286. Rituals kinships refers to:

(A) Legally defined marriage relationship

(B) Kins related to one another through blood

(C) Kinship on the basis of functional social relationship

(D) Persons related to each other through rituals

1287. The process by which ideas and beliefs become part of the self is called:
(A) Conformity
(B) Internalization
(C) Imitation
(D) Institutionalisation

1288. Consider the following two statements:
Assertion (A): Although controls human behaviour yet we can't call it repressive
Reason (R): It often does repress drives, but it does not do so completely.
Select the correct answer:
(A) Both A and R are true and R is the correct explanation of A
(B) Both A and R are true by R is not the correct explanation of A
(C) A is true, R is false
(D) A is false, T is true

1289. The process by which national integration is achieved by recognizing the separate but equal status of the various constituents of a nation is called:
(A) Multiculturalism
(B) Secularism
(C) Separatism
(D) Two-nation theory

1290. A system under which a woman is permitted to marry a person of the lower caste but vice-versa is not allowed is known as:
(A) Hypergamy (B) Hypogamy
(C) Preniscuily (D) Polygamy

1291. If a married couple lives with a brother of the groom's mother, then the couple follow which one of the following residency rules?
(A) Uxorilocal (B) Neolocal
(C) Avunculocal (D) Virilocal

1292. Which of the following relationships are revealed through the?
(A) Polygyny and cross cousin marriage
(B) Polyandry and sororate
(C) Matrilineal joint family
(D) Polyandry and cross cousin marriage

1293. A marriage always occurring outside a certain group of relations is called:
(A) Hypergamy (B) Endogamy
(C) Exogamy (D) Hypogamy

1294. According to Hindu traditional belief, marriage is a:
(A) Contract (B) Sacrament
(C) Bond (D) Formality

1295. Which of the following is the characteristic feature of classification of family into matrilineal and patrilineal family?
(A) Residence of spouse
(B) Sexual division of labour
(C) Decendants of same ancestor
(D) Differential power of male and female

1296. A group organized on the basis of the common ancestry of its members is called:
(A) Primary Group
(B) Secondary Group
(C) In Gr up (D) Descent Group

1297. The term 'Couvade' refers to a:
(A) Husband practicing magic during the pregnancy period of his wife
(B) Husband looking after the first child during the second delivery of his wife
(C) Husband leading the life of an invalid during the post delivery period of his wife
(D) Husband heading the life of a priest performing rituals for safe delivery for his wife

1298. The correct order of the main objectives of marriage in the traditional Hindu Society is:
(A) Praja, Rati, Dharma
(B) Rati, Dharma, Praja
(C) Dharma, Rati, Praja
(D) Dharma, Praja, Rati

1299. Changes in intelligence are usually quite small after:
(A) Upper childhood
(B) Lower childhood
(C) Middle childhood
(D) All of these

1300. Techniques for improving memory are termed as:
(A) Mnemonics (B) Loci
(C) Imaging (D) Chunking

1301. Storage function of sensory channels are called:
(A) Iconic image (B) Sensory register
(C) Retrieval (D) All of these

1302. When ethnic group maintain their cultural identies and differences but coexist peacefully is known as:
(A) Amalgamation
(B) Assimilation
(C) Isolation
(D) Cultural pluralism

1303. 'Cultural ethos' stands for the world-view of:
(A) Society (B) A people
(C) Culture (D) Transaction

1304. The mutual impact of British and Indian cultures on each other termed as:
(A) Accommodation (B) Acculturation
(C) Westernization (D) Assimilation

1305. Newborn individual is molded into social being and find their fulfillment with in the society through the process of:
(A) Acculturation (B) Socialization
(C) Amalgamation (D) Assimilation

1306. Consider the following two statements:
Assertion (A): Many tribal communities are said to have merged with Hindu communication around them
Reason (R): In the process of assimilation it is the relative power of the groups that is the deciding factor
Select the correct answer:
(A) Both A and R are true and R is the correct explanation of A
(B) Both A and R are true by R is not the correct explanation of A
(C) A is true, R is false
(D) A is false, T is true

1307. Study of population is called:
(A) Dermatology (B) Demagogy
(C) Anthropology (D) Demography

1308. Sociometry was developed by:
(A) Schramm (B) Rogers
(C) Moreno (D) Shoemaker

1309. Match list I with list II and select the correct answer:
List I
a. Accomodation b. Acculturation
c. Assimilation
List II
1. Mutual cultural diffusion through which persons or groups come to share a common culture
2. Acquisition by a group of individual of the traits of another culture
3. Peaceful coexistence between hostile or competing groups
4. The effort to monolise rewards by eliminating or weakening the competitions

Codes:

	a	b	c
(A)	4	2	3
(B)	3	2	1
(C)	2	4	3
(D)	2	3	1

1310. The process whereby the thought and behaviour of individuals in a given society are influenced by its culture is called:
(A) Cultural speciality
(B) Cultural standard
(C) Cultural norms
(D) Cultural conditioning

1311. The technique which is used in short term memory experiment:
(A) Free recall (B) Recall
(C) Recognition (D) Rehearsal

1312. Working together to accomplish a goal that all desire, implies:
(A) Competition (B) Conflict
(C) Co-operation (D) Compromise

1313. Which of the following is informal type of sanction:
(A) Ridicule (B) Imprisonment
(C) Fines (D) Torture

1314. In which one of the following cities specialization in production occurs in the product, not in the process?
(A) Industrial
(B) Pre-Industrial
(C) Modern Western City
(D) Post-Industrial

1315. The term assimilation is defined:
(A) By reference to some other object or system, to their compatibility within a common process
(B) Qualitatively with respect to its conformity to a standard of value
(C) Qualitatively with respect to the size of groups and its norms
(D) Qualitatively with respect to structural or functional differentiation

1316. Objects and things which are endowed with a special meanng or significance by people is called:
(A) Cultural symbols
(B) Cultural universals
(C) Cultural standard
(D) Cultural speciality

1317. The process in which complete merging of group with separate cultures into group with a common culture and identity takes place is called:
(A) Amalgamation
(B) Assimilation
(C) Integration
(D) Accommodation

1318. The process of accommodation is adopted in society to avoid:
(A) Competition between people
(B) Competition between ideologies
(C) Conflict between equals and non-equals
(D) Integration between formal and informal

1319. The role of statistical methods in the study of sociology is to increase:
(A) Subjectivity (B) Complexity
(C) Objectivity (D) Neutrality

1320. The storage capacity of the short term memory can be increased through:
(A) Rehearsal (B) Perception
(C) Chunking (D) Scanning

1321. A person's functional significance is indicated by:
(A) Status (B) Role
(C) Act (D) Culture

1322. The symbolic trait of an economic institution is:
(A) Emblem (B) Will
(C) Flag (D) Charter

1323. If a gang is involved in conflict with arrival gang, the intragroup conflict usually tends to:
(A) Increase
(B) Decrease
(C) Disappear
(D) Remain unaffected

1324. Division of power between higher and lower levels of government is called:
(A) Horizontal distribution
(B) Parallel distribution
(C) Vertical division
(D) Diagonal division

1325. Power is shared among different Political Parties that represent:
(A) Different candidates
(B) Same ideologies
(C) Different ideologies and social groups
(D) None of these

1326. Community government is:
(A) A government for the entire country
(B) A government elected by the people belonging to one community
(C) A social division based on shared culture
(D) A government which literally means "rule by the people"

1327. Urbanisation in India:
(A) Started as a result of industrialization
(B) Has been the cause of industrialization
(C) Started as a result of modern education
(D) Existed in India before industrialization and western impact

1328. Power sharing is good because:
(A) It increases the conflict between social groups
(B) It ensures the instability of political order
(C) It reduces the possibility of conflict between social groups
(D) It leads to violence

1329. Which one of the following was the most important condition for the growth of early cities?
(A) Increase in production
(B) Surplus food production
(C) Technological innovations
(D) Discoveries of new trader

1330. The social process in which individuals directly challenges the antagonist:
(A) Competition
(B) Conflict
(C) Assimilation
(D) Acculturation

1331. Which one of the following systems of power sharing is called checks and balances?
(A) Horizontal distribution of powers
(B) Federal division of powers
(C) Separation of powers
(D) Power shared among different levels of government (vertical division of powers)

1332. . major group of people who are biologically inbred is:
(A) Caste (B) Race
(C) Class (D) Religion

1333. Metropolitan region is:
(A) A congregation of cities
(B) A centre of a large ecological community
(C) An urban area that is highly industrialised
(D) A capital city

1334. Which of the following is an example of horizontal sharing of power?
(A) Power sharing between different states
(B) Power sharing between different organs of the government
(C) Power sharing between different levels of the government
(D) Power sharing between different political parties

1335. The distinction between upper class, middle class and lower class is based on:.
(A) Wealth (B) Birth
(C) Education (D) Politics

1336. The term ı espondent is used in:
(A) Social survey research
(B) Statistics
(C) Psychology (D) Interactionism

1337. A city is distinguished from a village by:
(A) Preponderance of secondary occupation
(B) Occupational unmobility
(C) High birth rate
(D) Informal social control

1338. Under vertical division of power sharing of power is:
(A) Among government at different levels
(B) Among different organs of the government
(C) Among different social groups
(D) Among different political parties

1339. The other name for regional government in India is:
(A) Provincial Government
(B) State Government
(C) Community Government
(D) None of these

1340. Social control in the city is different from that of village because:
(A) The city furnishes anonymous to those who desire it
(B) The formal social controls are absent in villages
(C) Primary groups do not have effective social control in the city
(D) Secondary groups do not enforce particularistic orientation

1341. A person who does not discriminate other on the basis of religious beliefs:
(A) Communalist
(B) Feminist
(C) Secularist
(D) Patriachal

1342. Which one of the following statements regarding the representation of wor n in the Parliament is not correct
(A) In India, the proportion of woman in legislature has been very low
(B) In Nordic countries the proportion of woman in legislature has been right
(C) The percentage of elected women members in Lok Sabha has never reached even 10%
(D) The share of women in the state assemblies is high

1343. The major share of a city's population growth in India is due to:
(A) Migration
(B) Natural growth
(C) Redefinition of municipal limits
(D) Industrialisation

1344. Which one of the following is not appropriate reason for power sharing?
(A) It leads to economic development
(B) It reduces the possibility of social conflict
(C) It ensures political stability
(D) It is the very spirit of democracy

1345. What is casteism?
(A) Organisation of people into social groups for the purpose of marriage, work and diet
(B) Using different regions in politics
(C) Both (A) and (B)
(D) None of these

1346. What are family laws?
(A) Those laws that deal with family related matters such as marriage, divorce, adoption, inheritance
(B) Allows freedom to practice any work
(C) Both (A) and (B)
(D) None of these

1347. The distribution of human settlement in terms of rural and urban constitutes:
(A) A dichotomy
(B) A continuum
(C) A dialetial
(D) Urban development

1348. Which of the following is the most visible institution in a democracy?
(A) Pressure Group
(B) Public Interest Group
(C) Sectional Interest Group
(D) Political Parties

1349. The aim of sociology according to Comte is:
(A) Scientific analysis of society
(B) Coordination among the various sciences
(C) Social reorganization
(D) All of these

1350. "Religion can never be separated from politics". Who said these words?
(A) Jawaharlal Nehru
(B) W.C. Bannerji
(C) Mahatma Gandhi
(D) Indira Gandhi

1351 Consider the following two statements:

Assertion (A): According to WHO, in 1993, there were 16.5 million deaths due to infections diseases world wide

Reason (R): Urbanisation without health planning leads to spread of diseases

Select the correct answer:

(A) Both A and R are true and R is the correct explanation of A

(B) Both A and R are true by R is not the correct explanation of A

(C) A is true, R is false

(D) A is false, T is true

1352. Under distributions of power sharing is shared among legislature, executive and judiciary:

(A) Diagonal (B) Parallel

(C) Horizontal (D) Vertical

1353. Consider the following two statements:

Assertion (A): Most of the time the urban individual participates and fulfills his needs in secondary groups

Reason (R): The primary groups have ceased to exist in modern societies

Select the correct answer:

(A) Both A and R are true and R is the correct explanation of A

(B) Both A and R are true by R is not the correct explanation of A

(C) A is true, R is false

(D) A is false, T is true

1354. It is a group of people who come together to contest elections and hold power in the government:

(A) Pressure Group

(B) Public Interest Group

(C) Sectional Interest Group

(D) Political Parties

1355. Consider the following two statements:

Assertion (A): Cities are the coenfluence of races, cultures and people

Reason (R): Cities are centres of innovation

Select the correct answer:

(A) Both A and R are true and R is the correct explanation of A

(B) Both A and R are true by R is not the correct explanation of A

(C) A is true, R is false

(D) A is false, T is true

1356. Which of the following is not a feature of communalism?

(A) Communal politics is based on the idea that religion is the principal basis of social community

(B) Religion is seen as the basis of nation

(C) Beliefs of one religion are presented as superior to those of other religions

(D) The followers of a particular religion can belong different communities

1357. A Person, who believes in equal rights and opportunities for women and men:

(A) Communalist (B) Feminist

(C) Secularist (D) Patriarchal

Chapter 13

Training manual

1. The process of designing and maintaining an environment in which individuals working together in groups effectively accomplish selected aims:
(A) Administration (B) Management
(C) Policy making (D) Planning

2. The term "management" was derived from:
(A) Latin (B) French
(C) Greek (D) Sanskrit

3. The literal meaning of the word "Management" is:
(A) House keeping
(B) Personal administration
(C) Store keeping (D) Staff placement

4. Management is called a process because it is:
(A) Applicable to manufacturing process
(B) Relevant for social organization
(C) Involves a series of action
(D) None of these

5. Social units, or human grouping, deliberately considered to seek specific goals:
(A) House keeping
(B) Personal administration
(C) Institution (D) Organization

6. Output-input ratio within a time period with due consideration for quality:
(A) Productivity (B) Profitability
(C) Quality control (D) Sustainability

7. 'POSCORB' as management function was developed by:
(A) Gullick & Urwick
(B) Urwick
(C) L. Gulick (D) H. Fayol

8. The acronym 'POSCORB' symbolizes:
(A) The activities of government
(B) The principles of administration
(C) The activities of an administrator
(D) The objectives of organization

9. Management interest or practice followed for a period or time with exaggerated zeal:
(A) Management fad
(B) Management zeal
(C) Management outlook
(D) Management culture

10. Work tends to expand to fill the time available:
(A) Parkison's law (B) Marconi law
(C) Management by objective
(D) Commitment principle

11. The management skill most needed for people working at top position:
(A) Technical Skills
(B) Conceptual Skills
(C) Clerical Skills (D) Leadership Skills

12. Middle management is:
(A) Consists of departmental managers
(B) Transmit orders down
(C) Motivate lower management
(D) All of these

13. Selecting missions and objectives and the actions to achieve them:
(A) Planning (B) Organizing
(C) Leading (D) Coordinating

14. General programs of action and development or resources to attain comprehensive objectives:
(A) Plan (B) Strategy
(C) Rule (D) Policies

15. A statement of expected results expressed in numerical terms:
(A) Budget (B) Programme
(C) Plan (D) Strategy

16. The inputs for formulating policies are otherwise called as:
(A) Objectives (B) Procedures
(C) Programmes (D) Projects

17. Plans should cover a period of time that is long enough to fulfill commitments involved in decisions made today:
(A) Commitment principle
(B) Principles of planning
(C) Management strategy
(D) Management by objective

18. A comprehensive management system that integrates many key management activities in a systematic manner and that is consciously directed towards efficient and effective accomplishment of organizational and individual objectives:
(A) Key Performance Method
(B) Critical Programme Method
(C) Management Information System
(D) Management by Objective

19. The concept of Management by Objective was first put forward by:
(A) Edward Weidner (B) Montgomery
(C) Peter F Druner (D) F.W. Riggs

20. A modern tool for analysis of the situation:
(A) Management by objective
(B) Win-loose game
(C) 'TOWS' matrix
(D) Brainstorming

21. When a company plans to liquidate or retrench, the business strategy is more of:
(A) 'SO' strategy (B) 'WO' strategy
(C) 'ST' strategy (D) 'WT' strategy

22. Planning Research and Action Institute is located at:
(A) New Delhi (B) Hyderabad
(C) Lucknow (D) Bhopal

23. Selection of a course of action from among the alternatives:
(A) Planning (B) Decision making
(C) Controlling (D) Leading

24. The core of planning is:
(A) Span of control (B) Coordination
(C) Hierarchy (D) Decision making

25. Any decision or action to be realistic has to be based on:
(A) Facts (B) Opinion
(C) Value judgement (D) Moral grounds

26. Ability to make prompt and good decisions is known as:
(A) Decisiveness (B) Initiative
(C) Cohesiveness (D) Social insight

27. The decisions concerned with structured or routine problems:
(A) Programmed decisions
(B) Non-programmed decisions
(C) Natural decisions
(D) Illogical decisions

28. Strategic decisions in general are:
(A) Programmed decisions
(B) Non-programmed decisions
(C) Natural decisions
(D) Illogical decisions

29. Which among the following is a quantitative tool used for decision making:
(A) Break even analysis
(B) Finite element analysis
(C) Factor analysis
(D) None of these

30. 'With or without principle' in-cision making approach related to:
(A) Coast benefit analysis
(B) Factor analysis
(C) Differential analysis
(D) Marginal analysis

31. The character that distinguish a non programmed decision from programmed decision:
(A) Routine nature (B) Subjectivity
(C) Structured (D) Precedence

32. The process of establishing an internal structure of roles for people to fill in an organization:
(A) Supervision (B) Controlling
(C) Managing (D) Organizing

33. The process of grouping of activities into units for the purpose of administration is characteristic of:
(A) Span of control
(B) Hierarchy
(C) Unity of command
(D) Dapartmentation

34. The cases of organizations working in various shifts is a clear example of departmentation based on:
(A) Territory (B) Function
(C) Product (D) Time

35. The span of control depends on:
(A) Personality of the supervisors
(B) Age of the organization
(C) Element of space (D) All of these

36. In which of the following cases span of control is smaller:
(A) Well planned work schedule
(B) Utilizing the staff assistance
(C) Effective communication system
(D) Heterogeneous work environment

37. Which of the following can widen the span of control:
(A) Delegation
(B) Simple task
(C) Ineffective meetings
(D) Both (A) and (B)

38. The informal organization is characterized by:
(A) Emotional (B) Customary
(C) Personal (D) All of these

39. The ability of an individual to induce or influence the beliefs or action of other persons of group:
(A) Power (B) Leadership
(C) Authority (D) Control

40. The right in a position to exercise discretion in making decisions affecting others;
(A) Responsibility (B) Hierarchy
(C) Authority (D) Unity of command

41. The concept of Zone of Indifferent is associated with:
(A) Decision making (B) Leadership
(C) Authority (D) Motivation

42. Which of the following is legitimate power:
(A) Intimidation (B) Authority
(C) Manipulation (D) Suggestion

43. The most logical criterion to distinguish a line function from staff is:
(A) Authority relationship
(B) Functional relationship
(C) Grouping of functions
(D) Departmentalization

44. Taking decision and acting upon them are the main responsibility of:
(A) Line authority (B) Staff authority
(C) Super authority
(D) Functional authority

45. Which one of the following is not a function of staff agencies:
(A) Planning
(B) Advising
(C) Consultation
(D) Achieving goals

46. Which of the following is not true of line activity:
(A) It has operational ability
(B) It stands for though and facilitation
(C) It decides the organizational goal
(D) It has authority

47. Which of the following factors could place limitations on the authority of a superior:
(A) Group behaviour
(B) Social relationships
(C) Organizational norms
(D) All of these

48. The chief reason for delegation of authority from headquarters to field officers is:
(A) Inculcate skill in the later
(B) Facilitate local decision making
(C) Decongest the headquarter
(D) Control the field officer

49. The concept of unity of command is complimentary to the principle of:
(A) Supervision (B) Scalar chain
(C) Span of control (D) Administration

50. When the delegation goes by a climate of understanding between the superior and subordinate, it is known as:
(A) General delegation
(B) Specific delegation
(C) Formal delegation
(D) Informal delegation

51. For field observation method preferred is:
(A) Conference (B) Primary source
(C) Correspondence (D) Interview

52. Sociology are primarily interested in:
(A) Similarities between human behaviours and the behaviour of other animals
(B) Way in which human beings mutually influence each other
(C) Individual since he is convinced that there really is no such thing as group
(D) Anything which is related in a casual way to the behaviour of mankind

53. Correct chronology of study method suggested by A. Comte is:
(A) History, Observation, Experiment and Comparison
(B) Experiment, Observation, History and Comparison
(C) Observation, Experiment, History and Comparison
(D) Experiment, Observation, Comparison and History

54. Procedures which cannot prove useful for ensuring objective fact finding is:
(A) Senior people should be sent to the field
(B) Some counter sample-checking should be done
(C) Trained people should be set to the field
(D) Respondents should be given monetary incentives

55. Hinderance in objective finding is:
(A) Investigator is unbiased
(B) Respondent is aware of the magnitude of the problem
(C) Respondent is aware of the magnitude of the problem
(D) Investigator is serious in investigating a problem

56. Dependable method for obtaining objectivity is:
(A) Questionnaire method
(B) Interview method (C) Field survey
(D) Group discussion

57. Primary methodology of functionalists is:
(A) Library work (B) Positivism
(C) Conjectural history (D) Field work

58. Objectivity in investigating social problem can be possible only when investigator has:
(A) Close mind
(B) Pre-conceived notions
(C) Scientific mind
(D) Pre-determined ideas

59. Experiments are:
(A) Quite common in sociology
(B) Rare but possible in sociology
(C) Impossible in sociology
(D) Easiest in sociology

60. Pareto termed the rational explanations for non-logical action as:
(A) Conceits
(B) Archi-types
(C) Derivations
(D) Pseudo-judgements

61. Social research is:
(A) Collection of facts on a given subject
(B) The relevant facts put together
(C) Collection of the required data, maps, charts, observations etc.
(D) Logical arrangement of facts which mutually explain each other

62. For collecting information about Indian society method preferred is:
(A) Participant observation method
(B) Social survey method
(C) Questionnaire method
(D) Interview method

63. Basic difference between E. Durkheim and M. Weber in terms of methodology is:
(A) Durkheim believed in subjective approach and Weber in the objective approach
(B) Weber believed in objective approach and the Durkheim in the subjective approach
(C) Both adopted each other method
(D) None of these

64. Error regarding an abstraction as a real phenomenon is called:
(A) Fallacy of abstraction
(B) Fallacy of confusion
(C) Fallacy of ratification
(D) Fallacy of verification

65. Objectivity in social investigation can be possible when the data:
(A) Has been manipulated
(B) Has been collected in particular manner
(C) Is sufficient
(D) Is unrelated to the problem

66. While investigating a social problem objective finding is:
(A) Very much possible
(B) Very difficult
(C) Possible without great difficulty
(D) None of these

67. By objectivity we generally mean:
(A) Impersonal attitude
(B) Unbiassed collection of data
(C) Unconditional motion of tangible and intangible facts
(D) All of these

68. According to Weber, values influence our judgement about the ...... of a specific event:
(A) Viability (B) Cause
(C) Repercussions (D) Effect

69. Quality for an investigator of social problem is that he should:
(A) Have a closed mid
(B) Have capacity to twist the facts
(C) Be capable of observing facts impartially
(D) Be guided by emotions

70. Which of the following methods is usually not used for objective fact finding?
(A) Historical method
(B) Library method
(C) Questionnaire method
(D) All of these

71. Individuals with a combination of low-ascribed and high-achieved statuses often act "pushy" in their attempts to emphasize their achieved status. This conceptualized as:
(A) Mediocracy (B) Double jeopardy
(C) Status inconsistency
(D) Relative mobility

72. Holistic approach in sociology is associated with:
(A) Evolutionary approach
(B) Neo diffusionist approach
(C) Diffusionist approach
(D) Structural-functional approach

73. In which of the following methods of data collection where maximum danger of losing one's sharpeness of observation is possible because of too much familiarity?
(A) Interview (B) Questionnaire
(C) Observation
(D) Participant observation

74. An observation, uninfluenced by personal bias is called:
(A) Functional (B) Critical
(C) Objective (D) Subjective

75. By which of the following means objective facts cannot be secured or established by the researcher?
(A) By observation (B) By research
(C) By objective analysis of data
(D) From material collected on the basis of oath and affirmation

76. Objectivity cannot be introduced at the time of:
(A) Data collection
(B) Data interpretation
(C) Field survey
(D) Meeting non-respondents

77. Which method is very frequently used by the sociologist?
(A) Ideal type method
(B) Sample method
(C) Sociometry (D) Field research method

78. Which of the following is define as a method of measuring individuals social behaviour?
(A) Sociometry (B) Sample method
(C) Verstehen method
(D) Interview method

79. First hand study of a small community is called:
(A) Ethnology (B) Ethnography
(C) Ethics (D) Aesthetics

80. Participant observation has a clear advantage over other methods of data collection in the sense that it:
(A) More uniform (B) More gratifying
(C) Yields in-depth information
(D) More precise

81. Sometimes mailed questionnaire is preferable to interview method because:
(A) Face-to-face contact is possible
(B) People immediately return it
(C) Illiterates can also be covered
(D) It works out cheaper and easy

82. Which of the following is a social fact?
(A) Tangible fact (B) Intangible fact
(C) Social custom (D) All of these

83. The scientific study of animal behaviour is:
(A) Ethymology (B) Ethology
(C) Pathology (D) Socio-biology

84. Second stage of scientific or experimental method is:
(A) Classification (B) Recording
(C) Observation (D) Prediction

85. According to Durkheim, biology has three principal division, viz.:
(A) Social morphology, social biology and general sociology
(B) Social morphology, social psychology and general sociology
(C) Social statistics, social dynamics and general sociology
(D) None of these

86. Correct sequence of the processes in the scientific method is:
(A) Hypotheses, observation, recording, classification, prediction and verification
(B) Observation, recording, classification, hypotheses, verification and prediction
(C) Recording, observation, classification, verification, hypotheses and prediction
(D) None of these

87. Recording data is called:
(A) Concept (B) Fact
(C) Hypotheses (D) Label

88. A man who studies effects of the social structure of a society on its form of government is most likely a/an:
(A) A sociology
(B) A physical anthropology
(C) An economist (D) A historian

89. Which of the following method is a form of qualitative analysis involving the very useful and complete observation of a person, a situation or an institution?
(A) Inverse deductive
(B) Case study
(C) Social survey
(D) Historical method

90. A study of relations between organism and their environment is called:
(A) Topography (B) Geography
(C) Ecology (D) Biology

91. Social research is:
(A) Logical arrangement of facts which mutually explain each other
(C) Collection of the required data-maps, charts, observations, etc
(C) Collection of facts on a given subject
(D) All the relevant facts put together

92. In the social sciences:
(A) Statistical methods are not applicable
(B) Social survey method is the only reliable method of data collection
(C) Prediction is possible
(D) None of these

93. Which of the following theories refutes the assumption that richness in itself bring down birth rate and in poverty the rate is bound to go up?
(A) Optimum theory of population
(B) Theory of increasing prosperity
(C) Theory of rationalism
(D) Theory of voluntarism

94. Sociology cannot make use of:
(A) Historical method
(B) Ideal types method
(C) Scientific method
(D) Social survey method

95. Method of participant observation is associated with:
(A) Psychology (B) Social Anthropology
(C) Archaeology (D) Biology

96. Technique more suitable for study in depth is:
(A) Mailed questionnaire
(B) Interview guide
(C) Interview schedule
(D) Structured interview

97. Maximum subjectivity in social investigation can come when:
(A) Problem is being identified
(B) Problem has been identified
(C) Questions to be asked are being finalised
(D) Report is being finalized

98. Study of Human society involves the study of:
(A) Man (B) Group
(C) Culture (D) Family

99. Field study is a method of research in which the observations are taken in the:
(A) Natural setup (B) Observed set up
(C) Laboratories (D) Field

100. The comparative method was used by:
(A) Positivists (B) Neo-positivist
(C) Diffusionist (D) Evolutionist

101. The role of statistical methods in the study:
(A) Objectivity (B) Complexity
(C) Subjectivity (D) None of these

102. An interviewer can collect good material when the relationship between interviewer and the interviewee is:
(A) Indifferent (B) Uninterested
(C) Cordial (D) Ethical

103. The nature of objectivity is:
(A) Always based on the mental cognition of the person concerned
(B) To come upon objectivity consistent thinking with logical approach is to be maintained
(C) Degree of objectivity differs from person to person in different or same situations
(D) All of these

104. Social survey studies:
(A) Economic problem
(B) Social problem
(C) Historical problems
(D) Political problems

105. Which of the following statements carry the correct meaning of the objectivity approach?
(A) Selection of facts should not be subject to one's like or dislikes
(B) Analysis should not be influenced by subjective approach
(C) Approach should be impersonal
(D) All of these

106. Objectivity exists in:
(A) Partially interpretating data
(B) Scientific attitude
(C) Highlighting one's own view point
(D) Creating ambiguities

107. Objective finding is hindered because:
(A) Investigators are educated
(B) Respondents are usually biased
(C) Investigators are illiterate
(D) Respondents are illiterate

108. Objectivity is:
(A) Seeks scientific temperature
(B) Is biased on cause-effect relationship
(C) In collecting facts, it fulfils the manifestation of problems in time and space relationship
(D) All of these

109. Objectivity is:
(A) Requires a detached impersonal view
(B) Requires a person to give secondary importance to his beliefs
(C) Requires a person to discard his own beliefs and moral convictions
(D) None of these

110. Objective facts refer to facts when they have been:
(A) Theoretically established
(B) Rationally determined
(C) Logically derived
(D) Empirically verifies

111. An important criterion of a good sample is its:
(A) Small size
(B) Large size
(C) Representativeness
(D) Ambiguities

112. Attitude is – in nature and character:
(A) Vague (B) Clear
(C) Subjective (D) Objective

113. Main goal of sociological study is:
(A) Social reform
(B) Social construction
(C) Social reconstruction
(D) Study society in its existing form

114. The interview guide will be used during interview by:
(A) Interviewee
(B) Interviewer
(C) Different people
(D) Both interviewer and interviewee

115. Most widely used method of data collection among sociologists is:
(A) Participant observation
(B) Interview method
(C) Mailed questionnaire
(D) Simple observation

116. Sociology is a science because it is:
(A) Objective (B) Quantitative
(C) Methodical (D) Qualitative

117. Which one of the following is a thought or, an idea that sample for a phenomenon or, a group of facts under a common lable?
(A) Concept (B) Definition
(C) Sign (D) Theory

118. Comparative method was generally used by:
(A) Positivests (B) Anti-positivists
(C) Evolutionists (D) Empericist

119. Which is the following means a looking at things?
(A) Observation (B) Recording
(C) Prediction (D) Verification

120. Crucial feature of scientific data is its:
(A) Reliability (B) Quality
(C) Quantifiability
(D) Universal applicability

121. Most frequently used method of research in sociology is/are:
(A) Experiment method
(B) Explorative method
(C) Descriptive method
(D) Explorative and descriptive methods

122. Social scientists deal with ideas:
(A) So that they can reinforce them
(B) To ascertain their origin
(C) To ascertain whether they are or can be turned into knowledge
(D) To prove them false

123. Objectivity requires a:
(A) Person to give secondary importance to his own biases and preferences
(B) Person to discard his own belief and moral conviction
(C) Person's detached and an empersonal view
(D) None of these

124. While conducting a social survey, the questionnaire technique is not applicable if:
(A) There is widespread literacy
(B) Area to be covered is vast
(C) There is male literacy
(D) There is over-population

125. A social researcher can use interview than a questionnaire to collect good and more material in getting information in a short time when he meets persons of:
(A) A particular area
(B) Different areas
(C) A bigger areas
(D) A wider areas

126. A set of techniques used to measure attractions and repulsions during interpersonal relations in quantitative and diagrammatic terms is called:
(A) Sociometry (B) Social survey
(C) Quantitative method
(D) Social statistics

127. Emotions, sensations, perceiving, desiring, etc come under the category of:
(A) Integrative events (B) Freudian events
(C) Objective events (D) Subjective events

128. The reaction of a child when scolded can be studied under controlled conditions by applying the technique of:
(A) Observation (B) Interview
(C) Questionnaire (D) Field survey

129. Case study aims to:
(A) Establish statistical correlation
(B) Bring out the structure of the units as a whole
(C) Expose persons danger to society
(D) Treats cases requiring self-help

130. A tentative proposition, the authenticity of which is yet to be ascertained is:
(A) Index (B) Hypothesis
(C) Introduction (D) Conclusion

131. Observation means to see things:
(A) With a purpose
(B) With many purposes
(C) Without a purpose
(D) For important discussion

132. Which of the following method consists in the collection of data concerning the living, the working conditions of people in a given area with a view to formulate practical social measures for their betterment and welfare?
(A) Ideal types (B) Social survey
(C) Sociometry (D) Statistics

133. If wishes or value do not consciously or unconsciously distort reporter's description of the situation, then report is:
(A) Subjective (B) Objective
(C) Absolute (D) Original

134. Objectivity cannot be introduced at the time of:
(A) Selection of respondents
(B) Selecting field
(C) At the time of giving publicity through media
(D) Report writing

135. According to Weber, the subject-matter we select for study determine:
(A) Material benefits
(B) Long-term perspective
(C) Prestige associated with the work
(D) Values

136. Objective finding is hindered because:
(A) Investigators are educated
(B) Respondents are usually biased
(C) Investigators are illiterate
(D) Respondents are illiterate

137. Objective fact finding in social research means:
(A) Some specific object in view
(B) Demands of a particular organization
(C) Impartially
(D) Some pre-conceived objectives

138. Which of the following is a social fact?
(A) Tangible fact (B) Unknown fact
(C) Intangible fact (D) All of these

139. Things external to the mind, material objects and overt bodily behaviour are:
(A) Physical events (B) External events
(C) Objective events (D) Subjective events

140. Schedule is a list of questions which will be answered in an interview by a/an:
(A) Interviewer (B) Respondent
(C) Surveyor (D) Researcher

141. Questionnaire method is not applicable if:
(A) There is overpopulation
(B) There is wide spread literacy
(C) There is illiteracy
(D) Area is vast

142. Use of an interview method by a social researcher is expensive because it covers:
(A) More group of an area
(B) A few group of an areas
(C) More group of different areas
(D) A group of different areas

143. Social research can be conducted to study the problems related to:
(A) An individual (B) A group
(C) All groups (D) Some groups

144. Case study involves:
(A) Very careful observation of a person
(B) Complete observation of a person
(C) Careful observation of a person
(D) Very careful and complete observation of a person

145. Sociologist in study of human behaviour:
(A) Treats values as data which he studies
(B) Determines which values are of greatest importance
(C) Leaves values completely along since they have no place in human behaviour
(D) Attempts to justify the values of his society if he is a good sociologist

146. In an experimental group dynamic situation and new social norms arise, when:
(A) The experimenter gives am instruction
(B) Individuals are faced with ambiguous situations
(C) Instrumental roles are performed
(D) Expressive roles are performed

147. Thurston scale is a type of:
(A) Interview schedule (B) Attitude scale
(C) Opinion poll (D) Questionnaire

148. Training Performance Index does not includes:
(A) Training Effectiveness Index
(B) Training Utility Index
(C) Training Cost Index
(D) Training Participation Index

149. A science is to be judged by three criteria, reliability of its knowledge, its organization and:
(A) Its method
(B) Its capacity to expose universal laws
(C) Its quantifiability
(D) Bulk of qualitative data it can produce

150. Respondent is the term used in:
(A) Formal sociology
(B) Sociology of knowledge
(C) Kinship studies
(D) Social survey research

151. Which of the following statements carry the correct meaning of the objective approach?
   (A) Selection of facts should not be subject to one's likes and dislike
   (B) Analysis should not get influenced by subjective approach
   (C) Approach should be impersonal and unbiased
   (D) All of these

152. Participant observation refers to:
   (A) Working with a group
   (B) Studying a people by participating in their activities
   (C) Observing a people's activities closely
   (D) Observation of the participants

153. Objectivity in sociology requires a person:
   (A) To negate his own culture entirely
   (B) To discard his beliefs and values
   (C) To give secondary importance to his own biases and preferences
   (D) To take a detached, impersonal view

154. Set a techniques to measure attractions and repulsions in inter-personal relations in quantitative and diagrammatic terms is:
   (A) Social statistics
   (B) Sociometry
   (C) Social survey
   (D) Quantitative method

155. The prime concern of a sociologist with human being and their relations to one another is as a/an:
   (A) Disciplined inquirer
   (B) Advocate
   (C) Preacher (D) Reformer

156. When area of research will be large, the appropriate research method is:
   (A) Participant observation method
   (B) Sample survey method
   (C) Questionnaire method
   (D) Mailing questionnaire

157. Objectivity means:
   (A) Absence of personal values
   (B) Lack of personal values and bias in one's work
   (C) Using scientific method of study
   (D) Full of personal value

158. Which of the following is important for the study of sociology?
   (A) It helps in solving the social problems
   (B It enriches our knowledge
   (C) It helps in raising the standards of the people
   (D) All of these

159. Correct chronology steps involved in sociological investigation are:
   (A) Formulation of problem, planning research design, field work, analyzing data, drawing the conclusion
   (B) Analysing data, planning research design, field work, formulation of problem, drawing conclusion
   (C) Formulation of problem, analyzing data, planning research design, drawing conclusion
   (D) Field work, planning research design, formulation of problem, field work and drawing conclusion

160. The method which consists of observation, recording, classification, hypothesis, verification and prediction is:
   (A) Historical method
   (B) Scientific method
   (C) Inverse-deductive method
   (D) Ideal-type method

161. Which of the following fall under the scope of sociology?
   (A) Study of society
   (B) Study of social relationship
   (C) Study of social action
   (D) All of these

162. Which of the following method involves comparisons of various kinds or groups of people in order to find out the differences as well as similarities in their ways of life and thus to find out clues to man's social behaviour?
(A) Historical method
(B) Comparative method
(C) Experimental method
(D) Inverse-deductive method

163. Experimentation is:
(A) Impossible is sociology
(B) Most important method in sociology
(C) Though it is possible not preferred in sociology
(D) Rarely possible in sociology

164. According to which method, an ideal in constructed from concrete cases and then a particular case is evaluated according to the degree of its approximation to the ideals?
(A) Inverse deductive
(B) Comparative
(C) Ideal type (D) Experimental

165. Which method has limitation that cannot be used among illiterates?
(A) Participant observation
(B) Non-participant observation
(C) Questionnaire
(D) Interview

166. A man who studies effects of the social structure os a society on its form of government is most likely a/an:
(A) Historian (B) Sociologist
(C) Economist (D) Anthropologist

167. The field of sociology is distinguished from other disciplines by:
(A) Its method
(B) Types of questions it asks and the particular perspectives it uses
(C) Its social phenomenon measurement scales
(D) The special interest in explaining human behaviour

168. An important feature of a good sample is:
(A) Smaller quantity
(B) Big quantity
(C) Correct representativeness
(D) None of these

169. Case study involves:
(A) Careful observation of a person
(B Complete observation of a person
(C) Very careful observation of a person
(D) Very careful and complete observation of a person

170. Sometimes mailed questionnaire is preferable to interview because:
(A) It works out cheaper
(B) Face –to-face contact is possible
(C) People immediately return it
(D) None of these

171. To obtain entirely reliable knowledge of the exact value of a parameter, observations must be made on:
(A) All the elements in a single stratum of the stratified population
(B) A stratified sample of the population
(C) A complete census of the population
(D) A simple random sampling of the population

172. Which of the following method is less expensive?
(A) Case study
(B) Mailing questionnaire
(C) Interview (D) Questionnaire

173. In participant observation method, observes can collect material from a group to a large extent if he acts:
(A) Emotionally (B) Passively
(C) Objectively (D) Subjectively

174. Correct chronology of Comte's method is:
(A) Experiment, observation, History and Sociology
(B) Observation, Experiment, Comparison and History
(C) History, Political Science, Biology and Sociology
(D) History, Physics, Biology and Sociology

175. There is maximum danger of loosing one's sharpness while gathering data due to excessive rapport in:
(A) Interview (B) Questionnaire
(C) Participant observation
(D) Mailing questionnaire

176. Which of the following is most appropriate?
(A) All social sciences study society
(B) Only sociology and social anthropology study society
(C) Only sociology, social anthropology and history study society
(D) Only sociology studies society

177. Filling and keeping filled, the positions in the organization structure:
(A) Planning (B) Organizing
(C) Staffing (D) Leading

178. Discussion papers based on in-depth study and research is presented and discussed under the guidance of experts:
(A) Colloquium (B) Seminar
(C) Buzz session (D) Workshop

179. The activities of staffing include:
(A) Recruitment (B) Promotion
(C) Separation (D) All of these

180. Detailed biographical information about a job applicant is referred to as:
(A) Questionnaire (B) Bio-data
(C) Interview (D) Discussion

181. The last stage in the process of recruitment is:
(A) Appointment (B) Orientation
(C) Placement (D) Promotion

182. Managers tends to be promoted to the level of their incompetence is said in:
(A) Parkinson's law
(B) Marconi law (C) Peter principle
(D) Commitment principle

183. The essence of promotion is:
(A) Change of title (B) Change of pay
(C) Change of assignment
(D) Change of duties and punishment

184. The basic objective of an interview is to assess the candidate's:
(A) Intellectual capacity
(B) General knowledge
(C) Personal qualities
(D) Power of expression

185. The technique that gives managers a clear idea of their path of development:
(A) Managerial appraisal
(B) Assignment centres
(C) Planned progression
(D) None of these

186. Purpose of time study is:
(A) To remove wastage of time
(B) To make workers punctual
(C) To determine fair day's work
(D) To give timely assistance

187. To identify exactly what workers do and what skills or knowledge they need to succeed in a job we go for:
(A) Job analysis (B) Item analysis
(C) Factor analysis (D) None of these

188. The situation that an employee must be able to cope with if he is going to succeed in a particular job:
(A) Critical incidents (B) Work behaviour
(C) Work motivation (D) Work environment

189. A lower job satisfaction will lead to:
(A) High absenteeism
(B) Low moral
(C) High employee turn over
(D) Both (A) and (B)

190. An expression of discontent or dissatisfaction with any aspect of organization is:
(A) Stress (B) Conflict
(C) Grievance (D) Competition

191. Which of the following is a source of stress in official situation:
(A) Overload of work
(B) Lack of adjustment
(C) Lack of autonomy
(D) All of these

192. Organizational conflict does not occur when there exist:
(A) Autocratic supervision
(B) Incompatible goals
(C) Unlimited resources
(D) Heterogeneous group

193. Which of the following is an informal means of coordination:
(A) Planning (B) Leadership
(C) Conferences (D) Budgeting

194. Formation of cliques in an organization ultimately creates:
(A) Damaging effect
(B) Performing effect
(C) Stabilizing effect (D) No effect

195. Measuring and correcting of activities of subordinates to ensure the events confirm to plans:
(A) Supervision (B) Controlling
(C) Managing (D) Leading

196. The principle of supervision is inherent in the principle of:
(A) Span of control (B) Coordination
(C) Hierarchy (D) Unity of command

197. The technique of supervision include:
(A) The promulgation of service standard
(B) Budgetary limitations upon the operations
(C) Reporting system on work progress
(D) All of these

198. The aspect of supervision does not include:
(A) Institutional aspect
(B) Personal aspect
(C) Political aspect
(D) Technical aspect

199. Each individual in an organization must receive his or her instructions for a particular operation from only one person is known as:
(A) Span of control (B) Unity of command
(C) Hierarchy (D) Power

200. The essence of the scaler principle is:
(A) Unity of command (B) Span of control
(C) Communication (D) Coordination

201. "If it is violated, authority is under mind, discipline is in jeopardy, order disturbed and stability threatened" refers mainly to the principle of:
(A) Hierarchy (B) Span of control
(C) Coordination (D) Unity of control

202. Discharging of duties according to the letter and spirit of the command signifies the principles of:
(A) Control (B) Authority
(C) Accountability (D) Responsibility

203. The span of control depends on:
(A) Personality of superior
(B) Nature of work to be supervised
(C) Age of the organization
(D) All of these

204. The principle of unity of command is complimentary to the principle of:
(A) Span of control (B) Authority
(C) Scaler chain (D) Supervision

205. The principles that is directly concerned with preventing overlap, conflict and friction in an organization:
(A) Direction (B) Supervision
(C) Coordination (D) Span of control

206. Which of the following is the limiting factor in the development of coordination:
(A) Time (B) Size
(C) Informal organization
(D) Both (A) and (B)

207. The most important means of coordination is:
(A) Hierarchy (B) Consultation
(C) Planning (D) Centralization

208. Allocation of funds for different sectors of programme:
(A) Controlling (B) Budgeting
(C) Financing (D) Supervising

209. Dividing a large group into much smaller ones, in which a topic is discussed within a limited period of time:
(A) Colloquium (B) Seminar
(C) Buzz session (D) Symposium

210. Which of the following is not an empirical method of assessing training needs:
(A) Job analysis
(B Performance evaluation
(C) Buzz session
(D) Questionnaire method

211. A cooperative gathering of individuals who discuss, learn and apply practical skills under expert supervision:
(A) Colloquium (B) Seminar
(C) Buzz session (D) Workshop

212. The characteristic feature of the fourth generation of evaluation is:
(A) Measurement oriented
(B) Description oriented
(C) Judgement oriented
(D) Collaboration oriented

213. The element in the evaluation fra ework that covers the trainees opinion about the programme:
(A) Reaction (B) Learning
(C) Behaviour (D) Result

214. The extent of change in the behvaiour interms of job performance as a result of training is determined by:
(A) Process evaluation
(B) Impact evaluation
(C) Plan evaluation
(D) Terminal evaluation

215. Objective-training evaluation model was given by:
(A) Ralph W Tyler (B) Daniel Stufflebeam
(C) Robert Stake (D) Michael Scriven

216. In a situation of abundant supply and low demand the village industry should thrive on:
(A) Export promotion
(B) Value addition (C) Distress sale
(D) Consumer education

217. The evaluation model that uncovers the utility and overall impact of the training programme without any set criteria:
(A) Objective model (B) Goal free model
(C) Expert model (D) Naturalistic model

218. A systematic, integrated and planned approach for improving enterprise effectiveness:
(A) Organizational development
(B) Futures organization
(C) Management grid
(D) Management development

219. The highest quality organization is:
(A) Transcendental (B) Mutualistic
(C) Parasitic (D) Accidental

220. The relationship wherein one gains from an association at the expense of the other:
(A) Transcendental(B) Mutualistic
(C) Parasitic (D) Accidental

221. The term "entrepreneur" is coined by:
(A) Richard Cantinlon (B) Tony Pears
(C) Curt Lewin (D) Gilford Pinchot

222. The most important for an entrepreneur to have is:
(A) High need for achievement
(B) High knowledge about the business
(C) High operational skills
(D) High financial resources

223. Technology recommended for an enterpreneurship must be:
(A) Appropriate (B) Innovative
(C) Risky (D) Costly

224. Enterpreneur are people who take:
(A) High risk (B) Low risk
(C) No risk (D) Moderate risk

225. Successful entrepreneurs are:
(A) Risky people (B) Rational people
(C) Power hungry (D) Empire builders

226. National Institute of Entrepreneurship and Small Business Development is located at:
(A) New Delhi (B) Jaipur
(C) Kolkata (D) Bangalore

227. An entrepreneur who worked within the organizational environment:
(A) Entrepreneur (B) Intrapreneur
(C) Intrepreneur (D) Entrapreneur

228. One of the basic elements in the leadership relationship is:
(A) Task (B) Faith
(C) Motivation (D) Education

229. The Path-Goal approach of leadership theory was formulated by:
(A) Robert House (B) Fred E Fiedler
(C) B F Skinner (D) L W Porter

230. Take the odd man out from the following:
(A) Country club leadership
(B) Middle road leadership
(C) Lassez-faire leadership
(D) Team leadership

231. Leader are born and not made:
(A) Behavioural theory of leadership
(B) Great man theory of leadership
(C) Situational theory of leadership
(D) None of these

232. A basically lazy approach that avoids as much work as possible:
(A) Impoverished management
(B) Country Club management
(C) Authority-compliance
(D) Team management

233. People define roles for themselves and others based on social learning and reading:
(A) Role theory of leadership
(B) Great man theory of leadership
(C) Situational theory of leadership
(D) None of these

234. Successful leadership is based in definable, learnable behavior:
(A) Behavioural theory
(B) Situational theory
(C) Trait theory
(D) Contingency theory

235. People are more committed to action where they have involved in the relevant decision making:
(A) Transactional leadership
(B) Situational leadership
(C) Participative leadership
(D) Contingency leadership

236. Contingency model of leadership was formulated by:
(A) Robert House (B) Fred E. Fiedler
(C) B F Skinner (D) L W Porter

237. The leader has a low concern for people and uses such methods as threats and other fear-based methods to achieve conformance:
(A) Exploitive authoritative
(B) Benevolent authoritative
(C) Consultative leadership
(D) Participative leadership

238. The best approach when the following are expert and their advice is both needed and they expect to be able to give it:
(A) Supportive leadership
(B) Directive leadership
(C) Participative leadership
(D) Achievement-oriented leadership

239. When people have agreed to do a job, a part of the deal is that they precede all authority to their manager:
(A) Transactional leadership
(B) Situational leadership
(C) Transformational leadership
(D) Contingency leadership

240. A leadership style in which the leader himself develops a plan of action and achieves the challenging goal is called:
(A) Instrumental (B) Supportive
(C) Participative
(D) Achievement-oriented

241. According to which theory, leadership is multidimentional:
(A) Behavioural theory
(B) Situational theory
(C) Trait theory (D) None of these

242. Benevolent autocrat un Linkert's four system of management comes under:
(A) System I (B) System 2
(C) System 3 (D) System 4

243. Transformational leadership theory was formulated by:
(A) Robert House (B) Bass, B.M.
(C) B F Skinner (D) L W Porter

244. The term "Motivation" means:
(A) To move or incite into action
(B) To bring forward
(C) To energize inner self
(D) To act according to wishes

245. Motivation is a general term which includes:
(A) Drives and desire
(B) Desire and needs
(C) Emotions and feelings
(D) Both (A) and (B)

246. Which of the following is true regarding Maslow's theory:
(A) Based on psychoanalysis
(B) First theory of motivation
(C) Published in 1942
(D) Based on prepotency of needs

247. The behavioral type embedded in th basic concept of motivation:
(A) Risk taking behavior
(B) Goal satisfying behavior
(C) Decision making behavior
(D) None of these

248. The process theory of motivation does not includes the models of:
(A) Victor Vroom (B) Herald Kelly
(C) Mc Gregor (D) Porter Lawler

249. The middle level need in Maslow's need hierarchy is:
(A) Social needs (B) Security needs
(C) Esteem needs
(D) Self actualization needs

250. The need for power, prestige, status and self-confidence belongs to:
(A) Esteem needs
(B) Security needs
(C) Social needs
(D) Self actualization needs

251. Motivation in Herzberg's need classification belong to which level of Maslow's need hierarchy:
(A) Social needs (B) Security needs
(C) Physiological needs
(D) Self actualization needs

252. Two factor theory of motivation was proposed by:
(A) F Herzberg (B) V. H. Vroom
(C) J. S. Adams (D) L. W. Porter

253. The satisfiers according to Herzberg does not include:
(A) Achievement
(B) Recognition
(C) Challenging work
(D) Work condition

254. The strength of a person's motivation is a function of:
(A) Valence (B) Expectancy
(C) Both (A) and (B) (D) None of these

255. According to Maslow:
(A) Human needs are always conscious
(B) Man is perpetually wanting animals
(C) A satisfied need does not motivate
(D) Both (A) and (C)

256. Latest theory of motivation among the following is:
(A) Expectancy theory (B) ERG theory
(C) Theory X and Y
(D) Maturity immaturity theory

257. Valence expectancy theory of motivation was proposed by:
(A) C. Argyris (B) V.H. Vroom
(C) A.H. Maslow (D) Mc Gregor

258. ERG theory of motivation was put forward by:
(A) F. Herzberd (B) V. H. Vroom
(C) Clayton Alderfer (D) L.W. Porter

259. The term self actualization was first coined by:
(A) Abraham Maslow
(B) Kurt Goldstein
(C) Mc Gregor (D) Herzberg

260. The first major exponent of the classical theory of motivation:
(A) Maslow (B) Fayol
(C) Taylor (D) Mc Gregor

261. The most important hygiene factor according to Herzberg is:
(A) Policy and administration
(B) Achievement
(C) Salary (D) Working condition

262. Which of the following are correct regarding Herzberg hygiene factors:
(A) They are external to the job
(B) They reflect the Adam nature of man
(C) They are related to the job content
(D) Both (A) and (B)

263. The book "An introduction to Motivation" is written by:
(A) J. W. Atkinson (B) D C Mc Clelland
(C) R A Clark (D) E L Lowell

264. Which of the following is not included under the hygiene factors of motivation by Herzberg:
(A) Pay and job security
(B) Working conditions
(C) Advancement and growth
(D) Interpersonal relations

265. The need theory of motivation was developed by:
(A) David Mc Clelland
(B) A H Maslow
(C) Douglas Mc Gregor
(D) None of these

266. Management grid was developed by:
(A) Samuel David
(B) B F Skinner
(C) Blake and Mouton
(D) Peter and Lawter

267. The most important of all human needs:
(A) Physiological need
(B) Safety needs
(C) Esteem needs
(D) Self actualization need

268. Carrot and stick approach of motivation can be effectively linked with:
(A) Natural learning methods
(B) Behavior control methods
(C) Controlled exposure methods
(D) Expository methods

269. Reinforcement theory of motivation was put forward by:
(A) F Herzberg (B) V H Vroom
(C) B F Skinner (D) L W Porter

270. The equity theory of motivation was put forward by:
(A) F Herzberg (B) V H Vroom
(C) J S Adams (D) L W Porter

271. The term "Theory X" was coined by:
(A) Mc Gregor (B) Marks
(C) Abrams (D) Jewel

272. Workers enjoy autonomy and are willing to accept responsibility is in assumption of:
(A) Theory X (B) Theory Y
(C) Theory Z (D) Balance theory

273. Which one of the following is not a proposition in Theory X:
(A) Man wished to avoid responsibility
(B) Man has inherent liking for work
(C) Man must be coerced for working
(D) Man must be threatened with punishment

274. Workers must be goaded or guided into being productive is the assumption of:
(A) Theory X (B) Theory Y
(C) Theory Z (D) Balance theory

275. The central principle which derive from Mc Gregor Theory Y is:
(A) Direction of control (B) Coordination
(C) Scaler principle (D) Integration

276. According to path goal theory the effectiveness of leadership style is contingent upon:
(A) Evolutional characters
(B) Characteristics of subordinates
(C) Work environment
(D) Both (B) and (C)

277. Administration is:
(A) Policy making
(B) Execution of policy
(C) Giving direction to things
(D) None of these

278. The term "Development Administration" was coined by:
(A) Edward Weidner
(B) Montegomery
(C) George Gant (D) E W Riggs

279. The hallmark of public administration is:
(A) Consistency of treatment
(B) Public responsibility
(C) Community service
(D) External financial control

280. The word watch of public administration is:
(A) Public accountability
(B) Uniformity of treatment
(C) Service orientation
(D) Political direction

281. The gangplank idea is found in the works of:
(A) Fayol (B) Follett
(C) Barnard (D) Simon

282. To whom administration and authority are intimately linked:
(A) Fayol (B) Barnard
(C) Weber (D) Simon

283. Which one of the following is, according to Max Webber, the defining characteristic of authority:
(A) Power (B) Coercion
(C) Tradition (D) Legitimacy

284. Who among the following authors attached more importance to structure rather than persons in organizations:
(A) Luther Gulick (B) Elton Mayo
(C) Chester Barnard
(D) Mary Porter Follett

285. The influence of socio-psychological factors on worker's output was first proved by:
(A) M P Follett (B) F W Raylor
(C) E Mayo (D) R Linkert

286. Who among the following does not belong to the new human relations approach:
(A) C Argyris (B) E Mayo
(C) R Linkert (D) W Bennis

287. The theory of bureaucracy was mainly developed by:
(A) Psychologists (B) Sociologists
(C) Political scientists (D) Economists

288. The emphasis of scientific management theory is on:
(A) Scholarly view
(B) Explaining behavior
(C) Improving practice
(D) None of these

289. The classical theory of organization was developed during the period:
(A) 1900-1950 (B) 1800-1850
(C) 1700-1750 (D) 1600-1650

290. The classical theory is chiefly concerned with:
(A) Formal organizational structure
(B) Efficiency at the operational level
(C) Basic management process
(D) Both (A) and (C)

291. The most important exponent of the classical theory is:
(A) Luther Gulick (B) Henry Fayol
(C) JD Mooney (D) Lyndall Urwick

292. The Hawthorne studies were conducted by:
(A) Elton Mayo (B) Hugo
(C) FW Taylor (D) Fayol

293. According to human relations theory, the productivity of employees mainly depends on:
(A) Physical condition at work
(B) Relationship of superiors and colleagues
(C) Amount of salary received
(D) None of these

294. According to Marx, class membership was determined by ones:
(A) Family (B) Relationship to the state
(C) Education
(D) Relation to the means of production

295. The greatest resistance to scientific management came from:
(A) Human relationists (B) Workers
(C) Managers (D) Trade unions

296. According to human relations theory, an organization consist of which of the following elements:
(A) Informal groups
(B) Formal groups
(C) Intergroup relations
(D) All of these

297. Which of the following is a characteristic of classical theory:
(A) Specialization (B) Impersonality
(C) Hierarchy (D) All of these

298. Scientific management means:
(A) One best way of doing things
(B) Scientific division of work
(C) Functional foremanship
(D) Mental revolution

299. The idea of scientific management was introduced by:
(A) Gilbreth (B) Mayo
(C) Lewin (D) Taylor

300. Scientific management uses:
(A) Time and motion studies
(B) Task analysis
(C) Job specialization (D) All of these

301. Which of the following theories views leadership as a manipulative function:
(A) Human relations theory
(B) Scientific management theories
(C) Classical theory
(D) Decision making theory

302. Which of the following have advocated positional theory of authority:
(A) Weber and Barnard
(B) Fayol and Simon
(C) Barnard and Simon
(D) Fayol and Weber

303. The office of district collector of India is based on the principle of:
(A) Delegation (B) Decetralization
(C) Devolution (D) Deconcentration

304. The Indian Institute of Public Administration is located at:
(A) Mumbai (B) New Delhi
(C) Kolkata (D) Kanpur

305. The scientific management stressed upon:
(A) Rationality (B) Specialization
(C) Predictability (D) All of these

306. The experiment conducted by Taylor was intended to understand the:
(A) Human behavior in organization
(B) Process of mental revolution
(C) Motivation of workers
(D) Best way to carry out specific tasks

307. The human relations theory lays emphasis on:
(A) People (B) Motivation
(C) Informal group (D) All of these

308. A hierarchical arrangement among the different parts of an organization with the divisibility of functions and authority is an important feature of:
(A) Span of control
(B) Bureaucracy
(C) Management by objective
(D) Organizational climate

309. Which one the following is the major drawback of hierarchical organization:
(A) Defined structure
(B) Red tapism
(C) Rule orientation
(D) Fixed decision making levels

310. Taylor's concept of functional foremanship leads to:
(A) Multiple command
(B) Specialization
(C) Negation unity of command
(D) All of these

311. Weber's analysis of bureaucracy was from:
(A) Political perspective
(B) Sociological perspective
(C) Economic perspective
(D) Cultural perspective

312. The behavioural characteristics of bureaucracy are:
(A) Rationality and role specificity
(B) Role specificity and impersonality
(C) Impersonality and role orientation
(D) Rationality and system of rules

313. Theory X, or scientific management, is concerned primarily with improving:
(A) Output (B) Productivity
(C) Work efficiency
(D) Work environment

314. The structural characteristics of bureaucracy includes:
(A) Hierarchy (B) Rule orientation
(C) Role specificity (D) All of these

315. According to Max Webber, bureaucracy is superior to any other form of organization in:
(A) Precision (B) Stability
(C) Discipline (D) All of these

316. Which of the following is not a behavioral characteristics of bureaucracy:
(A) Impersonality (B) Role specificity
(C) Rationality (D) Rule orientation

317. The essential focus of the human relations theory was on:
(A) Physiological factors
(B) Social factors
(C) Economic factors
(D) Both (A) and (B)

318. The democratic and participatory style of leadership become popular during the era of:
(A) Behavioural theory
(B) Classical theory
(C) Systems theory
(D) Human relations theory

319. Plans and actions should be evaluated by their consequences:
(A) Utilitarian theory
(B) Theory of rights
(C) Theory of justice
(D) Parkinson's law

320. Administration will differ from one field to another according to the subject matter:
(A) Administrative view
(B) Wider view
(C) Integral view
(D) Narrower view

321. Which of the following is described as anti-theorist, anti-positivist and anti-hierarchical:
(A) Development communication
(B) New public administration
(C) Comparative communication
(D) Administrative development

322. The principal disadvantage of hierarchy is:
(A) Rigidity in organization
(B) Creates superiority or inferiority feeling
(C) Delay in disposal of case
(D) Decentralization of decision making

323. Which of the following Is not a feature of system approach:
(A) Adaptive (B) Deterministic
(C) Multi-variate (D) Descriptive

324. Behavioural approach in management is not concerned with:
(A) Prescription (B) Quantification
(C) Mathematization (D) Theorization

325. Which of the following approaches gave rise to contingency management:
(A) Behavioural approach
(B) Classical approach
(C) Systems approach
(D) Human Relations approach

326. Systems approach in management is also known as:
(A) Contingency theory
(B) Integral theory
(C) Ecological theory (D) Modern theory

327. Training is basically is:
(A) Product (B) Process
(C) Activity (D) Programme

328. Training, looking in its true sense:
(A) Stimulate and create potential
(B) Stimulate potential
(C) Create potential (D) None of these

329. Training basically is not a:
(A) Overt process (B) Covert process
(C) Communication process
(D) Learning process

330. The focus of training and learning are respectively on:
(A) Events and self
(B) Events and outcome
(C) Outcome and self
(D) Outcome and Process

331. Training is a function of:
(A) Knowledge only (B) Attitude only
(C) Skill only (D) All of these

332. The most emphasized behaviour dimension in training is:
(A) Knowledge (B) Attitude
(C) Skill (D) None of these

333. The behavior domain that gets the highest emphasis in training and education are:
(A) Cognitive and Connotative
(B) Cognitive and Effective
(C) Affective and Cognitive
(D) Affective and Psychomotive

334. Training differs from development in the sense the former is:
(A) Long term (B) Short term
(C) Equal term (D) None of these

335. Training is related to learning in the same way:
(A) Sowing to seed (B) Soil to seed
(C) Seed to soil (D) None of these

336. The learning curve in training follows a:
(A) S-shape (B) U-shape
(C) V-shape (D) C-shape

337. The place in which a training is conducted is technically called:
(A) Training institute
(B) Training organization
(C) Training association
(D) Training group

338. The training skill that need to be emphasized for top level management:
(A) Conceptual
(B) Technical
(C) Design
(D) Both (A) and (B)

339. Training programme for VLW's should essentially focus on:
(A) Technical skill (B) Conceptual skill
(C) Design skill (D) Leadership skill

340. Extension education Institutes focus their training for functionaries working under:
(A) Grass root level
(B) Middle level
(C) Top level (D) Both (A) and (B)

341. Institution share in the training course is the highest for:
(A) ICAR (B) SAUs
(C) ACTs (D) CGO

342. Foundation training is otherwise known as:
(A) Portal (B) Vestibule
(C) Inductive (D) All of these

343. The training type that prepares an employee to take a new assignment is:
(A) Re-training (B) Refresher training
(C) Inductive training
(D) Orientation training

344. Training and visit system essentially make use of:
(A) Refresher training
(B) Inductive training
(C) Re training (D) Development training

345. The extension worker trains the farmer at their farm situation is termed as:
(A) Pipeline training
(B) Vestibule training
(C) Peripatatic training
(D) Inservice training

346. Action research programme is considered as an example for:
(A) Formal training
(B) Non formal training
(C) Direct training (D) Distant training

347. The training type that can also be grouped under pre service training:
(A) Inductive training
(B) Development training
(C) Orientation training
(D) Overseas training

348. The NAARM training for freshly recruited ARS trainees falls under:
(A) Pre-training (B) Induction training
(C) Orientation training
(D) Refresher training

349. Summer institute is a good example for:
(A) Induction training
(B) Developmental training
(C) Orientation training
(D) Refresher training

350. Learning by doing is called:
(A) Formal training
(B) Informal training
(C) In-service training
(D) Post-entry training

351. Training principle that best explain the use of analogue is:
(A) Association (B) Readiness
(C) Capacity (D) Intensity

352. Academic orientation is the characteristic feature of:
(A) Process approach
(B) System approach
(C) Traditional approach
(D) Experimental approach

353. The training approach developed from engineering concept:
(A) Discrete approach
(B) System approach
(C) Process approach
(D) Performance approach

354. The concept of training is considered very loose and less precise in:
(A) Discrete approach (B) Process approach
(C) Experimental approach
(D) System approach

355. Training model widely used by policy makers and managers:
(A) Decision model (B) Expert model
(C) Experimental model
(D) Objective model

356. Cyclic model of training was developed by:
(A) Lynton and Pareek (B) R D Miller
(C) A York (D) DG Reay

357. System concept of training was developed by:
(A) A York (B) R D Miller
(C) DG Reay (D) D Kolb

358. The number of steps in the discrete approach developed by RD Miller is:
(A) Three (B) Four
(C) Five (D) Six

359. System approach of training is basically a:
(A) Diagnostic approach
(B) Productive approach
(C) Descriptive approach
(D) Experimental approach

360. Training model that shares an empathetic attitude towards training institution:
(A) Elaborated model (B) Discrete model
(C) Simple model (D) Cyclic model

361. The training stage that does not form a component of cyclic model:
(A) Preparatory training
(B) Pre-training
(C) Training (D) Post training

362. Moving phase of Curt Lewin matches with:
(A) Preparatory training phase
(B) Pre training phase
(C) Training phase
(D) Post training phase

363. The training phase that is otherwise known in the adjustment phase:
(A) Preparatory phase
(B) Pre training phase
(C) Training phase
(D) Post training phase

364. The motivational phase in training is otherwise called:
(A) Preparatory phase
(B) Pre-training phase
(C) Training phase
(D) Post training phase

365. Receptability of organization plays a crucial role in:
(A) Preparatory phase
(B) Pre-training phase
(C) Training phase (D) Post training phase

366. The probable outcome at the end of post training phase is:
(A) Enhanced motivation
(B) Internalization of new knowledge
(C) Expectancy building
(D) Institutionalization of change

367. The rational training need analysis method among the following is:
(A) Observation (B) Check list
(C) Critical incident method
(D) Card sort method

368. The first step in SATNA need assessment method is:
(A) Job description
(B) Problem identification
(C) People identification
(D) KAS identification

369. "A need is something whose absence or deficiency proves harmful" represent:
(A) Democratic view
(B) Autocratic view
(C) Diagnotic view
(D) Discrepancy view

370. Logical analysis is the characteristic feature of:
(A) Rational method (B) Empirical method
(C) Discovery method
(D) Encounter method

371. Internal training strategy covers everything except:
(A) Individual receptability
(B) Training scheduling
(C) Logistical arrangement
(D) Curriculum development

372. Training methods differ from training modality in the sense:
(A) Method is broader
(B) Modality is broader
(C) Both are same (D) None of the same

373. The analogue of heart and blood vessels in training best represent to:
(A) Methods and Curriculum
(B) Strategy and Objective
(C) Curriculum and Methods
(D) Need and Methods

374. The level in the training need where a person can perform a job effectively but takes more time:
(A) Precision level (B) Perfection level
(C) Sufficiency level
(D) Deficiency level

375. Phenomena which shows continuous change over time is termed as:
(A) Systematic
(B) Process
(C) Erratic
(D) None of these

376. "Training for Organizational Transformation" is a book written by:
(A) Lynton and Pareek
(B) Berger and Berger
(C) PN Singh
(D) Pareek, Rao and Parimjee

377. Curriculum literacy means:
(A) Keep on running
(B) Content in training
(C) Theory in training
(D) Printed material

378. The S in the SMART objective refers to:
(A) Simple (B) Specific
(C) Sincere (D) Secure

379. What is the first step in SQ3R method:
(A) Study (B) Select
(C) Speculate (D) Survey

380. Training objective does not contain:
(A) Criteria (B) Behaviour
(C) Condition (D) Goal

381. The latest priority component in the concentric approach in curriculum development is:
(A) Could know (B) Should know
(C) Must know (D) Would know

382. Outer and inner core of human brains are respectively the seats of:
(A) Emotional behavior
(B) Logical behavior
(C) Emotional and logical behavior
(D) Logical and Emotional behavior

383. The curriculum development approach that equate with the principle of thermostat:
(A) Reductionlist approach
(B) Constructionalist approach
(C) Cybernetic approach
(D) Goal management approach

384. Kolb is associated with:
(A) Programmed instruction
(B) Learning style
(C) Role play
(D) ELC cycle

385. The learning style wherein people like hand on experience is:
(A) Accommodates (B) Diverges
(C) Assimilators (D) Convergers

386. People with high imaginative ability and sensitivity falls under:
(A) Accommodates
(B) Diverges
(C) Assimilators (D) Convergers

387. Integraton of ideas and formulation of theories is the characteristic feature of:
(A) Accommodates (B) Diverges
(C) Assimilators (D) Convergers

388. The learning style with more emphasis on pragmatism rather than conceptualization is:
(A) Accommodates (B) Diverges
(C) Assimilators (D) Convergers

389. The stage in learning cycle where the trainees become aware about the changes around them and develop interest about the advances in the subject:
(A) Unconscious incompetence
(B) Conscious incompetence
(C) Conscious competence
(D) Unconscious competence

390. Transfer of learning from concrete experience to reflective observation needs which mode of learning:
(A) Accommodative (B) Divergent
(C) Assimilative (D) Convergent

391. The first stage in ELC cycle is:
(A) Experiencing (B) Publishing
(C) Generalizing (D) Applying

392. The fulcrum of ELC cycle is:
(A) Experiencing (B) Processing
(C) Generalizing (D) Applying

393. Theory can be integrated in the experiential learning cycle at:
(A) Experiencing (B) Processing
(C) Generalizing (D) Applying

394. The last phase in experiential learning cycle is:
(A) Application (B) Generalization
(C) Publication (D) Experience

395. The first and last step in the seven step model of ELC are:
(A) Climate setting and closure
(B) Experience and closure
(C) Experiencing and Application
(D) Objective clarification and closure

396. The role of trainer in experiential approach is:
(A) Facilitator (B) Contributor
(C) Controller (D) Enabler

397. Sensitivity training normally needs a time period of:
(A) 15-20 minutes (B) 1-2 hours
(C) 1-2 days (D) 5-10 days

398. "He is passive in laboratory culture, take it in with no struggle or fight and back home will put quickly to the old shell":
(A) Tourist (B) Expatriate
(C) Missionary (D) Easy convert

399. The best bet for transfer of learning for a training programme is:
(A) Easy convert (B) Self-mystic
(C) Learner critic (D) Expatriate

400. The basic training orientation that gives equal importance to content and concept:
(A) Laboratory (B) Academic
(C) Activity
(D) Person development

401. Going from specific to general in a training programme make use of:
(A) Inductive reasoning
(B) Deductive reasoning
(C) Horizontal reasoning
(D) Vertical reasoning

402. Relating class room learning with real life experience takes place in;
(A) Experiencing (B) Processing
(C) Generalizing (D) Applying

403. Transfer of learning to work situation does not take place in:
(A) Tourist model (B) Missionary model
(C) Self mystic model
(D) Learner critic model

404. The amount of learning that takes place through non-verbal communication is:
(A) 10 per cent (B) 20 per cent
(C) 30 per cent (D) 40 per cent

405. Participants role in field training is:
(A) Outsider – Learner
(B) Player – Observer
(C) Initiator – Contributor
(D) Contributor – Reporter

406. The focus of trainer in a training sessions is essentially on:
(A) Learner (B) Subject matter
(C) Learning climate
(D) Learning outcome

407. The responsibility for the outcome of training rest solely on:
(A) Learner (B) Trainer
(C) Institution (D) Both (A) and (B)

408. The training needs of individual employees in an organization has to be assessed by:
(A) Employee (B) Colleagues
(C) Employer
(D) Training organization

409. The process of transforming the experiences into knowledge as a basic for action is called:
(A) Continued learning
(B) Supplemental learning
(C) Topical learning
(D) Experimental learning

410. Human beings continues to learn till:
(A) Adulthood (B) Childhood
(C) College days (D) Life end

411. Institutional aids used in a training session can:
(A) Provide verbal experience
(B) Replace trainer
(C) Includes worksheet
(D) Decorates learning situation

412. Instructional aids used in a training for:
(A) Supplement learning
(B) Supplant learning
(C) Both (A) and (B)
(D) None of these

413. The phase of interactive demonstration wherein the participants make attempt to intellectualize the skill:
(A) Cognitive phase
(B) Fixation phase
(C) Autonomous phase
(D) Association phase

414. The behavior domain that acquires the most prominence in simulation method:
(A) Knowledge (B) Skill
(C) Attitude (D) All of these

415. Compared to education training show variation in all except:
(A) Vocational component
(B) Facilitator role
(C) Inductive process
(D) Delayed usage

416. Training by feeling is:
(A) Concrete experience
(B) Reflective observation
(C) Abstract conceptualization
(D) Active experimentation

417. The number of trainees attending an open learning programme will be:
(A) Very small (B) Small
(C) Medium (D) Large

418. Which type of training centre is Krishi Vigyan Kendra:
(A) Trainers Training Centre
(B) Extension Workers Training Centre
(C) Vocational Training Centre
(D) Leadership Training Centre

419. The apex body to offer training for extension functionaries:
(A) MANAGE (B) NAARM
(C) NCAP (D) NIRD

420. The normal pace of delivery in the lecture method is:
(A) 120 words per minutes
(B) 160 words per minutes
(C) 200 words per minutes
(D) 240 words per minutes

421. The behavior domain that assumes significance in lecture presentation:
(A) Cognitive (B) Effective
(C) Connative (D) All of these

422. Printed or typesdsheet of paper used for quick communication before lecture:
(A) Handout (B) Lead letter
(C) Circular letter (D) Notice

423. Question following inductive reasoning:
(A) Divergent question
(B) Rhetorical question
(C) Lead question
(D) Convergent question

424. The stage in group development characterized by uncertainty:
(A) Forming (B) Storming
(C) Norming (D) Performing

425. Use of psychological tools in training is otherwise known as:
(A) Instrumental exercise
(B) Simulation exercise
(C) Sensitivity exercise
(D) Laboratory exercise

426. The training method best suites to all situation:
(A) Simulation method
(B) Lecture method
(C) Sensitive method
(D) None of these

427. Simulation method does not include:
(A) In basket exercise (B) Role play
(C) Business game
(D) Brainstorming

428. Win loose game is staged to demonstrate:
(A) Intra-team collaboration
(B) Inter-team collaboration
(C) Inter-personal collaboration
(D) Intra-personal collaboration

429. Microlab exercise in a training programme is conducted at its:
(A) Beginning (B) Middle
(C) End (D) Any time

430. The purpose of microb exercise in a training programme is:
(A) Concept building
(B) Rapport building
(C) Strategy building
(D) Experience building

431. Win-loose game when made to play with the same group for a second time will gives:
(A) Better results (B) Average result
(C) Best result (D) No result

432. T-group training is grouped under:
(A) Transactional method
(B) Sensitivity method
(C) Experiential method
(D) Laboratory method

433. The training type that standardized data from real life for systematic step by step learning of certain type:
(A) Encounter method
(B) Inspiratory method
(C) Expository method
(D) Individualized method

434. Generation of data during training from real life situation is not the characteristic feature of:
(A) Individualized method
(B) Simulation method
(C) Laboratory method
(D) Field method

435. The co-trainer concept of training is very much related to:
(A) T group (B) L group
(C) O group (D) Q group

436. The training method that is liked most by people with divergent learning style:
(A) Role play (B) Lecture
(C) Clinical study (D) Library review

437. "Handbook of Structural Experience" – a collection of training methods is published by:
(A) Pfeiffer and Jones
(B) David G Reay
(C) D Kolb (D) Lynton and Pareek

438. The training method that gives more importance to cognitive domain:
(A) Lecture (B) Role play
(C) Brainstorming (D) Workshop

439. The stage of brainstorming in which free flow of ideas takes place:
(A) Red stage (B) Green stage
(C) Yellow stage (D) Orange stage

440. Evaluation of ideas in brainstorming takes place in:
(A) Red stage (B) Green stage
(C) Yellow stage (D) Orange stage

441. Brainstorming method was developed by:
(A) Osgood (B) Gorden
(C) Andrews (D) Kolb

442. The training method best suits to large audience:
(A) I iscussion method
(B) Simulation method
(C) Sensitivity method
(D) Transactional method

443. A programmed instruction takes participants:
(A) Outside social context
(B) Nearer to the participants
(C) Nearer to the trainer
(D) Both (A) and (B)

444. Computer based training is classified under:
(A) Expository method
(B) Encounter method
(C) Discovery method
(D) Individualized method

445. The training method that work on the principle of empathy:
(A) Role play (B) Case study
(C) Transactional analysis
(D) Sensitivity method

446. Transactional analysis is classified under:
(A) Laboratory method
(B) Simulation method
(C) Discussion method
(D) Individualized method

447. Sensitivity training includes:
(A) T group (B) L group
(C) Process work (D) All of these

448. Sensitivity training is a:
(A) Self-disclosure method
(B) Trainer dominated method
(C) Process work method
(D) Both (A) and (C)

449. The essential difference between games and exercise in a training session rest on:
(A) Games can be repeated
(B) Exercise can be repeated
(C) Games do not have rule
(D) None of these

450. "Showing method among the following is:
(A) Lecture (B) Seminar
(C) Brainstorming (D) Demonstration

451. The number of participants is minimum in:
(A) Group discussion (B) Huddle method
(C) Buzz session (D) Symposium method

452. Training method that samples real life is:
(A) Incidence and Case
(B) Role play and Case
(C) Incidence and Craftmanship
(D) Study tours and Case

453. Counseling as a training method comes under:
(A) Simulation method
(B) Laboratory method
(C) Individualized method
(D) Discussion method

454. The training method that emphasis on problem diagnosis:
(A) Case method (B) Role play
(C) Group discussion
(D) Project method

455. The focus of sensitivity training essentially is on:
(A) Blind self (B) Unknown self
(C) Open self (D) Hidden self

456. The training method that gets its name from a particular behavior of honey bee:
(A) Brainstorming (B) Huddle method
(C) Buzz method (D) Role play

457. The training method that does not represent on the job training:
(A) Craftmanship
(B) Process work
(C) Apprenticeship
(D) Both (A) and (C)

458. Syndicate is essentially a:
(A) Telling method
(B) Doing method
(C) Showing method
(D) Reading method

459. Reciprocal communication between trainees is most effective in:
(A) Lecture (B) Role play
(C) Group discussion (D) Seminar

460. Role play wherein an individual plays his own role is called:
(A) Socio-drama (B) Culturo-drama
(C) Psycho-drama (D) Arto-drama

461. T group training with trainers from different discipline:
(A) Stranger group (B) Family group
(C) Cousin group (D) Informal group

462. Buzz session is normally conducted for:
(A) 5-10 minutes (B) 15-20 minutes
(C) 25-30 minutes (D) 55-60 minutes

463. Computer based training make use of:
(A) Digital image (B) Analogue image
(C) Iconic image (D) Enactive image

464. The loss of productive days is on a minimum for:
(A) Laboratory training
(B) Field training
(C) Simulated training
(D) Library training

465. Trainer role in brainstorming session is:
(A) Stimulator (B) Facilitator
(C) Concepualiser (D) Controller

466. Data in role play method is:
(A) Self generated
(B) Brought from outside
(C) Standardized (D) General

467. Behaviour conditioning in interactive demonstration takes place in:
(A) Cognitive phase
(B) Fixation phase
(C) Autonomous phase
(D) Association phase

468. 'Covedale' methodology of training is the characteristic methodology followed in:
(A) IARI (B) MANAGE
(C) IIM (D) SIETI

469. To make people to reach precision level the type of training method best suits is:
(A) Operational training
(B) Behavioural training
(C) Sensitivity training
(D) Applied training

470. Data is very abstract and general in:
(A) Lecture (B) Business game
(C) Case (D) Brainstorming

471. Trainer-trainee ratio is maximum in:
(A) Syndicate (B) Brainstorming
(C) Case (D) Lecture

472. Trainer-trainee ratio is minimum in:
(A) Syndicate (B) Brainstorming
(C) Role play (D) Lecture

473. The training setup under Training and Visit system can be considered as:
(A) Short arm long finger
(B) Long arm short finger
(C) Short arm short finger
(D) Long arm long finger

474. Training programme for VLWs in training and visit system is organized on a:
(A) Weekly basis (B) Fortnightly basis
(C) Monthly basis (D) Half yearly basis

475. Interaction and involvement of participants is in the highest degree in:
(A) Field method (B) Case method
(C) Laboratory method
(D) Lecture method

476. Feedback from the participants is the highest in:
(A) Instrumental method
(B) Discussion method
(C) Simulation method (D) Field method

477. Very intensive learning is a characteristic feature of:
(A) Field method (B) Simulation method
(C) Laboratory method
(D) Instructional method

478. Cost of training is the highest for:
(A) Field training
(B) Laboratory training
(C) Simulation training
(D) Instructional training

479. Role play can be considered as a:
(A) Projective method
(B) Simulation method
(C) Encounter method
(D) All of these

480. "Doing" method among the following is:
(A) Seminar (B) Brainstorming
(C) Movie (D) Role play

481. The Lynton and Pareek classification of training method is basically on:
(A) Involvement of trainee
(B) Cognitive emphasis
(C) Sensory capacity
(D) Involvement of trainer

482. The "Showing, Doing, Telling' classification of training methods was done by:
(A) Lynton and Pareek
(B) D Kolb (C) D G Reay
(D) Pfeiffer and Jones

483. The tower building exercise was developed by:
(A) Alfred York
(B) Danieel Stufflebeam
(C) Robert Stake (D) Michael Scriven

484. The concept of microlab was first used in:
(A) USA (B) Canada
(C) Japan (D) India

485. When trainers for the T-group are selected from the same department:
(A) Cousin group (B) Father group
(C) Family group (D) Stranger group

486. When trainers in the T-group are selected from only social scientists:
(A) Cousin group (B) Father group
(C) Family group (D) Stranger group

487. A role-play wherein a person directs a blind fold man to perform best represents:
(A) Role rotation (B) Role reversal
(C) Role multiplication
(D) Role simplification

488. The three dimentional classification of training method by Pareek and Rao is not based on:
(A) Cognitive emphasis
(B) Learner activity
(C) Sensory perception
(D) Use of experience

489. To remove the afternoon sleep from the participants, we use:
(A) Ice breaks (B) Microlab
(C) Climate setting (D) All of these

490. Experiential games and exercise are conducted during:
(A) Beginning of the session
(B) Middle of the session
(C) End of the session
(D) Both (A) and (C)

491. Hypnosis as a training method is classified under:
(A) Expository method
(B) Individualized method
(C) Controlled exposure method
(D) Behaviour control method

492. Entrepreneurship development training does not contain:
(A) Stimulus phase
(B) Supplementary phase
(C) Sustenance phase
(D) Support phase

493. Low learner activity and high cognitive emphasis is the characteristic feature of:
(A) Encounter method(B) Discovery method
(C) Expository method(D) Enactive method

494. Use of experience, learner activity and cognitive emphasis is high in:
(A) Discovery method (B) Enactive method
(C) Encounter method
(D) Inspiratory method

495. The learner activity and use of experimentation in encounter method are:
(A) High-Low (B) Low-High
(C) High-High (D) Low-Low

496. For large group we can adopt:
(A) Buzz session (D) Brainstorming
(C) Debate (D) Symposium

497. The adult ego state in transitional analysis best represent:
(A) Though concept of life
(B) Taught concept of life
(C) Felt concept of life
(D) Both (A) and (B)

498. The concept of transitional analysis was first proposed by:
(A) Eric Berne (B) David G Reay
(C) Pfeiffer and Ballow
(D) Lynton and Pareek

499. The number of human birth in transitional analysis is:
(A) One (B) Two
(C) Three (D) Four

500. The child ego state in transitional analysis best represent:
(A) Thought concept of life
(B) Taught concept of life
(C) Felt concept of life
(D) Both (A) and (B)

501. A child who holds his cry when his mother beats is a:
(A) Normal child (B) Adaptive child
(C) Creative child (D) Dominant child

502. The part of personality of a trainee that is known only to him:
(A) Blind self (B) Unknown self
(C) Open self (D) Hidden self

503. The basic life position that denotes over dominance:
(A) I am OK You are OK
(B) I am OK You are not OK
(C) I and You are not OK
(D) I am not OK but You are OK

504. Collection of bad and negative stamps in transactional analysis is otherwise called:
(A) Gold stamps (B) Silver stamps
(C) Black stamps (D) Grey stamps

505. The basic life scripts are recorded before as age of:
(A) One (B) Three
(C) Five (D) Ten

506. The unspoken psychological flow of communication that runs in parallel:
(A) Transactions (B) Perception
(C) Empathy (D) Opinion

507. A state in which people behave, feel and think in response to what is going on in the 'here-and-now', using all of their resources as an adult human being with many years of life experience to guide them:
(A) Neopsychic (B) Exteropsychic
(C) Archaepsychic (D) None of these

508. The modified form of panel discussion in which three or more resource persons discuss a specific topic is called:
(A) Symposium (B) Colloquium
(C) Workshop (D) Brainstorming

509. Pooling of experience and opinion among group of people who have special qualifications in the area is called:
(A) Symposium (B) Conference
(C) Workshop (D) Brainstorming

510. A discussion method, usually more than one day in duration and giving emphasis on lectures, individual conferences with emphasis on small working groups:
(A) Conference (B) Panel discussion
(C) Symposium (D) Workshop

511. In which of the following training methods do the trainees expose their idiosyncrasies to their fellow-trainees who rub them off through mutual interaction:
(A) Syndicate method
(B) Case study method
(C) Lecture method (D) T-Group method

512. The process through which a trainer communicate to a group of participants through the spoken words supplemented by audio-visual aids and discussion is:
(A) Group discussion (B) Lecture
(C) Brainstorming (D) Workshop

513. The training type wherein speakers covers sub-divisions of a specific topic for a definite period of time:
(A) Seminar (B) Symposium
(C) Panel discussion (D) Lecture

514. The method of verbal presentation on a topic by a speaker to a group of audience:
(A) Lecture (B) Seminar
(C) Panel discussion (D) Workshop

515. A meeting in which a small group of resource persons present short prepared papers on a given topic:
(A) Colloquium (B) Seminar
(C) Buzz session (D) Symposium

516. Which of the following directly measures competence?
(A) Personality tests
(B) Paper pencil tests
(C) Standardized tests
(D) Performance tests

517. Motivation is meant to:
(A) Create interest in the task
(B) Reduce anxiety
(C) Enhance task related anxiety
(D) Develop personality

518. Fill in the blank:
In asking higher order questions, teacher must ask ———————— questions.
(A) Closed ended (B) Fact based
(C) Concept based (D) Convergent type

519. Pickout the odd one:
(A) Development proceeds from centre to periphery
(B) Development proceeds from head to tail
(C) Childhood reflects the future adult's personality
(D) Development of mathematical concepts starts with number value

520. Which practice in questioning brings about more classroom interaction?
(A) Asking questions before calling on a student
(B) Focussing on divergent questions
(C) Focussing on convergent questions
(D) Asking questions after calling on a student

521. Greatest weakness of intelligenc tests is that they:
(A) Are not reliable
(B) Are not fair to persons of different background
(C) Do not provide motivation to different task takers
(D) Measure intelligence rather than ability

Chapter 14

AGRICULTURAL STATISTICS

1. Who is considered as father of statistics:
(A) R.A. Fisher (B) Pearson
(C) A.L. Bowley (D) Boddington

2. When statistics is used in plural sense it means:
(A) Statistical methods
(B) Collect statistics
(C) Both (a) and (b) (D) None of these

3. Which measure of central tendency is applied for calculation of regression and correlation:
(A) Arithmetic mean (B) Mode
(C) Median
(D) Geometric mean

4. The problem of effect of the observer is minimized by:
(A) Concealing the behavior
(B) Concealing the observer
(C) Concealing the observed
(D) None of these

5. Coefficient of variation cannot find a place in interval level of measurement because of the problem in:
(A) Standard Deviation (B) Mean
(C) Both (A) and (B) (D) None of these

6. Select the two dimension diagrams of data it means:
(A) Squares (B) Rectangular bodies
(C) Cubes (D) Pictures

7. A hypothesis when tested true becomes a:
(A) Principle (B) Theory
(C) Philosophy (D) Law

8. When a test measure what it is intended to measure, we calls it:
(A) Reliable (B) Valid
(C) True (D) None of these

9. Linkert scale for measurement of attitude falls under:
(A) Nominal (B) Ordinal
(C) Interval (D) Ratio

10. In a symmetrical distribution of data which pair is correct
(A) Mean=Median=Mode
(B) Mean>Median>Mode
(C) Mean<Median<Mode
(D) Mean=Median>Mode

11. The first psychological test constructed was:
(A) Binet-Simon intelligence scale
(B) Army Alpha test
(C) Thorndike Handwritting scale
(D) Woodworths scale

12. Reliability essentially is a property of:
(A) Test score (B) Test as such
(C) Norms of test (D) Sample of test

13. Which measure of central tendency requires data arrangement in ascending or descending order for its estimation:
(A) Arithmetic mean
(B) Median (C) Mode
(D) Harmonic mean

14. If any items of the series is zero, which mean of central tendency become zero:
(A) Harmonic mean (B) Geometric mean
(C) Arithmetic mean
(D) Weighted harmonic mean

15. The term ex post facto was first coined by:
(A) Chapin and Greenwood
(B) Festinger and Katz
(C) Goode and Hatt
(D) Campbell and Katona

16. To calculate the average speed, which item of central tendency is most suited:
(A) Mode (B) Median
(C) Weighted HM (D) GM

17. Freedom from variable errors is known as:
(A) Reliability (B) Validity
(C) Credibility (D) Sustainability

18. Which is the simplest measure of dispersion:
(A) Mean deviation
(B) Standard deviation
(C) Range (D) Variance

19. The first central moment (μ_1) is always:
(A) 0 (B) 2 (C) 4 (D) 6

20. A pilot study to give more insight into the research problem form the characteristic feature of:
(A) Explorative design
(B) Diagnostic design
(C) Descriptive design
(D) Experimental design

21. The curves which are very highly peaked and have the value of $\beta_2>3$ are called:
(A) Mesokurtic (B) Leptokurtic
(C) Platykurtic (D) All of these

22. The mean of the binomial distribution is:
(A) $\sqrt{np}$ (B) $\sqrt{npq}$
(C) $\sqrt{npq}$ (D) None of these

23. A hypothesis is a must in:
(A) Explorative design
(B) Diagnostic design
(C) Descriptive design
(D) Experimental design

24. The first and third moments about the mean of normal distribution is
(A) 0 (B) δ^2
(C) $3\delta^4$ (D) None of these

25. To study the effect of independent variable:
(A) Factorial design (B) CRD
(C) LSD (D) RBD

26. β_1 and β_2 constants of the normal distribution are ...., respectively:
(A) 0, 3 (B) 3, 0
(C) 0, 0 (D) 3, 3

27. Experimental and non-experimental research shares in common which of the following characteristics:
(A) Control (B) Manipulation
(C) Measurement (D) All of these

28. Student's t test is used when sample size is .... and population standard deviation is not known:
(A) Large (B) Small
(C) Both (D) None of these

29. Opinion polls falls under:
(A) Explorative design
(B) Diagnostic design
(C) Descriptive design
(D) Experimental design

30. For comparison of two means from independent sample which test is applicable:
(A) t-test (B) x^2test
(C) F-test (D) z-test

31. Experimenter has the highest control in manipulation of variable in:
(A) Laboratory method
(B) Library method
(C) Simulation method
(D) Field method

32. x^2 is applied for
(A) Testing of independence of attributes
(B) Genetic problem and detection of linkages
(C) Testing the expectation of a ratio
(D) All of these

33. The following description "I will bet this would happen if I did this" best represent to:
(A) Explorative design
(B) Confirmatory experiment
(C) Crucial experiment
(D) Pilot experiment

34. The range of multiple correlation coefficient lies between:
(A) 0 to +1 (B) -1to +1
(C) 0 to ∞ (D) $-\infty$ to $+\infty$

35. A research design without hypothesis among the following is:
(A) Explorative design
(B) Diagnostic design
(C) Descriptive design
(D) Experimental design

36. The principle of making use of greater homogeneity in groups of experimental unit to reduce the experimental error is referred as:
(A) Replication
(B) Randomization
(C) Local control
(D) None of these

37. Which of the following is not a characteristic feature of a good research design:
(A) High external validity
(B) Low external validity
(C) High internal validity
(D) None of these

38. The completely randomized design is appropriate when the experimental material is
(A) Limited and homogenous
(B) Unlimited and heterogeneous
(C) Limited and heterogeneous
(D) Unlimited and homogenous

39. A sample of 12 specimen taken from a normal population is expected to be 50 kg. To test Ho: μ= 50Vs: Hi not μ= 50, the test used is:
(A) t- test (B) X^2test
(C) Z test (D) F-test

40. When the fertility gradient of the field goes in two directions, which design is most appropriate:
(A) Augmented design (B) CRD
(C) LSD (D) SPD

41. Equality of two regression coefficient can be tested with:
(A) 't' test (B) 'X^2' test
(C) Z test (D) ANOVA

42. In randomized block design, how many treatments can be adopted without any loss of efficiency:
(A) 10 (B) 15 (C) 18 (D) 20

43. When the experimental material is very less and number of genotypes are very large, which design is most suited:
(A) Split Plot Design (B) LSD
(C) RBD
(D) Augmented Design

44. Analysis of variance utilizes:
(A) t- test (B) X^2test
(C) Z test (D) F-test

45. Hypothesis that the population variance has a specified value can be tested by:
(A) t- test (B) X^2test
(C) Z test (D) F-test

46. The critical difference (CD) is a form of t-test and its formula is given by:
(A) CD=t.SE(d) (B) CD=t+SE
(C) CD=t/SE(d) (D) CD=F.SE(d)

47. The ANOVA is a tool by which total variation may be spit up into several physically assignable components was defined by:
(A) Karl Pearson (B) RA Fisher
(C) Horace Secrist (D) AL Bowley

48. The degree of freedom for statistics for paired t test of n pairs of observations is:
(A) 2 (n-1) (B) n – 1
(C) 2n – 1 (D) None of these

49. The value of the sum of deviations from mean is:
(A) Zero (B) Positive
(C) Negative (D) Different from zero

50. The correlation coefficient remains unaffected by change of:
(A) Origin (B) Scale
(C) Both (a) and (b) (D) None of these

51. The manipulated variable in an experiment is:
(A) Independent variable
(B) Dependent variable
(C) Active variable
(D) Both (A) and (C)

52. The variance that reflects the systematic differences between groups of measures:
(A) Experimental variance
(B) Extraneous variance
(C) Error variance
(D) None of these

53. Assignment of values to alternatives is called:
(A) Observation (B) Judgment
(C) Measurement (D) Evaluation

54. The basic focus of Thematic Apperception Test is on:
(A) Product (B) Process
(C) Input (D) Both (A) and (B)

55. The regression coefficient are independent of change of
(A) Origin (B) Scale
(C) Both (a) and (b) (D) None of these

56. Questionnaire involves presentation of:
(A) Oral-verbal stimuli
(B) Written-verbal stimuli
(C) Oral-written stimuli
(D) None of these

57. On the spot reliability method among the following is:
(A) Test re-test method
(B) Odd even method
(C) Parallel foams method
(D) Comparable method

58. When the value of correlation coefficient is 1 then two variable are:
(A) Uncorrelated
(B) Partially correlated
(C) Perfectly correlated
(D) Highly correlated

59. Index of discrimination is otherwise known as:
(A) Item validity index
(B) Item of reliability index
(C) Item difficulty index
(D) Item credibility index

60. Cluster sample are more effective if the population taken is:
(A) Cluster of small size
(B) Clusters with largest within variation
(C) Clusters with least within variation
(D) None of these

61. The measure of association or relationship between two or more variables:
(A) Dispersion (B) Correlation
(C) Central tendency (D) Regression

62. To test the hypothesis about proportions of items in a class, we use:
(A) t- test (B) X^2test
(C) Z test (D) F-test

63. Which transformation is most appropriate for percentages:
(A) Square root
(B) Arc sine
(C) Logarithmic
(D) All of these

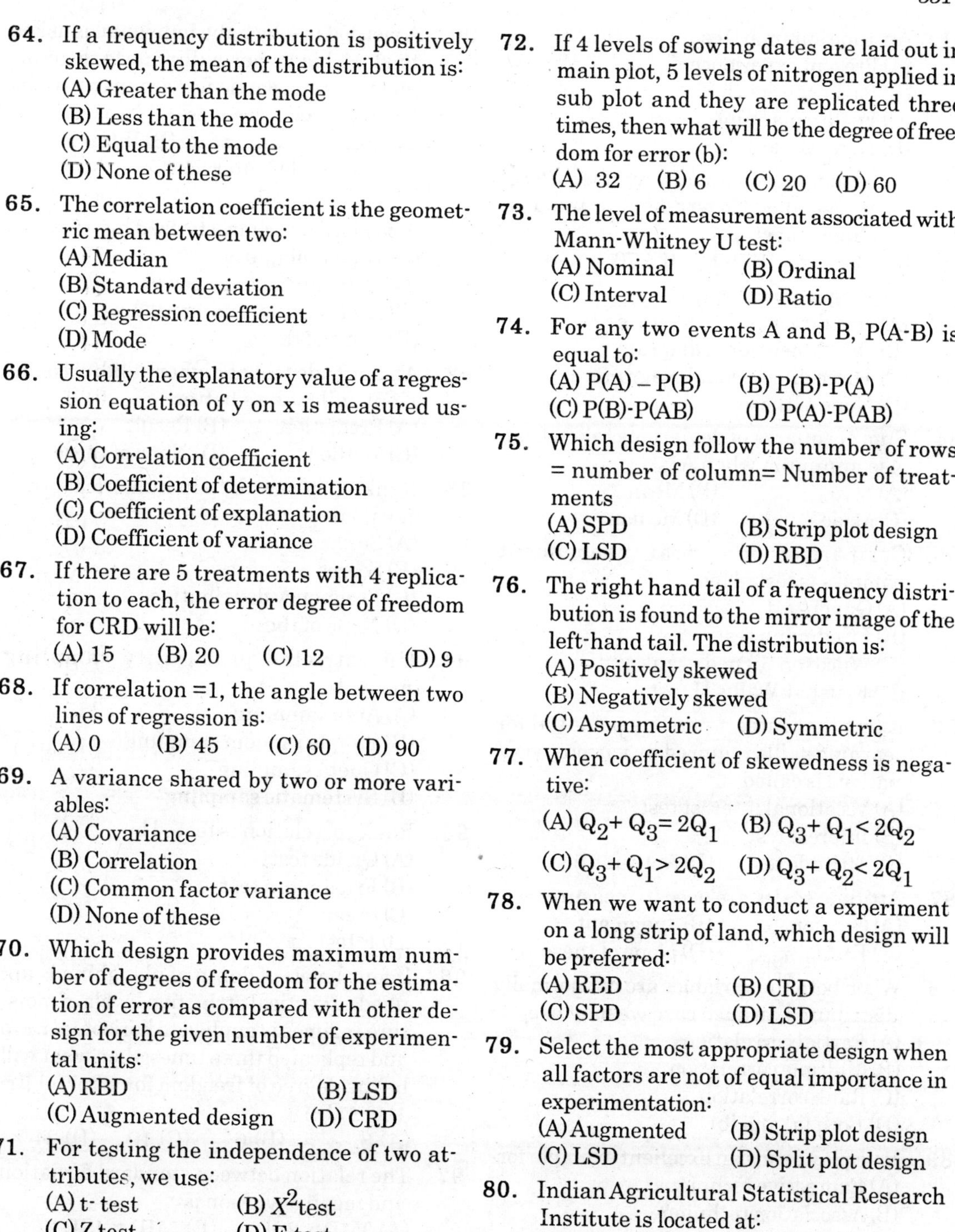

64. If a frequency distribution is positively skewed, the mean of the distribution is:
(A) Greater than the mode
(B) Less than the mode
(C) Equal to the mode
(D) None of these

65. The correlation coefficient is the geometric mean between two:
(A) Median
(B) Standard deviation
(C) Regression coefficient
(D) Mode

66. Usually the explanatory value of a regression equation of y on x is measured using:
(A) Correlation coefficient
(B) Coefficient of determination
(C) Coefficient of explanation
(D) Coefficient of variance

67. If there are 5 treatments with 4 replication to each, the error degree of freedom for CRD will be:
(A) 15 (B) 20 (C) 12 (D) 9

68. If correlation =1, the angle between two lines of regression is:
(A) 0 (B) 45 (C) 60 (D) 90

69. A variance shared by two or more variables:
(A) Covariance
(B) Correlation
(C) Common factor variance
(D) None of these

70. Which design provides maximum number of degrees of freedom for the estimation of error as compared with other design for the given number of experimental units:
(A) RBD (B) LSD
(C) Augmented design (D) CRD

71. For testing the independence of two attributes, we use:
(A) t- test (B) X^2test
(C) Z test (D) F-test

72. If 4 levels of sowing dates are laid out in main plot, 5 levels of nitrogen applied in sub plot and they are replicated three times, then what will be the degree of freedom for error (b):
(A) 32 (B) 6 (C) 20 (D) 60

73. The level of measurement associated with Mann-Whitney U test:
(A) Nominal (B) Ordinal
(C) Interval (D) Ratio

74. For any two events A and B, P(A-B) is equal to:
(A) P(A) – P(B) (B) P(B)-P(A)
(C) P(B)-P(AB) (D) P(A)-P(AB)

75. Which design follow the number of rows = number of column= Number of treatments
(A) SPD (B) Strip plot design
(C) LSD (D) RBD

76. The right hand tail of a frequency distribution is found to the mirror image of the left-hand tail. The distribution is:
(A) Positively skewed
(B) Negatively skewed
(C) Asymmetric (D) Symmetric

77. When coefficient of skewedness is negative:
(A) $Q_2+ Q_3= 2Q_1$ (B) $Q_3+ Q_1< 2Q_2$
(C) $Q_3+ Q_1> 2Q_2$ (D) $Q_3+ Q_2< 2Q_1$

78. When we want to conduct a experiment on a long strip of land, which design will be preferred:
(A) RBD (B) CRD
(C) SPD (D) LSD

79. Select the most appropriate design when all factors are not of equal importance in experimentation:
(A) Augmented (B) Strip plot design
(C) LSD (D) Split plot design

80. Indian Agricultural Statistical Research Institute is located at:
(A) Lucknow (B) Jaipur
(C) New Delhi (D) Kolkata

81. Attitude summarizes:
(A) Present experience
(B) Past experience
(C) Future planning
(D) None of these

82. At present time how many number of fore-casts issued by Government of India on 34 important crops:
(A) 50 (B) 60 (C) 70 (D) 80

83. Pie diagram is a type of:
(A) One dimensional diagram
(B) Two dimensional diagram
(C) Three dimensional diagram
(D) None of these

84. Sum of squares of deviations taken from A is minimum when A is:
(A) Zero (B) Mean
(C) Mode (D) Median

85. Comparison of several independent samples is tested by:
(A) Quade test
(B) Friedmann test
(C) Wilcoxon Signed –rank test
(D) Kruskal-Wallis H test

86. A test that rates a persons potential for learning skills required by various occu-pations is called:
(A) Vocational interest test
(B) Interest test
(C) Attitude test (D) Aptitude test

87. Attitude scales are used to measure:
(A) Opinion (B) Judgment
(C) Fact (D) None of these

88. When both the variables are not normally distributed, in such case we may use:
(A) Partial correlation
(B) Multiple correlation
(C) Rank correlation
(D) Both (a) and (b)

89. Role play forms an excellent example for:
(A) Expressive technique
(B) Association technique
(C) Constructive technique
(D) Completion technique

90. A method in which the individual is asked to state his or her willingness to admit members of a particular group to various levels of social closeness:
(A) Social distance scale (B) Interview
(C) Open ended interview
(D) Attitude scale

91. A good item is characterized by:
(A) High ambiguity
(B) High discrepancy power
(C) High discriminatory items
(D) None of these

92. An aptitude test of higher validity coeffi-cient will make a better:
(A) Prediction (B) Result
(C) Method (D) None of these

93. Ranking of raw data in numerical order is called:
(A) Series
(B) Scores
(C) Frequency distribution
(D) None of these

94. Unrestricted probability sampling method:
(A) Area sampling
(B) Simple random sampling
(C) Quota sampling
(D) Systematic sampling

95. Rank correlation is tested by:
(A) Quade test
(B) Friedmann test
(C) t-test
(D) F-test

96. If four levels of depth of ploughing is ap-plied to vertical strip, five levels of mois-ture regimes is applied to horizontal strip and replicated three times, then what will be the degree of freedom for error (c).[(r-1),(a-1)(b-1)]
(A) 8 (B) 6 (C) 10 (D) 24

97. The relation between standard deviation and mean deviation is:
(A) 3MD = 2SD (B) 5MD = 4SD
(C) 6MD = 5SD (D) MD = SD

98. Which of the following is correct:

(A) $r_{12.34} = r_{13.24}$

(B) $r_{12.3} = r_{21.3}$

(C) $r_{13} = r_{23}$

(D) $r_{13.2} = r_{13.4}$

99. The square of the standard deviation is known as the:

(A) Variance (B) Standard deviation
(C) Mean deviation
(D) Coefficient of variance

100. A subset or portion of a population is called:

(A) Population (B) Sample
(C) Representative (D) All of these

101. Rating scales are used to measure:

(A) Actual behavior
(B) Perceived behavior
(C) Both (A) and (B)
(D) None of these

102. When 4 levels of sowing dates is applied to vertical strip, five levels of irrigation is applied to horizontal strip, four levels of nitrogen randomly allocated within the intersection plots of vertical and horizontal strip and replicated three times, then what will be the degree of freedom for error (d).[ab(r-1).(c1)] (Strip split plot design)

(A) 72 (B) 120 (C) 24 (D) 8

103. A set of different measures of an individual or group, each of which is expressed in the same unit of measure:

(A) Profile (B) Biodata
(C) Set (D) Character

104. A measure of the peakedness or convexity of a curve is known as :

(A) Skewness (B) Kurtosis
(C) Ogive (D) Histogram

105. With the help of ogive one can determine:

(A) Mean (B) Median
(C) Mode (D) Geometric mean

106. Greater importance to bigger items of a given series is given by:

(A) Median (B) Mode
(A) Arithmetic mean (D)None of these

107. The basic unit containing the elements of the population to be sampled:

(A) Sampling unit
(B) Sampling frame
(C) Sampling element
(D) Nonc of thcsc

108. Which one of the following is not a measure of central tendency?

(A) Mean (B) Median
(C) Mode (D) Variance

109. An example for two way non-parametric analysis of variance among the following is:

(A) F test (B) Kvuskal Wallis test
(C) Friedman test (D) Z test

110. Geometric mean is better than other mean when:

(A) Data are positive as well as negative
(B) Data are in ratios or percentage
(C) Data are binary
(D) Data are on interval scale

111. Mean of ten observations is 20. If a value 11 is removed, what is the mean now?

(A) 19 (B) 20
(C) 21 (D) 22

112. Greater the distance of Lorenz curve from the line of equal distribution:

(A) More is inequality
(B) Less in equality
(C) Nothing can be inferred
(D) None of these

113. Mean of eight observations is 5. When one value x is added, mean remains same. What is the value of x?

(A) 3 (B) 5 (C) 8 (D) 13

114. The technique employed to find the degree of correlation is:

(A) Experimental (B) Observational
(C) Statistical (D) None of these

115. Which one of the following is not a dependent variable:

(A) Chronological age (B) Rate of response
(C) REM rebound (D) Reaction time

116. Mean of a frequency distribution is m. If every frequency is doubled . What is new mean?
(A) m (B) m/2
(C) 2m (D) m+2

117. The logical process that operates when a researcher moves from hypothesis testing to making of prediction:
(A) Vertical logic (B) Horizontal logic
(C) Inductive logic (D) Deductive logic

118. Standard error of number of successes is given by:
(A) $\sqrt{pq/n}$ (B) $\sqrt{npq}$
(C) npq (D) n^2p^2

119. Nominal level of measurement can categories objects:
(A) Qualitatively and Quantitatively
(B) Only Qualitatively
(C) Only Quantitatively
(D) None of these

120. Validity of a measuring instrument in simple terms is the freedom from:
(A) Variable errors (B) Systematic errors
(C) Random errors
(D) Non-systematic errors

121. The correlation type in which an increase in one measure is matched by an increase in the other:
(A) Positive (B) Negative
(C) Both (A) and (B) (D) None of these

122. If for a distribution, coefficient of kurtosis $\gamma_2<0$, the frequency curve is:
(A) Leptokurtosis (B) Platykurtosis
(C) Mesokurtosis (D) None of these

123. Range of the values 15, 12, 27, 6, 9, 18, 21 is:
(A) 6 (B) 21 (C) 4.5 (D) 3

124. From the cumulative frequency diagram which one of the following types of average can be estimated?
(A) Geometric mean (B) Harmonic mean
(C) Median (D) Mode

125. While testing the significance of the difference of two sample means in case of small samples, the degrees of freedom is calculated by:
(A) $v = n_1 + n_2 - 2$ (B) $v = n_1 + n_2 - 1$
(C) $v = n_1 - n_2 - 2$ (D) $v = n_1 + n_2 + 2$

126. — error are made when we accept a hypothesis which is not true:
(A) Type I (B) Type II
(C) Both (A) and (B) (D) None of these

127. The standard distribution of Poisson's distribution is: (m=mean)
(A) m (B) $\sqrt{m}$
(C) m^2 (D) $1/\sqrt{m}$

128. Correct formula for variance is:
(A) $1/n\ \Sigma X_1{}^2 – X^{-2}$
(B) $1/n\ \Sigma X^2 – nX^{-2}$
(C) $1/n\ \Sigma X_1{}^2 – X^{-2}\ /2$
(D) None of these

129. Theoretically the best average in the construction of index number is:
(A) Median (B) Geometric mean
(C) Mode (D) Arithmetic mean

130. Variance of twenty observations is 8 if every observation is multiplied by 2, what is the value of variance now?
(A) 4 (B) 8 (C) 16 (D) 32

131. The coefficient of correlation is independent of change of:
(A) Origin (B) Scale
(C) Both (A) and (B) (D) None of these

132. During one year the ratio of prices of two commodities is 3, whereas during the next year it is 2. For averaging these ratios:
(A) Arithmetic mean is most appropriate
(B) Harmonic mean is most appropriate
(C) Geometric mean is most appropriate
(D) All the three, arithmetic, harmonic and geometric means are equally appropriate

133. Total number of possible out comes in a cricket match is:
(A) 1 (B) 2 (C) 3 (D) 11

134. Measurement in social science is concerned with:
(A) Objects
(B) Properties of objects
(C) Phenomena
(D) Properties of phenomena

135. Consistency and reproductibility of test score are indicated by:
(A) Reliability (B) Validity
(C) Credibility (D) Sustainability

136. Reliability is found to increase when we go for:
(A) Less number of items
(B) Homogeneous items
(C) High discriminatory items
(D) Less difficult items

137. Attitude can be considered as a mixture of:
(A) Belief and emotion
(B) Emotion and feeling
(C) Thought and emotion
(D) Belief and thought

138. If A contains B, what can be said about the value of r = P (A)/ P (B):
(A) $r < 1$ (B) $r = 0$
(C) $r > 1$ (D) $0 < r < 1$

139. Each item in the attitude scale is given a scale value in:
(A) Guttman scale
(B) Thurstone and Clave scale
(C) Adorne scale (D) Likert scale

140. The primary elements of the universe of possible outcome is called:
(A) Sample point (B) Sample space
(C) Sample event (D) None of these

141. Which among the following is not a measure of central tendency:
(A) Median (B) Mode
(C) Arithmetic mean (D) Variance

142. Correlation coefficient is considered as that of two regression coefficients:
(A) Mean (B) Geometric mean
(C) Square (D) None of these

143. If standard deviation is greater than variance, the value of standard deviation is:
(A) 1 (B) Less than 1
(C) Greater than 1
(D) Nothing definite can be said

144. The tendency to act in ways that influence experimental results so that we obtain results that are consistent with our expectations:
(A) Self fulfilling prophecy
(B) Cock tail party effect
(C) Sapir-Whorf hypothesis
(D) None of these

145. If A and B are mutually exclusive events. P(A) = ¼ and P (A U B) = ¾, what is the value of P (B)?
(A) 0 (B) ½ (C) 3/8 (D) 3/16

146. "After this, therefore caused by this" is usually denoted to indicate:
(A) Experimental design
(B) Survey design
(C) Ex Post facto design
(D) Explorative design

147. Which among the following is the testable foam in a scientific research:
(A) Research problem
(B) Research hypothesis
(C) Research variable
(D) All of these

148. To discriminate groups from one another on the basis of sets of measures, we use:
(A) Profile analysis
(B) Factor analysis
(C) Discriminate analysis
(D) Disaster analysis

149. When compared with reality, measurement procedure and number system are:
(A) Isomorphic (B) Polymorphic
(C) Monomorphic (D) Oligomorphic

150. A die is rolled once. What is the probability of obtaining two or four points:
(A) 1/2 (B) 1/3 (C) 1/4 (D) 1/6

151. For statistical stability and reliability the ideal number of cards needed in Q distribution is:
(A) 1-30 (B) 31-60
(C) 61-90 (D) 91-120

152. Rorschach projection tests foams an example for:
(A) Association test
(B) Construction technique
(C) Completion technique
(D) Ordering technique

153. The measurement of temperature falls under:
(A) Nominal level (B) Ordinal level
(C) Interval level (D) Ratio level

154. In order to study fertility fluctuations, which one of the following is used?
(A) Randomization (B) Replication
(C) Local control (D) Uniformity trials

155. Information gained through direct observation and measurement is called:
(A) Empirical evidence
(B) Experimental variance
(C) Observational evidence
(D) None of these

156. The first step in summarizing the data is:
(A) Tabulation (B) Classification
(C) Interpretation (D) Observation

157. The hypothesis type stated for the purpose of rejection:
(A) Null hypothesis
(B) Alternate hypothesis
(C) Parallel hypothesis
(D) Both (B) and (C)

158. The underlying philosophy of scientific method followed in different disciplines of science are:
(A) Exactly the same (B) Almost same
(C) Some what the same
(D) Entirely different

159. In a 2 x 2 contingency table, what is the degree of freedom in x^2– test?
(A) 0 (B) 1 (C) 2 (D) 4

160. What is the main purpose of design of experiments?
(A) To compare means only
(B) To compare variances only
(C) To compare means and variances both
(D) To compare neither means nor variances

161. If there are 8 blocks and 6 treatments in RBD, what are the degrees of freedom corresponding to total sum of squares?
(A) 5 (B) 7 (C) 3 (D) 47

162. What is the geometric mean of 1, 4 and 16?
(A) 2 (B) 4 (C) 6 (D) 8

163. Age is classified under:
(A) Discontinuous variable
(B) Discrete variable
(C) Continuous variable
(D) Intervening variable

164. A test used to compare the variance of two independent samples:
(A) t- test (B) X^2test
(C) Z test (D) F-test

165. If the correlation coefficient between X and Y is r, the correlation between x^2 and y^2 is:
(A) r (B) r^2
(C) 0 (D) None of these

166. Any categorization scheme must be on:
(A) Four level of discourse
(B) Three level of discourse
(C) Two level of discourse
(D) One level of discourse

167. If correlation between two variables (x and y) is negative, the regression coefficient of y on x is:
(A) Positive (B) Negative
(C) Not certain (D) Zero

168. Which one of the following is not a fundamental principle of experimental design?
(A) Randomization (B) Replication
(C) Validation (D) Local control

169. If r is the correlation coefficient, r^2 is:
(A) Coefficient of determination
(B) Coefficient of non determination
(C) Coefficient of alienation
(D) None of these

170 If all the variate values are the same (e.g. 8, 8, 8), the standard deviation will be:
(A) 4 (B) 8 (C) 16 (D) 0

171. What are the degrees of freedom in testing the difference between means by t-test?
(A) $n_1 + n_2 - 1$ (B) $n_1 + n_2 + 1$
(C) $n_1 + n_2 - 2$ (D) $n_1 + n_2 + 2$

172. The ratio of regression sum of squares to the total sum of squares is:
(A) Error means square
(B) Coefficient of determination
(C) Coefficient of non determination
(D) Both (A) and (B)

173. If correlation coefficient is 0.7, then coefficient of determination is:
(A) 0.7 (B) 0.07
(C) 0.49 (D) None of these

174. To study average rate of change in population, we use:
(A) Median (B) Mode
(C) Arithmetic mean (D) Geometric mean

175. Which measure of dispersion is least affected by extreme values:
(A) Range (B) Quartile deviation
(C) Standard deviation
(D) Mean deviation

176. The simplest way of visualizing a correlation is to construct a:
(A) Graph (B) Table
(C) Scatter diagram (D) None of these

177. Local control is a device to maintain homogeneity:
(A) Among blocks (B) Within Blocks
(C) (A) and (B) both (D) None of these

178. To test the hypothesis about proportions of items in a class, we use:
(A) t- test (B) X^2 test
(C) Z test (D) F-test

179. Ranking of raw data in numerical order is called:
(A) Series (B) Scores
(C) Frequency distribution
(D) None of these

180. Index numbers used in social science measure are having:
(A) Arbitrary zero point
(B) Absolute zero point
(C) Natural zero point
(D) Both (A) and (C)

181. In a series of values if one value is zero, G.M. is:
(A) Positive (B) Zero
(C) Indeterminate (D) None of these

182. Probable error is:
(A) 0.6745 S.E. (B) 0.6457 S.E.
(C) 0.6753 S.E. (D) 0.7645 S.E.

183. The first step in analysis of data is:
(A) Manipulation (B) Categorization
(C) Interpretation (D) Summarization

184. If two events A and B are dependent, the conditional probability of A given B, i.e., P(A/B) is calculated as:
(A) P (A) + P (B) (B) P (A) x P (B)
(C) P (A) – P (B) (D) P (AB) / P (B)

185. If a constant 50 is subtracted from each of the value of X and Y, the regression coefficient is:
(A) Reduced by 50 (B) $1/50^{th}$ of original
(C) Increased by 50 (D) Not changed

186. If all the variate values are negative, the standard deviation will be:
(A) Negative (B) Positive
(C) No sign (D) Cannot calculate

187. The level of measurement attached to categorical variable is:
(A) Nominal (B) Ordinal
(C) Interval (D) Ratio

188. When neither variable is the cause of the other, the relationship is called:
(A) Symmetrical (B) Reciprocal
(C) Asymmetrical (D) None of these

189. An interval in which scores are grouped is called:
(A) Point (B) Value
(C) Class interval (D) None of these

190. If there is an increase in a series at constant rate, the graph will be:
(A) Convex curve (B) Concave curve
(C) Parabola (D) Straight line

191. If both the regression coefficients are negative, the correlation coefficient would be——:
(A) Negative (B) Positive
(C) Zero (D) None of these

192. A sub set or portion of a population is called:
(A) Population
(B) Sample
(C) Representative
(D) All of these

193. The primary elements of the universe of possible outcome is called:
(A) Sample point (B) Sample space
(C) Sample event (D) None of these

194. The under root of two——— coefficient gives us the value of correlation coefficient:
(A) Regression
(B) Standard Error
(C) Standard Deviation
(D) None of these

195. Survey method would most typically used in:
(A) Clinical assessment
(B) Opinion poll
(C) Laboratory experiment
(D) Psychoanalysis

196. Sampling frame is a must in the case of:
(A) Random sampling
(B) Systematic sampling
(( ) Stratified sampling
(D) Cluster sampling

197. The number of possible samples of size two from a population of four units in a simple random sample without replacement is:
(A) 2 (B) 4 (C) 6 (D) 8

198. A constant value of 50 is subtracted from each observation of a set, the mean of the set is:
(A) Increased by 50 (B) Decreases by 50
(C) Not affected (D) Zero

199. Take the odd one out:
(A) Area sampling (B) Snowball sampling
(C) Quota sampling
(D) Incidental sampling

200. Cross section of population is represented less in:
(A) Simple random sampling
(B) Multistage sampling
(C) Cluster sampling
(D) Stratified random sampling

201. If sum of the product of deviations of x and y series from their means, i.e. Σ xy is zero, the coefficient of correlation shall be:
(A) +1 (B) 0
(C) -1 (D) None of these

202. There will be only one regression line in case of two variables if:
(A) $r = 0$ (B) $r = \pm 1$
(C) $r - 0.5$ (D) None of these

203. To have an accurate picture of how large segments of the population feel about current issues:
(A) Case study (B) Lab experiment
(C) Field survey (D) None of these

204. Unrestricted probability sampling method:
(A) Area sampling
(B) Simple random sampling
(C) Quota sampling
(D) Systematic sampling

205. The sample type wherein similarity between immediate neighbours and difference between distinct neighbours are both taken care of:
(A) Random sample
(B) Cluster sample
(C) Systematic sample
(D) Stratified sample

206. If each observation of a set is multiplied by 10, the mean of the new set of observation:
(A) Remains the same
(B) 10 times the original mean
(C) $1/10^{th}$ of the original mean
(D) Increased by 10

207. When information is to be collected from a geographically dispersed population we go for:
(A) Stratified sampling
(B) Systematic sampling
(C) Cluster sampling
(D) Multipurpose sampling

208. Regression coefficient is independent of change of:
(A) Origin but not of scale
(B) Both origin and scale
(C) Scale but not of origin
(D) None of these

209. The best sample size from a perfectly homogenous population in respect of a character:
(A) A large sample (B) A small sample
(C) A single item (D) No item

210. The regression lines cut each other at the point of:
(A) Average of x and y
(B) Average of x only
(C) Average of y only
(D) None of these

211. Sum of measure divided by the number of the measure:
(A) Mean (B) Mode
(C) Median (D) None of these

212. A constant of 5 is subtracted from each observation of a set, the variance is:
(A) Reduced by 5 (B) Reduced by 25
(C) Unaltered (D) Increased by 25

213. Greater importance to bigger items of a given series is given by:
(A) Median (B) Mode
(C) Arithmetic mean (D) None of these

214. Both the regression coefficients cannot be:
(A) Less than one (B) Greater than one
(C) Equal to one (D) None of these

215. Geometric mean is better than other mean when:
(A) Data are positive as well as negative
(B) Data are in ratios or percentage
(C) Data are binary
(D) Data are on interval scale

216. To study the average of quantity prices, we use:
(A) Median (B) Harmonic mean
(C) Arithmetic mean
(D) Geometric mean

217. If each of a set is divided by 10, the standard deviation of anew observation is:
(A) $1/10^{th}$ of SD of originals
(B) $1/100^{th}$ of SD of originals
(C) Not altered
(D) 10 times of SD of originals

218. To study the average height of plants, we use:
(A) Mode (B) Harmonic mean
(C) Arithmetic mean
(D) Geometric mean

219. Mention the correct answers. The ranks according to two attributes in a sample are given below:

R1	1	2	3	4	5
R2	5	4	3	2	1

The rank correlation between them is:
(A) O (B) +1
(C) -1 (D) None of these

220. The AM of the age of 20 students in a class is 15 years. When the teachers age also is added, the mean increase by one year. The age of the teacher is:
(A) 16 years (B) 21 years
(C) 36 years (D) 25 years

221. Which among the following is not a measure of central tendency:
(A) Mean (B) Median
(C) Mode (D) Variance

222. Formula for calculating coefficient of variation is:
(A) SD/Mean x 100
(B) Mean/SD x 100
(C) Mean x SD/100
(D) 100/Mean x SD

223. The method of assembling correspondence without experiment control over them is called:
(A) Dispersion (B) Correlation
(C) Central tendency
(D) None of these

224. Mean deviation is minimum when taken about:
(A) Mean (B) Median
(C) Mode (D) Zero

225. The relation between arithmetic mean, geometric mean and harmonic mean is:
(A) AM = GM = HM (B) AM>GM>HM
(C) AM<GM<HM (D) AM<HM<GM

226. The most frequent score in a given distribution:
(A) Mean (B) Mode
(C) Median (D) None of these

227. In case of positively skewed distribution the relation between mean,, median and mode that hold is:
(A) Median > Mean > Mode
(B) Mean > Median > Mode
(C) Mean = Median = Mode
(D) None of these

228. Which mean of dispersion is most affected by extreme values:
(A) Range (B) Quartile deviation
(C) Standard deviation
(D) Mean deviation

229. Experimental method of data collection among the following is:
(A) ield study (B) Sample survey
(C) Library method (D) Field experiment

230. Interview is the widely used technique in:
(A) Local study (B) Public studies
(C) Empirical studies
(D) Non-scientific studies

231. The measure of dispersion which ignores signs of deviation from the central value:
(A) Range
(B) Quartile deviation
(C) Standard deviation
(D) Mean deviation

232. Which measure of dispersion is least affected by extreme values:
(A) Range
(B) Quartile deviation
(C) Standard deviation
(D) Mean deviation

233. The range of simple correlation coefficient is:
(A) 0 to ∞ (B) $-\infty$ to $+\infty$
(C) -1 to +1 (D) 0 to 1

234. Range is not a reliable measure of:
(A) Mean (B) Dispersion
(C) Median (D) Mode

235. Empirical relation between quartile deviation and standard deviation is:
(A) 3QD = 2 SD (B) 4 QD = 3 SD
(C) 6 QD = 5SD (D) 5 QD = 4 SD

236. Correlation coefficient is independent of:
(A) Scale (B) Origin
(C) Both (A) and (B)
(D) Neither (A) nor (B)

237. The relation between standard deviation and mean deviation is:
(A) 3 MD = 2 SD (B) 5 MD = 4 SD
(C) 6 MD = 5 SD (D) MD = SD

238. The average of sum of squares of deviation about the mean is called:
(A) Absolute deviation
(B) Variance
(C) Standard deviation
(D) Mean deviation

239. Which of the following is the unit less measure of dispersion:
(A) Standard deviation
(B) Mean deviation
(C) Coefficient of variance
(D) Range

240. The most commonly used method in a social survey is:
(A) Observation (B) Interview
(C) Questionnaire (D) Case study

241. The value of co-efficient of concordance ranges between:
(A) 0 to + 1 (B) ∞ ? to + ∞
(C) -1 to +1 (D) -1 to 0

242. The degree of linear relationship between variables are expressed through:
(A) Coefficient of correlation
(B) Standard deviation
(C) Arithmetic mean (D) None of these

243. A constant of 5 is subtracted from each observation of a set, the variance is:
(A) Reduced by 5 (B) Reduced by 25
(C) Unaltered (D) Increased by 25

244. Survey research employs which of the following techniques:
(A) Panel technique
(B) Personal interview
(C) Controlled observation (D) All of these

245. If each of a set is divided by 10, the standard deviation of a new observation is:
(A) $1/10^{th}$ of SD of originals
(B) $1/100^{th}$ of SD of originals
(C) Not altered
(D) 10 times of SD of originals

246. The best control for practice effects in an experiment is:
(A) Random selection
(B) Counter balancing
(C) Equivalent group
(D) Split half presentation

247. The unit of correlation coefficient is:
(A) Kg/cc (B) Per cent
(C) No unit (D) None of these

248. In a normal distribution the points of inflexion occur at:
(A) $X + \sigma$ (B) $x + 2\delta$
(C) $X + 3\sigma$ (D) None of these

249. If an event cannot take place, probability will be:
(A) +1 (B) -1 (C) 0 (D) None of these

250. The circular test is satisfied when:
(A) $P_{12} x P_{23} P_{31} = 0$
(B) $P_{12} x P_{32} x P_{31} = 1$
(C) $P_{12} x P_{23} x P_{31} = 1$
(D) $P_{12} x P_{23} P_{13} = 1$

251. If β xy and β yx are two regression coefficient, they have:
(A) Same sign (B) Opposite sign
(C) Either same or opposite sign
(D) Nothing can be said

252. An example for non parametric test from among the following is:
(A) t- test (B) X^2 test
(C) Z test (D) F-test

253. When degree of freedom of X^2 test is more than 100, t test is approximated to:
(A) t- test (B) X^2 test
(C) Z test (D) F-test

254. Correlation coefficient is independent of:
(A) Scale (B) Origin
(C) Both (A) and (B)
(D) Neither (A) nor (B)

255. Degree of freedom from X^2 in case of 4 x 3 contingency table is:
(A) 12 (B) 9 (C) 6 (D) 8

256. The measure of association or relationship between two or more variables:
(A) Dispersion (B) Correlation
(C) Central tendency (D) Regression

257. Time reversal test is satisfied when:
(A) $P_{01} x P_{23} = 0$
(B) $P_{01} x P_{10} = 1$
(C) $P_{02} x P_{10} > 1$
(D) $P_{01} x P_{10} < 1$

258. Addition theorem states that if two events A and B are mutually exclusive the probability of occurrence of either A or B is given by:
(A) P (A) + P (B) (B) P (A) x P (B)
(C) P (A) – P (B) (D) None of these

259. The normal distribution with x = 0 and α = 1 is known as:
(A) Standard normal distribution
(B) Poisson's distribution
(C) Both (A) and (B)
(D) None of these

260. Which among the correlation coefficient represent the strongest relationship:
(A) -0.86 (B) +0.86 (C) +0.10 (D) +0.09

261. The most important factors causing seasonal variations are:
(A) Growth of population
(B) Technological improvement
(C) Weather and social customs
(D) Change in fashions

262. If correlation =1, the angle between two lines of regression is:
(A) 0 (B) 45 (C) 60 (D) 90

263. Which one of the following average can be computed from a cumulative frequency curve:
(A) Geometric mean (B) Harmonic mean
(C) Mode (D) Median

264. —— asserts that there is no true difference in the sample and the population in the particular matter under consideration:
(A) Null hypothesis
(B) Alternate hypothesis
(C) Both (A) and (B)
(D) None of these

265. Coefficient of determination is defined as:
(A) r^3 (B) $1 - r^2$
(C) $1 + r^2$ (D) r^2

266. If X and Y are independent, the value of regression coefficient β xy is equal to:
(A) 0 (B) 1
(C) a (D) Any positive value

267. The farther the two regression lins cut each other:
(A) The greater will be degree of correlation
(B) The lesser will be degree of correlation
(C) Does not really matter
(D) None of these

268. Historically the first index was constructed in:
(A) 1764 (B) 1844 (C) 1850 (D) 1744

269. If the correlation coefficient between X and Y is r, the correlation between x^2 and y^2 is:
(A) r (B) r^2
(C) 0 (D) None of these

270. If two events A and B are independent, the probability that they will both occur is given by:
(A) P (A) + P (B) (B) P (A) x P (B)
(C) P (A) – P (B) (D) P(A)+P(B)-P(AB)

271. — error is committed when the hypothesis is true but out test rejects it:
(A) Type I (B) Type II
(C) Both (A) and (B) (D) None of these

272. If the standard deviation of a variable x is s and if y = a + bs, where a and b are constants, the standard deviation of y would be:
(A) a + bs (B) bs
(C) √b.s (D) b^2

273. If correlation between two variables (x and y) is negative, the regression coefficient of y on x is:
(A) Positive (B) Negative
(C) Not certain (D) Zero

274. "Hit and run" style of interview is adopted in:
(A) Invasion method
(B) Immersion method
(C) Case method (D) None of these

275. The most ubiquitous kin of variance in research is:
(A) Experimental variance
(B) Extraneous variance
(C) Error variance (D) None of these

276. In an experiment measuring the effect of temperature on dilation of blood vessels, the level of temperature is the:
(A) Control variable
(B) Antecedent variable
(C) Dependent variable
(D) Independent variable

277. Scientific data must be:
(A) Objective (B) Measurable
(C) Predictable (D) All of these

278. The value of correlation coefficient lies in between:
(A) 0 (B) -1 to +1
(C) 1 to 2 (D) 0 to -1

279. Which among the following accounts for existing data, predict new observation and guide further research:
(A) Hypothesis (B) Theory
(C) Definition (D) Method

280. Research design is not essentially a:
(A) Plan (B) Product
(C) Structure (D) Strategy

281. If correlation coefficient is 0.7, then coefficient of determination is:
(A) 0.7 (B) 0.07
(C) 0.49 (D) None of these

282. The classification of variable into independent and dependent is based on:
(A) Uses (B) Kinds
(C) Foam (D) Both (A) and (C)

283. The word hypothesis literally means:
(A) Less than generally held view
(B) More than generally held view
(C) Same as generally held view
(D) None of these

284. To penetrate deep into the inner sanctum of the group we go for:
(A) Case study
(B) Participatory observation
(C) Formal discussion (D) Interview

285. If r is the correlation coefficient, r^2 is:
(A) Coefficient of determination
(B) Coefficient of non determination
(C) Coefficient of alienation
(D) None of these

286. The most widely used method of measuring seasonal variations is:
(A) Ratio-to-moving average method
(B) Ratio-to trend method
(C) Link relative method
(D) None of these

287. The ratio of regression sum of squares to the total sum of squares is:
(A) Error means square
(B) Coefficient of determination
(C) Coefficient of non determinant
(D) Both (A) and (B)

288. To study the average size of shoes sold in the market, we use:
(A) Median (B) Mode
(C) Arithmetic mean (D) Geometric mean

289. Correlation coefficient is considered as that of two regression coefficients:
(A) Mean (B) Geometric mean
(C) Square (D) None of these

290. When we want average of rates of change or ratios or index number, which measure of central tendency is suitable:
(A) Geometric mean (B) Mode
(C) Median (D) Arithmetic mean

291. Under non-parametric the comparison of several related samples is tested by:
(A) Quade test (B) Friedmann test
(C) F-test (D) Both (A) and (B)

292. The correlation type wherein more than one dependent and independent variables are included in the regression model:
(A) Illusory correlation
(B) Spurious correlation
(C) Cannonical correlation
(D) Hallsinative correlation

293. In Confounding, generally which order of interaction is confounded:
(A) Lower order (B) First order
(C) Higher order (D) All of these

294. In a experiment, a treatment requires large area and b treatment requires smaller area (different precision), which design is most appropriate:
(A) Split Plot Design (B) LSD
(C) RBD (D) Strip Plot Design

295. The technique employed to find the degree of correlation is:
(A) Experimental (B) Observational
(C) Statistical (D) None of these

296. Local control is not applied for the design:
(A) RBD (B) CRD
(C) LSD (D) SPD

297. Which test is used to test of significance of the difference between two means:
(A) t-test (B) F-test
(C) z-test (D) x^2test

298. Normal distribution is due to the work of
(A) James Bernoulli (B) SD Poisson
(C) Laplace and Gauss(D) Demoivre

299. Mean and variance of a standard normal distribution are respectively:
(A) 0 and 1 (B) 1 and 0
(C) Both (A) and (B) (D) None of these

300. A normal variate takes values in the range of:
(A) 0 to 1 (B) 0 to ∞
(C) ∞ ? to + ∞ (D) -11.96 to +1.96

301. To study the average height of plants, we use:
(A) Median (B) Mode
(C) Arithmetic mean
(D) Geometric mean

302. The population for sampling is defined in terms of:
(A) Time (B) Element
(C) Extent (D) All of these

303. "Items on a mathematical ability are examined to see whether they seem to be related to this ability" is an example for:
(A) Face validity (B) Content validity
(C) Concurrent validity
(D) Predictive validity

304. Correlation coefficient is:
(A) More than +1
(B) Less than -1
(C) Lies between -1 and +1
(D) Lies between +1 and -1

305. The correlation type in which an increase in one measure is matched by an increase in the other:
(A) Positive (B) Negative
(C) Both of these (D) None of these

306. Quartile deviation is:
(A) $\frac{4}{5}\alpha$ (B) $\frac{3}{2}\alpha$
(C) $\frac{2}{3}\alpha$ (D) $\frac{5}{4}\alpha$

307. Standard error can be:
(A) 0 (B) 1 (C) >1 (D) All of these

308. A relationship in correlation coefficient that does not form a straight line is:
(A) Linear relationship
(B) Negative relationship
(C) Curvilinear relationship
(D) Positive relationship

309. The value of x^2always lies between:
(A) 0 to 1 (B) 0 to ∞
(C) -1 to +1 (D) - ∞ to + ∞

310. Pick the odd man out:
(A) F test (B) Kvuskal Wallis test
(C) Friedman test (D) Z test

311. .... is known as the 'Ideal' formula for constructing index numbers:
(A) Fisher's Ideal Index
(B) Laspeyre's Index
(C) Paasche's Index (D) None of these

312. Mean $\pm 3\sigma$ covers
(A) 99.37 % items (B) 90 % items
(C) 99.9 % items (D) 99.73 % items

313. The degrees of freedom for statistics for paired t test of n pairs of observations is:
(A) 2 (n-1) (B) n-1
(C) 2n-1 (D) none of these

314. Ogive for more than type and less than type distribution intersect at:
(A) Mean (B) Median
(C) Mode (D) None of these

315. Measures of dispersion includes the range and:
(A) Standard deviation (B) Mode
(C) Median (D) None of these

316. Unit of variance of a data in cm is:
(A) No unit (B) Cm^2
(C) cm (D) cm^3

317. The range of multiple correlation coefficient R is:
(A) 0 to 1 (B) 0 to ∞
(C) -1 to +1 (D) - ∞ to + ∞

318. Coefficient of quartile deviation is calculated by the formula:
(A) $\frac{Q_3+Q_1}{4}$ (B) $\frac{Q_3+Q_1}{2}$
(C) $\frac{Q_3-Q_1}{Q_3+Q_1}$ (D) $\frac{Q_3+Q_1}{Q_3-Q_1}$

319. Which of the following is correct:
(A) $r_{12.34}=r_{13.24}$ (B) $r_{12.3}-r_{21.3}$
(C) $r_{13}=r_{23}$ (D) $r_{13.2}=r_{13.4}$

320. If the correlation coefficient between x and y is r, the correlation coefficient of y and x is:
(A) -r (B) r^2 (C) r (D) 1/r

321. For testing the significance of difference between two means, we use:
(A) t- test (B) X^2test (C) Z test (D) F-test

322. Usually the explanatory value of a regression equation of y on x is measured using:
(A) Correlation coefficient
(B) Coefficient of determination
(C) Coefficient of explanation
(D) Coefficient of variance

323. The difference of two means in case of small samples is tested by the formula:
(A) $T=\frac{X_1-X_2}{S}$
(B) $T=\frac{X_1+X_2}{S}\sqrt{\frac{n_1+n_2}{n_1-n_2}}$
(C) $T=\frac{X_1+X_2}{S}\sqrt{\frac{n_1 n_2}{n_1+n_2}}$
(D) None of these

324. The correlation type wherein more than one dependent and independent variables are included in the regression model:
(A) Illusory correlation
(B) Spurious correlation
(C) Cannonical correlation
(D) Hallsinative correlation

325. Equality of two regression coefficient can be tested with:
(A) t-test (B) x^2test
(C) Z test (D) ANOVA

326. To test the hypothesis about proportions of items in a class, we use:
(A) F test (B) x^2test
(C) Z test (D) t test

327. For testing the significance of correlation coefficient, we use:
(A) F test (B) ANOVA
(C) x^2test (D) t test

328. Standard deviation of sample mean is given by:
(A) σ (B) n
(C) σ/root n (D) None of these

329. Who is considered as father of statistics:
(A) Ronald Fisher
(B) Emile Dukeman
(C) Francis Galton (D) Carl Pearson

330. The non-parametric substitute for the parametric t test:
(A) x^2test
(B) Kruskal Wallis test H test
(C) Friedman test
(D) Mann-Whitney U test

331. To study the average of quantity prices, we use:
(A) Median (B) Harmonic mean
(C) Arithmetic mean
(D) Geometric mean

332. To get a short hand summary of large number of observation we go for:
(A) Pure statistics
(B) Descriptive statistics
(C) Numerical statistics
(D) None of these

333. Statistics serves to provide a short hand description of large amounts of:
(A) Scores (B) Data
(C) Points (D) None of these

334. Guttman scale is basically a:
(A) Summated rating scale
(B) Equal appearing interval scale
(C) Cumulative scale
(D) None of these

335. Which among the concepts is mathematically considered the biggest:
(A) Reliability (B) Validity
(C) Both (A) and (B) (D) None of these

336. The internal and external validity for laboratory experiment are respectively:
(A) High and High (B) High and Low
(C) Low and Low (D) Low and High

337. Test scores can be used for scientific purposes if they are:
(A) True and effective
(B) Reliable and true
(C) Reliable and valid
(D) None of these

338. An example for two way non-parametric analysis of variance among the following is
(A) X^2test
(B) Kruskal Wallis test H test
(C) Friedman test (D) Z test

339. Opinion level of difficulty of a test corresponds to:
(A) Maximum item inter-correlation
(B) Maximum item total correlation
(C) Higher reliability of test scores
(D) All of these

340. To discriminate groups from one another on the basis of sets of measures, we use:
(A) Profile analysis (B) Factor analysis
(C) Discriminate analysis
(D) Distracter analysis

341. Harmonic mean gives weightage to:
(A) Small values (B) Positive values
(C) Large values (D) Negative values

342. An example for one way non-parametric analysis of variance among the following is:
(A) X^2test
(B) Kruskal Wallis test H test
(C) Friedman test (D) Z test

343. The level of measurement associated with Mann-Whitney U test:
(A) Nominal (B) Ordinal
(C) Interval (D) Ratio

344. The value of coefficient of kurtosis can be:
(A) Less than 3 (B) Greater than 3
(C) Equal to 3 (D) All of these

345. If there are ten treatments and three replications in a CRD experiment, the error degrees of freedom is:
(A) 18 (B) 20 (C) 27 (D) 29

346. A family of parametric distribution with mean is equal to variance is:
(A) Binomial (B) Normal
(C) Poisson (D) Negative binomial

347. Sum of square divided by the number of the measure:
(A) Mean (B) Mode
(C) Median (D) None of these

348. When information is to be collected from a geographically dispersed population we go for:
(A) Stratified sampling
(B) Systematic sampling
(C) Cluster sampling
(D) Multipurpose sampling

349. If $a<b$ and $b<c$, then $a<c$ denotes:
(A) Transitive Property of Inaquality
(B) Addition Property of Equality
(C) Commutative Property of Addition
(D) Multiplication Property of Equality

350. In an RBD with 10 treatment and 4 replications, what will be the error degree of freedom if two observations are lost:
(A) 25 (B) 2 (C) 27 (D) 28

351. The test used in selection process to measure a persons interests, existing skills, and potential for acquiring skills:
(A) Intelligence tests (B) Aptitude tests
(C) Vocation tests (D) Personality test

352. Which of the following is not an ability test:
(A) Aptitude test (B) Social intelligence test
(C) Gottshehold test
(D) General intelligence test

353. If a=b, then a+C = b+C denotes:
(A) Transitive Property of Inequality
(B) Addition Property of Equality
(C) Commutative Property of Addition
(D) Multiplication Property of Equality

354. Distribution wherein mean is always greater than variance is:
(A) Binomial (B) Poisson
(C) Normal (D) Both (A) and (B)

355. The units within and between each stratum in a stratified random sampling are respectively:
(A) Homogenous and homogenous
(B) Homogenous and Heterogeneous
(C) Heterogeneous and Heterogeneous
(D) Heterogeneous and Homogenous

356. Degree of freedom of X2 test for 4 x 3 contingency table is:
(A) 12 (B) 6 (C) 1 (D) 7

357. Harmonic mean is better than other means if data are for:
(A) Speeds or rates
(B) Heights and lengths
(C) Binary values 0 and 1
(D) Ratio and proportion

358. The variance of a binomial distribution with n and p as parameters is:
(A) n (B) n(1-p)
(C) np (D) np (1-p)

359. The sampling procedure most commonly used in order to ensure a representative sample:
(A) Random sampling
(B) Systematic sampling
(C) Cluster sampling(D) Quota sampling

360. A normal curve is all but not:
(A) Symmetrical (B) Continuous
(C) Unimodal (D) Asymbiotic

361. If X is binomial variant with parameters n and p. If n=1, the distribution is:
(A) Poison (B) Normal
(C) Bernoulli (D) Binomial

362. Theoretical value of test statistics for five per cent when compared with one per cent is:
(A) Same (B) Greater
(C) Lesser (D) None of these

363. The value of probability ranges from:
(A) -1 to +1 (B) 0 to ∞
(C) 0 to 1 (D) None of these

364. Probability of two persons being born on the same day is:
(A) 1/49 (B) 1/365
(C) 1/7 (D) None of these

365. Mean Deviation is minimum when taken about:
(A) Median (B) Mode
(C) Arithmetic mean (D) Zero

366. Average performance of a representative sample on a given test is called:
(A) Norm (B) Validity
(C) Reliability (D) Objectivity

367. Sampling size is influenced by:
(A) Heterogeneity of population
(B) Sampling method
(C) Degree of accuracy
(D) All of these

368. Sampling frame is a must in the case of:
(A) Random sampling
(B) Systematic sampling
(C) Stratified sampling
(D) Quota sampling

369. The relation between arithmetic, geometric mean and harmonic mean is:
(A) AM=GN=HM (B) AM>GM>HM
(C) AM<GM<HM (D) AM<HM<GM

370. Probability of a leap year will have 53 Sundays is:
(A) 1/7 (B) 2/7 (C) 2/53 (D) 52/53

371. The limiting relative frequency approach of probability is known as:
(A) Statistical probability
(B) Classical probability
(C) Mathematical probability
(D) Asymmetrical probability

372. The most frequent score in a given distribution:
(A) Median
(B) Mode
(C) Arithmetic mean
(D) Geometric mean

373. If it is known an event A has already occurred, the probability of an event E given A is:
(A) Statistical probability
(B) Classical probability
(C) Mathematical probability
(D) Asymmetrical probability

374. Accepting a hypothesis when it is false:
(A) Size of Type I error
(B) Size of Type II error
(C) Both of the these (D) None of these

375. The probability that the throw of two dice yields a total of 5 or 8 is:
(A) ¼ (B) 1/5
(C) 2/5 (D) None of these

376. The method used to identify the units or dimensions behind the measures:
(A) Multiple regression analysis
(B) Discriminate analysis
(C) Profile analysis
(D) Factor analysis

377. For testing the significance of correlation coefficient, we use:
(A) X^2 test (B) F test
(C) Z test (D) t test

378. A structural observations is used:
(A) For testing a causal hypothesis
(B) In exploratory analysis
(C) In formulative studies
(D) All of these

379. According to which process the cone of experience classify the instructural methods:
(A) Experimental correctness
(B) Difficulty in understanding
(C) Ease in use and application
(D) Level of abstraction in thinking

380. After this therefore caused by this is usually denoted to indicate:
(A) Experimental design
(B) Survey design
(C) Ex post facto design
(D) Exploratory design

381. Age is classified under:
(A) Discontinous variable
(B) Discrete variable
(C) Continuous variable
(D) Intervening variable

382. An exact test for testing the independence of attributes is a contingency table of order (2 x 2) was given by:
(A) Karl Pearson (B) Pascal
(C) De moivre (D) R A Fisher

383 .If a card is drawn from a pack of cards the probability of getting either a king or a queen is:
(A) 2/13 (B) 5/15
(C) ¼ (D) None of these

384. In 1933, the theory of testing of hypothesis was propounded by:
(A) R A Fisher (B) J Nijman
(C) E L Lehman (D) Karl Pearson

385. In case of a positively skewed distribution:
(A) Median > mean > mode
(B) Mean > median > mode
(C) Mean = median = mode
(D) None of these

386. In case of conditional probability (A and B) = ?
(A) P (B) x P (A/B)
(B) P (B) x P (A)
(C) P (A) + P (B)
(D) None of these

387. In Sequential Probability Ratio Test (SPRT), the sample size is:
(A) Fixed
(B) Fixed but small
(C) Fixed but large
(D) A random variable

388. Mention the type of correlation that operated between rurality and population density:
(A) Positive (B) Negative
(C) Accidental (D) Scattered

389. Out of two errors, the most severe is:
(A) Type I error (B) Type II error
(C) Both of these (D) None of these

390. Student's t test is applicable in case of:
(A) Small sample
(B) For sample size 5 – 30
(C) Large samples (D) None of these

391. Student's t test was invented by:
(A) R A Fisher (B) G W Snesecor
(C) W S Gosset (D) W.G. Cochran

392. Student's t test is applicable only when:
(A) Variate values are independent
(B) The variate is distributed normally
(C) The sample is not large
(D) All of these

393. The coefficient of reliability at a perfectly homogenous experimental condition will be:
(A) 0 (B) 1 (C) 0.5 (D) 2

394. The measure of dispersion which ignores signs of deviation from the central value is:
(A) Range
(B) Quartile deviation
(C) Standard deviation
(D) Mean deviation

395. The most ubiquitous kind of variance in research is:
(A) Experimental variance
(B) Sample variance
(C) Population variance
(D) Error variance

396. The probability of an impossible event is:
(A) 0 (B) 1
(C) ∞ (D) None of these

397. The number of possible samples of size two from a population of four units in a simple random sample without replacement is:
(A) 2 (B) 4 (C) 6 (D) 8

398. The range of multiple correlation coefficient R is:
(A) 0 to 1 (B) 0 to ∞
(C) -1 to +1 (D) -∞ to +∞

399. The ratio of difference between sample variance and with in sample variance follows:
(A) F- distribution (B) X^2– distribution
(C) Z – distribution (D) t- distribution

400. The suitable measure to find the average speed when time for each speed is fixed would be:
(A) Arithmetic mean
(B) Geometric mean
(C) Harmonic mean
(D) None of these

401. To control variance is the major purpose of:
(A) Statistical test (B) Research design
(C) Ex post facto method of research
(D) Interview schedule

402. Any categorization scheme must be on:
(A) One level of discourse
(B) Three level of discourse
(C) Two level of discourse
(D) Four level of discourse

403. Take the odd one out:
(A) Area sampling (B) Snowball sampling
(C) Quota sampling
(D) Incidental sampling

404. To findout the average height of plants we should use:
(A) Arithmetic mean
(B) Median
(C) Mode
(D) Harmonic mean

405. To find the average of quality prices, the measure of central tendency to be used is:
(A) Harmonic mean (B) Geometric mean
(C) Arithmetic mean (D) None of these

406. To findout the average size of shoes sold in the market, we should be used:
(A) Arithmetic mean
(B) Geometric mean
(C) Median (D) Mode

407. To test the hypothesis about proportions of items in a class, the usual test is:
(A) T- test (B) Z -test
(C) F -test (D) All of these

408. When information is to be collected from a geographically dispersed population we go for:
(A) Stratified sampling
(B) Systematic sampling
(C) Cluster sampling
(D) Multipurpose sampling

409. When degree of freedom for $x^2$is 100 or more, $x^2$is approximated to:
(A) t-distribution
(B) F- distribution
(C) Z -distribution
(D) None of these

410. Whether a test is one sided of two sided depends on:
(A) Alternate hypothesis
(B) Composite hypothesis
(C) Null hypothesis
(D) Simple hypothesis

411. Classification of research data should follow which among the following rules:
(A) Categories should be exhaustive
(B) Mutually exclusive
(C) Independent (D) All of these

412. Which of the following is the unit less measure of dispersion:
(A) Standard deviation
(B) Mean deviation
(C) Coefficient of variation
(D) Range

413. Which of the following statistics are used to divide a normal adoptor distribution into categories:
(A) Mean and Variance
(B) Mean and Standard deviation
(C) Mean and Regression
(D) Correlation and Standard deviation

414. Which among the following accounts for existing data, predict new observations and guide further research:
(A) Hypothesis (B) Theory
(C) Definition (D) Method

415. The process of drawing conclusions or inferences after a careful analysis of the collected data:
(A) Interpretation (B) Tabulation
(C) Classification (D) Evaluation

416. The basic research difference between classical research and participatory research lies in their:
(A) Control (B) Content
(C) Power (D) Substance

417. The summary and conclusion chapter in a research thesis is otherwise known as:
(A) Epilogue (B) Prologue
(C) Alphalogue (D) Symatologue

418. Opinion polls falls under:
(A) Descriptive design
(B) Explorative design
(C) Diagnostic design
(D) Experimental design

419. If all values of a variable are multiplied by a given number, which one of the following will NOT be affected?
(A) Mean (B) Median
(C) Range (D) Coefficient of variation

420. A symbol to which numerals or values are assigned:
(A) Mean (B) Mode
(C) Variable (D) Median

421. Variables are essentially a:
(A) Property (B) Symbol
(C) Constructs (D) All of these

422. A quantity which can vary from one individual to another individual is called:
(A) Mean (B) Mode
(C) Variable (D) Median

423. **Assertion (A):** Variance is always greater than the standard deviation
Reason (R): is the square of the standard deviation
(A) Both A and R are individually true and R is the correct explanation of A
(B) Both A and R are individually true but R is not the correct explanation of A
(C) A is rue but R is false
(D) A is false but R is true

424. The condition under investigation that act as potential cause of some change in behaviour:
(A) Dependent variable
(B) Independent variable
(C) Intervening variable
(D) None of these

425. Independent variables are all but not:
(A) Presumed cause
(B) Predicted to
(C) Manipulated
(D) Antecedent

426. The variable that reflects changes in the independent variable is:
(A) Dependent variable
(B) Independent variable
(C) Extraneous variable
(D) None of these

427. The manipulate variable in an experiment is:
(A) Independent variable
(B) Dependent variable
(C) Active variable
(D) Both (A) and (C)

428. All variables that can be measured in the smallest degree of finness is called:
(A) Continuous variable
(B) Discrete variable
(C) Qualitative variable
(D) Moderator variable

429. Age is classified under:
(A) Dicotonimous variable
(B) Discrete variable
(C) Continuous variable
(D) Intervening variable

430. Sex falls under which of the following type of variable:
(A) Dichotomies variable
(B) Polytomies variable
(C) Continuous variable
(D) Qualitative variable

431. Response variable are a function of:
(A) Orgasmic variable
(B) Stimulus variable
(C) Both (A) and (B) (D) None of these

432. If r is the correlation coefficients between the two variables, then:
(A) $0 < r < 1$ (B) $-1 < r < 0$
(C) $-1 < r < 1$ (D) $1 < r < 2$

433. The price index that uses base year quantities as weights is called:
(A) Passche's index
(B) Laspeyere's index
(C) Fisher's ideal index
(D) Wholesale price index

434. An example for quantitative variable from among the following is:
(A) Sex (B) Age
(C) Race (D) Religion

435. An example for discrete variable from among the following is:
(A) Age (B) Sex
(C) Intelligence (D) Height

436. The word hypothesis literary means:
(A) Less than generally held view
(B) More than generally held view
(C) Same as generally held view
(D) None of these

437. A hypothesis is essentially a:
(A) Interrogative statement
(B) Empirical statement
(C) Relational proposition
(D) Generally held view

438. Which one of the following statistical measures is based on all observation?
(A) Quartile deviation
(B) Harmonic mean
(C) Range (D) Mode

439. Which among the following is the testable form in a scientific research:
(A) Research problem
(B) Research hypothesis
(C) Research variable (D) All of these

440. The hypothesis type stated for the purpose of rejection:
(A) Null hypothesis
(B) Alternate hypothesis
(C) Parallel hypothesis(D) Both (A) and (C)

441. A hypothesis when tested true become:
(A) Principle (B) Theory
(C) Philosophy (D) Law

442. The accuracy of a measurement signifies the closeness with which a estimate approaches the:
(A) True value (B) Average value
(C) Both (A) and (B) (D) None of these

443. The collection of data in research falls under:
(A) Statistical design
(B) Observational design
(C) Operational design
(D) Sampling design

444. In the field experimentation, local control is used for:
(A) Validity of estimate of error
(B) Validity of treatment effects
(C) Diminution of error
(D) All of these

445. Experimental method of data collection among the following is:
(A) Field study (B) Sample survey
(C) Library method (D) Field experiment

446. Survey method would typically used in:
(A) Clinical assessment
(B) Opinion poll
(C) Laboratory experiment
(D) Psychoanalysis

447. Formula of linear correlation coefficient (r_{yx}) is:

(A) $r_{yx} = \dfrac{B_{yx} + b_{xy}}{2}$

(B) $r_{yx} = \dfrac{\sqrt{b_{yx} \cdot b_{xy}}}{2}$

(C) $r_{yx} = \left(\dfrac{1}{Byx} + \dfrac{1}{bxy}\right)$

(D) $r_{yx} = \dfrac{W1.b_{yx} + W2.b_{xy}}{W_1 + W_2}$

448. Survey research employs which of the following techniques:
(A) Panel technique
(B) Personal interview
(C) Controlled observation
(D) All of these

449. The best control for practice effects in an experiment is:
(A) Random selection
(B) Counter balancing
(C) Equivalent group
(D) Split half presentation

450. The R rank correlation was suggested by:
(A) James Bernoulli
(B) SD Poisson
(C) C Spearman
(D) Laplace and Gauss

451. If some of the counts are very small or zero, the most appropriate transformed variate will be
(A) $\sqrt{x+1/2}$ (B) $\sqrt{x}$
(C) $\sqrt{x+1/2}$ (D) $\sqrt{x+1}$

452. The most commonly used method in a social survey is:
(A) Observation (B) Interview
(C) Questionnaire (D) Case study

453. Interview is widely used technique in:
(A) Local studies
(B) Public studies
(C) Empirical studies
(D) Non-scientific studies

454. Assignment of values to alternatives is called:
(A) Observation (B) Judgment
(C) Measurement (D) Evaluation

455. Questionnaire involves presentation of:
(A) Oral verbal stimuli
(B) Written-verbal stimuli
(C) Oral written stimuli
(D) None of these

456. When $\beta 2 < 3$ the distribution is:
(A) Leptokurtic (B) Platykurtic
(C) Mesokurtic (D) None of these

457. In a normal distribution:
(A) Mean and median are equal
(B) Median and mode are equal
(C) All the three are equal
(D) None of these

458. If fertility variation in the field is in two direction at right angles, which of the experimental design is suitable:
(A) CRBD (B) RBD
(C) Latin square (D) CRD

459. As a general rule, the number of classes in the frequency distribution should be about:
(A) 30 (B) 15 (C) 6 (D) 20

460. In case of positively skewed distribution, the relation between mean, median and mode that hold is:
(A) Median>Mean>Mode
(B) Mean>Median>Mode
(C) Mean=Median=Mode
(D) None of these

461. Range of distribution is:
(A) The difference between largest and smallest observations
(B) Averages of all observations
(C) Middle of the series
(D) All of these

462. The second moment about mean is:
(A) $\frac{\Sigma (x - x)^2}{N}$ (B) $\frac{(x - x)^2}{N}$
(C) $\frac{(x + x)^2}{N}$ (D) None of these

463. In the regression line $Y = \alpha + \beta x$, β is called:
(A) Intercept (B) Slope of the line
(C) Mean of the X values
(D) Mean of the Y values

464. Which mean of dispersion is most affected by extreme values:
(A) Range
(B) Quartile deviation
(C) Standard deviation
(D) Mean deviation

465. During experimentation, we lost some information and we want to get idea about these values, which technique will be useful:
(A) Field plot technique
(B) Missing plot technique
(C) Seed plot technique
(D) Uniformity trial

466. F-test can be used for testing the significance of:
(A) Several differences
(B) A single difference
(C) Not used for testing the difference
(D) Two differences

467. When the fertility gradient of the field is in one direction, which design is appropriate:
(A) CRD (B) LSD
(C) RBD (D) SPD

468. X^2test is applied to:
(A) Quantitative observations
(B) Qualitative observations
(C) In both the cases
(D) In none of the cases

469. The probability of committing type I error is known as:
(A) Test of significance
(B) Level of significance
(C) Composite hypothesis
(D) None of these

470. Which type of average would be suitable to calculate the size of agricultural holding:
(A) Mode
(B) Median
(C) Weighted HM
(D) Weighted AM

471. When a frequency distribution is not symmetrical about the mean it is said to be:
(A) Kurtosis
(B) Moment
(C) Skewed
(D) None of these

472. The randomization of treatments in the field experiments is essential:
(A) The statistical procedures hold good only when randomization is done
(B) To average out the effects of environmental differences
(C) For comparison of treatment effects
(D) For comparison of plot wise differences

473. Which test is applied for the test of significance between the two variances:
(A) T-test (B) Z-test
(C) x^2test (D) F-test

474. Local control helps in:
(A) Reducing the experimental error
(B) Homogeneity of experimental units
(C) Both of these
(D) None of these

475. Which of following is not a measure of dispersion?
(A) Range
(B) Coefficient of variation
(C) Mean deviation
(D) Standard deviation

476. What is the correct formula of coefficient of variation?
(A) $\frac{SD}{AM} \times 100$ (B) $\frac{AM}{SD} \times 100$
(C) $\frac{SD}{AM}$ (D) $\frac{AM}{SD}$

477. For testing the independence of two attributes, the test used is:
(A) F-test (B) T-test
(C) x^2test (D) Z-test

478. For testing the significance of correlation coefficient, the test used is:
(A) F-test (B) T-test
(C) x^2test (D) Z-test

479. Latin square design is suitable for comparing:
(A) 2 to 4 treatment
(B) 3 treatment
(C) 15 treatment (D) 5-12 treatment

480. The reasoning process wherein the researcher collect observation and formulate hypothesis Based upon it:
(A) Inductive reasoning
(B) Deductive reasoning
(C) Both (A) and (B) (D) None of these

481. For the field experimentation the accurate idea of the fertility variation calculated with the help of :
(A) Experimental field
(B) Unformity trial
(C) Coordinated trials
(D) All of these

482. An abstract formed from generalization from particular is called:
(A) Principle (B) Hypothesis
(C) Theory (D) Concept

483. Concepts are powerful tools that make us to think:
(A) Absurdly (B) Abstractly
(C) Positively (D) Negatively

484. Do such and such things in so and so manner is the characteristic feature of:
(A) Operative definition
(B) Constitutive definition
(C) Conceptual definition
(D) Observational definition

485. Defining intelligence in terms of verbal reasoning analytical ability, comprehension and numerical capacity forms the characteristic feature of:
(A) Operative definition
(B) Constitutive definition
(C) Conceptual definition
(D) Observational definition

486. The variation due to uncontrolled factors is spoken as
(A) Standard error
(B) Experimental error
(C) Treatment effects (D) All of these

487. Pure research differs from applied research in the sense the former:
(A) Illuminates theory
(B) Uses highly technical language
(C) Shares multipurpose view
(D) Concentrate on the "Why" aspect

488. The basic difference between classical research and participatory research lies in their:
(A) Control (B) Content
(C) Power (D) Substance

489. The basic function of a good design is to:
(A) Decrease experimental variance
(B) Maximize extraneous variance
(C) Minimize error variance
(D) Explain metaphysical observation

490. Poisson distribution deals with
(A) Discrete variables
(B) Continuous variables
(C) Both (A) and (B (D) None of these

491. A symbol to which numerals or values are assigned:
(A) Mean (B) Mode
(C) Variable (D) Median

492. Variables are essentially a:
(A) Property (B) Symbol
(C) Constructs (D) All of these

493. A quantity which can vary from one individual to another individual is called:
(A) Mean (B) Mode
(C) Variable (D) Median

494. Correlation does not demonstrate:
(A) Causation (B) Effect
(C) Reason (D) None of these

495. If two factors are involved in experimentation and both the factors requires long plot sizes, the design suited for this will be:
(A) Split plot (B) Strip plot
(C) CRD (D) RBD

496. Median of seven observations 9, 3, 5, 4, 2, 3, 2 is:
(A) 2 (B) 3 (C) 4 (D) 5

497. The condition under investigation that act as potential cause of some change in behavior:
(A) Dependent variable
(B) Independent variable
(C) Intervening variable (D) None of these

498. Two unbiased coins are tossed once What is the probability of getting two heads?
(A) 0 (B) ¼ (C) ½ (D) 1

499. If a<b and b<c, then a<c denotes:
(A) Transitive Property of Inequality
(B) Addition Property of Equality
(C) Cumulative Property of Addition
(D) Multiplication Property of Equality

500. Independent variables are all but not:
(A) Presumed cause (B) Predicted to
(C) Manipulated (D) Antecedent

501. Harmonic mean gives weightage to:
(A) Small values (B) Positive values
(C) Large values (D) Negative values

502. A constant value of 50 is subtracted from each observation of a set, the mean of the set is:
(A) Increased by 50 (B) Decreased by 50
(C) Not affected (D) Zero

503. A relationship in correlation coefficient that does not form a straight line is:
(A) Linear relationship
(B) Negative relationship
(C) Curvilinear relationship
(D) Positive relationship

504. The variable that reflects changes in the independent variable is:
(A) Dependent variable
(B) Independent variable
(C) Extraneous variable
(D) None of these

505. The rule of correspondent between a dependent variable and an independent variable:
(A) Function (B) Equation
(C) Argument (D) Premise

506. All variables that can be measured in the smallest degree of fitness is called:
(A) Continuous variable
(B) Discrete variable
(C) Qualitative variable
(D) Moderator variable

507. The most important measure of dispersion used in industry for quality control:
(A) Standard deviation
(B) Mean deviation
(C) Coefficient of variation
(D) Range

508. What is the mean of 5, 13, 0, 6?
(A) 3 (B) 4 (C) 6 (D) 8

509. The x^2(chi-square) test for goodness of fit and independence of attributes come under:
(A) Parametric tests
(B) Non-parametric test
(C) Both (A) and (B)
(D) Not classified

510. When the same interaction is confounded in all the replications, it is known as:
(A) Partial confounding
(B) Complete confounding
(C) Both (A) and (B)
(D) None of these

511. A variable selected to prevent alteration of the relationship between primary independent variable with the dependent variable:
(A) Active variable
(B) Intervening variable
(C) Attribute variable
(D) Moderator variable

512. The values of regression coefficient lies between:
(A) $-\infty$ to $+\infty$ (B) 0 to 1
(C) -1to +1 (D) 0 to ∞

513. How many number of treatments can be adopted in LSD:
(A) 5 to 8 (B) 5 to 12
(C) Both (A) and (B) (D) Not estimated

514. For x^2 no theoretical cell frequency should be small, it should be at least:
(A) 2 (B) 5 (C) 8 (D) 12

515. Sex falls under which of the following type of variable:
(A) Dichotomies variable
(B) Polytomies variable
(C) Continuous variable
(D) Qualitative variables

516. When the distribution of data is continuous, which distribution is applicable:
(A) Normal distribution (B) Binomial
(C) Poisson
(D) None of these

517. Degree of freedom of x^2test for 4 x 3 contingent table is:
(A) 12 (B) 6 (C) 1 (D) 7

518. If all the variate values are negative, the standard deviation will be:
(A) Negative (B) Positive
(C) No sign (D) Cannot calculate

519. A true dicotominous variable among the following is:
(A) Male-female
(B) Literate-illiterate
(C) Intelligent-unintelligent
(D) Introvert-extrovert

520. Take the odd man out:
(A) Independent variable
(B) Active variable
(C) Stimulus variable
(D) Orgasmic variable

521. Response variable are a function of:
(A) Orgasmic variable
(B) Stimulus variable
(C) Both (A) and (B)
(D) None of these

522. A variance produced in the dependent variable by the manipulation of the active variable by the experimenter:
(A) Experimental variance
(B) Extraneous variance
(C) Error variance (D) None of these

523. Which among the following is not a character of error variance:
(A) Self compensation
(B) Unpredictability
(C) Generalizability (D) Skewedness

524. A hypothesis is essentially a:
(A) Interrogative statement
(B) Empirical statement
(C) Relational proposition
(D) Generally held view

525. Hypothesis that states the negation of the assertion is commonly called as:
(A) Null hypothesis
(B) Parallel hypothesis
(C) Alternate hypothesis
(D) Both (B) and (C)

526. The bridge between theory and empirical enquiry is connected through:
(A) Research principle
(B) Research hypothesis
(C) Research problem
(D) Research output

527. Finding the cause of behavior from number of observation is called;
(A) Inductive reasoning
(B) Horizontal reasoning
(C) Vertical reasoning
(D) Deductive reasoning

528. When we want average of rates of change or ratios or index number, which measure of central tendency is suitable:
(A) Geometric mean (B) Mode
(C) Median
(D) Arithmetic mean

529. Under non-parametric the comparison of several related samples is tested by:
(A) Quade test (B) Friedmann test
(C) F-test (D) Both (A) and (B)

530. In Confounding, generally which order of interaction is confounded:
(A) Lower order (B) First order
(C) Higher order (D) All of these

531. A tentative explanation of an event or observation:
(A) Variable (B) Hypothesis
(C) Argument (D) Principle

532. In a experiment, a treatment requires large area and b treatment requires smaller area (different precision), which design is most appropriate:
(A) Split Plot Design (B) LSD
(C) RBD
(D) Strip Plot Design

533. Whether a test is one tailed or two tailed depends on:
(A) Alternate hypothesis
(B) Null hypothesis
(C) Composite hypothesis
(D) Simple hypothesis

534. The research method which focus independently on different aspects of a single subject:
(A) Field studies (B) Case studies
(C) Experimental studies
(D) None of these

535. Local control is not applied for the design:
(A) RBD (B) CRD
(C) LSD (D) SPD

536. Which test is used to test of significance of the difference between two means:
(A) T-test (B) F-test
(C) Z-test (D) x^2test

537. Normal distribution is due to the work of
(A) James Bernoulli
(B) SD Poisson
(C) Laplace and Gauss
(D) Demoivre

538. Which among the following can be thought as natural clinical tests of the effects of unusual variables:
(A) Observational variables
(B) Empirical studies
(C) Experimental studies
(D) Case studies

539. Which among the following is the main disadvantage of mailed questionnaire:
(A) Dishonest response
(B) Low return rate
(C) Time consuming and costly
(D) Postal lose

540. If a constant 50 is subtracted from each of the value of X and Y, the regression coefficient is:
(A) Reduced by 50 (B) 1/50th of original
(C) Increased by 50 (D) Not changed

541. The method of assembling correspondence without experimental control over then is called:
(A) Dispersion (B) Correlation
(C) Central tendency (D) None of these

542. The intelligency, ability or efficiency can be measured by use of:
(A) Median (B) Mode
(C) Weighted HM (D) Geometric mean

543. When there are several factors with different levels to be experimented simultaneously with the same level of precision, which design is most appropriate:
(A) CRD (B) LSD
(C) RBD (D) Factorial design

544. The collection of data in research falls under:
(A) Statistical design
(B) Observational design
(C) Operational design
(D) Sampling design

545. The interview type where in conversation are informal and the interviewee determines the subjects for discussion:
(A) Direct observation
(B) Projective test
(C) Unstructured interview
(D) Structured interview

546. What is/are the basic principles of field experimentation:
(A) Randomization (B) Replication
(C) Local control (D) All of these

547. For paired observation, which test is used for testing the significance of a mean difference:
(A) Paired t-test (B) T-test
(C) F-test (D) x^2test

548. Binomial distribution is a very useful distribution for dealing with ———— variates:
(A) Continuous (B) Discrete
(C) Both (A) and (B) (D) None of these

549. The idea of randomization was put forwarded by:
(A) Chapman and Greenwood
(B) Kerlinger
(C) Ronald A Fisher
(D) Sukathmae

550. The assignment of numbers to certain characteristics of objects according to rules:
(A) Measurement (B) Evaluation
(C) Observation (D) Survey

551. The most frequent occurred value of data or whose frequency is maximum is known as:
(A) Arithmetic mean (B) Mode
(C) Median (D) GM

552. The degree of linear relationship between variables are expressed through:
(A) Coefficient of correlation
(B) Standard deviation
(C) Arithmetic mean
(D) None of these

553. If X is multiplied by 10 and Y by 20, b_{xy}, the regression coefficient is:
(A) Same as b_{xy} (B) Half by b_{xy}
(C) 4 times of b_{xy} (D) Double of b_{xy}

554. The exact procedures used to represent a concept:
(A) Operational definition
(B) Constitutive definition
(C) Conceptual definition
(D) Observational definition

555. A functional variate for establishing a parameter is called:
(A) An estimate (B) An estimator
(C) A frame (D) A statistic

556. The most eliminatory level of measurement is:
(A) Nominal (B) Ordinal
(C) Interval (D) Ratio

557. In case of a positively skewed distribution:
(A) Median> mean > mode
(B) Mean > Median > mode
(C) Mean = median = mode
(D) None of these

558. Transitivity postulate find its application right from:
(A) Nominal level (B) Ordinal level
(C) Interval level (D) Ratio level

559. The lowest level of measurement having an arbitrary zero is:
(A) Nominal level (B) Ordinal level
(C) Interval level (D) Ratio level

560. The level of measurement that goes smooth with non-parametric statistics is:
(A) Nominal-Ordinal
(B) Ordinal – Interval
(C) Nominal - Interval
(D) Ordinal – Ratio

561. Intelligence, aptitude and personality measures strictly speaking falls under:
(A) Nominal level
(B) Ordinal level
(C) Interval level
(D) Ratio level

562. If the measure of kurtosis is greater than 3, then the curve is:
(A) Platykrurosis (B) Mesokurtosis
(C) Leptokurtosis (D) Normal

563. The experimental design wherein the heterogeneity in the experimental units are in two directions:
(A) CRD (B) RBD
(C) LSD (D) BIBD

564. Q methodology was put forwarded by:
(A) Fisher (B) Osgood
(C) Mereno (D) William Stephenson

565. Reflective, transitive and symmetrical relationship is the characteristic feature of:
(A) Nominal level (B) Ordinal level
(C) Interval level (D) Ratio level

566. To test the agreement between observed frequencies and expected frequencies, we use:
(A) F test (B) ANOVA
(C) X^2 test (D) t- test

567. Height and weight are obvious examples for:
(A) Nominal level (B) Ordinal level
(C) Interval level (D) Ratio level

568. The error degree of freedom. for a RBD experiment with 5 treatments and 3 replications is:
(A) 15 (B) 12 (C) 8 (D) 14

569. Paper pencil test is otherwise called as:
(A) Verbal test (B) Non verbal test
(C) Performance test
(D) Non-language test

570. To judge the capacity of individuals in synthesis of ideas and critical thinking:
(A) New type test (B) Paper pencil test
(C) Free answer test (D) Selection type test

571. The item type best suited for measuring individual achievement is:
(A) Matching item
(B) Multiple choice item
(C) Two alternative item (D) None of these

572. Degree of freedom is related to the number of:
(A) People involved
(B) Hypothesis under set
(C) Independent observations
(D) None of these

573. The chance for guessing from a two alternative item is:
(A) 100 per cent (B) 75 per cent
(C) 50 per cent (D) 25 per cent

574. Thematic Apperception Test is used to study:
(A) Leadership behavior
(B) Individual creativity
(C) Conflict management
(D) Achievement motivation

575. When the experimental units are homogenous, the design used is:
(A) CRD (B) RBD
(C) LSD (D) None of these

576. An example for constructive techniques from among the following is:
(A) Rorschach Projection Test
(B) Thematic Apperception Test
(C) Sentence Completion Test
(D) Word Association Test

577. Projetcile test are:
(A) Objective test (B) Subjective test
(C) Attitude test (D) Both (B) and (C)

578. Which design is most appropriate for the laboratory experiments:
(A) CRBD (B) CRD
(C) LSD (D) Factorial RBD

579. The stimulus used in a projectile test must be:
(A) Structured ambiguous
(B) Unstructured ambiguous
(C) Structured unambiguous
(D) Unstructured unambiguous

580. For testing the significance of difference between two means, we use:
(A) F- test (B) Z- test
(C) X^2 - test (D) t -test

581. The lowest level of measurement that facilitate us to go for addition and subtraction is:
(A) Nominal level (B) Ordinal level
(C) Interval level (D) Ratio level

582. Expressive test gives basic emphasis on:
(A) Product (B) Process
(C) Input (D) Both (A) and (B)

583. The principle of experimental design include:
(A) Randomization (B) Local control
(C) Both (A) and (B) (D) None of these

584. The method used to observe and measure the psychological meaning of concept:
(A) Somatic differentiation
(B) Projectile technique
(C) Content analysis
(D) Q methodology

585. The method of analyzing communication in a systematic, objective and quantitative manner to measure variable is called:
(A) Content analysis
(B) Process analysis
(C) Hypothesis analysis
(D) Variable analysis

586. Correlation coefficient lies between:
(A) - ¥ < r < 0 (B) 0 £ r £ 1
(C) -1 < r < + 1 (D) -¥ £ r £ ¥

587. Which measure of central tendency is appropriate for index numbers:
(A) Harmonic mean (B) Arithmetic mean
(C) Mode (D) Geometric mean

588. If there are 5 varieties and 4 replications to each, the error degree of freedom for RBD will be:
(A) 20 (B) 19 (C) 12 (D) 9

589. Ogive is a method of:
(A) Diagrammatic representation
(B) Graphic representation
(C) Both (A) and (B)
(D) None of these

590. Find the modal value in 9, 13, 18, 14, 21, 18 and 13:
(A) 9 (B) 13 (C) 18 (D) 21

591. Who has developed the technique of analysis of variance?
(A) Cox (B) Fisher
(C) Kempthorne (D) Rao

592. To study the average size shoes sold in the market, we use:
(A) Mode (B) Harmonic mean
(C) Arithmetic mean (D) Geometric mean

593. The smallest unit of content analysis is:
(A) Theme (B) Words
(C) Characters (D) Items

594. The t test tends to be Z test when sample size becomes more than:
(A) 100 (B) 50 (C) 30 (D) 5

595. The computer programme used for content analysis:
(A) General inquirer (B) Dream weaver
(C) Mind mapper (D) Coral draw

596. To quantify readability in content analysis we go for:
(A) Fog index (B) Log index
(C) Content index (D) Variable index

597. Who is the author of the book Sampling Techniques':
(A) W G Cochran (B) A L Bowley
(C) R L Ackoff (D) Wison M.

598. Which one of the following is generally used for testing the equality of two variances?
(A) T - test (B) Z- test
(C) x^2 – test (D) F – test

599. Sum of absolute deviations about median is:
(A) Least (B) Greatest
(C) Zero (D) Equal

600. Naturalistic observation only provides description of:
(A) Experience (B) Feeling
(C) Behaviour (D) None of these

601. The non-parametric test used for two independent samples is:
(A) T-test (B) Z-test
(C) F-test (D) Mann-Whitney –test

602. Which design gives precision high enough to reduce the standard error to less than 1%:
(A) SPD (B) Strip plot design
(C) RBD (D) LSD

603. In confounding, the precision on the main effects and certain interactions of lower order:
(A) Increases (B) Decreases
(C) Remain same (D) None of these

604. For the comparison of two correlated samples under non-parametric alternative to paired F-test is:
(A) Z-test
(B) F-test
(C) Wilcoxon Signed –rank test
(D) Kruskal-Wallis H test

605. The three R's in the SQ3R strategy of comprehension does not include:
(A) Read (B) Recite
(C) Review (D) Reward

606. When 5 is added to every observation, which of the one of the following does:
(A) Mean (B) Median
(C) Mode (D) Variance

607. A method in which people are asked to respond to questions by expressing their personal evaluations:
(A) Attitude scale (B) Interview
(C) Self report method (D) All of these

608. Mode can be calculated by the formula:
(A) Mode = 4 median – 3 mean
(B) Mode = 3 median – 2 mean
(C) Mode = median –mean
(D) None of these

609. For testing HO : m = 0 in a normal distribution N(m, 1), which one of the following test is used?
(A) T – test (B) Normal – test
(C) x^2 – test (D) F – test

610. Statistical conclusion are always:
(A) Absolutely true (B) Absolutely wrong
(C) Either (A) or (B) (D) True on average

611. The observational procedure in explorative studies is more likely:
(A) Structured (B) Unstructured
(C) Partially structured
(D) None of these

612. A low error variance results in a:
(A) High validity (B) Low validity
(C) High reliability (D) Both (A) and (C)

613. The alpha coefficient of variability indicates:
(A) Temporal stability
(B) Internal consistency
(C) Both (A) and (B)
(D) None of these

614. F test is used to test:
(A) Ratio of two variances
(B) Ratio of two means
(C) Square of normal variate
(D) None of these

615. Equality of several variance is done by:
(A) F (B) X^2
(C) Z (D) None of these

616. The concept of semantic differentiation was discovered by:
(A) Fisher (B) Osgood
(C) Mereno (D)William Stephenson

617. Which of the following is not a unit of content analysis:
(A) Themes (B) Readability
(C) Characters (D) Items

618. The lowest common denominator of a test is otherwise called:
(A) Stem (B) Item
(C) Root (D) Base

619. A high inter-correlation among different items of a test indicates:
(A) High validity (B) Low validity
(C) Low reliability (D) High reliability

620. The arithmetic mean of the Poisson distribution is:
(A) 3m2+m (B) √ m
(C) m (D) 1/m

621. For test of goodness of fit, which statistical test is applied:
(A) x^2test (B) F-test
(C) Z-test (D) T-test

622. The allocation of the treatments to the different experimental units in a random manner is known as:
(A) Randomization (B) Replication
(C) Local control (D) Sampling

623. The hypothesis of no difference is known as:
(A) Simple hypothesis
(B) Null hypothesis
(C) Composite hypothesis
(D) Alternative hypothesis

624. The external validity of the findings are the lowest in:
(A) Experimental design
(B) Explorative design
(C) Diagnostic design
(D) Survey design

625. The reliability of test with a high error score will be:
(A) Constant (B) High
(C) Low (D) Zero

626. The type of validity attached to an instrument when it appears to measure what it sets out to measure:
(A) Face validity (B) Content validity
(C) Criteria related validity
(D) Construct validity

627. Hypothesis that stated the negation of the assertion is commonly called as:
(A) Null hypothesis
(B) Parallel hypothesis
(C) Alternate hypothesis
(D) Both (A) and (C)

628. Coefficient of variation is useful for comparing:
(A) Comparison between different populations
(B) Same population
(C) Both of these
(D) None of these

629. Attitude scale giving emphasis for intensity of altitude expression:
(A) Cumulative scale
(B) Equal appearing interval scale
(C) Summated rating scale
(D) None of these

630. The research method which focus independently on different aspects of a single subject:
(A) Field studies (B) Case studies
(C) Experimental studies
(D) None of these

631. The type of validity that assume most significance when some set standards are compared with the measuring score:
(A) Face validity (B) Content validity
(C) Criteria related validity
(D) Construct validity

632. The coefficient of reliability at a perfectly homogenous experimental condition will be:
(A) Zero (B) One
(C) 0.5 (D) Two

633. The parent population in analysis of variance is assumed to follow:
(A) Normal distribution
(B) Binomial distribution
(C) Poisson distribution
(D) None of these

634. When degree of freedom of x^2test is more than 100, t test is approximated to:
(A) F-test (B) X^2test
(C) Z- test (D) None of these

635. The statement "All that is reliable is valid but all those is valid is not reliable" is:
(A) Completely false
(B) Partially true
(C) Completely true (D) Partially false

636. The range of correlation coefficient lies between
(A) -1to+1 (B) 0 to 1
(C) 0 to ∞ (D) 0 to 100

637. Arithmetic mean and variance are always equal in:
(A) Normal distribution
(B) Poisson distribution
(C) Binomial (D) All of these

638. The property of a test that gives us the same score when corrected by different people:
(A) Validity (B) Reliability
(C) Objectivity (D) Credibility

639. Compared to laboratory experiment, the internal and external validity of field experiment are respectively:
(A) High-Low (B) Low-High
(C) Low-Low (D) High-High

640. The value of x^2ranges from .... and is always positive:
(A) -1 to +1 (B) 0 to 2
(C) 0 to ∞ (D) None of these

641. External validity is the highest for:
(A) Field experiment (B) Field study
(C) Sample survey (D) Case study

642. A variance shared by two or more variables:
(A) Covariance (B) Correlation
(C) Common factor variance
(D) None of these

643. Jury opinion is used to establish:
(A) Reliability (B) Validity
(C) Creativity (D) Adjustability

644. A sample of 12 specimens taken from a normal population is expected to be 50 kg. To test H0:χ = 50 Vs Hi: :χ not = 50, the test used is:
(A) Z-test (B) T-test
(C) x^2test (D) F-test

645. An example for summated rating scale from among the following:
(A) Guttman scale (B) Thurstone scale
(C) Adorno scale (D) Likert scale

646. A test used to compare the variance of two independent samples:
(A) F- test (B) X^2 - test
(C) Z- test (D) None of these

647. The area under standard normal curve beyond Z= + 1.96 is:
(A) 95 per cent (B) 99 per cent
(C) 5 per cent (D) 1 per cent

648. Reduction in the size of the test results in:
(A) Decrease in its power
(B) Increase in its power
(C) No change in power
(D) All of these

649. To test the hypothesis about proportions of items in a class, we use:
(A) F- test (B) X^2 -test
(C) Z -test (D) None of these

650. Analysis of variance utilizes:
(A) F- test (B) x^2test
(C) Z- test (D) t- test

651. The ratio of difference between sample variance and with in sample variance follows:
(A) Z distribution (B) t distribution
(C) F distribution (D) X^2distribution

652. Measures of dispersion includes the range and:
(A) Standard deviation
(B) Mode
(C) Median (D) None of these

653. An observable phenomena substituted for an less observable phenomena is called:
(A) Index (B) Ratio
(C) Linkage (D) Concept

654. An example for summated rating scale from among the following:
(A) Guttman scale (B) Thurstone scale
(C) Adorno scale (D) Likert scale

655. A tentative explanation of an even or observation:
(A) Variable (B) Hypothesis
(C) Argument (D) Principle

656. The word 'scale' is correctly used in which of the following cases:
(A) Achievement scales (B) Attitude scales
(C) Intelligent scales (D)Both (A) and (B)

657. The basic emphasis while placing item in the cumulative scale is on:
(A) Scalability of set of items
(B) Items and their placement
(C) Scale position of individuals
(D) Both (A) and (C)

658. For calculation of x^2, N should be at least:
(A) 20 (B) 30 (C) 40 (D) 50

659. Attitude scale giving emphasis for intensity of attitude expression:
(A) Cumulative scale
(B) Equal appearing interval scale
(C) Summated rating scale
(D) None of these

660. Guttman scale is basically a:
(A) Summated rating scale
(B) Equal appearing interval scale
(C) Cumulative scale (D) None of these

661. The difficulty indices of items producing maximum variance among examinees lies between:
(A) 0.2 to 0.4 (B) 0.4 to 0.6
(C) 0.6 to0.8 (D) 0.8 to 1.0

662. The probability of an impossible event is:
(A) 0 (B) 1 (C) ∞ (D) None of these

663. Analysis of variance utilizes:
(A) F- test (B) X^2test
(C) Z- test (D) T-test

664. A lower scale value in Thurstone and Clave scale indicates:
(A) Positive attitude
(B) Negative attitude
(C) Neutral attitude (D) Zero attitude

665. Tests designed to measure individual potential and to predict future achievements:
(A) Attitude tests (B) Achievement tests
(C) Aptitude tests (D) Personality tests

666. Kruskall Wallis analysis of data is meant for:
(A) No classification
(B) One way classification
(C) Two way classification
(D) None of these

667. The first step in summarizing the data is:
(A) Tabulation (B) Classification
(C) Interpretation (D) Observation

668. The decision criterian in SPRT depends in the function of:
(A) Type I error (B) Type II error
(C) Type I and II error (D) None of these

669. An exact test for testing the independence of attributes in a contingency table of order (2 x 2) is based on the calculation of:
(A) The value of x^2
(B) Probability of configuration
(C) Value of Z (D) None of these

670. Area of critical region depends upon:
(A) Size of type I error
(B) Size of type II error
(C) Value of statistics
(D) Number of observation

671. For any two evens A and B, P (A-B) is equal to:
(A) P (A) – P (B) (B) P (B) – P (A)
(C) P (B) – P (AB) (D) P (A) – P (AB)

672. The measure of central tendency to be used to study average rate of change in the population is:
(A) Arithmetic mean (B) Mode
(C) Geometric mean (D) Median

673. An interval in which scores are grouped is called:
(A) Point (B) Value
(C) Class interval (D) None of these

674. The first step in analysis of data is:
(A) Manipulation (B) Categorization
(C) Interpretation (D) Summarization

675. Equality of several normal population means can be tested by:
(A) F- test (B) X^2test
(C) Z- test (D) t- test

676. The t test is used to test the significance of:
(A) Two means of large sample
(B) Two means of small sample
(C) Correlation coefficient
(D) Both (B) and (C)

677. The reliability method that take into consideration both temporal stability and internal consistency dimensions:
(A) Test re-test method
(B) Odd even method
(C) Parallel foams method
(D) Split half method

678. If there is an increase in a series at constant rate, the graph will be:
(A) Convex curve (B) Concave curve
(C) Parabola (D) Straight line

679. Assessment of numbers of the football players of a team is an example of:
(A) Ratio measurement
(B) Internal measurement
(C) Ordinal measurement
(D) Nominal measurement

680. The hypothesis under test is:
(A) Simple hypothesis
(B) Alternate hypothesis
(C) Null hypothesis (D) None of these

681. The key of accurate sampling is:
(A) Randomness (B) Opinion
(C) Prediction (D) All of these

682. Cross section of population is represented less in:
(A) Simple random sampling
(B) Multistage sampling
(C) Cluster sampling
(D) Stratified random sampling

683. The sample type wherein similarity between immediate neighbours and difference between distinct neighbours are both taken care of:
(A) Stratified sampling
(B) Systematic sampling
(C) Cluster sampling
(D) Multipurpose sampling

684. If A and B are two events which have no points in common, the events A and B are:
(A) Mutually exclusive
(B) Independent
(C) Complimentary to each other
(D) None of these

685. The best sample size from a perfectly homogenous population in respect of a character:
(A) A large sample (B) A small sample
(C) A single item (D) No item

686. Mean and variance of a standard normal distributing respectively:
(A) 0 and 1 (B) 1 and 0
(C) Both (A) and (B) (D) None of these

687. If A is an event the probability of A given A is equal to:
(A) Zero (B) One
(C) Infinite (D) Indeterminate

688. When r = +1, the two regression lines are:
(A) Coincident
(B) Parallel
(C) Perpendicular to each other
(D) Perpendicular to X axis

689. For testing the independence of two attributes, we use:
(A) F- test (B) Z- test
(C) x^2-test (D) t- test

690. The sampling type that had developed from the principles of sociometry:
(A) Area sampling
(B) Snowball sampling
(C) Quota sampling
(D) Incidental sampling

691. Square of a standard normal variate is:
(A) F (B) X^2
(C) Z (D) None of these

692. Pick the odd man out:
(A) x^2 - test
(B) Kruskal Wallis test H test
(C) Friedman test
(D) Z - test

693. Statistical conclusions are always:
(A) Absolutely true
(B) Absolutely wrong
(C) Either (A) or (B)
(D) True on average

694. Which of the following is not a discrete distribution:
(A) Binomial (B) Poisson
(C) Normal (D) Both (A) and (B)

695. Area of critical region depends upon:
(A) Size of Type I error
(B) Size of Type II error
(C) Both of the these
(D) None of these

696. Mean deviation is minimum when devi-ants are taken from:
(A) Mean (B) Median
(C) Mode (D) None of these

697. Who is the author of the book 'Sampling Techniques':
(A) W.G. Cochran (B) A.L.Bowley
(C) R.L. Ackoff (D) Wilson M

698. The AM of the age of 20 students in a class is 15 years. When the teachers age also added, the mean increase by one year. The age of the teacher is:
(A) 16 years (B) 26 years
(C) 36 years (D) 46 years

699. The sampling type that had developed from the principles of sociometry:
(A) Area sampling (B) Snowball sampling
(C) Quota sampling
(D) Incidental sampling

700. Sampling size is influenced by:
(A) Heterogeneity of population
(B) Sampling method
(C) Degree of accuracy
(D) All of these

701. The measure of dispersion which ignores signs of deviation from the central value:
(A) Range
(B) Quartile deviation
(C) Standard deviation
(D) Mean deviation

702. Range is not a reliable measure of:
(A) Mean (B) Dispersion
(C) Median (D) Mode

703. The key of accurate sampling is:
(A) Randomness (B) Opinion
(C) Prediction (D) All of these

704. The most important measure of dispersion used in industry for quality control:
(A) Standard deviation
(B) Mean deviation
(C) Coefficient of variation (D) Range

705. The population for sampling is defined in terms of:
(A) Time (B) Element
(C) Extent (D) All of these

706. The best measure of dispersion among the following is:
(A) Standard deviation
(B) Mean deviation
(C) Coefficient of variation (D)Range

707. The range of variable ratio F is:
(A) -1 to +1 (B) $-\infty$ to $+\infty$
(C) 0 to ∞ (D) 0 to 1

708. Empirical relation between quartile deviation and standard deviation is:
(A) 3QD = 2SD (B) 4QD = 4SD
(C) 6QD = 5SD (D) 5QD = 4SD

709. The average of sum of squares of deviation about the mean is called:
(A) Absolute deviation (B) Variance
(C) Standard deviation
(D) Mean deviation

710. The sampling procedure most commonly used in order to ensure a representative sample:
(A) Random sampling
(B) Systematic sampling
(C) Cluster sampling
(D) Quota sampling

711. Which of the following is the unit less measure of dispersion:
(A) Standard deviation (B) Mean deviation
(C) Coefficient of variation (D) Range

712. An abstract formed generalization for particular is called:
(A) Principle (B) Hypothesis
(C) Theory (D) Concept

713. To find the average speed of the vehicle, when the same distance is covered with different speeds, the suitable measure is:
(A) AM (B) HM
(C) GM (D) None of these

714. The most frequent score in a given distribution:
(A) Mean (B) Mode
(C) Median (D) None of these

715. The first step is summarizing the data is:
(A) Tabulation (B) Classification
(C) Graphical representation
(D) Collection

716. The value of coefficient of concordance ranges between:
(A) 0 + 1 (B) $-\infty$ to $+\infty$
(C) - 1 to +1 (D) -1 to 0

717. The unit of correlation coefficient is:
(A) Kg/cc (B) Per cent
(C) No unit (D) None of these

718. Level of significance is the probability of:
(A) Type I error
(B) Type of II error
(C) Non committing error
(D) All of these

719. Classification of research data should follow which among the following rules:
(A) Categories should be exhaustive
(B) Mutually exclusive
(C) Independent (D) All of these

720. Which among the correlation coefficient represent the strongest relationship:
(A) - 0.86 (B) +0.86
(C) +0.10 (D) +0.09

721. The variance is the square of :
(A) Mean deviation
(B) Standard deviation
(C) Range (D) Correlation

722. Regression coefficient is independent of:
(A) Origin
(B) Scale
(C) Both origin and scale
(D) Neither origin nor scale

723. When is the normality of populations is used in analysis of experiments?
(A) At the time of planning of experiments
(B) At the time of computing various sums of squares
(C) At the time of constructing ANOVA table
(D) At the time of testing of hypothesis

724. The basic unit containing the elements of the population to be sampled:
(A) Sampling unit (B) Sampling frame
(C) Sampling element
(D) None of these

725. The process of drawing of conclusions or inferences after a careful analysis of the collected data:
(A) Interpretation (B) Tabulation
(C) Classification (D) Evaluation

726. What is the main purpose of local control?
(A) To reduce the number of plots
(B) To reduce the number of replications
(C) To reduce the experimental error
(D) To reduce the degrees of freedom

727. Regression coefficient takes values:
(A) $-\infty$ to 1 (B) 1 to ∞
(C) $-\infty$ to $+\infty$ (D) -1 to +1

728. Mean deviation is the least when calculated about (minimal property) :
(A) Arithmetic mean (B) Median
(C) Mode (D) Geometric mean

729. If X and Y are independent, the value of regression coefficient β_{xy} is equal to:
(A) 0 (B) 1
(C) a (D) Any positive value

730. Which mean of dispersion is most affected by extreme values:
(A) Range
(B) Quartile deviation
(C) Standard deviation
(D) Mean deviation

731. An example for quantitative variable from among the following is:
(A) Sex (B) Age
(C) Race (D) Religion

732. The word statistics is used as:
(A) Singular
(B) Plural
(C) Singular and plural both
(D) None of these

733. What is the probability that the sum of points in a toss of two dice is 14?
(A) 0 (B) 1/6
(C) 1/36 (D) 1

734. In CRD, number of treatments are 5 and number of replications are 6. What is the error degrees of freedom:
(A) 20 (B) 25 (C) 29 (D) 4

735. The critical value for a t-distribution with 120 degrees of freedom at 5% level of significance in a two tailed test is:
(A) 1.24 (B) 1.98
(C) 2.54 (D) 3.13

736. If p is the probability of any event, which one of the following is true?
(A) $-x < p < 0$ (B) $0 < p < 1$
(C) $1 < p < x$ (D) $-x < p < x$

737. The class interval of this grouped data 10-19, 20-29, 30-39, 40-49 is:
(A) 9 (B) 10 (C) 14.5 (D) 39

738. An example of discrete variable from among the following is:
(A) Age (B) Sex
(C) Intelligence (D) Height

739. Scientific data must be:
(A) Objective (B) Measurable
(C) Predictable (D) All of these

740. If the reliability coefficient of a test is one, the standard error will be:
(A) +1 (B) -1
(C) 0.5 (D) 0

741. All the three principles of randomization, replication and local control are used in:
(A) CRD only
(B) RBD only
(C) CRD and RBD both
(D) None of these

742. If X is multiplied by 10 and Y by 20, b_{xy}, the regression coefficient is:
(A) Same as b_{xy} (B) Half by b_{xy}
(C) 4 times of b_{xy} (D) Double of b_{xy}

743. Mean and variance of a standard normal distribution are respectively:
(A) 0 and 1
(B) 1 and 0
(C) Both (A) and (B)
(D) None of these

744. The second central moment (about the arithmetic mean) is always equal to:
(A) Variance
(B) Mean deviation
(C) Standard deviation
(D) None of these

745. A normal variate takes values in the range of:
(A) 0 to 1 (B) 0 to ∞
(C) $-\infty$ to $+\infty$ (D) -11.96 to +1.96

746. Correlation does not demonstrate:
(A) Causation (B) Effect
(C) Reason (D) None of these

747. Unit less measure of dispersion is
(A) Standard deviation (B) Mean deviation
(C) Range (D) CV

748. Standard Error can be:
(A) 0 (B) 1 (C) >0 (D) All of these

749. The range of multiple correlation coefficient R is:
(A) 0 to 1 (B) 0 to ∞
(C) -1 to +1 (D) $-\infty$ to $+\infty$

750. The coef..cient of variation can be calculated by using the formula:
(A) Standard deviation/arithmetic mean x 100
(B) Arithmetic mean/standard deviation x 100
(C) Square root of standard deviation/arithmetic mean x 100
(D) Standard deviation/geometric mean x 100

751. When r = +1, the two regression lines are:
(A) Coincident (B) Parallel
(C) Perpendicular to each other
(D) Perpendicular to X axis

752. If the correlation coefficient between x and y is r, the correlation coefficient of y and x is:
(A) -r (B) r^2
(C) r (D) 1/r

753. The value of X^2 always lies between:
(A) 0 to 1 (B) 0 to
(C) -1 to +1 (D) ∞ ? to $+\infty$

754. For testing the significance of correlation coefficient, we use:
(A) t- test (B) X^2 - test
(C) Z- test (D) F-test

755. To test the agreement between observed frequencies and expected frequencies, we use:
(A) t- test (B) X^2 - test
(C) Z- test (D) F-test

756. The t test tends to be Z test when sample size becomes more than:
(A) 100 (B) 50 (C) 30 (D) 5

757. The parent population in analysis of variance is assumed to follow:
(A) Normal distribution
(B) Binomial distribution
(C) Poisson distribution
(D) None of these

758. Kruskall Wallis analysis of data is meant for:
(A) No classification
(B) One way classification
(C) Two way classification
(D) None of these

759. Ogives for more than type and less than type distributions intersect at:
(A) Mean (B) Median
(C) Mode (D) None of these

760. Hypothesis that the population variance has a specified value can be tested by:
(A) F- test (B) Z- test
(C) x^2 -test (D) t- test

761. The simplest way of visualizing a correlation is to construct a:
(A) Graph (B) Table
(C) Scatter diagram (D) None of these

762. Equality of several normal population means can be tested by:
(A) t- test (B) X^2test
(C) Z - test (D) F-test

763. Formula for calculating coefficient of variance is:
(A) SD/Mean x 100 (B) Mean/SD x 100
(C) Mean x SD/100 (D) 100/Mean x SD

764. Regression coefficient is independent of:
(A) Origin (B) Scale
(C) Both origin and scale
(D) Neither origin nor scale

765. In the regression line $Y = \alpha + \beta x$, β is called:
(A) Intercept
(B) Slope of the line
(C) Mean of the X values
(D) Mean of the Y values

766. The 't' test is used to test the significance of:
(A) Two means for large sample
(B) Two means for small sample
(C) Correlation coefficient
(D) Both (B) and (C)

767. Degree of freedom from x^2in case of 4 x 3 contingency table is:
(A) 12 (B) 9 (C) 6 (D) 8

768. Greater the distance of Lorenz curve from the line of equal distribution:
(A) More is inequality
(B) Less is inequality
(C) Nothing can be inferred
(D) None of these

769. If for a distribution, coefficient of kurtosis ,2 <0, the frequency curve is:
(A) Leptokurtosis
(B) Platykurtosis
(C) Mesokurtosis
(D) None of these

770. F – test is used to test:
(A) Ratio of two variance
(B) Ratio of two means
(C) Square of normal variate
(D) None of these

771. Square of a standard normal variate is:
(A) F (B) X^2 (C) Z (D) None of these

772. If P (AEB) = 0.8 and P(A)) = P(B) = 0.3, what is the value of P(A c B)?
(A) 0.1 (B) 0.2 (C) 0.3 (D) 0.5

773. 'Annawari' procedure is a:
(A) Traditional method for estimation of crop yield
(B) Method for weather forecasting
(C) Method for rainfall forecasting
(D) All of these

774. For the expression of standard error of mean which formula is applicable:
(A S/√n (B) 2S/√n
(C) √S/n (D) n/√S

775. The statistical procedure for estimating whether the difference under study is significant or non significant is known as:
(A) Test of significance
(B) Standard error
(C) Level of significance
(D) Sampling

776. For application of t – test, it is assumed that population is:
(A) Negatively skewed (B) Normal
(C) Positively skewed (D) None o these

777. Equality of several variance is done by:
(A) F (B) X^2
(C) Z (D) None of these

778. Standard deviation of sample mean is given by:
(A) σ (B) n
(C) σ/root n (D) None of these

779. The ratio of difference between sample variance and with in sample variance follows:
(A) Z distribution (B) t distribution
(C) F distribution (D)X^2distribution

780. The area under standard normal curve beyond Z =+ 1.96 is:
(A) 95 per sent (B) 99 per cent
(C) 5 per cent (D) 1 per cent

781. Who is considered as father of statistics:
(A) Ronald A. Fisher (B) Emile Dukeman
(C) Francis Galton (D) Carl Pearson

782. Statistics serves to provide a short hand description of large amounts of:
(A) Scores (B) Data
(C) Points (D) None of these

783. The non parametric substitute for the parametric 't' test:
(A) X^2-test (B) Kvuskal Wallis test
(C) Friedman test
(D) Mann-Whitney U test

784. An example for one way non-parametric analysis of variance among the following is:
(A) F -test (B) Kvuskal Wallis test
(C) Friedman test (D) Z- test

785. The value of coefficient of kurtosis can be:
(A) Less than 3 (B) Greater than 3
(C) Equal to 3 (D) All of these

786. If the measure of kurtosis is greater than 3, then the curve is:
(A) Platykurtic (B) Mesokurtic
(C) Leptokurtic (D) Normal

787. In case of a positively skewed distribution:
(A) Median > Mean > Mode
(B) Mean > Median > Mode
(C) Mean = Median = Mode
(D) None of these

788. A functional variate for establishing a parameter is called:
(A) An estimate (B) An estimator
(C) A frame (D) A statistic

789. The idea of randomization was put forwarded by:
(A) Chapman and Greenwood
(B) Kerlinger
(C) Ronald A Fisher
(D) Sukathamae

790. The principle of experimental design include:
(A) Randomization (B) Local control
(C) Both (A) and B) (D) None of these

791. When the experimental units are homogeneous, the design used is:
(A) CRD (B) RBD
(C) LSD (D) None of these

792. Degree of freedom is related to the number of:
(A) People involved
(B) Hypothesis under test
(C) Independent observations
(D) None of these

793. The experimental design wherein the heterogeneity in the experimental units are in two directions:
(A) CRD (B) RBD
(C) LSD (D) BIBD

794. The error d.f. for a RBD experiment with 5 treatments and 3 replications is:
(A) 15 (B) 12
(C) 8 (D) 14

795. If there are ten treatments and three replications in a CRD experiment, the error degrees of freedom is:
(A) 18 (B) 20
(C) 27 (D) 29

796. If ten treatment are there, how many blocks are needed to have a minimum error degree of freedom of 12 in RBD:
(A) 1 (B) 2 (C) 3 (D) 4

797. In an RBD with 10 treatment and 4 replications, what will be the error degree of freedom if two observations are lost:
(A) 25 (B) 26 (C) 27 (D) 28

798. A family of parametric distribution with mean is equal to variate is:
(A) Binomial (B) Normal
(C) Poisson (D) Negative binomial

799. The variance of a binomial distribution with n and p as parameters is:
(A) n (B) n (1-p)
(C) np (B) np(1-p)

800. Which of the following is not a discrete distribution:
(A) Binomial (B) Poisson
(C) Normal (D) Both (A) and (B)

801. Distribution wherein mean is always greater than variance is:
(A) Binomial (B) Poisson
(C) Normal (D) Both (A) and (B)

802. A normal curve is all but not:
(A) Symmetrical (B) Continuous
(C) Unimodal (D) Asymbiotic

803. If X is binomial variant with parameters n and p. If n=1, the distribution is:
(A) Binomial (B) Poisson
(C) Normal (D) Bernoulli

804. Theoretical value of test statistics for five per cent when compared with one per cent is:
(A) Same (B) Greater
(C) Lesser (D) None of these

805. The value of probability ranges from:
(A) -1 to +1 (B) 0 to ∞
(C) 0 to 1 (D) None of these

806. Probability of two persons being born on the same day is:
(A) 1/49 (B) 1/365
(C) 1/7 (D) None of these

807. Probability of a leap year will have 53 Sundays is:
(A) 1/7 (B) 2/7
(C) 2/53 (D) 52/53

808. If A and B are two events which have no points in common, the events A and B are:
(A) Mutually exclusive (B) Independent
(C) Complementary to each other
(D) None of these

809. The probability of an impossible event is:
(A) 0 (B) 1 (C) 2 (D) None of these

810. The limiting relative frequency approach of probability is known as:
(A) Statistical probability
(B) Classical probability
(C) Mathematical probability
(D) Asymmetric probability

811. If it is know that an event A has already occurred, the probability of an event E given A is:
(A) Statistical probability
(B) Conditional probability
(C) Mathematical probability
(D) Asymmetric probability

812. If A is an event the probability of A given A is equal to:
(A) Zero (B) One
(C) Infinite (D) Indeterminate

813. The data reveals that 10 patient die in a particular type of operation. A doctor performed 9 operations and all of them survived. What is the chance of the 10th patient die on being operated:
(A) 100% (B) >10% but <100%
(C) <10% (D) 10%

814. The basis of sampling theory is:
(A) Probability (B) Randomization
(C) Local control (D) Relative frequency

815. List of sampling units is:
(A) Sample population
(B) Sampling frame
(C) Sample element (D) Sampling unit

816. The nature of relationship between sample size and sampling error is:
(A) Positive (B) Negative
(C) Zero (D) Constant

817. Difference between estimated value and actual value is measured as:
(A) Non sampling error
(B) Bias
(C) Sampling error
(D) None of these

818. Level of significance is the probability of:
(A) Type I error (B) Type II error
(C) Not committing error
(D) Any one of these

819. Rejecting a hypothesis when it is true:
(A) Type I error
(B) Type II error
(C) Not committing error
(D) Any one of these

820. Area beyond $Z = \pm 1.96$ in a standard normal distribution:
(A) 95 (B) 5 (C) 10 (D) 1

821. Rejecting a hypothesis when it is true:
(A) Type I error (B) Type II error
(C) Not committing error
(D) Any one of the above

822. Area of critical region depends upon:
(A) Size of Type I error
(B) Size of Type II error
(C) Both of these (D) None of these

823. Accepting a hypothesis when it is false:
(A) Type I error (B) Type II error
(C) Both of these (D) None of these

824. The method used to identify the units or dimensions behind the measures:
(A) Multiple regression analysis
(B) Discriminant analysis
(C) Profile analysis (D) Factor analysis

825. Which of the following method consists in the collection of data concerning the living, the working conditions of people in a given area with a view to formulate practical social measures for their betterment and welfare?
(A) Ideal types (B) Social survey
(C) Sociometry (D) Statistics

826. Level of significance is the probability of:
(A) Type I error (B) Type II error
(C) Not committing error
(D) Any one of the above

827. Consider the following statements:
1. Median is not useful while dealing with skewed distribution such as income distribution or price distribution
2. Median can be determined graphically by drawing only one give
3. Mode can be used to describe qualitative phenomenon

Which of the above statements are correct?
(A) 1 and 2 (B) 1 and 3
(C) 2 and 3 (D) 1, 2 and 3

828. The following equation indicates the relationship among average median and mode:
(A) Mode = 3 Median – 2 Mean
(B) Mean = 2 Mode – 3 Median
(C) Median = 3 Mean – 2 Mode
(D) None of these

829. What is the median of:
36, 72, 49, 45, 60, 33, 61, 72, 52?
(A) 33 (B) 52 (C) 61 (D) 60

830. The data reveals that 10 patient die in a particular type of operation. A doctor performed 9 operations an all of them survived. What is the chance of the 10th patient die on being operated:
(A) 100% (B) >10% but <100%
(C) <10% (D) 10%

831. In a moderately asymmetrical distribution:
(A) AM>GM>HM (B) AM<GM<HM
(C) AM<GN>HM (D) None of these

832. The geometric mean of 8, 64 and 512 is:
(A) 194 (B) 64 (C) 16 (D) None of these

833. If the reliability coefficient of a test is one, the standard error will be:
(A) +1 (B) -1 (C) 0.5 (D) 0

834. The basis of sampling theory is:
(A) Probability (B) Randomization
(C) Local control (D) Relative frequency

835. The range of simple correlation coefficient is:
(A) 0 to ∞ (B) $-\infty$ to $+\infty$
(C) -1 to +1 (D) 0 to 1

836. The best measure of dispersion among the following is:
(A) Standard deviation
(B) Mean deviation
(C) Coefficient of variation
(D) Range

837. Cluster sample are more effective if the population taken is:
(A) Cluster of small size
(B) Clusters with largest within variation
(C) Clusters with least within variation
(D) None of these

838. The points of intersection of the 'less than' and the 'more than' gives correspond to:
(A) Mean (B) Median
(C) Geometric mean (D) None of these

839. In CRD, which one of the following is the alternative hypothesis?
(A) All the means are unequal
(B) At least one mean is different
(C) At least half of the means are unequal
(D) All the means are equal

840. A distribution is said to be symmetrical if the values are uniformly distributed around the:
(A) Mean (B) Mode
(C) Median (D) Mean deviation

841 The validity of regression coefficient is tested by:
(A) z -test (B) t-test
(C) x^2-test (D) Both (A) and (B)

842. An example for non parametric test from among the following is:
(A) Z- test (B) t- test
(C) F- test (D) x^2 -test

843. If ten treatments are there, how many blocks are needed to have a minimum error degree of freedom of 12 in RBD:
(A) 1 (B) 2 (C) 3 (D) 4

844. Histogram is useful to determine graphically the value of:
(A) Arithmetic mean
(B) Median
(C) Mode (D) None of these

845. Area beyond Z = ±1.96 in a standard normal distribution:
(A) 95 (B) 5 (C) 10 (D) 1

846. Coefficient of variation is calculated by the formula:
(A) $\frac{X}{\Sigma} \times 100$ (B) $\frac{\sigma}{x} \times 100$
(C) $\frac{X}{\Sigma}$ (D) $\frac{\sigma}{X}$

847. If a=b, then a+c=b+c denotes:
(A) Transitive Property of Inequality
(B) Addition Property of Equality
(C) Commutative Property of Addition
(D) Multiplication Property of Equality

848. Mean of numbers 0, 1, 2, 3, ...10:
(A) 10 (B) 11 (C) 5.5 (D) 5

849. The sum of squared deviations is minimum when taken from:
(A) Mean (B) Median
(C) Mode (D) None of these

850. The validity of correlation coefficient is tested by:
(A) Z-test (B) T-test
(C) x^2 test (D) F-test

851. Which measure of central tendency is especially suited in the field of business:
(A) Median (B) Arithmetic mean
(C) Harmonic mean (D) Mode

852. Karl Pearson's coefficient of skewednes is ...... Bowley's coefficient of skewed for any skewed distribution:
(A) Equal to (B) Less than
(C) Greater than (D) Not related to

853. Geometric mean of a given series is always less than its:
(A) Harmonic mean (B) Weighted mean
(C) Arithmetic mean
(D) Weighted harmonic mean

854. $x^2$is calculated with the help of formula:
(A) $x^2=\Sigma(O-E)^2/E$ (B) $x^2=\Sigma(O-E)/(E)^2$
(C) $x^2=\Sigma(O-E)^2/\varepsilon O$ (D) $x^2=\Sigma(O-E)^2/\Sigma(O)^2$

855. Coefficient of determination is a square of
(A) Correlation coefficient
(B) Regression coefficient
(C) Both (A) and (B)
(D) None of these

856. If there are 6 levels of moisture regime and 6 replications each, the error degree of freedom for LSD will be:
(A) 25 (B) 35
(C) 36 (D) 20

857. Harmonic mean of two observations 4 and 6 is:
(A) 12/5 (B) 12/10
(C) 24/5 (D) 28/10

858. To get a short hand summary of large number of observation we go for:
(A) Pure statistics
(B) Descriptive statistics
(C) Numerical statistics
(D) None of these

859. The measure of variation that is least affected by extreme observation is:
(A) Range
(B) Mean deviation
(C) Standard deviation
(D) Quartile deviation

860. Difference between estimated value and actual value is measured as:
(A) Non sampling error (B) Bias
(C) Sampling error (D) None of these

861. When coefficient of skewed is zero the distribution is:
(A) J-shaped (B) U-shaped
(C) Symmetrical (D) L-shaped

862. In a negatively skewed distribution is:
(A) Mode>Median>Mean
(B) Median>Mode>Mean
(C) Mode<Median<Mean
(D) None of these

863. The nature of relationship between sample size and sampling error is:
(A) Positive (B) Negative
(C) Zero (D) Constant

864. The coefficient of correlation:
(A) Has no limits
(B) Can be less than one
(C) Can be more than one
(D) Varies between ± 1

865. List of sampling units is:
(A) Sample population
(B) Sampling frame
(C) Sample element
(D) Sampling unit

866. In the context of normal distribution N (m, 1), which one of the following is a simple hypothesis?
(A) H : m <0 (B) H : m >0
(C) H : m = 0 (D) H : -1 < m < 1

867. The relationship between independent and dependent variables is computed by:
(A) Regression coefficient
(B) Correlation coefficient
(C) Both (A) and (B)
(D) None of these

868. The significance of multiple correlation (R) is tested by:
(A) T-test (B) F-test
(C) Z-test (D) Paired t-test

869. Crop cutting survey on sugarcane recommended the plot size for estimation of crop yield is:
(A) 20 m x 10 m (B) 10 m x 10 m
(C) 10 m x 5 m (D) 5 m x 5 m

870. The minimum error degree of freedom should be at least:
(A) 6 (B) 12
(C) 18 (D) 20

871. The repetition of the treatment under investigation is known as:
(A) Randomization (B) Replication
(C) Local control (D) None of these

872. The optimum plot size in the experiment is/are determined by:
(A) Maximum curvative method
(B) Fair field Smith's variance law
(C) Both (A) and (B)
(D) Uniformity trial

873. Which one of the following is true for n (degrees of freedom) in large sample tests?
(A) n <5 (B) 5 £ n <20
(C) $n^{2}30$ (D) no restrictions

874. When one regression coefficient is negative the other would be:
(A) Negative (B) Positive
(C) Zero (D) None of these

875. Where r is zero the regression lines but each other making an angle of:
(A) 45^0 (B) 60^0
(C) 90^0 (D) None of these

876. If with a rise of 10% in prices the wages are increased by 20%, the real wage increased is by:
(A) 10%
(B) More than 10%
(C) Less than 10%
(D) 20%

877. Fisher's ideal index is:
(A) Arithmetic mean of Laspeyre's and Paasche's index
(B) Median of Laspeyre's and Paasche's index
(C) Geometric mean of Laspeyre's and Paasche;s index
(D) None of these

878. Laspeyre's index is based on:
(A) Base year quantities
(B) Current year quantities
(C) Average of current and base year
(D) None of these

879. A good index number is one that satisfies:
(A) Unit test (B) Time test
(C) Time Reversal test and Factor Reversal test
(D) None of these

880. Cyclical fluctuations are caused by:
(A) Wars (B) Earth quakes
(C) Floods (D) None of these

881. Indian Livestock Census is conducted after every:
(A) 2 years (B) 5 years
(C) 10 years (D) 20 years

882. In India the available acreage statistics for the total geographical area is:
(A) 51% (B) 61% (C) 71% (D) 100%

883. If E(X) is the mathematical expectation of the variable X and C is constant then E(CX) =?
(A) E(C) + E(X) (B) E(C) x E(X)
(C) CE(X) (D) None of these

884. E (x + y) = ?
(A) E(x) + E(y) (B) E(x) x E(y)
(C) E(xy) (D) None of these

885. Which of the following is not a characteristic feature of a good research design:
(A) High external validity
(B) Low external validity
(C) High internal validity
(D) None of these

886. If A and B are mutually exclusive events, P(AB) = ?
(A) Zero (B) P (A) + P (B)
(C) P (A) (D) None of these

887. A research design without hypothesis among the following is:
(A) Explorative design
(B) Diagnostic design
(C) Descriptive design
(D) Experimental design

888. If two events A and B are dependent, the conditional probability of B given A, i.e. P(A/B) is calculated as:
(A) P (A) + P (B) (B) P (A) x P (B)
(C) P (A) – P (B) (D) P (AB) / P (A)

889. To pilot study to gain more insight into the research problem form the characteristic feature of:
(A) Explorative design
(B) Diagnostic design
(C) Descriptive design
(D) Experimental design

890. ${}^5C_2$ is equal to:
(A) 20 (B) 10 (C) 30 (D) 100

891. Two cards are drawn from a well – shuffled pack of 52 cards. Find the probability that they are both aces if the first is replaced:
(A) $\frac{1}{169}$ (B) $\frac{2}{52}$
(C) $\frac{1}{221}$ (D) None of these

892. The standard deviation of Binomial distribution is:
(A) $\sqrt{npq}$ (B) npq
(C) $n^2p^2q^2$ (D) np

893. The basic concept of a good research design is to:
(A) Decrease experimental variance
(B) Maximize extraneous variance
(C) Minimize error variance
(D) Explain metaphysical observation

894. In case of normal distribution $x + 2\sigma$ covers:
(A) 95.45% (B) 95.54%
(C) 94.45% (D) 99.73%

895 .If in Binomial distribution mean is 10 and standard deviation 2, q will be:
(A) 0 (B) 1 (C) 0.4 (D) 0.8

896. After this, therefore caused by this" is usually denoted to indicate:
(A) Experimental design
(B) Survey design
(C) *Ex post facto* design
(D) Exploratory design

897. The normal distribution is an approximation to:
(A) Binomial distribution
(B) Poisson's distribution
(C) Both (A) and (B)
(D) None of these

898. The mea plus minus 1.96 standard deviation includes —— per cent of the items of normal distribution:
(A) 75 (B) 47.5
(C) 74.5 (D) None of these

899. The standard deviation of sampling distribution is called:
(A) Standard error (B) Probable error
(C) Both (A) and (B) (D) None of these

900. Research design is not essentially a:
(A) Plan (B) Product
(C) Structure (D) Strategy

901. A hypothesis is a must in:
(A) Experimental design
(B) Explorative design
(C) Descriptive design
(D) Both (A) and (B)

902. Dispersions are NOT calculated about:
(A) Mean (B) Median
(C) Mode (D) Geometric mean

903. The following description " I will bet this would happen if I did this" best represent to:
(A) Explorative design
(B) Confirmatory experiment
(C) Crucial experiment
(D) Pilot experiment

904. To study the effect of independent variable on a single experiment we go for:
(A) Factorial design (B) CRD
(C) LSD (D) RBD

905. Large sample theory is applicable when:
(A) N>30 (B) N<30
(C) N=30 (D) None of these

906. If the mode of frequency distribution is Mo = 16 and its mean = 16 then median is:
(A) Zero (B) 16
(C) 32 (D) <16

907. To study average rate of change in population, we use:
(A) Median (B) Mode
(C) Arithmetic mean (D) Geometric mean

908. Student's 't' Distribution was discovered by:
(A) Karl Pearson (B) Laplace
(C) Fisher (D) Gosset

909. Experimenter has the highest control in manipulation of variable in:
(A) Laboratory method
(B) Library method
(C) Simulation method
(D) Field experiment

910. Consider the following measures:
1. Correlation Coefficient
2. Covariance
3. Coefficient of variation
4. Index number
Which of these are unit-free?
(A) 1 and 2 (B) 1 and 3
(C) 2 and 4 (D) 1, 3 and 4

911. The units within and between each stratum in a stratified random sampling are respectively:
(A) Homogenous and homogenous
(B) Homogenous and Heterogenous
(C) Heterogenous and Homogenous
(D) Heterogenous and Heterogenous

912. Harmonic mean is better than other means if data are for:
(A) Speeds or rates
(B) Heights and lengths
(C) Binary values 0 and 1
(D) Ratio and proportion

913. Regression coefficient takes values:
(A) $-\infty$ to 1 (B) 1 to ∞
(C) $-\infty$ to $+\infty$ (D) -1 to +1

914. If each observation of a set is multiplied by 10, the mean of the new set of observation:
(A) Remains the same
(B) 10 times the original mean
(C) $1/10^{th}$ of the original mean
(D) Increased by 10

915. Standard deviation : Variability : : Mean : ________.
(A) Central tendency
(B) Discrimination
(C) Level of difficult
(D) Correlation

916. Which out of the following is correct about median?
(A) It is a measure of variability
(B) It is the 50th percentile
(C) It is most stable measure of central tendency
(D) It is significantly affected by extreme scores

Chapter 15

APPENDIX

ICAR RESEARCH INSTITUTES

AGRICULTURAL ENGINEERING

1. CIAE	Central Institute of Agricultural Engineering, Nabi Bagh, Berasia road, Bhopal- 462038 (MP)
2. CIPHET	Central Institute of Post Harvest Engineering & Technology, Ludhiana- 141004 (Punjab)
3. CICT	Central Institute of Cotton Technology, P.B. No. 16640, Mumbai – 400019 (Maharashtra)
4. ILRI	Indian Lac Research Institute, Ranchi- 834010 (Jharkhand)
5. NIRJAFT	National Institute of Research on Jute & Allied Fibre Technology, Ranikuthi, Kolkata -700120 (W.B.)

AGRICULTURE EXTENSION

1. NRCWA	National Research Centre for Women in Agriculture, 1199 Jagmara, Bhubneshwar-751030 (Orrisa)
2.DIPA	Directorate of Information and Publication of Agriculture, KAB-I, Pusa, New Delhi-110012

ANIMAL SCIENCES

1. CARI	Central Avian Research Institute, Izzatnagar , Bareili-243122(U.P)
2. CIRB	Central Institute for Research on Buffaloes, Sirsa Road, Hisar- 125001 (Haryana)
3. CIRG	Central Institute for Research on Goats, Makhdoom, Mathura-281122 (U.P)
4. IVRI	Indian Veterinary Research Institute, Izzatnagar, Bareili-243122 (U.P)
5. NAINP	National Institute of Animal Nutrition & Physiology, Adugodi, Bangalore-560030 (Karnataka)
6. CSWRI	Central Sheep and Wool Research Institute, Avikanagar- 304501 (Rajasthan)
7. NBAGR	National Bureau of Animal Genetic Resources, P.B. No. 129, Kernal-132001

(Haryana)

8. NDRI National Dairy Research Institute, Karnal-132001(Haryana)
9. NRCC National Research Centre on Camel, Jobber, P.B. No. 07, Bikhaner-334001 (Rajasthan)
10. NRCM National Research Centre for Mithun, Jharnapani, Medziphema, Nagaland-500059 (A.P.)
11. NRCM National Research Centre on Meat, Saidabad, CRIDA campus, Hyderabad-500059 (A.P.)
12. NRCY National Research Centre on Yak, Dirang, West Kameng-790101 (Arunachal Pradesh)
13. NRCP National Research centre on Pigs, Panjabari Road, 6th mile, Guwahati-785037 (Assam)
14. PDADMS Project Directorate on Animal Disease Monitoring and Surveillance, Hebbal, Bangalore-560024 (Karnataka)
15. PDC Project Directorate on Cattle, Grass farm Road, P.B. No. 17, Meerut Cant, Meerut- 250001 (U.P.)
16. PDFMD Project Directorate on Foot and Mouth Disease, IVRI, Mukteshwar, Kumaon-263138 (Uttaranchal)
17. PDP Project Directorate on Poultry, Hyderabad-500030(A.P.)
18. NRCE National Research Centre for Equines, Hisar - 125001

CROP SCIENCES

1. CICR Central Institute for Cotton Research, P.B. No. 2, Nagpur-440010 (Maharashtra)
2. CRIJAF Central Research Institute for Jute & Allied Fiber, Nilganj, Barrackpore-700120 (W.B.)
3. CRRI Central Rice Research Institute, Cuttak, 753006 (Orrisa)
4. CTRI Central Tobacco Research Institute, Rajamundry-533105(A.P.)
5. DMR Directorate of Maize Research, IARI Pusa Campus, New Delhi -110012
6. DOR Directorate of Oilseed Research, Rajendranagar, Hyderabad-500030 (A.P.)
7. DRR Directorate of Rice Research, Hyderabad-500030 (A.P.)
8. DWR Directorate of Wheat Research, PB No.158, Karnal -132001 (Haryana)
9. IARI Indian Agricultural Research Institute, New Delhi-110012
10. IGFRI Indian Grassland and Fodder Research Institute, Jhansi-284003 (U.P.)
11. IIPR Indian Institute of Pulses Research, Kanpur-208024
12. IISR Indian Institute of Sugarcane Research, Lucknow-226602
13. NBAIMO National Bureau of Agriculturally Important Microorganisms, Kusmaur, Mau Nath Bhanjan-275101 (U.P.)
14. NBPGR National Bureau of Plant Genetic Resources, New Delhi-110012
15. NCIPM National Centre for Integrated Pest Management, LBS Building, Pusa campus, New Delhi -110012
16. NRPB National Research Centre on Plant Biotechnology, Pusa campus, New Delhi-110012
17. DGR Directorate of Groundnut Research, PB No. 5, Ivnagar Road, Junagarh-362001 (Gujarat)

18. DRMR Directorate of Rapeseed and Mustard Research, Bharatpur - 321303 (Rajasthan)
19. DSR Directorate of Sorghum Research, Rajendranagar, Hyderabad-500030 (A.P.)
20. DSR Directorate of Soybean Research, Indore-452017 (M.P.)
21. NABII National Bureau of Agriculturally Important Insect, PB.No. 2491, Hebbal, Banglore-560024 (Karnataka)
22. VPKAS Vivekanand Parvatiya Krishi Anusandhan Sansthan, Almora-263601 (Uttaranchal)
23. DSR Directorate of Seed Research, P.B. No. 11, Village Kushmour, P.O. Kaithuli, Mau – 275901
24. SBI Sugarcane Breeding Institute, Coimbatore – 641007

FISHERIES

1. CIBA Central Institute for Brackish Water Aquaculture, 75 Santhome High Road, RA Puram, Chennai-600028 (T.N.)
2. CIFA Central Institute for Fresh Water Aquaculture, Kausalyaganga, Bhubneshwar-751002 (Orissa)
3. CIFA Central Institute for Fisheries Education, Mumbai-400061 (Maharashtra)
4. CIFRI Central Inland Fisheries Research Institute, Barrackpore- 700120 (W.B)
5. CIFT Central Institute of Fisheries Technology, Cochin-682029 (Kerala)
6. CMFRI Central Marine Fisheries Research Institute, PB.No. 1603, Tatpuram P.O. Kochi-682018 (Kerala)
7. NBFGR National Bureau of Fish Genetic Resources, P.O. Dilkusha, Lucknow-226002 (U.P)
8. DCFR Directorate of Coldwater Fisheries Research, Saurabh Cottage, Bhimtal, Nainital-263136 (Uttaranchal)

HORTICULTURE

1. CARI Central Agricultural Research Institute, PB No. 181, Port Blair- 744101(A&N Island)
2. CIAH Central Institute of Arid Horticulture, Sri Ganganagar Highway, Bikaner-334006(Rajasthan)
3. CISH Central Institute for Subtropical Horticulture, Rehmankheda, PO Kakori, Lucknow-227107 (U.P)
4. CITH Central Institute of Temperate Horticulture, Old Air Field, Rangreth-190007 (J&K)
5. CPCRI Central Plantation Crop Research Institute, Kasargod-671124 (Kerela)
6. CPRI Central Potato Research Institute, Shimla-171001 (H.P.)
7. CTCRI Central Tuber Crops Research Institute, Sreekariam, Thiruvanthapuram-695017 (Kerela)
8. IIHR Indian Institute of Horticultural Research, Hassarghatta Lake post, Banglore-560089 (KN)
9. IISR Indian Institute of Spices Research, PB.No. 1701, Marikunnu PO, Kozhikode-673012 (Kerela)
10. IIVR Indian Institute of Vegetable Research, PB No. 01, PO.Jakini, Sahansapuram, Varanasi- 221305 (U.P.)

11. NRCB	National Research Centre for Banana, Thogamalai Main Road, Thayar PO Thiruchirapalli -620102 (TN)
12. DCR	Directorate of Cashew Research, Puttur-574202 Dakshina Kannada (KN)
13. NRCC	National Research Centre for Citrus, PB No. 464 Shankarnagar PO, Nagpur-440010
14. NRCSS	National Research Centre for Seed Spices, Tabiji, Ajmer-305206 (Rajasthan)
15. NRCG	National Research Centre for Grapes, P B No. 3, Manjri Farm Post, Solapur Road, Pune-412307 (MH)
16. NRCL	National Research Centre for Litchi, Manchi House, Club Road, Muzaffarpur-842002 (Bihar)
17. NRCM	National Research Centre for Mushroom, Chambaghat, Solan-173213 (H.P)
18. DOR	Directorate of Oil Palm Research, Near Jawahar Navodaya Vidyalaya Pedavegi-534450, West Godavari dist. (A.P)
19. NRCO	National Research Centre for Orchids, Pakyang, Sikkim- 737106
20. NRCOG	National Research Centre for Onion & Garlic, Ragurunagar-410505, Pune (MH)

MANAGEMENT AND STATISTICS

1. IASRI	Indian Agricultural Statistics Research Institute, Library Avenue, Pusa, New Delhi -110012
2. NAARM	National Academy of Agricultural Research and Management, Hyderabad-500030 (A.P.)
3. NCAEPR	National Centre for Agricultural Economics & Policy Research, Library Avenue, Pusa, New Delhi-110012

NATURAL RESOURCES MANAGEMENT

1. CAZRI	Central Arid Zone Research Institute, Jodhpur-342003 (Rajasthan)
2. CRIDA	Central Research Institute for Dryland Agriculture, Santhoshnagar, Hyderabad-500059 (A.P.)
3. CSSRI	Central Soil Salinity Research Institute, Kernal- 132001 (Haryana)
4. CSWCRTI	Central Soil & Water Conservation Research & Training Institute, 218 Kaulagarh Road, Dehradun -248195 (Uttaranchal)
5. ICARRCE	ICAR Research Complex of Eastern Region, Walmi Complex, PO Phulwari Sharif, Patna-801505 (Bihar)
6. ICARRCG	ICAR Research Complex for Goa, Ela , Old Goa-403402
7. ICARRCNEH	ICAR Research Complex for NEH Region, Barapani-793103 (Meghalaya)
8. IISS	Indian Institute of Soil Science, Nabi bagh, Berasia Road, Bhopal-462038 (M.P.)
9. NBSSLUP	National Bureau of Soil Survey and Land Use Planning, Nagpur-440010 (MH)
10. NRCAF	National Research Centre for Agroforestry, Near Pahujdam, Jhansi-284003 (UP)
11. DWRS	Directorate of Weed Science Research, Maharajpur, Adhartal, Jabalpur-482004 (M.P.)
12. PDFSR	Project Directorate on Farming Systems Research, Meerut-250110
13. DWM	Directorate of Water management, Chandra Sekharpur, PO SE Railway, Project Complex, Bhubneswar-751023 (Orissa)

14. NIASM	National Institute of Abiotic Stress Management, Malegaon, Baramati – 413115 (Pune- Maharashtra)

OTHER IMPORTANT NATIONAL ORGANISATIONS

1. APEDA	Agricultural and Processed Food Export Development Authority, NCU 1 Building, 3 Siri Industrial Area , August Kranti marg, Hauzkhas, New Delhi-110016
2. BARC	Bhabha Atomic Research Centre, Trombay, Mumbai
3. CFTRI	Central Food Technological Research Institute, Mysore
4. CIMAP	Central Medicinal and Aromatic Plants, Lucknow
5. CPPTI	Central Plant Protection Training Institute, Hyderabad
6. CSIR	Council of Scientific and Industrial Research, New Delhi
7. CSMCRI	Central Salt and Marine Chemical Research Institute, Bhavnagar
8. FRI	Forest Research Institute, Dehradun
9. IAS	Indian Academy of Science, Pune
10. ICFRI	Indian Council of Forestry Research and Education, Dehradun
11. ICM	Institute of Cooperative Management, Ahmedabad
12. IFGTB	Institute of Forest Genetics and Tree Breeding, Coimbatore
13. IGNOU	Indira Gandhi National Open University, New Delhi
14. IIM	Indian Institute of Management, Ahmedabad
15. INSA	Indian National Science Academy, Bahadur Shah Zafar Marg, NewDelhi
16. IWST	Institute of Wood Science & Technology, Banglore
17. JARI	Jute Agricultural Research Institute, Barrackpore
18. JTRI	Jute Technology Research Institute, Kolkata
19. MANAGE	National Institute of Agricultural Extension Management, Hyderabad
20.	MS Swaminathan Research Foundation, 3rd Cross Street, Taramani Institutional Area, Chennai-600133(TN)
21. NABARD	National Bank of Agriculture and Rural Development
22. NAFED	National Agricultural Cooperative Marketing Federation of India Ltd., NAFED House, Siddartha Enclave, Asharm Chowk , New Delhi
23. NBRI	National Botanical Research Institute, Lucknow
24. NCDC	National Cooperative Development Corporation, New Delhi
25. NCERT	National Council of Education Research & Training, New Delhi
26. NCL	National Chemical Laboratory, Pune
27. NDDB	National Dairy Development Board, Safadarjung Enclave, New Delhi
28. NIAG	National Institute of Animal Genetics, Kernal (Harayana)
29. NIN	National Institute on Nutrition, Hyderabad
30. NIRD	National Institute for Rural Development, Hyderabad
31. NOVDB	National Oil Seed & Vegetable Development Board, New Delhi
32. NPL	National Physical Laboratory, New Delhi
33. NRL	Nuclear Research Laboratory, IARI, New Delhi
34. NSC	National Seed Corporation, Bee Bhawan, Pusa, New Delhi
35. NSI	National Sugar Institute, Kalyanpur, Kanpur
36. PPII	Patash Research Institute of India, Gurgaon (Haryana)
37. RSAC	Remote Sensing Application Centre, Lucknow
38. UGC	University Grant Commission, Bahadur Shah Zafar Marg, New Delhi

39. WLMTRI Water and Land Management Training Research Institute, Hyderabad
40. WTC Water Technology Centre, IARI, New Delhi

INTERNATIONAL ORGANISATIONS

1. AIT Asian Institute of Technology, P.O. Box 2734, Bankok, Thailand-10501
2. AVRDC Asian Vegetable Research and Development Centre, P.O.Box 42 , Shanhua, Taiwan 74199, Taiwan
3. BARC Bangladesh Agricultural Research Council, Dhaka
4. CGIAR Consultative Group on International Agriculture Research, Washington (USA)
5. CIAT Centro International de Agricultural Tropical (International Centre for Tropical Agriculture, Coli, Columbia), South Africa
6. CIDA Canadian International Development Agency, Cuba
7. CIMMYT Centro International deMajormiento de Maizy Trigo (International Centre for Improvement of Maize, Wheat and Barley),
8. CIP International Potato Centre (Centro Internationally de la Papa), Peru, Mexico
9. CSIRO Commonwealth Scientific and Industrial Research Organization, Near Perth, Australia
10. FAO Food and Agriculture Organization, Rome, Italy
11. IBPGR International Plant and Genetic Resources, Italy
12. IBSNAT International Benchmark Soil Network for Agrotechnology Transfer, Hawaii
13. ICARDA International Centre for Agricultural Research for Dryland Areas, Syrea
14. ICRAF International Centre for Research in Agroforestry, Nairobi, Kenya
15. ICRISAT International Crop Research Institute for the semi Arid Tropics, Patancheru, Hyderabad
16. IDRC International Development Research Centre, Canada
17. IFAD International Fund for Agriculture Development, Rome
18. IFDC International Fertilizer Development Centre, Muscale Shoala, Alobama (USA)
19. IFPRI International Food Policy Research Institute, Washington DC (USA)
20. IIASA International Institute for Applied Systems Analysis, Luxemberg, Vienna, Austria
21. IITA International Institute for Tropical Agriculture, Ibadan, Nigeria
22. IIASA International Institute for Applied Systems Analysis, Luxemberg, Vienna, Austria
23. INSFFER International Network on Soil Fertility and Fertilizer Evaluation of Rice, Manila, Philippines
24. INSNAR International Service for National Agricultural Research, The Hague, Netherland
25. IPGRI International Plant Genetic Research Institute
26. IRRI International Rice Research Institute, Manila Philippines
27. ISRIC International Soil Reference and Information Centre, Netherland
28. IWMI International Water Management Institute
29. LURI Land Use Research Institute, Aberdeen, Scotland
30. NIAS National Institute on Agricultural Science, Japan
31. PPIC Potash and Phosphate Institute, Canada
32. TPRI Tropical Pesticides Research Institute, Arushe, Tanzania
33. WARDA West Africa Rice Development Association, Manrovia, Liberia

34. WMO World Meteorological Organization, Geneva, Switzerland

STATE AGRICULTURAL UNIVERSITIES

1. AAU Anand Agricultural University, Anand (Gujarat)
2. AAU Assam Agricultural University, Jorhat (Assam)
3. ANGRAU Acharya NG Ranga Agricultural University, Rajendranagar, Hyderabad (A.P.)
4. BAU Birsa Agricultural University, Kanke, Ranchi (Jharkhand)
5. BCKV Bidhan Chandra Krishi Vishwavidhalaya, Mohanpur, Nadia (WB)
6. BSKKV Dr. Babasaheb Sawant Kokan Krishi Vidyapeeth, Dapoli, Ratnagiri, (Maharashtra)
7. CAU Central Agricultural University, P O Box No. 23, Imphal, (Manipur)
8. CCSHAU Chaudhary Charan Singh Haryana Agricultural University, Hisar (Haryana)
9. CSAUA&T Chandra Shekhar Azad University of Agriculture & Technology, Kanpur, (U.P.)
10. CSKKV Ch. Sarwan kumar Krishi Vishwavidyalaya, Palampur (HP)
11. CVV&GAS U.P. Pandit Deen Dayal Upadhyaya Pashu Chiktsha Vigyan Vishwavidhayala evam Go Anushandhan Sansthan, Mathura, (U.P.)
12. GBPUA&T Gobind Ballabh Pant University of Agriculture & Technology, Pantnagar-263145, Udam Singhnagar (Uttaranchal)
13. IGKV Indira Gandhi Krishi Vishwavidyalaya, Krishak negar, Raipur-492012 (Chhattisgarh)
14. JAU Junagarh Agricultural University, Junagarh-362001 (Gujarat)
15. JNKVV Jawaharlal Nehru Krishi Vishwa Vidyalaya, Jabalpur – 482004 (MP)
16. KAU Kerala Agricultural University, Vellanikara, Trichur-680656 (Kerala)
17. MAS&FU Maharastra Animal Science & Fishery University, Seminary Hillls, Nagpur-440006 (Maharashtra)
18. MAU Marathwada Agricultural University, Parbhani-431 402 (Maharastra)
19. MPKV Mahatma Phule Krishi Vidyapeeth, Rahuri-413 402 (Maharastra)
20. MPUA&T Maharana Pratap University of Agriculture & Technology, Udaipur-3133001 (Rajasthan)
21. NAU Navsari Agricultural University, Navsari (Gujarat)
22. NDAU&T Narendra Deo University of Agriculture & Technology, Kumar Ganj, Faiza bad-224 229 (UP)
23. OUA&T Orissa University of Agriculture & Technology, Bhubneshwar-751003 (Orissa)
24. PAU Punjab Agriculture University, Ludhiana -141004 (Punjab)
25. PDKV Dr. Panjabrao Deshmukh Krishi Vidyapeeth, Krishinagar, Akola-444 104 (Maharashtra)
26. RAU Rajasthan Agriculture University, Bikaner-33400 (Rajasthan)
27. RAU Rajendra Agriculture University, Pusa, Samastipur-848 125 (Bihar)
28. SBBPUA&T Sardar Ballabh Bhai Patel University of Agriculture & Technology, Modipuram, Meerut-250 110 (UP)
29. SKDAU Sardar Krushinagar Dantiwada Agricultural University, Sardar Krushinagar, Banskantha-385506 (Gujarat)
30. SKUAC&T Sher-E-Kashmir University of Agricultural Sciences & Technology, Jammu -188812 (J&K)
31. SKUAC&T Sher-E-Kashmir University of Agricultural Sciences & Technology, Kashmir -191121 (J&K)

32.TAU	Tamilnadu Agricultural University, Coimbattor-641003 (Tamilnadu)
33. TV&ASU	Tamilnadu Veterinary & Animal Sciences University, Chennai-600051 (Tamilnadu)
34. UAS	University of Agricultural Sciences GKVK Bangalore-560065 (Karnataka)
35. UAS	University of Agricultural Sciences, Dharwad-580005 (Karnataka)
36. UBKV	Uttar Banga Krishi Vishwavidyalaya, Pundibari, Cooch behar – 736 165 (WB)
37. WUA&FS	West Bengal University of Animal & Fishery Sciences, 68 K Sahani, Kolkata-700037 (WB)
38. YSPUH&F	Dr. Yashwant Singh Parmar University of Horticulture & Forestry, Solan, Nauni-173 230 (HP)
39. UHF	University of Horticulture and Forestry, Pauri, Uttarakhand
40. MKJUAT	Manyawar Kansiram Ji University of Agriculture and Technology, Banda (UP)
41. SVVU	Sri Venkateswara Veterinary University, Trupati – 517502 (AP)
42. APUH	Andhra Pradesh University of Horticulture, Venkataramanagudem, West Godawari – 534001 (AP)
43. RVSKV	Rajmata Vijayaraje Scindia Krishi Vishwavidyalaya, Gwalior – 474002 (MP)
44. MARFSU	Maharashtra Animal & Fisheries Sciences University, Nagpur – 4400006 (MH)
45.GADVASU	Guru Angad Dev Veterinary And Animal Science University, Ludhiana – 141004 (Punjab)
46.SHIATS	Sam Higginbottom Institute of Agriculture Technology and Sciences, Naini, Allahabad – 211007
47. KUFOS	Kerala University of Fisheries and Ocean Studies, Pananagad, Kochi – 682506
48. KV&ASU	Kerala Veterinary and Animal Sciences University, Pookot, Wayanand – 673576
49.LLRUVAS	Lala Lajpat Rai University of Veterinary & Animal Sciences, Hisar (Haryana)
50.RUV&AS	Rajasthan University og Veterinary and Animal Sciences, Bijay Bhawan Palace Complex, Near Pt. Deendayal Circle, Bikaner – 334006 (Rajasthan)
51. BAU	Bihar Agriculture University, Sabour, Bhagalpur – 813210 (Bihar)
52. MPPCVV	Madhya Pradesh Pashu Chikitsa Vigyan Vishwavidhyalaya, Jabalpur – 482004 (MP)
53.UHS	University of Horticultural Sciences, Sector-60, Navanagar, Bagalkot – 587102
54. UAS	University of Agriculture Science, Raichur – 584102 (Karnataka)

ICAR INSTITUTES DEEMED TO BE UNIVERSITY

IARI	Indian Agricultural Research Institute, New Delhi – 110012
IVRI	Indian Veterinary Research Institute, Izatnagar, Berrreili - 243122 (UP)
NDRI	National Dairy Research Institute, Kernal – 132001 (Haryana)
CIFE	Centrral Institute of Fisheries Education, Mumbai – 400061

CENTRAL UNIVERSITIES RELATED TO AGRICULTURE

AMU	Aligarh Muslim University, Aligarh
BHU	Banaras Hindu University, Varanasi
VISWA	Viswa Bharti (Palli Siksha Bhawan), P.O. Sriniketan – 731236

NU	Nagaland University (School of Agricultural Sciences and Rural Development), Modziphema - 797106

ABBREVIATIONS/ACRONYMS

A3P	Accelerated Pulse Production Programme
ACIAR	Australian Council on International Agriculture Research
AGRIS	International Information System for Agricultural Science and Technology
AICRP	All India Coordinated Research Project
AOAC	Association of Official Analytical Chemists
AP Cess Fund	Agricultural Product Cess Fund
APAS	Andhra Pradesh Academy of Science
APEDA	Agriculture & Processed Products Export Authority
APWSS	Asian Pacific Weed Science Society
ARIC	Agricultural Research Information Centre
ARS	Agricultural Research Services
ASA	American Society of Agronomy
ASRB	Agricultural Scientist Recruitment Board
ASSOD	Assessment of the Status of Human –included Soil Degradation
ATIC	Agricultural Technology Information Centre
ATMA	Agriculture Technology Management Agency
BARC	Bhabha Atomic Research Centre
BHC	Benzene Hexa chloride
CAB	Commonwealth Agricultural Bureau
CCIS	Comprehensive Crop Insurance Scheme
CDD	Community Development Directorate
CEP	Central Exchange Programme
CFB	Corrugated Fibre Board
CIDA	Canadian International Development Agency
CSC	Central Seed Committee
CSSA	Crop Science Society of America
CVRC	Central Variety Release Committee
DAP	Di Ammonium Phosphate
DARE	Department of Agricultural Research and Education
DBT	Department of Bio Technology
DDP	Desert Development Programme
DDT	Dichloro Diphenyle Trichloro Ethane
DPAP	Drought Prone Area Programme
DPD	Directorate of Pulses Development
DST	Department of Science and Technology
EAS	Employment Assurance Scheme
EDB	Ethylene Di Bromide
EDCIL	Educational Development Consultant India Ltd
ESCAP	Economic and Social Commission for Asia and the Pacific
FAI	Fertilizer Association of India
FAO	Food and Agriculture Organisation
FCI	Food Corporation of India

FDA	Food and Drug Adulteration Act
FPC	Fertilizer Promotion Campaign
FPTC	Food Processing and Training Centre
FRC	Food Research Council
FSC	Farm Science Centre
FYM	Farm Yard Manure
GLASOD	Global Assessment of the Status of Human-included Soil Degradation
GOI	Government of India
GSFC	Gujarat State Fertilizer Corporation Ltd
HCN	Hydro Cyanic Acid
HETA	Horticulture Experiment & Training Centre
HIL	Hindustan Insecticide Limited
HSDC	Haryana Seed Development Corporation
HYVP	High Yielding Variety Programme
IAAP	Intensive Agriculture Area Programme
IADP	Intensive Agriculture District programme
IARC	International Agricultural Research Centre
IBRD	International Bank for Reconstruction and Development
ICDS	Integrated Child Development Scheme
ICID	International Commission on Irrigation and Drainage
IEL	Indian Explosive Limited
IFFCO	Indian Farmers Fertilizers Corporation Ltd
IFPC	Intensive Fertilizer Promotion Campaign
IFWA	International Federation for Woman Development
IJO	International Jute Organisation
ILCA	International Livestock Centre for Africa
IRDC	International Research and Development Council
IRDP	Integrated Rural Development Programme
IRTP	International Rice Testing Programme
ISA	Indian Society of Agronomy
ISOPOM	Integrated Scheme for Oilseeds, Pulses, Oilpalm and Maize
ISRIC	International Soil Reference and Information Centre
ISSS	Indian Society of Soil Science
ISSS	International Soil Science Society
ISTS	Indian Society of Tobacco Science
ITEC	Indian Technical and Economic Corporation
IUFRO	International Union of Forest Research Organisation
IWSS	Indian Weed Science Society
KGK	Krishi Gyan Kendra
KRIBHCO	Krishak Bharti Corporation Limited
KVK	Krishi Vigyan Kendra
LCU	Lac Coated Urea
LEISA	Low External Input and Sustainable Agriculture
LLP	Lab to Land Programme
LTPC	Long Term Programme of Cooperation

MCFC	Maharastra Cooperative Fertilizer and Chemical Limited
MEPDA	Marine Product Export Development Authority
MFAL	Marginal Farmers and Agricultural Labour Programme
MGREGS	Mahatma Gandhi Rural Employment Guarantee Scheme
MIS	Management Information System
NAAS	National Academy of Agriculture Science
NACA	Network of Aquaculture Centre in Asia –Pacific
NAFED	National Agriculture Cooperative Marketing Federation of India Limited
NAIP	National Agriculture Innovative Project
NARP	National Agricultural Research Project
NARS	National Agricultural Research System
NATP	National Agricultural Technology Project
NCA	National Commission on Agriculture
NCAER	National Council of Applied Economic Research
NCDC	National Cooperative Development Corporation
NCIPM	National Centre for Integrated Pest Management
NCL	National Chemical Laboratory
NCU	Neem coated Urea
NDDB	National Dairy Development Board
NDP	National Development Programme
NET	National Eligible Test
NFL	National Fertilizer Limited
NFSM	National Food Security Mission
NGO	Non Government Organisation
NIPMA	North Indian Pesticide Manufacture Association
NICRA	National Initiative on Climate Resilient Agriculture
NOVDB	National Oilseed and Vegetable Development Board
NPC	National Productivity Council
NPL	National Physical Laboratory
NRDC	National Research Development Corporation
NREGA	National Rural Employment Guarantee Agency
NREP	National Rural Employment Programme
NSKS	Neem Kernel Seed Extract
NWDPRA	National Watershed Development Project for Rainfed Area
OFC	Orissa Fertilizer and Chemical Limited
ORP	Oilseeds Research Project
PARC	Pakistan Agriculture Research Council
PCARR	Philippines Council of Agriculture and Resources Research
PCARRD	Philippines Council for Agriculture Forestry and Natural Resources Research and Development
PCNB	Penta Chloro Nitro Benzene
PIRCOM	Project for Intensification of Regional Research on Cotton, Oilseeds and Millets
PPCL	Pyrites Phosphate and Chemical Limited
PRA	Participatory Rural Appraisal

PRAI	Planning Research and Action Institute
PSSC	Punjab State Seed Corporation
PTD	Participatory Technology Development
RAPI	Rural Agro Processing Industries
RKVY	Rastriya Krishi Vikas Yojna
RLEGP	Rural Landless Employment Guarantee Programme
RNE	Renewable Natural Energy
RRA	Rapid Rural Appraisal
SAARC	South Asian Association for Regional Cooperation
SADCC	Southern African Development Coordination Conference
SAMETI	State Agricultural Management and Extension Training Institute
SAU	State Agricultural Universities
SFCI	State Farms Corporation of India
SFDA	Small Farmers Development Agency
SIDA	Swedish International Development Agency
SISSTA	South Indian Sugarcane and Sugar Technology Association
SITE	Satellite Instructional Television Experiment
SOTER	Soils and Terrain Digital Database
SRI	System of Rice Intensification
SSDC	State Seed Development Corporation
SSP	Single Super Phosphate
SURE	Society to Uplift Rural Economy
SVRC	State Variety Release Committee
TAS	Time Allocation Study
TDC	Terai Development Corporation
TMO	Technology Mission on Oilseeds
TRIMS	Trade Related Investment Measures
TRIPS	Trade Related Intellectual Property Rights
TRYSEM	Training of Rural Youth Self Employment Programme
UGC	University Grand Commission
UNDP	United Nations Development Programme
UNEP	United Nations Environment Programme
UNESCO	United Nations Educational Scientific and Cultural Organisation
UNICEF	United Nations International Children Emergency Fund
UPCAR	Uttar Pradesh Council of Agricultural Research
USAID	United States Agency for International Development
USDA	United State Department of Agriculture
USIF	United State Indian Fund
WAPCCS	Water and Power Consultative Service India Limited
WMO	World Meteorological Organisation

Table 1 : Year wise area, production and productivity of food grain in India

Year	Area (m ha)	Production (m tonnes)	Productivity (kg/ha)
1950-51	97.32	50.82	522
1960-61	115.58	82.02	710
1970-71	124.32	108.42	872
1980-81	126.67	129.59	1023
1990-91	127.84	176.39	1380
1991-92	121.87	168.38	1382
1992-93	123.15	179.48	1457
1993-94	122.75	184.26	1501
1994-95	123.86	191.50	1546
1995-96	121.01	180.42	1491
1996-97	123.58	199.44	1614
1997-98	123.85	192.26	1552
1998-99	125.17	203.61	1627
1999-2000	123.10	209.80	1704
2000-01	121.05	196.81	1626
2001-02	122.78	212.85	1734
2002-03	113.86	174.77	1535
2003-04	123.45	213.19	1727
2004-05	120.00	198.36	1652
2005-06	121.60	208.60	1715
2006-07	123.71	217.28	1756
2007-08	124.07	230.78	1860
2008-09	122.83	234.47	1909
2009-10	121.33	218.11	1798
2010-11	126.67	244.49	1930
2011-12	125.03	257.44	2059

Table1.1:Area, production and productivity of food grain producing states in India (2011-12)

States	Area (mha)	Production (mtonnes)	Productivity (kg/ha)
Uttar Pradesh	80.16	50.29	2495
Punjab	6.53	28.15	4339
Andhra Pradesh	7.29	18.40	2525
West Bengal	6.09	16.29	2675

Haryana	4.63	17.96	3878
Rajasthan	14.41	18.96	1316
Madhya Pradesh	13.50	19.05	1411
Maharashtra	10.71	12.32	1150
Bihar	6.69	14.05	2102
Karnataka	7.45	12.20	1637
Tamilnadu	3.59	9.64	2687
Orissa	4.96	6.43	1297
Gujarat	4.62	9.07	1961
Chhattishgarh	4.84	6.84	1415
Jharkhand	2.68	4.66	1741
Assam	2.43	4.19	1723
Uttarakhand	0.95	1.85	1942
Others	3.49	6.88	-
All India	125.03	257.44	2059

Table 2 : Year wise area, production and productivity of rice in India

Year	Area (m ha)	Production (m tonnes)	Productivity (kg/ha)
1950-51	30.81	20.58	668
1960-61	34.13	34.58	1013
1970-71	37.59	42.22	1123
1980-81	40.15	53.63	1336
1990-91	42.69	74.29	1740
1991-92	42.65	74.68	1751
1992-93	41.78	72.86	1744
1993-94	42.54	80.30	1888
1994-95	42.81	8.81	1911
1995-96	42.84	76.98	1797
1996-97	43.43	81.74	1882
1997-98	43.45	82.53	1900
1998-99	44.80	86.08	1921
1999-2000	45.16	89.68	1986
2000-01	44.71	84.98	1901
2001-02	44.90	93.34	2079
2002-03	41.18	71.82	1744
2003-04	42.59	88.53	2077

2004-05	41.91	83.13	1984
2005-06	43.66	91.79	2102
2006-07	43.81	93.35	2131
2007-08	43.91	96.69	2202
2008-09	45.54	99.18	2178
2009-10	41.92	89.09	2125
2010-11	42.86	95.98	2239
2011-12	43.97	104.32	2372

Table 2.1 : Area, production and productivity of rice producing states in India (2011-12)

States	Area (mha)	Production (mtonnes)	Productivity (kg/ha)
West Bengal	5.46	14.85	2719
Andhra Pradesh	4.10	12.89	3146
Uttar Pradesh	5.95	14.03	2358
Punjab	2.82	10.54	3741
Orissa	4.02	5.82	1448
Tamilnadu	2.01	6.89	3423
Chhattisgarh	3.77	6.03	15.97
Bihar	3.34	7.20	2158
Karnataka	1.39	4.04	2897
Haryana	1.24	3.76	3044
Jharkhand	1.69	3.42	2018
Assam	2.17	4.01	1849
Maharashtra	1.54	2.81	1821
Gujarat	0.82	1.76	2162
Madhya Pradesh	1.66	1.84	1106
Kerala	0.21	0.56	2668
Others	1.79	3.89	-
All India	43.97	104.32	2372

Table 3: Year wise area, production and productivity of wheat in India

Year	Area (m ha)	Production (m tonnes)	Productivity (kg/ha)
1950-51	9.75	6.46	663
1960-61	12.93	11.00	851
1970-71	18.24	23.83	1307
1980-81	22.28	36.31	1630
1990-91	24.17	55.14	22.81
1991-92	23.26	55.69	23.94
1992-93	24.59	57.21	23.27
1993-94	25.15	59.84	23.80
1994-95	25.70	65.77	25.59
1995-96	25.01	62.10	24.83
1996-97	25.89	69.35	26.79
1997-98	26.70	66.35	24.85
1998-99	27.52	71.29	25.90
1999-2000	27.49	76.37	2778
2000-01	25.73	69.68	2708
2001-02	26.34	72.77	27.62
2002-03	25.20	65.76	2610
2003-04	26.60	72.16	2713
2004-05	26.38	68.64	2602
2005-06	26.48	69.35	2619
2006-07	27.99	75.80	2708
2007-08	28.04	78.57	2802
2008-09	27.75	80.68	2907
2009-10	28.46	80.80	2839
2010-11	29.07	86.87	2988
2011-12	29.90	93.90	3140

Table 3.1 : Area, production and productivity of wheat producing states in India (2011-12)

States	Area (mha)	Production (mtonnes)	Productivity (kg/ha)
Uttar Pradesh	9.73	30.29	3113
Punjab	3.51	17.21	4898
Haryana	2.52	12.68	5030

Madhya Pradesh	4.89	10.58	2164
Rajasthan	2.94	9.32	3175
Bihar	2.17	4.79	2206
Gujarat	1.35	4.10	3035
Maharashtra	0.84	1.31	1558
Uttarakhand	0.37	0.87	2369
West Bengal	0.32	0.88	2800
Himanchal Pradesh	0.36	0.60	1671
Jammu & Kashmir	0.29	0.41	1404
Karnataka	0.23	0.19	843
Jharkhand	0.18	0.34	1876
Assam	0.05	0.06	1135
Others	0.16	0.27	-
All India	29.90	93.90	3140

Table 4: Year wise area, production and productivity of maize in India

Year	Area (m ha)	Production (m tonnes)	Productivity (kg/ha)
1950-51	3.16	1.73	547
1960-61	4.41	4.08	926
1970-71	5.85	7.49	1279
1980-81	6.01	6.96	1159
1990-91	5.90	8.96	1518
1991-92	5.86	8.06	1376
1992-93	5.96	9.99	1676
1993-94	6.00	9.60	1602
1994-95	6.14	8.88	1570
1995-96	5.98	9.53	1595
1996-97	6.26	10.77	1720
1997-98	6.32	10.82	1711
1998-99	6.20	11.15	1797
1999-2000	6.42	11.51	1792
2000-01	6.61	12.04	1822
2001-02	6.58	13.16	2000
2002-03	6.64	11.15	1681
2003-04	7.34	14.98	2041
2004-05	7.43	14.17	1907

2005-06	7.59	14.71	1938
2006-07	7.89	15.10	1912
2007-08	8.12	18.96	2335
2008-09	8.17	19.73	2414
2009-10	8.26	16.72	2024
2010-11	8.55	21.73	2540
2011-12	8.71	21.57	2476

Table 4.1 : Area, production and productivity of maize producing states in India (2011-12)

States	Area (mha)	Production (mtonnes)	Productivity (kg/ha)
Karnataka	1.36	4.09	3018
Andhra Pradesh	0.86	3.76	4354
Bihar	0.60	1.52	2507
Uttar Pradesh	0.78	1.31	1668
Maharashtra	0.84	2.30	2745
Rajasthan	1.06	1.67	1579
Madhya Pradesh	0.86	1.29	1492
Tamilnadu	0.30	1.57	5173
Himanchal Pradesh	0.29	0.72	2433
Jammu & Kashmir	0.31	0.51	1608
Punjab	0.13	0.50	3984
Gujarat	0.55	0.77	1532
Jharkhand	0.26	0.41	1602
West Bengal	0.09	0.35	3825
Others	0.46	0.82	-
All India	8.71	21.57	2476

Table 5: Year wise area, production and productivity of bajra in India

Year	Area (m ha)	Production (m tonnes)	Productivity (kg/ha)
1950-51	9.02	2.60	288
1960-61	11.47	3.28	2.86
1970-71	12.91	8.03	622
1980-81	11.66	5.34	458
1990-91	10.48	6.89	658

1991-92	10.03	4.67	465
1992-93	10.62	8.88	836
1993-94	9.55	4.97	521
1994-95	10.22	7.16	700
1995-96	9.32	5.38	577
1996-97	9.98	7.87	788
1997-98	9.67	7.64	791
1998-99	9.30	6.96	748
1999-2000	8.90	5.78	650
2000-01	9.83	6.76	688
2001-02	9.53	8.28	869
2002-03	7.74	4.72	610
2003-04	10.61	12.11	1141
2004-05	9.23	7.93	859
2005-06	9.58	7.68	802
2006-07	9.51	8.42	886
2007-08	9.57	9.97	1042
2008-09	8.75	8.89	1015
2009-10	8.90	6.51	731
2010-11	9.61	10.37	1079
2011-12	8.69	10.05	1156

Table 5.1 : Area, production and productivity of bajra producing states in India (2011-12)

States	**Area (mha)**	**Production (mtonnes)**	**Productivity (kg/ha)**
Rajasthan	5.00	4.16	832
Uttar Pradesh	0.89	1.63	1839
Maharashtra	0.84	0.84	1000
Gujarat	0.80	1.42	1771
Haryana	0.58	1.18	2040
Madhya Pradesh	0.18	0.34	1924
Karnataka	0.28	0.28	979
Tamilnadu	0.06	0.12	21488
Andhra Pradesh	0.04	0.06	1419
Jammu & Kashmir	0.02	0.01	589
Others	0.01	0.02	-
All India	8.69	10.05	1156

Table 6: Year wise area, production and productivity of sorghum in India

Year	Area (m ha)	Production (m tonnes)	Productivity (kg/ha)
1950-51	15.57	5.50	353
1960-61	18.41	9.81	533
1970-71	17.37	8.11	466
1980-81	15.81	10.43	660
1990-91	14.36	11.68	814
1991-92	12.36	8.10	655
1992-93	13.04	12.81	982
1993-94	12.71	11.41	895
1994-95	11.51	8.97	779
1995-96	11.33	9.33	823
1996-97	11.43	10.93	956
1997-98	10.80	7.53	697
1998-99	9.79	8.42	859
1999-2000	10.25	8.68	847
2000-01	9.86	7.53	764
2001-02	9.80	7.56	771
2002-03	9.30	7.01	754
2003-04	9.33	6.68	716
2004-05	9.09	7.24	797
2005-06	8.67	7.24	880
2006-07	8.47	7.15	844
2007-08	7.76	7.93	1021
2008-09	7.53	7.25	962
2009-10	7.79	6.70	860
2010-11	7.38	7.00	949
2011-12	6.32	6.03	954

Table 6.1 : Area, production and productivity of food grain producing states in India (2011-12)

States	Area (mha)	Production (mtonnes)	Productivity (kg/ha)
Maharashtra	3.27	2.69	818
Karnataka	1.15	1.24	1079
Madhya Pradesh	0.39	0.61	1558

Andhra Pradesh	0.28	0.38	1365
Rajasthan	0.56	0.41	735
Tamilnadu	0.28	0.31	1095
Uttar Pradesh	0.19	0.21	1104
Gujarat	0.12	0.15	1185
Haryana	0.07	0.03	508
Orissa	0.01	0.01	616
Others	0.01	0.01	-
All India	6.32	6.03	954

Table7 : Year wise area, production and productivity of coarse cereals in India

Year	Area (m ha)	Production (m tonnes)	Productivity (kg/ha)
1950-51	37.67	15.38	408
1960-61	44.96	23.74	528
1970-71	45.95	30.55	665
1980-81	41.78	29.02	695
1990-91	36.32	32.70	900
1991-92	33.42	25.99	778
1992-93	34.42	36.59	1063
1993-94	32.82	30.82	939
1994-95	32.17	29.88	929
1995-96	30.88	29.03	940
1996-97	31.81	34.10	1072
1997-98	30.83	30.40	986
1998-99	29.34	31.34	1068
1999-2000	39.34	30.33	1034
2000-01	30.26	31.08	1027
2001-02	29.52	33.38	1131
2002-03	26.99	26.07	966
2003-04	30.80	37.60	1221
2004-05	29.03	33.47	1153
2005-06	29.04	34.07	1172
2006-07	28.71	33.92	1182
2007-08	28.48	40.75	1413
2008-09	27.45	40.04	1459

2009-10	27.68	33.55	1212	
2010-11	28.34	43.40	1531	
2011-12	26.37	42.01	1593	

Table 7.1 : Area, production and productivity of coarse cereals producing states in India (2011-12)

States	Area (mha)	Production (mtonnes)	Productivity (kg/ha)
Maharashtra	5.14	5.98	1163
Rajasthan	6.89	7.03	1020
Karnataka	3.50	6.91	1975
Uttar Pradesh	2.03	3.55	1747
Andhra Pradesh	1.25	4.26	3397
Madhya Pradesh	1.77	2.47	1397
Bihar	0.64	1.56	2434
Gujarat	1.51	2.42	1606
Tamilnadu	0.79	2.35	2959
Haryana	0.69	1.39	2001
Himanchal Pradesh	0.32	0.75	2318
Punjab	0.14	0.55	3901
Jammu & Kashmir	0.36	0.53	1480
Uttarakhand	0.25	0.33	1335
Jharkhand	0.27	0.42	1559
West Bengal	0.10	0.36	3505
Chhattisgarh	0.15	0.19	1267
Orissa	0.19	0.26	1383
Others	0.38	0.71	-
All India	26.37	42.01	1593

Table 8: Year wise area, production and productivity of total oilseeds in India

Year	Area (m ha)	Production (m tonnes)	Productivity (kg/ha)
1950-51	10.73	5.16	481
1960-61	13.77	6.98	507
1970-71	16.64	9.63	579
1980-81	17.60	9.37	532
1990-91	24.15	18.61	771

1991-92	25.89	18.60	719
1992-93	25.24	20.11	797
1993-94	26.90	21.50	799
1994-95	25.30	21.34	843
1995-96	25.96	22.11	851
1996-97	26.34	24.38	926
1997-98	26.12	21.32	816
1998-99	26.23	24.75	944
1999-2000	24.28	20.72	853
2000-01	22.77	18.44	810
2001-02	22.64	20.66	913
2002-03	21.49	14.84	691
2003-04	23.06	25.19	1064
2004-05	27.52	24.25	885
2005-06	27.86	27.98	1004
2006-07	26.51	24.29	916
2007-08	26.69	29.76	1115
2008-09	27.56	27.72	1006
2009-10	26.96	24.88	958
2010-11	27.22	32.48	1193
2011-12	24.44	30.01	1135

Table 8.1 : Area, production and productivity of total oilseeds producing states in India (2011-12)

States	**Area (mha)**	**Production (mtonnes)**	**Productivity (kg/ha)**
Madhya Pradesh	7.20	7.72	1071
Rajasthan	4.62	5.73	1241
Maharastra	3.70	4.54	1228
Gujarat	3.08	4.93	1603
Andhra Pradesh	1.95	1.33	683
Karnataka	1.44	0.95	658
Tamilnadu	0.54	1.14	2103
Uttar Pradesh	1.13	0.94	828
Haryana	0.55	0.89	1601
West Bengal	0.69	0.75	1076
Orissa	0.25	0.16	639

Bihar	0.13	0.14	1111
Assam	0.27	0.15	558
Punjab	0.05	0.07	1364
Others	0.85	0.59	-
All India	24.44	30.01	1135

Table 9: Year wise area, production and productivity of groundnut in India

Year	Area (m ha)	Production (m tonnes)	Productivity (kg/ha)
1950-51	4.49	3.48	775
1960-61	6.46	4.81	745
1970-71	7.33	6.11	834
1980-81	6.80	5.01	736
1990-91	8.31	7.51	904
1991-92	8.67	7.09	818
1992-93	8.17	8.56	1049
1993-94	8.32	7.83	941
1994-95	7.85	8.06	1027
1995-96	7.52	7.58	1007
1996-97	7.60	8.64	1138
1997-98	7.09	7.37	1040
1998-99	7.40	8.98	1214
1999-2000	6.87	5.26	766
2000-01	6.56	6.41	977
2001-02	6.24	7.03	1127
2002-03	5.94	4.12	694
2003-04	5.99	8.13	1357
2004-05	6.64	6.77	1020
2005-06	6.74	7.99	1187
2006-07	5.61	4.86	866
2007-08	6.29	9.18	1459
2008-09	6.16	7.17	1163
2009-10	5.48	5.43	991
2010-11	5.86	8.26	1411
2011-12	5.31	6.93	1305

Table 9.1 : Area, production and productivity of ground nut producing states in India (2011-12)

States	Area (mha)	Production (mtonnes)	Productivity (kg/ha)
Gujarat	1.65	2.64	1603
Tamilnadu	0.45	1.07	2382
Andhra Pradesh	1.31	0.85	650
Rajasthan	0.41	0.80	1930
Maharastra	0.29	0.36	1247
Karnataka	0.70	0.50	705
Madhya Pradesh	0.21	0.34	1618
Orissa	0.07	0.07	1147
Uttar Pradesh	0.09	0.09	1000
Others	0.13	0.20	-
All India	5.31	6.93	1305

Table 10 : Year wise area, production and productivity of rapeseed and mustard in India

Year	Area (m ha)	Production (m tonnes)	Productivity (kg/ha)
1950-51	2.07	0.76	368
1960-61	2.88	1.35	467
1970-71	3.32	1.98	594
1980-81	4.11	2.30	560
1990-91	5.78	5.23	904
1991-92	6.55	5.86	895
1992-93	6.19	4.80	776
1993-94	6.29	5.33	847
1994-95	6.01	5.76	958
1995-96	6.55	6.00	916
1996-97	6.55	6.66	1017
1997-98	7.04	4.70	668
1998-99	6.51	5.66	869
1999-2000	6.03	5.79	960
2000-01	4.48	4.19	935
2001-02	5.07	5.08	1002
2002-03	4.54	3.88	854

2003-04	5.43	6.29	1159
2004-05	7.32	7.59	1038
2005-06	7..28	8.13	1117
2006-07	6.79	7.44	1095
2007-08	5.83	5.81	1001
2008-09	6.30	7.20	1143
2009-10	5.59	6.61	1183
2010-11	6.90	8.08	1185
2011-12	5.92	6.78	1145

Table 10.1 : Area, production and productivity of rapeseed and mustard producing states in India (2011-12)

States	Area (mha)	Production (mtonnes)	Productivity (kg/ha)
Rajasthan	2.50	2.97	1187
Uttar Pradesh	0.64	0.72	1125
Haryana	0.54	0.86	1609
Madhya Pradesh	0.79	0.87	1108
Gujarat	0.21	0.33	1577
West Bengal	0.42	0.44	1049
Assam	0.24	0.13	558
Bihar	0.09	0.10	1151
Punjab	0.03	0.04	1333
Others	0.47	0.31	-
All India	5.92	6.78	1145

Table 11: Year wise area, production and productivity of soyabean in India

Year	Area (m ha)	Production (m tonnes)	Productivity (kg/ha)
1970-71	0.03	0.01	426
1980-81	0.61	0.44	728
1990-91	2.56	2.60	1015
1991-92	3.18	2.49	7.82
1992-93	3.79	3.39	894
1993-94	4.37	4.75	1086
1994-95	4.32	3.93	911
1995-96	5.04	5.10	1012

1996-97	5.45	5.38	987
1997-98	5.99	6.46	1079
1998-99	6.49	7.14	1100
1999-2000	6.22	7.08	1138
2000-01	6.42	5.28	822
2001-02	6.34	5.96	940
2002-03	6.11	4.65	762
2003-04	6.56	7.82	1193
2004-05	7.57	6.87	908
2005-06	7.71	8.27	1073
2006-07	8.32	8.85	1063
2007-08	8.88	10.97	1135
2008-09	9.51	9.91	1041
2009-10	9.73	9.96	1024
2010-11	9.60	12.74	1327
2011-12	10.18	12.28	1207

Table 11.1 : Area, production and productivity of soybean producing states in India (2011-12)

States	**Area (mha)**	**Production (mtonnes)**	**Productivity (kg/ha)**
Madhya Pradesh	5.67	6.28	1108
Maharastra	3.07	4.03	1312
Rajasthan	0.90	1.39	1544
Andhra Pradesh	0.13	0.21	1618
Karnataka	0.20	0.18	900
Others	0.21	0.20	-
All India	10.18	12.28	1207

Table 12: Year wise area, production and productivity of sunflower in India

Year		**Area (m ha)**	**Production (m tonnes)**	**Productivity (kg/ha)**
1970-71		0.12	0.08	653
1980-81		0.12	0.07	555
1990-91		1.63	0.87	535
1991-92		2.11	1.19	565
1992-93		2.09	1.18	567

1993-94	2.67	1.35	505
1994-95	2.00	1.22	610
1995-96	2.12	1.26	593
1996-97	1.93	1.25	646
1997-98	1.74	0.89	548
1998-99	1.82	0.94	517
1999-2000	1.29	0.69	538
2000-01	1.07	0.65	602
2001-02	1.18	0.68	577
2002-03	1.64	0.87	531
2003-04	2.01	0.93	464
2004-05	2.17	1.19	549
2005-06	2.34	1.44	615
2006-07	2.16	1.23	567
2007-08	1.91	1.46	765
2008-09	1.81	1.16	639
2009-10	1.48	0.85	576
2010-11	0.93	0.65	701
2011-12	0.72	0.50	692

Table 12.1 : Area, production and productivity of sunflower producing states in India (2011-12)

States	Area (mha)	Production (mtonnes)	Productivity (kg/ha)
Karnataka	0.38	0.19	503
Andhra Pradesh	0.16	0.12	785
Maharashtra	0.09	0.05	586
Tamilnadu	0.02	0.03	1742
Bihar	0.01	0.01	1429
Haryana	0.01	0.02	1800
Uttar Pradesh	0.01	0.02	2143
Others	0.05	0.06	-
All India	0.72	0.50	692

Table 13: Year wise area, production and productivity of total pulses in India

Year	Area (000 ha)	Production (000 tonnes)	Productivity (kg/ha)
1950-51	19090.0	8410.0	441
1960-61	23560.0	12700.0	539
1970-71	22534.0	11817.8	524
1980-81	22456.9	10626.8	473
1990-91	24662.2	14265.3	578
1991-92	22542.5	12014.7	533
1992-93	22359.8	12814.5	573
1993-94	22250.1	13304.8	598
1994-95	23027.7	14037.3	610
1995-96	22283.3	12309.9	552
1996-97	22447.1	14243.9	635
1997-98	22871	12979.3	567
1998-99	23500.7	14907.3	634
1999-2000	21116.2	13418.3	635
2000-01	20348.3	11075.6	544
2001-02	22008.4	13368.1	607
2002-03	20496.2	11125.0	543
2003-04	23462.8	14909.7	635
2004-05	22473.0	13381.0	595
2005-06	22391.0	13391.0	598
2006-07	23760.0	14200.0	594
2007-08	23633.9	14761.5	625
2008-09	22099.9	14570.0	659
2009-10	23280.0	14660.0	630
2010-11	26400.0	18240.0	691
2011-12	24780.0	17210.0	694

Table 13.1 : Area, production and productivity of total pulses producing states in India (2011-12)

States	Area (million ha)	Production (million tonnes)	Productivity (kg/ha)
Andhra Pradesh	1.93	1.25	646
Bihar	0.54	0.51	946
Chhattisgarh	0.80	0.50	617

Gujarat	0.95	0.78	823
Haryana	0.18	0.13	706
Jharkhand	0.54	0.49	914
Karnataka	2.33	1.06	455
Madhya Pradesh	5.18	4.16	803
Maharashtra	3.19	2.22	695
Orissa	0.76	0.36	471
Rajasthan	4.45	2.36	530
Tamilnadu	0.78	0.40	512
Uttar Pradesh	2.45	2.43	991
West Bengal	0.21	0.19	911
Others	0.50	0.39	-
All India	24.78	17.21	694

Table A14: Year wise area, production and productivity of chickpea in India

Year	Area (000 ha)	Production (000 tonnes)	Productivity (kg/ha)
1950-51	7570.0	3650.0	482
1960-61	9280.0	6250.0	674
1970-71	7838.7	5199.2	663
1980-81	6584.1	4328.0	657
1990-91	7521.3	5356.4	712
1991-92	5579.9	4121.0	739
1992-93	6453.7	4416.7	684
1993-94	6358.9	4980.8	783
1994-95	7540.0	6460.0	853
1995-96	7120.0	4980.0	700
1996-97	6850.0	5570.0	813
1997-98	7560.0	6130.0	811
1998-99	8470.0	6800.0	803
1999-2000	6146.3	5118.1	833
2000-01	5185.3	3855.4	744
2001-02	6416.2	5473.0	853
2002-03	5906.4	4236.8	717
2003-04	7052.3	5717.2	811
2004-05	6706.0	5630.0	815
2005-06	6934.0	5603.0	808

2006-07	7632.0	6330.0	782
2007-08	7543.7	5748.6	762
2008-09	7890.0.0	7060.0	895
2009-10	8170.0	7480.0	915
2010-11	9190.0	8220.0	895
2011-12	8320.0	7580.0	912

Table 14.1 : Area, production and productivity of chickpea producing states in India (2011-12)

States	Area (mha)	Production (mtonnes)	Productivity (kg/ha)
Andhra Pradesh	0.57	0.52	920
Bihar	0.06	0.06	1134
Chhattisgarh	0.24	0.24	995
Gujarat	0.24	0.27	1138
Haryana	0.08	0.07	911
Karnataka	0.80	0.28	473
Madhya Pradesh	3.04	3.29	1081
Maharashtra	1.05	0.82	775
Orissa	0.04	0.03	695
Rajasthan	1.43	0.99	691
Uttar Pradesh	0.58	0.72	1248
West Bengal	0.02	0.03	1116
Others	0.17	0.17	-
All India	8.32	7.58	912

Table15 : Year wise area, production and productivity of pigeonpea in India

Year	Area (000 ha)	Production (000 tonnes)	Productivity (kg/ha)
1950-51	2180.0	1720.0	788
1960-61	2430.0	2070.0	849
1970-71	2655.0	1883.3	709
1980-81	2842.2	1957.3	689
1990-91	3593.0	2416.8	673
1991-92	3627.0	2133.1	588
1992-93	3575.6	2330.8	652
1993-94	3531.7	2692.2	762

1994-95	3309.4	2135.0	644
1995-96	3450.0	2310.0	670
1996-97	3510.0	2660.0	756
1997-98	3360.0	1850.0	551
1998-99	3440.0	2710.0	789
1999-2000	3427.0	2694.0	786
2000-01	3632.3	2246.5	618
2001-02	3327.7	2259.8	679
2002-03	3358.9	2185.8	651
2003-04	3515.6	2356.4	670
2004-05	3518.5	2346.9	667
2005-06	3632.0	2574.0	709
2006-07	3528.0	2640.0	712
2007-08	3725.8	3075.9	826
2008-09	3380.0	2270.0	671
2009-10	3470.0	2460.0	711
2010-11	4370.0	2860.0	655
2011-12	4040.0	2650.0	656

Table 15.1 : Area, production and productivity of pigeonpea producing states in India (2011-12)

States	Area (mha)	Production (mtonnes)	Productivity (kg/ha)
Andhra Pradesh	0.48	0.15	307
Bihar	0.03	0.05	1897
Gujarat	0.24	0.26	1057
Jharkhand	0.14	0.13	914
Karnataka	0.77	0.36	466
Madhya Pradesh	0.53	0.33	625
Maharashtra	1.21	0.85	704
Orissa	0.14	0.12	813
Tamilnadu	0.05	0.04	839
Uttar Pradesh	0.32	0.29	891
Others	0.05	0.04	-
All India	4.04	2.65	659

Table 16: Year wise area, production and productivity of urdbean in India

Year	Area (000 ha)	Production (000 tonnes)	Productivity (kg/ha)
1964-65	1980.0	640.0	321
1970-71	2066.5	656.2	318
1980-81	2829.8	959.1	339
1990-91	3482.5	1645.5	473
1991-92	3417.0	1497.5	438
1992-93	3023.3	1530.7	506
1993-94	3049.7	1403.1	460
1994-95	2800.0	1140.0	407
1995-96	2800.0	1320.0	471
1996-97	3010.0	1460.0	484
1997-98	3060.0	1290.0	421
1998-99	3120.0	1350.0	433
1999-2000	2939.4	1330.9	453
2000-01	3011.3	1296.4	431
2001-02	3303.0	1499.1	454
2002-03	3550.1	1473.5	415
2003-04	3424.1	1471.1	430
2004-05	3169.3	1326.6	419
2005-06	2968.8	1250.0	419
2006-07	3176.0	1440.0	453
2007-08	3243.0	1520.0	469

Table 16.1 : Area, production and productivity of urdbean producing states in India (2007-08)

States	Area (000 ha)	Production (000 tonnes)	Productivity (kg/ha)
Andhra Pradesh	503.0	259.0	515
Bihar	25.0	20.0	800
Chhattisgarh	113.0	34.0	301
Gujarat	104.0	71.0	683
Haryana	2.0	1.0	500
Himanchal Pradesh	12.0	6.0	500
Jammu & Kashmir	14.0	6.0	429
Jharkhand	97.0	61.0	629

Karnataka	162.0	63.9	395
Madhya Pradesh	472.0	166.1	352
Maharastra	576.0	327.1	568
Orissa	150.0	46.9	313
Punjab	4.0	2.0	500
Rajasthan	145.0	76.9	531
Tamilnadu	341.0	121.0	355
Uttar Pradesh	391.0	172.0	440
West Bengal	70.0	49.9	714
Others	62.0	36.2	-
All India	3243.0	1520.0	469

Table 17: Year wise area, production and productivity of lentil in India

Year	**Area (000 ha)**	**Production (000 tonnes)**	**Productivity (kg/ha)**
1970-71	753.8	374.6	497
1980-81	934.9	465.2	498
1990-91	1187.8	851.4	717
1991-92	1189.1	801.7	674
1992-93	1203.9	790.8	657
1993-94	1183.4	747.9	632
1994-95	1160.0	780.0	675
1995-96	1260.0	710.0	566
1996-97	1370.0	960.0	702
1997-98	1290.0	800.0	623
1998-99	1390.0	940.0	675
1999-2000	1461.6	1078.9	738
2000-01	1477.8	915.2	619
2001-02	1466.4	974.4	664
2002-03	1377.0	873.2	634
2003-04	1396.4	1034.8	741
2004-05	1468.0	990.4	675
2005-06	1505.4	946.2	629
2006-07	1470.0	910.0	621
2007-08	1310.0	810.0	622
2008-09	1380.0	950.0	693
2009-10	1480.0	1030.0	697
2010-11	1600.0	940.0	591

Table 17.1 : Area, production and productivity of lentil producing states in India (2010-11)

States	Area (mha)	Production (mtonnes)	Productivity (kg/ha)
Assam	0.02	0.01	500
Bihar	0.24	0.21	875
Madhya Pradesh	0.59	0.18	305
Rajasthan	0.04	0.04	1000
Uttar Pradesh	0.59	0.41	695
West Bengal	0.06	0.05	833
Others	0.06	0.04	-
All India	1.60	0.94	591

Table 18: Year wise area, production and productivity of field pea in India

Year	Area (000 ha)	Production (000 tonnes)	Productivity (kg/ha)
1967-68	926.9	661.0	842
1970-71	926.9	780.9	842
1980-81	423.2	291.2	688
1990-91	554.8	604.9	1090
1991-92	578.6	552.2	954
1992-93	608.5	549.5	903
1993-94	690.6	640.3	927
1994-95	760.0	670.0	879
1995-96	760.0	640.0	836
1996-97	730.0	720.51	987
1997-98	740.0	660.0	891
1998-99	830.0	710.0	845
1999-2000	791.5	814.8	1029
2000-01	654.7	536.2	819
2001-02	670.3	607.6	906
2002-03	663.6	591.5	891
2003-04	709.5	725.2	1022
2004-05	791.2	785.6	993
2005-06	777.8	709.9	913

Table 18.1 : Area, production and productivity of field pea producing states in India (2005-06)

States	Area (000 ha)	Production (000 tonnes)	Productivity (kg/ha)
Assam	1.9	11.3	631
Bihar	24.1	21.5	892
Chattisgarh	18.2	6.4	352
Haryana	1.4	1.7	1214
Himanchal Pradesh	5.1	10.5	2059
Jhharkhand	7.7	6.2	805
Madhya Pradesh	229.2	111.9	488
Maharastra	19.7	8.7	442
Rajasthan	13.4	31.7	2366
Uttar Pradesh	413.0	483.3	1170
Uttara khand	5.5	4.8	873
West Bengal	13.1	11.5	878
Others	25.5	0.40	-
All India	777.8	709.9	913

Table 19 : Year wise area, production and productivity of mungbean in India

Year	Area (000 ha)	Production (000 tonnes)	Productivity (kg/ha)
1964-65	1990.0	600.0	302
1970-71	2065.6	700.4	339
1980-81	2843.4	979.1	344
1990-91	3355.0	1384.2	413
1991-92	3489.0	1283.9	368
1992-93	2986.7	1406.7	471
1993-94	2938.8	1232.5	419
1994-95	2900.0	1070.0	369
1995-96	2710.0	820.0	303
1996-97	3060.0	1320.0	430
1997-98	3020.0	950.0	316
1998-99	3050.0	1160.0	380
1999-2000	2905.8	1077.4	371
2000-01	3008.4	1023.2	340
2001-02	3086.8	1110.6	360

2002-03	3014.5	867.4	288
2003-04	3547.8	1702.1	475
2004-05	3341.2	1057.8	317
2005-06	3108.8	946.3	304
2006-07	3237.0	1120.0	346
2007-08	3771.0	1561.2	414

Table 19.1 : Area, production and productivity of mungbean producing states in India (2007-08)

States	Area (000 ha)	Production (000 tonnes)	Productivity (kg/ha)
Andhra Pradesh	440.0	216.9	493
Bihar	177.0	95.0	537
Chattisgarh	16.0	5.0	313
Gujarat	230.0	125.0	543
Haryana	21.0	6.0	286
Karnataka	528.0	108.7	206
Madhya Pradesh	80.0	27.0	33
Maharastra	671.0	371.7	554
Orissa	255.0	61.9	243
Rajasthan	1060.0	416.5	393
Tamilnadu	171..0	55.9	327
Uttar Pradesh	72.0	40.0	556
West Bengal	10.0	5.0	500
Others	40.0	26.6	-
All India	3771.0	1561.2	414

Table 20 : Year wise area, production and productivity of lathyrus in India

Year	Area (mha)	Production (m tonnes)	Productivity (kg/ha)
1970-71	1.67	0.84	504
1980-81	1.35	0.57	418
1990-91	1.06	0.56	530
1991-92	0.96	0.57	598
1992-93	0.88	0.51	580
1993-94	0.97	0.62	641
1994-95	0.99	0.60	605

1995-96	0.93	0.42	455
1996-97	0.88	0.55	623
1997-98	0.85	0.24	288
1998-99	0.66	0.41	621
1999-2000	0.76	0.57	756
2000-01	0.52	0.33	639
2001-02	0.69	0.44	640
2002-03	574.6	354.8	617
2003-04	709.5	492.7	698
2004-05	641.2	304.0	474
2005-06	626.1	345.8	552

Table 20.1 : Area, production and productivity of lathyrus producing states in India (2005-06)

States	Area (000 ha)	Production (000 tonnes)	Productivity (kg/ha)
Bihar	92.1	78.6	853
Chattisgarh	426.9	194.2	455
Madhya Pradesh	42.1	27.8	660
Maharastra	31.7	14.8	467
West Bengal	33.3	30.4	913
All India	626.1	345.8	552

Table 21 : Year wise area, production and productivity of cotton in India

Year	Area (m ha)	Production (m bales of 170 kg)	Productivity (kg/ha)
1950-51	5.88	3.04	88
1960-61	7.61	5.60	125
1970-71	7.61	4.76	106
1980-81	7.82	7.01	152
1990-91	7.44	9.84	225
1991-92	7.66	9.71	216
1992-93	7.54	11.40	257
1993-94	7.32	10.74	249
1994-95	7.87	11.89	257
1995-96	9.04	12.86	242
1996-97	9.12	14.23	265

1997-98	8.87	10.85	208
1998-99	9.34	12.29	224
1999-2000	8.71	11.53	225
2000-01	8.53	9.52	190
2001-02	9.13	10.00	186
2002-03	7.67	8.62	191
2003-04	7.60	13.73	307
2004-05	8.79	16.43	318
2005-06	8.68	18.50	362
2006-07	9.14	22.63	421
2007-08	9.41	22.88	467
2008-09	9.41	22.28	403
2009-10	10.13	24.02	403
2010-11	11.24	33.00	499
2011-12	12.18	35.20	491

Table 21.1 : Area, production and productivity of cotton producing states in India (2011-12)

States	Area (mha)	Production (m bales)	Productivity (kg/ha)
Gujarat	2.96	12.00	689
Maharastra	4.13	7.20	297
Punjab	0.56	2.30	698
Andhra Pradesh	1.88	4.90	443
Haryana	0.64	2.65	703
Madhya Pradesh	0.71	2.00	482
Rajasthan	0.47	1.34	483
Karnataka	0.55	1.20	368
Tamilnadu	0.13	0.45	575
Others	0.15	1.17	-
All India	12.18	35.20	491

Table 22 : Year wise area, production and productivity of jute and mesta in India

Year	Area (m ha)	Production (m bales of 180 kg)	Productivity (kg/ha)
1950-51	0.57	3.31	1043
1960-61	0.90	5.26	1049
1970-71	1.08	6.19	1032
1980-81	1.30	8.16	1130
1990-91	1.02	9.23	1634
1991-92	1.11	10.29	1662
1992-93	0.93	8.59	1658
1993-94	0.89	8.43	1713
1994-95	0.93	9.08	1760
1995-96	0.93	8.81	1712
1996-97	1.10	11.13	1818
1997-98	1.11	11.02	1792
1998-99	1.03	9.81	1722
1999-2000	1.04	10.56	1836
2000-01	1.02	10.56	1867
2001-02	1.05	1168	2007
2002-03	1.04	11.28	1960
2003-04	1.00	11.17	2008
2004-05	0.92	10.27	2019
2005-06	0.90	10.84	2173
2006-07	0.94	11.27	2170
2007-08	0.96	11.21	2101
2008-09	0.90	10.37	2071
2009-10	0.91	11.82	2349
2010-11	0.87	10.62	2192
2011-12	0.91	11.57	2283

Table 22.1 : Area, production and productivity of jute and mesta producing states in India (2011-12)

States	Area (mha)	Production (m bales)	Productivity (kg/ha)
West Bengal	0.60	8.62	2568
Bihar	0.15	1.89	2286
Assam	0.07	0.64	1593
Andhra Pradesh	0.03	0.20	1398
Orissa	0.02	0.10	1005
Meghalaya	0.01	0.05	1170
Maharastra	0.02	0.03	1272
Others	0.02	0.08	-
All India	0.91	11.57	2283

Table 23: Year wise area, production and productivity of sugarcane in India

Year	Area (m ha)	Production (m tonnes)	Productivity (kg/ha)
1950-51	1.71	57.05	33422
1960-61	2.42	110.00	45549
1970-71	2.62	126.37	48322
1980-81	2.67	154.25	57844
1990-91	3.69	241.05	65395
1991-92	3.84	254.00	66069
1992-93	3.57	228.03	63843
1993-94	3.42	229.66	67120
1994-95	3.87	275.54	71254
1995-96	4.15	281.10	67787
1996-97	4.17	277.56	66496
1997-98	3.93	279.54	71134
1998-99	4.05	288.72	71203
1999-2000	4.22	299.32	70935
2000-01	4.32	295.96	68577
2001-02	4.41	297.21	67370
2002-03	4.52	287.38	63576
2003-04	3.93	233.86	59380
2004-05	3.66	237.08	64752
2005-06	4.20	281.17	66928

2006-07	5.15	355.52	69022
2007-08	5.06	348.19	68871
2008-09	4.42	285.03	64553
2009-10	4.17	292.30	70020
2010-11	4.88	342.38	70091
2011-12	5.09	357.67	70317

Table 23.1 : Area, production and productivity of sugarcane producing states in India (2011-12)

States	Area (mha)	Production (mtonnes)	Productivity (kg/ha)
Uttar Pradesh	2.16	128.82	59583
Maharastra	1.02	81.86	80017
Tamilnadu	0.38	39.28	102837
Karnataka	0.43	38.81	90251
Andhra Pradesh	0.20	16.73	82000
Gujarat	0.20	14.78	70181
Haryana	0.10	6.96	73253
Uttarakhand	0.11	6.60	61074
Punjab	0.08	4.67	58375
Bihar	0.23	12.07	51457
Madhya Pradesh	0.07	2.68	38685
West Bengal	0.02	1.18	72981
Orissa	0.01	0.88	6 1014
Assam	0.03	0.97	38600
Others	0.04	1.99	-
All India	5.09	357.67	70317

Table 24: Year wise area, production and productivity of potato in India

Year	Area (m ha)	Production (m tonnes)	Productivity (kg/ha)
1950-51	0.24	1.66	6917
1960-61	0.38	2.72	7251
1970-71	0.48	4.81	9976
1980-81	0.73	9.67	13258
1990-91	0.94	15.21	16254
1991-92	1.03	16.39	15904

1992-93	1.05	15.23	14458
1993-94	1.05	17.39	16610
1994-95	1.07	17.40	16262
1995-96	1.11	18.84	16973
1996-97	1.25	24.22	19376
1997-98	1.21	17.65	14587
1998-99	1.32	23.61	17886
1999-2000	1.34	24.71	18443
2000-01	1.22	22.49	18404
2001-02	1.21	23.92	19806
2002-03	1.35	23.27	17300
2003-04	1.29	23.06	17887
2004-05	1.32	23.63	17923
2005-06	1.40	23.91	17058
2006-07	1.48	22.18	14943
2007-08	1.55	28.47	18331
2008-09	1.83	34.39	18810
2009-10	1.84	36.58	19951
2010-11	1.86	42.34	22724
2011-12	1.87	41.32	22074

Table 24.1 : Area, production and productivity of potato producing states in India (2010-11)

States	Area (000ha)	Production (000 tonnes)	Productivity (kg/ha)
Uttar Pradesh	556.5	13576.6	24396
West Bengal	406.3	13391.2	32955
Punjab	83.6	2088.4	24988
Gujarat	65.3	1881.8	28809
Bihar	314.2	5784.3	18410
Madhya Pradesh	62.0	743.0	11990
Karnataka	40.0	400.8	10010
Assam	66.7	739.6	8507
Haryana	26.8	598.2	22336
Jharkhand	43.2	655.5	15174
Others	178.5	2482.1	-
All India	1863.2	42339.4	22724

Table 25: Year wise area, production and productivity of tobacco in India

Year	Area (m ha)	Production (m tonnes)	Productivity (kg/ha)
1950-51	0.36	0.26	731
1960-61	0.40	0.31	766
1970-71	0.45	0.36	810
1980-81	0.45	0.48	1065
1990-91	0.41	0.56	1353
1991-92	0.43	0.58	1369
1992-93	0.42	0.60	1425
1993-94	0.38	0.56	1463
1994-95	0.38	0.57	1486
1995-96	0.39	0.54	1356
1996-97	0.43	0.62	1444
1997-98	0.46	0.64	1394
1998-99	0.51	0.74	1451
1999-2000	0.43	0.52	1211
2000-01	0.26	0.34	1318
2001-02	0.35	0.55	1565
2002-03	0.33	0.49	1506
2003-04	0.37	0.55	1486
2004-05	0.37	0.55	1498
2005-06	0.37	0.55	1481
2006-07	0.37	0.52	1409
2007-08	0.35	0.49	1417
2008-09	0.39	0.57	1456
2009-10	0.44	0.69	1559
2010-11	0.49	0.83	1687

Table 25.1 : Area, production and productivity of tobacco producing states in India (2010-11)

States	Area (000 ha)	Production (000 tonnes)	Productivity (kg/ha)
Andhra Pradesh	157.00	281.00	1790
Uttar Pradesh	26.00	79.80	3069
Gujarat	148.00	281.00	1425
Karnataka	125.00	134.00	1072

Bihar	10.99	20.90	1902
Tamilnadu	5.37	8.12	1512
Maharastra	2.00	2.00	1000
Others	18.61	25.74	-
All India	489.97	831.41	1687

Table 26: Area, production and yield of principle crops of various countries in 2010

Nam	Area (ha)	Production (t)	Yield (kg/ha)
1. Paddy			
World	159416542	672015587	43736
China	30116862	197212010	65482
India	36950000	120620000	32644
Indonesia	13244200	66411500	50144
Bangladesh	11800000	49355000	41826
Vietnam	7513700	39988900	53221
Myanmar	8051700	33204500	41239
Thailand	10990100	31597200	28751
Philippines	4354160	15771700	36222
Brazil	2722460	11236000	41272
Japan	1628000	10600000	65111
Pakistan	2365000	7235000	30592
USA	1462950	11027000	75375
Cambodia	2776510	8245320	29697
Egypt	459525	4329500	94217
Korea	892074	6136300	68787
Nepal	1481290	4023820	27164
Madagaskar	1808000	4737970	26206
Srilanka	1060360	4300620	40558
Nigeria	1788200	3218760	18000
Iran	563517	2288150	40605
Russia	200900	1060660	52795
2. Wheat			
World	217219395	653654525	30092
China	24256086	115180303	47486
India	28460000	80800000	28391
Russia	21639800	41507600	19181

USA	19270900	60062400	31167
France	5931000	40787000	68769
Canada	8268700	23166800	28017
Germany	3297700	24106700	73102
Pakistan	9131600	23310800	25528
Australia	13507000	22138000	16390
Ukraine	6284100	16851300	26816
Turkey	8053670	19660000	24411
Kazakhstan	13138000	96384000	7336
UK	1939000	14878000	76730
Iran	7035020	15028800	21363
Poland	2406100	9487800	39432
Egypt	1287630	7169000	55676
Argentina	4373440	14914500	34102
Italy	1830480	6849860	37421
Romania	2152520	5811810	27000
Spain	1907300	5610700	29417
Syrian	1599100	3083100	19280
Bangladesh	376256	901490	23959
3. Maize			
World	161765388	840308214	51946
U.S.A	32960400	316165000	95923
China	32517868	177540788	54598
Brazil	12683400	55394800	43675
Mexico	7148050	23301900	32599
Indonesia	4131680	18327600	44359
India	7180000	14060000	19582
France	1571000	13975000	88956
Argentina	2902750	22676900	78122
South Africa	2742000	12815000	46738
Ukraine	2647600	11953000	45147
Canada	1202900	11714500	97385
Romania	2094250	9042030	43176
Italy	925967	8827810	95336
Hungry	1060610	6967170	65690
Philippines	2499040	6376800	25517
Egypt	968519	7183000	74165

Serbia	1223580	7207190	58902
Germany	463600	4072900	87854
Turkey	593552	4310000	72814
Russian Fed	1025200	3084350	30085
Pakistan	939000	3341000	35580
4. Sugarcane			
World	23877378	1711087173	716614
Brazil	9076710	717462000	790443
India	4170000	292300000	700959
China	1695228	111454359	657459
Thailand	977956	68807800	703588
Pakistan	942870	49372900	523645
Mexico	703943	50421600	716274
Colombia	380000	38500000	1013158
Australia	405000	31457000	776716
Argentina	350000	25000000	714286
U.S.A.	355112	24820600	698951
Indonesia	336000	24450000	727679
Philippines	362834	34000000	937068
South Africa	267000	16015600	599835
Guatemala	234650	22216700	946802
Egypt	134538	15708900	1167618
Viet Nam	266300	15946800	598828
Cuba	431400	11300000	261938
Peru	76983	9660900	1254939
Venezuela	118000	8907670	754887
Ecuador	106928	8347180	780636
Bangladesh	121000	5304000	438347
Mauritius	58709	4365830	743639
5. Groundnut (in shell)			
World	24011537	37953949	15807
China	4547917	15709036	34541
India	4930000	5640000	11440
Nigeria	2636230	2636230	10000
U.S.A.	507890	1885510	37124
Myanmar	866460	1341000	15477
Senegal	1195570	1286860	10764

Sudan	1151640	762500	6621
Indonesia	620563	779228	12557
Argentina	218828	611040	27923
Vietnam	231284	485792	21004
Ghana	353376	530887	15023
Chad	504400	394400	7819
Tanzania	415000	300000	7229
United Republic			
Guinea	215000	291700	13567
Mali	336978	314458	9332
Burkina Faso	409922	340166	8298
Brazil	94329	261455	27717
Malawi	295236	297487	10076
Egypt	66764	202906	30392
Uganda	235000	172000	7319
Cameroon	300000	460000	15333
Thailand	28405	45344	15963
Congo	45200	29100	6438
Japan	7720	16200	20984
6.Tobacco Unmanufactured			
World	3946665	7037514	17832
China	1345703	3005753	22336
Brazil	449629	787617	17517
India	459600	755500	16438
U.S.A.	136582	325766	32576
Malawi	195000	220198	11292
Indonesia	193916	122276	6306
Argentina	75500	137000	18146
Italy	27829	89112	32021
Pakistan	55800	119323	21384
Zimbabwe	94175	109737	11652
Turkey	80977	55000	6792
Mozambique	59200	86000	14527
Zambia	59988	89700	14953
Thailand	30641	58912	19227
Tanzania	41000	50000	12195

United Republic			
Bulgaria	25161	41056	16317
Canada	15600	40200	25769
Korea Republic	14000	41100	29357
Bangladesh	38270	55288	14447
Poland	15500	31414	20267
Japan	15000	29300	19533
Philippines	29706	40530	13644
Spain	10600	32300	30472
Greece	15600	22000	14103
Lao People's	6500	29100	44769
France	7245	18428	25435
Iran, Islam Republic	9586	14145	14756

Table 27: India's position in world agriculture in 2010

Item	India	World	India's position		
			% share	Rank	Next to
1. Area (m ha)					
Total area	329	13445	2.4	Seventh	Russia, Canada, USA China, Brazil, Australia
Land area	297	13015	2.3	Seventh	Russia, Canada, USA China, Brazil, Australia
Arable land	160	1423	11.3	Second	USA
2. Population (Million)					
Total	1214	6909	17.6	Second	China
Agriculture	661	2617	25.2	Second	China
3. Economically active population (million)					
Total	491	3282	15.0	Second	China
Agriculture	267	1310	25.2	Second	China
4. Crop Production (million Tones)					
(A) Total	260	2458	10.6	Third	China, USA
Wheat	81	654	12.4	Second	China
Rice (Paddy)	121	672	17.9	Second	China
Total Pulses	17	67	25.5	First	
(B) Oilseeds					
Groundnut (in shell)	6	38	14.9	Third	China, Indonesia
Rapeseed	6.4	59.0	10.9	Third	China, Canada

5. Fruits and Vegetables (Million Tonnes)					
(A) Vegetables & melon	100	1036	9.7	Second	China
(B) Fruits excluding melon	75	603	12.5	Second	China
(C) Potato	37	324	11.3	Third	China, Russia
(D) Onion	15	79	19.3	Second	China
6. Commercial crops (Million Tonnes)					
(A) Sugarcane	292	1711	17.1	Second	Brazil
(B) Tea	0.99	4.5	22.1	Second	China
(C) Coffee (green)	0.29	8.23	3.5	Sixth	Brazil,Vientam, Colombia, Indonesia
(D) Jute & allied fibre	1.85	3.30	56.0	First	
(E) Cotton (lint)	5.70	23.30	24.4	Third	China, USA
(F) Tobacco leaves	0.76	7.04	10.7	Third	China, Brazil
7. Livestock					
(A) Cattle	210	1430	14.7	Second	Brazil
(B) Buffalo	111	194	57.3	First	
(C) Camel	0.45	25	1.8	Sixth	Sudan, Ethopia Kenya, Chand, Pakistan
(D) Sheep	74	1078	6.9	Third	China, Australia
(E) Goat	154	910	16.9	Second	China
(F) Chick	774	19571	4.0	Fifth	China, USA, Indonesia Brazil
8. Animal products					
(A) Total	121847	723143.31	16.8	First	
(B) Eggs	3378.1	69102.73	4.9	Third	China, USA
(C) Total	6190	295462	2.1	Fifth	China, USA, Brazli
9. Implements (Thousands numbers)					
Tractors	3149	29320	10.7	Second	USA

Table 28: Selected categories of land use from 1950-51 to 2005-06

000 hectares

Year	Net area sown	Total cropped area	area sown more than onece	Net irrigated area	Gross cropped area once	Area irrigated more than
1950-51	118.75	131.89	13.15	20.85	22.56	1.71
1960-61	133.20	152.77	19.57	24.66	27.98	3.32
1970-71	140.27	165.79	25.52	31.10	38.20	7.09
1980-81	140.00	172.63	32.63	38.72	49.78	11.06
1990-91	143.00	185.74	42.74	48.02	63.20	15.18
1991-92	141.63	182.24	40.61	49.87	65.68	15.81
1992-93	142.72	185.70	42.98	50.29	66.76	16.47
1993-94	142.34	186.58	44.25	51.34	68.26	16.92
1994-95	142.96	188.05	45.09	53.00	70.65	17.65
1995-96	142.20	187.47	45.27	53.40	71.35	17.95
1996-97	142.93	189.50	46.57	55.11	76.03	20.91
1997-98	141.95	190.02	48.07	55.17	75.42	20.25
1998-99	142.76	191.69	48.92	57.41	78.37	20.96
1999-2000	141.06	188.43	47.37	57.11	78.79	21.69
2000-01	141.40	185.37	43.97	55.08	76.57	21.50
2001-02	141.45	189.71	48.26	56.67	78.73	22.06
2002-03	132.69	175.62	42.93	53.78	73.55	19.77
2003-04	140.97	190.20	49.23	56.62	78.00	21.38
2004-05	141.14	190.42	49.28	58.87	80.00	21.13
2005-06	141.89	192.80	50.90	60.20	82.63	22.43

Table 29: All India consumption of fertilizers in terms of nutrients (NPK) from 1951-52 to 2011-12

000 tones

Year	N	P	K	Total
1951-52	58.7	6.9	-	65.6
1960-61	210.0	53.1	29.0	292.1
1970-71	1487.0	462.0	228.0	2177.0
1980-81	3678.1	1213.6	623.9	5515.6

1990-91	7997.2	3221.0	1328.0	12546.2
1991-92	8046.3	3321.2	1360.5	12728.0
1992-93	8426.8	2843.8	883.9	12154.5
1993-94	8788.3	2669.3	908.4	12366.0
1994-95	9507.1	2931.7	1124.7	13563.5
1995-96	9822.8	2897.5	1155.8	13876.1
1996-97	10301.8	2976.7	1029.6	14308.1
1997-98	10901.8	3913.6	1372.5	16187.9
1998-99	11353.8	4112.2	1331.5	16797.5
1999-2000	11592.7	4798.3	1678.7	18069.7
2000-01	10920.2	4214.6	1567.5	16702.3
2001-02	11310.2	4382.4	1667.1	17359.7
2002-03	10474.1	4018.8	1601.2	16094.1
2003-04	11077.0	4124.3	1597.9	16799.1
2004-05	11713.9	4623.8	2060.6	18398.3
2005-06	12723.3	5203.7	2413.3	20340.3
2006-07	13772.9	5543.3	2334.8	21651.0
2007-08	14419.1	5514.7	2636.3	22570.1
2008-09	15090.5	6506.2	3312.6	24909.3
2009-10	15580.0	7274.0	3632.4	26486.4
2010-11	16558.2	8049.7	3514.3	28122.2
2011-12	17300.3	7914.3	2575.5	27740.0

Table 30: Fertilizer consumption kg per hectare of agricultural land in selected countries in 2008 and 2009

Country	2008				2009			
	N	P	K	Total	N	P	K	Total
Africa								
Egypt	321.6	50.9	15.5	388.1	323.4	40.7	10.8	375.0
Morocco	6.5	3.7	1.5	11.7	6.7	3.8	1.6	12.0
South Africa	4.3	1.9	1.2	7.3	4.2	1.8	1.1	7.1
North & Central America								
Canada	26.9	8.5	2.6	38.0	28.4	9.3	4.0	41.7
Mexico	10.2	1.7	1.9	13.9	12.2	1.5	1.7	15.3
USA	25.7	7.0	6.9	39.6	27.4	9.2	10.1	46.7
South America								
Brazil	9.4	12.1	13.9	35.5	9.7	12.7	11.9	34.3

Chile	16.3	6.8	5.7	28.7	16.5	6.8	5.7	29.0
Asia								
Bangladesh	130.7	8.9	5.6	145.1	125.6	32.7	17.1	176.0
China	63.6	20.1	9.0	92.7	64.1	21.4	8.5	93.9
India	84.0	36.2	18.4	138.6	86.6	40.4	20.2	147.2
Indonesis	52.6	8.1	18.3	79.1	55.0	10.3	14.9	80.3
Japan	90.8	75.6	54.0	220.4	83.3	66.3	65.1	214.6
Jordan	167.1	49.9	55.4	272.4	169.9	48.5	56.6	275.1
Korea	63.7	22.5	89.4	175.6	66.3	29.2	88.9	184.5
Nepal	1.9	1.9	0.2	4.0	2.0	0.2	0.2	4.2
Pakistan	115.5	24.8	1.0	141.2	132.3	32.7	0.9	165.9
Srilanka	62.6	18.9	26.5	107.9	65.1	17.2	19.2	101.5
Thailand	47.3	15.0	18.5	80.8	62.0	12.1	8.4	82.5
Turkey	29.0	8.4	2.3	39.6	36.3	14.8	1.6	52.8
Vietnam	94.2	59.8	36.9	191.0	115.8	58.4	29.2	203.5
Europe								
Belarus	62.1	26.0	86.7	174.7	59.4	26.2	82.2	167.7
Denmark	71.2	6.4	12.0	89.6	72.9	6.5	13.3	92.6
France	71.6	10.1	13.3	95.0	70.7	13.9	14.2	988.8
Germany	91.6	10.3	10.6	112.5	92.9	13.9	21.5	128.3
Netherland	128.0	17.1	18.7	163.6	127.8	18.3	17.2	163.3
Poland	65.0	25.4	25.4	115.8	61.1	21.8	26.0	108.9
Russia	6.6	2.4	1.5	10.5	6.9	2.5	1.6	11.0
Spain	26.0	5.6	6.7	38.3	29.4	11.7	9.9	51.0
United Kingdom	51.9	7.5	11.7	71.1	58.6	10.3	15.2	84.1
Ukrain	17.8	4.2	3.8	25.8	17.0	4.6	1.8	23.3
Oceania								
Australia	2.0	2.0	0.5	4.5	2.1	1.6	0.4	
New Zealand	25.9	25.7	9.0	60.7	28.6			
World	20.1	6.9	4.8	31.8	7.7			

Table 31: All India production of milk, eggs and wool from 1985-86 to 2010-11

Year	Milk (Million Tonnes)	Eggs (Million Nos.)	Wool (Million kg)
1985-86	44.2	16.1	39.1
1986-87	46.1	17.3	40.0
1987-88	46.7	17.8	40.1
1988-89	48.4	19.0	40.8
1989-90	51.4	20.2	41.7
1990-91	53.9	21.1	41.2
1991-92	55.7	22.0	41.6
1992-93	58.0	22.9	38.8
1993-94	60.6	24.2	39.9
1994-95	64.0	26.0	40.6
1995-96	66.2	27.2	42.4
1996-97	69.1	27.5	44.4
1997-98	72.1	28.7	45.6
1998-99	75.4	29.5	46.9
1999-2000	78.3	30.4	47.9
2000-01	80.6	36.6	48.4
2001-02	84.4	38.7	49.5
2002-03	86.2	39.8	50.5
2003-04	88.1	40.4	48.5
2004-05	92.6	45.2	44.6
2005-06	87.1	46.2	44.9
2006-07	100.9	50.7	45.1
2007-08	104.8	53.5	44.0
2008-09	112.2	55.6	42.8
2009-10	116.4	60.3	43.1
2010-11	121.8	63.0	43.0

Table 32 : Fish production in India

Million tonnes

Year	Marine	Inland	Total
1950-51	5.34	2.18	7.52
1960-61	8.80	2.80	11.60
1970-71	10.86	6.70	17.56
1980-81	15.55	8.87	24.42

1981-82	14.45	9.99	24.44
1982-83	14.27	9.40	23.67
1983-84	15.19	9.87	25.06
1984-85	16.98	11.03	28.01
1985-86	17.16	11.60	28.76
1986-87	17.13	12.29	29.42
1987-88	16.58	13.01	29.59
1988-89	18.17	13.35	31.52
1989-90	22.75	14.02	36.77
1990-91	23.00	15.36	38.36
1991-92	24.47	17.10	41.57
1992-93	25.76	17.89	43.65
1993-94	26.49	19.95	46.44
1994-95	26.92	20.97	47.89
1995-96	27.07	22.42	49.49
1996-97	29.67	23.81	53.48
1997-98	29.50	24.38	53.88
1998-99	29.96	26.02	52.98
1999-2000	28.52	28.23	56.75
2000-01	28.11	28.45	56.56
2001-02	28.30	31.26	59.56
2002-03	29.90	32.10	62.00
2003-04	29.41	34.58	63.99
2004-05	27.79	35.25	63.05
2005-06	28.16	37.56	65.72
2006-07	30.24	38.45	68.69
2007-08	29.19	42.07	71.26
2008-09	29.77	46.60	76.37
2009-10	31.04	48.94	79.98
2010-11	32.25	51.98	84.23

Table 33: Statewise fish production in 2010-11

(000 tonnes)

State/UTs	Marine	Inland	Total
Andhra Pradesh	288637	1079565	1368202
Arunachal Pradesh	0.00	3035	3035
Assam	0.00	227242	227242
Bihar	0.00	299910	299910
Goa	89962	3308	93270
Gujarat	688930	85972	774902
Haryana	0.00	96195	96195
Himachal Pradesh	0.00	7381	7381
Jammu & Kashmir	0.00	19700	19700
Karnataka	295570	204271	499841
Kerala	560398	139475	699873
Madhya Pradesh	0.00	56451	56451
Maharashtra	444703	148546	595249
Manipur	0.00	20200	20200
Meghalaya	0.00	4557	4557
Mizoram	0.00	2901	2901
Nagaland	0.00	6585	6585
Orissa	133481	267284	400765
Punjab	0.00	97040	97040
Rajasthan	0.00	23708	23708
Sikkim	0.00	180	180
Tamilnadu	424842	189967	614809
Tripura	0.00	49231	49231
Uttar Pradesh	0.00	417479	417479
West Bengal	197108	1436466	1633574
A&N Island	33735	186	186
Chandigarh	0.00	242	242
D & N Haveli	0.00	50	50
Daman & Deu	16851	124	16975
Delhi	0.00	820	820
Lakshadweep	12372	0.00	12372
Pondichery	36100	5849	41949
Chhattisgarh	0.00	228207	228207
Uttarakhand	0.00	3818	3818
Jharkhand	0.00	71886	71886
Total	**3224689**	**5197831**	**8422520**

Table 34: Per capita net availability of food grains (per day) in India

(Grammes per day)

Years	Rice	Wheat	Other Cereals	cereals	Gram	Pulses	Food grains
1951	158.9	65.7	109.6	334.2	22.5	60.7	394.9
1956	187.7	61.5	111.2	360.4	29.0	70.3	430.7
1961	201.1	79.1	119.5	399.7	30.2	69.0	468.7
1966	161.9	95.4	102.6	359.9	18.3	48.2	408.1
1971	192.6	103.6	121.4	417.6	20.0	51.2	468.8
1976	187.2	79.5	107.4	373.8	20.2	50.5	424.3
1981	197.8	129.6	89.9	417.3	13.4	37.5	454.8
1985	188.8	138.6	87.9	415.3	12.9	38.1	453.4
1990	212.1	132.6	86.8	431.5	10.7	41.1	472.6
1991	221.7	166.8	80.0	468.5	13.4	41.6	510.1
1992	217.0	158.6	58.9	434.5	10.1	34.3	468.8
1993	201.1	140.2	86.6	427.9	10.7	36.2	464.1
1994	207.4	159.5	67.1	434.0	11.8	37.2	471.2
1995	220.0	172.7	64.9	457.6	14.9	37.8	495.5
1996	204.4	176.0	62.0	442.5	11.3	32.7	475.2
1997	214.0	179.1	72.9	466.0	12.4	37.1	503.1
1998	200.3	151.5	62.4	414.2	13.4	32.8	447.0
1999	203.4	162.3	63.4	429.2	14.6	36.5	465.7
2000	203.7	160.0	59.0	422.7	10.8	31.8	454.4
2001	190.5	135.8	56.2	386.2	8.0	30.0	416.2
2002	228.7	166.6	63.4	458.7	10.7	35.4	494.1
2003	181.4	180.4	46.7	408.5	8.5	29.1	437.6
2004	195.4	162.2	69.3	426.9	11.2	35.8	462.7
2005	177.3	154.3	59.4	390.9	10.6	31.5	422.4
2006	198.0	154.3	60.5	412.8	10.7	32.5	445.3
2007	196.8	156.1	58.9	409.9	11.9	29.4	439.3
2008	175.4	145.1	54.1	394.2	10.6	41.8	436.0
2009	188.4	154.7	63.9	407.0	12.9	37.0	444.0
2010	182.0	168.2	51.4	401.7	13.5	35.4	437.1
2011	188.8	164.6	70.0	423.5	14.6	39.4	462.9

Table 35 : Per capita net availability of food grains (per annum) in India

(kg per year)

Years	Rice	Wheat	Other Cereals	cereals	Gram	Pulses	Food grains
1951	58.0	24.0	40.0	122.0	8.2	22.1	144.1
1956	68.7	22.5	40.7	131.9	10.6	25.7	157.6
1961	73.4	28.9	43.6	145.9	11.0	25.2	171.1
1966	59.1	34.8	37.5	131.4	6.7	17.6	149.0
1971	70.3	37.8	44.3	152.4	7.3	18.7	171.1
1976	68.5	29.1	39.2	13.8	7.4	18.5	155.3
1981	72.2	47.3	32.8	152.3	4.9	13.7	166.0
1985	68.9	50.6	32.1	151.6	4.7	13.9	165.5
1990	77.4	48.4	31.7	157.5	3.9	15.0	172.5
1991	80.9	60.0	29.2	1710	4.9	15.2	186.2
1992	79.2	57.9	21.5	158.6	3.7	12.5	171.1
1993	73.4	51.2	31.6	156.2	3.9	13.2	169.4
1994	75.7	58.2	24.5	158.4	4.3	13.6	172.0
1995	80.3	63.0	23.7	167.0	5.4	13.8	180.8
1996	74.6	64.3	22.6	161.5	4.1	12.0	173.5
1997	78.1	65.4	26.6	170.1	4.5	13.5	183.6
1998	73.1	55.3	22.8	151.2	4.9	12.0	163.2
1999	74.2	59.2	23.1	156.7	5.3	13.3	170.0
2000	74.3	58.4	21.5	154.3	3.9	11.6	165.9
2001	69.5	49.6	20.5	141.0	2.9	10.9	151.9
2002	83.5	60.8	23.1	167.4	3.9	12.9	180.4
2003	66.2	65.8	17.1	149.1	3.1	10.6	159.7
2004	71.3	59.2	25.3	155.8	4.1	13.1	168.9
2005	64.7	56.3	21.7	142.7	3.9	11.5	154.2
2006	72.3	56.3	22.1	150.7	3.9	11.8	162.5
2007	71.8	57.0	20.8	149.6	4.3	10.7	160.4
2008	64.0	53.0	19.7	143.9	3.9	15.3	159.2
2009	68.8	56.5	23.3	148.6	4.7	13.5	162.1
2010	66.4	61.4	18.8	146.6	4.9	12.9	159.5
2011	68.9	60.1	25.6	154.6	5.3	14.4	169.0

Table 36:Three largest producing states of important crops during 2011-12

Crops/group of crops	States	Production (Million tonnes)	% share in India
1. Food grains			
Rice	West Bengal	14.85	14.23
	Uttar Pradesh	14.03	13.45
	Andhra Pradesh	12.89	12.36
	All India	104.32	
Wheat	Uttar Pradesh	30.29	32.26
	Punjab	17.21	18.33
	Haryana	12.68	13.50
	All India	93.90	
Maize	Karnataka	4.09	18.96
	Andhra Pradesh	3.76	17.43
	Maharashtra	2.30	10.66
	All India	21.57	
Total coarse cereal	Rajasthan	7.03	16.73
	Karnataka	6.91	16.45
	Maharastra	5.98	14.23
	All India	42.01	
Total pulses	Madhya Pradesh	4.16	24.17
	Uttar Pradesh	2.43	14.12
	Rajasthan	2.36	13.71
	All India	17.21	
Total food grains	Uttar Pradesh	50.29	19.53
	Punjab	28.35	11.01
	Madhya Pradesh	19.05	7.40
	All India	257.44	
Oilseeds			
Groundnut	Gujarat	2.64	38.10
	Tamilnadu	1.07	15.44
	Andhra Pradesh	0.85	12.27
	All India	6.93	
Rapeseed & Mustard	Rajasthan	2.97	43.81
	Madhya Pradesh	0.87	12.83
	Haryana	0.86	12.68

	All India	6.78	
Soyabean	Madhya Pradesh	6.28	51.14
	Maharastra	4.03	32.82
	Rajasthan	1.39	11.32
	All India	12.28	
Sunflower	Karnataka	0.19	38.00
	Andhra Pradesh	0.12	24.00
	Maharashtra	0.05	10.00
	All India	0.50	
Total Oilseeds	Madhya Pradesh	7.72	25.72
	Rajasthan	5.73	19.09
	Gujarat	4.93	16.42
	All India	30.01	
Other cash crops			
Sugar cane	Uttar Pradesh	128.82	36.02
	Maharastra	81.86	22.89
	Tamilnadu	39.28	10.98
	All India	357.67	
Cotton*	Gujarat	12.00	34.09
	Maharashtra	7.20	30.45
	Andhra Pradesh	4.90	13.92
	All India	35.20	
Jute & Mesta**	West Bengal	8.62	74.50
	Bihar	1.89	16.33
	Assam	0.64	5.53
	All India	11.57	
Potato	Uttar Pradesh	13.58	32.07
	West Bengal	13.39	31.62
	Punjab	2.88	6.80
	All India	42.34	
Onion	Gujarat	2.33	29.75
	Maharashtra	1.11	15.50
	Karnataka	0.85	11.87
	All India	7.16	

* Production in million bales of 170 kg each

** Production of million bales of 180 kg each

Table 37 : Distribution of agricultural land by use in India from 1950-51 to 2009-10

Classification	1950-51	1960-61	1970-71	1980-81	1990-91	1995-96	2000-01	2001-02	2002-03	2003-04	2004-05	2005-06	2007-08	2008-09	2009-10
Geographical area	328.73	328.73	328.73	328.73	328.73	328.73	328.73	328.73	328.73	328.73	328.73	328.73	328.73	328.73	328..73
Reporting area	284.32	298.46	303.76	304.16	304.86	304.88	305.17	305.11	305.34	305.40	305.31	305.27	305.61	305.59	305.61
Utilisation Statistics															
1. Forest	40.48	54.05	63.92	67.47	67.81	68.82	69.53	69.57	69.70	69.73	69.72	69.79	70.02	70.03	70.04
2. Net available for sultivation															
A. Area under cultivation (%)	9.36	14.84	16.48	19.66	21.09	22.38	23.86	24.06	24.27	24.66	24.76	25.03	25.71	26.06	26.17
B. Barren & Uncultivable land (%)	38.16	35.91	28.16	19.96	19.39	19.01	17.60	17.72	17.81	17.58	17.58	17.48	16.99	16.80	16.78
3. Other uncultivated land excluding fallow land															
A. Permanent other grazzland	6.68	13.97	13.26	11.97	11.40	11.06	10.66	10.58	10.50	10.44	10.42	10.42	10.20	10.18	10.15
B. Land under crops &	19.83	4.46	4.30	3.61	3.82	3.48	3.43	3.38	3.37	3.40	3.38	3.38	3.41	3.36	3.35
C. Culturable (%)	22.94	19.21	17.50	16.74	15.00	14.10	13.67	13.40	13.53	13.13	13.16	13.12	13.06	12.75	12.86
4. Fallow lands															
A. Fallow land current fallow	17.45	11.18	8.76	9.92	9.66	10.02	10.31	10.32	11.78	11.21	10.71	10.50	10.33	10.29	10.48
B. Current fallow	10.68	11.64	11.12	14.83	13.70	13.83	14.77	14.64	21.69	14.28	14.43	13.67	14.51	14.19	15.75
5. Net area sown (%)	118.75	133.20	140.27	140.00	143.00	142.20	141.40	141.45	132.69	140.97	141.14	141.89	141.38	141.93	140.02
6. Total cropped (Gross cropped area)	131.89	152.77	165.79	172.63	185.74	187.47	185.37	189.71	175.62	190.20	190.42	192.80	195.14	195.56	191.20
7. Area sown	13.15	19.57	25.52	32.63	42.74	45.27	43.96	48.26	42.93	49.23	49.28	50.90	53.76	53.43	52.18
8. Cropping Intensity	111.1	114.7	118.2	123.3	129.9	131.8	131.1	134.1	132.4	134.9	134.9	135.9	138.03	137.64	137.26
Net irrigated	20.85	24.66	31.10	38.72	48.02	53.40	55.08	56.67	53.78	56.62	58.87	60.20	63.29	63.74	63.26
Gross irrigated	22.56	27.98	38.20	49.78	63.20	71.35	76.57	78.73	73.55	78.00	80.00	82.63	87.98	88.87	86.42

Table 38: State wise per cent coverage of irrigated area under principle crops during 2009-10

State /Uts area	Rice	Jowar	Bajra	Maize	Wheat	Barley	Total cereals	Gram	Arhar	Total	Total Pulses grains	Sugrance food	Groundnut	Rapeseed mustard	Soyabeen &	Sunfl. Oowner	Total 36. eds	Cotton oilse-	Tobacco	Total under all crops
Andhra Pradesh	97.5	8.1	42.0	44.2	86.3	-	31.6	79.6	1.0	2.5	57.3	91.8	24.4	22.6	7.9	-	25.5	16.3	23.2	45.9
Arunachal Pradesh	40.3	-	-	-	46.1	-	0.0	26.8	-	-	25.7	-	-	-	-	-	-	-	-	20.3
Assam	7.1	-	-	-	0.6	-	0.0	6.9	-	1.0	6.6	-	-	-	-	82.9	0.3	-	-	5.5
Bihar	56.7	6.9	-	62.7	92.1	14.4	60.3	69.9	5.6	3.7	84.2	26.0	85.3	40.9	-	94.9	40.7	-	93.0	61.7
Chattisgarh	31.7	0.1	-																	
Goa	33.8	-	-	5.6	68.1	1.8	2.0	30.6	30.8	10.3	27.2	95.6	12.5	7.9	1.9	-	4.7	-	95.5	26.7
Gujrat	63.3	16.3	22.1	9.6	89.8	100.0	17.3	49.2	28.8	10.4	41.5	100.0	11.0	85.4	-	-	28.2	56.7	86.4	44.3
Haryana	99.9	71.3	3.1	24.0	99.4	96.4	44.9	90.7	17.2	30.7	89.0	99.5	93.1	79.1	-	-	79.2	99.8	100.0	87.3
Himachal Pradesh	64.9	-	100.0	9.6	20.2	14.1	9.9	20.3	10.6	13.3	20.0	50.7	7.1	13.0	30.8	-	18.9	7.9	60.0	20.2
Jammu & Kashmir	91.2	-	3.1	7.3	30.0	24.3	9.8	39.7	-	12.3	38.8	-	-	75.5	-	-	69.9	47.1	18.2	41.9
Jharkhand	2.3	-	-	3.2	91.5	1.1	1.8	6.0	12.8	4.7	5.9	88.0	-	73.5	-	-	15.2	-	-	11.1
Karnataka	75.2	9.8	11.2	40.3	56.7	-	18.9	36.4	13.5	7.3	27.3	100.0	29.2	6.0	14.3	22.2	28.2	17.5	19.7	31.8
Kerala	72.3	-	-	-	-	-	0.0	71.3	-	-	70.0	97.5	-	-	-	-	19.4	-	-	17.1
Madhya Pradesh	17.9	0.2	-	1.9	84.7	52.7	3.1	52.4	49.1	35.3	45.9	99.8	8.0	49.4	0.2	32.7	6.0	44.2	35.3	33.5
Maharastra	26.1	9.5	5.0	12.7	73.9	-	8.9	19.8	24.3	10.1	17.1	100.0	20.8	9.0	0.4	22.8	3.7	2.7	16.7	19.2
Manipur	30.7	-	-	-	-	-	0.0	29.9	-	-	27.6	-	-	-	-	-	-	-	-	22.3
Meghalaya	45.8	-	-	-	100.0	-	0.0	39.9	-	-	37.9	-	-	46.3	-	-	33.4	-	-	21.9
Mizoram	28.1	-	-	-	-	-	0.0	229	-	-	21.1	-	-	-	-	-	0.0	-	-	8.4
Nagaland	46.1	-	-	0.3	36.1	-	0.2	31.1	7.0	1.1	27.6	-	-	3.3	-	5.4	6.2	-	-	17.6
Orissa	46.8	-	-	16.9	100.0	-	15.1	44.2	-	7.7	33.6	100.0	40.3	23.0	-	97.0	18.7	3.6	100.0	35.1
Punjab	99.6	100.0	91.8	70.4	98.7	97.5	74.2	98.5	83.0	83.8	98.4	95.2	35.6	91.1	-	-	86.4	100.0	-	98.0
Rajasthan	66.7	0.5	4.5	1.0	99.6	98.8	6.5	30.1	46.5	13.9	25.9	97.8	75.0	83.8	0.7	-	53.5	91.9	92.5	33.6
Sikkim	8.5	-	-	3.0	17.0	12.1	3.1	20.4	-	3.0	17.5	-	-	20.0	1.5	-	12.5	-	-	12.8
Tamilnadu	92.8	10.3	11.1	44.8	44.4	-	23.0	74.5	2.0	6.1	62.4	100.0	37.8	14.9	92.8	61.0	58.7	28.6	100.0	58.1
Tripura	38.4	-	-	13.4	-	-	13.4	38.1	57.1	16.7	37.6	4.8	17.9	3.7	-	-	6.8	5.4		34.4
Uttarakh-and	69.5	-	-	1.3	57.5	2.4	0.5	46.4	70.6	8.5	44.4	98.0	10.8	49.3	2.3	18.5	29.2	-	100.0	76.3
Uttar Pradesh	79.0	0.8	6.	33.8	98.1	70.2	22.2	83.4	16.6	25.6	75.8	93.1	1.6	81.9	0.6	97.8	47.3	96.7	100..	76.3
West Bengal	48.2	19.6	94.6	14.4	95.9	53.0	19.2	50.1	30.6	23.3	49.3	59.3	-	88.1	-	-	82.4	79.2	97.5	58.0
All India	58.0	8.7	8.6	23.8	91.7	73.3	14.4	55.4	32.2	16.2	47.8	93.5	24.3	71.1	0.7	29.5	25.9	35.3	44.6	45.0

Table 39: Consumption and production of fertilizers from 1980-81 to 2011-12

Year	Consumption				Production			
	N	P	K	Total	N	P	K	Total
1981-82	40.69	13.22	6.73	60.64	31.44	9.49	-	40.93
1982-83	42.24	14.37	7.27	63.88	34.25	9.80	-	44.04
1983-84	52.05	17.30	7.75	77.10	34.85	10.48	-	45.33
1984-85	54.87	18.86	8.38	82.11	39.17	12.63	-	51.80
1985-86	56.61	20.05	8.08	84.74	43.28	14.28	-	57.56
1986-87	57.16	20.79	8.50	86.45	54.10	16.60	-	70.70
1987-88	57.17	21.87	8.80	87.84	54.66	16.65	-	71.31
1988-89	72.51	27.21	10.68	110.40	67.12	22.52	-	89.64
1989-90	73.86	30.14	11.68	115.68	67.47	17.96	-	85.43
1990-91	79.97	32.21	13.28	125.46	69.93	20.52	-	90.45
1991-92	80.46	33.21	13.61	127.28	73.01	25.62	-	98.63
1992-93	84.26	28.43	8.84	121.53	74.30	23.06	-	97.36
1993-94	87.88	26.69	9.09	123.66	72.31	18.16	-	90.47
1994-95	95.07	29.32	11.25	135.64	79.45	24.93	-	104.38
1995-96	98.23	28.98	11.56	138.77	87.77	25.58	-	113.35
1996-97	103.01	29.77	10.30	143.08	85.99	25.56	-	111.55
1997-98	109.01	39.14	13.73	161.88	100.86	29.76	-	130.62
1998-99	113.54	41.12	13.32	167.02	104.80	31.44	-	136.24
1999-2000	115.92	47.99	16.78	180.69	108.90	33.99	-	142.89
2000-01	109.20	42.15	15.67	167.02	109.61	37.43	-	147.04
2001-02	113.10	43.82	16.67	173.59	107.68	38.60	-	146.28
2002-03	104.74	40.19	16.01	160.94	105.61	39.04	-	144.65
2003-04	110.77	41.24	15.98	167.99	106.34	36.32	-	142.66
2004-05	117.14	46.24	20.61	183.99	113.39	40.64	-	154.03
2005-06	127.23	52.04	24.13	203.40	113.54	42.21	-	155.75
2006-07	137.73	55.43	23.35	216.51	115.78	45.18	-	160.96
2007-08	144.19	55.19	26.36	225.70	109.00	38.07	-	147.07
2008-09	150.91	65.06	33.12	249.09	108.70	34.64	-	143.34
2009-10	155.80	72.74	36.32	264.86	119.00	43.21	-	162.21
2010-11	165.58	80.50	35.14	281.22	121.54	45.22	-	163.78
2011-12	173.00	79.14	25.25	277.40	122.58	41.03	-	163.61

Table 40: State wise consumption of fertilizers in 2010-11 and 2011-12

(Thousand tones)

State/Zone	2010-11				2011-12			
	N	P	K	Total	N	P	K	Total
South zone								
Andhra Pradesh	1966.63	1031.98	498.18	3496.79	1977.29	1043.02	322.04	3342.35
Karnataka	1016.21	696.17	398.05	2110.43	1215.9	786.76	332.85	2335.55
Kerala	117.68	69.00	96.86	283.54	135.54	66.16	99.63	301.33
Tamilnadu	643.18	279.91	306.10	1229.19	684.56	316.39	263.96	1264.91
Pondicherry	19.14	4.81	5.43	29.38	14.61	3.90	3.06	21.57
A&N Island	0.39	0.33	0.19	0.91	0.37	0.37	0.17	0.91
Lakshadweep	0.00	0.00	0.00	0.00	0.00	0.00	0.00	0.00
Sub total (South zone)	**3763.23**	**2082.20**	**1304.81**	**7150.24**	**4028.31**	**2216.60**	**1021.71**	**7266.62**
West zone								
Gujarat	1241.22	518.00	179.94	1939.16	1183.30	417.02	132.74	1733.06
Madhya Pradesh	998.30	741.11	128.33	1867.74	1061.75	750.76	79.47	1891.98
Chhattisgarh	321.99	171.19	68.99	562.17	356.40	177.33	61.84	595.57
Maharashtra	1657.29	1126.37	671.68	3455.34	1610.91	1011.76	399.48	3022.15
Rajasthan	870.39	413.30	34.95	1318.64	913.49	416.11	26.18	1355.78
Goa	3.32	2.27	1.79	7.38	3.06	2.67	1.89	7.62
Daman & Diu	0.42	0.16	0.03	0.61	0.06	0.01	0.00	0.07
D&N Haveli	0.72	0.53	0.03	1.28	0.60	0.41	0.00	1.04
Sub total (West zone)	**5093.65**	**2972.93**	**1085.74**	**9152.32**	**5129.57**	**2776.07**	**701.60**	**8607.24**
North zone								
Haryana	974.04	335.95	47.63	1357.62	1020.90	369.62	37.53	1428.05
Punjab	1402.91	435.17	73.43	1911.51	1416.56	448.65	52.85	1918.06
Uttar Pradesh	2951.01	1097.49	267.39	4315.89	3067.10	1024.23	116.42	4207.75
Himanchal Pradesh	32.59	10.73	11.81	55.13	32.80	9.70	8.96	51.43
Jammu & Kashmir	72.82	37.30	11.15	121.27	66.13	28.65	5.29	100.07
Delhi	0.37	0.05	0.00	0.42	0.35	0.22	0.01	0.58
Chandigarh	1.00	0.00	0.00	0.00	0.00	0.00	0.00	0.00
Sub total (north zone)	**5545.66**	**1947.60**	**425.44**	**7918.70**	**5727.62**	**1913.38**	**231.37**	**7872.37**
East zone								
Bihar	907.28	289.59	155.84	1352.71	967.78	297.01	115.36	1380.15
Jharkhand	79.08	38.11	9.01	126.20	118.02	42.01	11.34	171.37
Orissa	294.72	153.97	89.15	537.84	323.41	135.48	55.80	514.89
West Bengal	712.37	495.58	363.87	1571.82	831.99	476.17	309.04	1617.20
Sub total (East zone)	**1993.45**	**977.25**	**617.87**	**3588.57**	**2241.20**	**950.67**	**491.54**	**3683.41**

North East zone								
Assam	142.76	59.51	74.79	277.06	151.05	49.08	75.52	275.65
Tripura	8.01	4.47	3.46	1594	10.42	5.49	2.73	18.64
Manipur	5.10	1.10	0.30	6.50	6.59	0.97	0.44	8.00
Meghalaya	3.03	1.52	0.48	5.03	3.27	1.24	0.25	4.76
Nagaland	0.77	0.48	0.17	1.42	0.75	0.49	0.20	1.44
Arunachal Pradesh	0.52	0.22	0.09					
Mizoram	2.05	2.43	1.12	5.60	0.92	0.21	0.06	1.19
Sikkim	0.00	0.00	0.00	0.00	0.00	0.00	0.00	0.00
Sub total (North east zone)	**162.24**	**69.73**	**80.41**	**312.38**	**173.55**	**57.58**	**79.23**	**310.36**
All India	**16558.23**	**8049.71**	**3514.27**	**28122.21**	**17300.25**	**7914.30**	**2525.45**	**27740.00**

Table 41: Production and use of agricultural inputs in India from 1999-2000 to 2011-12

Inputs	Unit	1999-2000	2000-01	2001-02	2002-03	2003-04	2004-05	2005-06	2006-07	2007-08	2008-09	2009-10	2010-11	2011-12
1. Seeds Production of Breeder seeds	Thousand Qtls	51.13	42.69	45.54	48.42	61.82	66.46	68.64	73.83	91.96	94.41	105.00	119.21	119.21
Production of foundation seeds	Lakh Qtls	4.66	5.91	5.44	6.14	6.50	6.90	7.40	7.96	8.22	9.69	10.50	17.53	21.86
Distribution of certified seeds	Lakh Qtls	87.98	86.27	91.80	98.03	108.59	120.26	126.75	155.01	179.05	215.81	257.11	277.34	283.85
2. Consumption of chemical fertilizers														
Nitrogenous (N)	Lakh Tonnes	115.92	109.20	113.10	104.74	110.77	117.13	127.23	137.73		144.19	150.91	165.58	173.00
Phosphatic (P)	Lakh Tonnes	47.99	42.15	43.82	40.19	41.24	46.24	52.04	55.43		35.15	65.06	80.50	79.14
Potassic (K)	Lakh Tonnes	16.78	15.67	16.67	16.01	15.98	20.61	24.13	23.35		26.36	33.12	35.14	25.26
Total	Lakh Tonnes	180.69	167.02	173.60	160.94	167.99	183.98	203.40	216.51		25.71	249.09	281.22	277.40
	Kg/ha	94.94	89.63	91.13	91.45	88.05	94.52	104.50	112.69		116.51	128.58	146.32	144.33
3. Consumption of pesticides	Thousand Tonnes	46.20	43.58	47.02	4.30	41.00	40.67	39.77	37.95		43.63	41.82	55.54	50.58
Area covered under soil conservation	Million hectares	4.70	4.30	5.55	7.37	8.67	11.41	7.34			6.90	5.32	7.49	4.72

Table 42: Livestock population in India (Million numbers)

Species	1951	1956	1961	1966	1972	1977	1982	1987	1992	1997	2003	2007
Cattle	155.3	158.7	175.6	176.2	178.3	180.0	192.5	199.7	204.6	198.8	185.2	199.08
Adult female cattle	54.4	47.3	51.0	51.8	53.4	54.6	59.2	62.1	64.4	63.6	64.5	70.95
Buffalo	43.4	44.9	51.2	53.0	57.4	62.0	69.8	76.0	84.2	89.9	97.9	105.34
Adult female buffalo	21.0	21.7	24.3	25.4	28.6	31.3	32.5	39.1	43.8	46.8	51.0	54.47
Total Bovine	198.7	203.6	226.8	229.2	235.7	242.0	262.4	275.8	289.0	289.0	283.4	304.42
Sheep	39.1	39.3	40.2	42.4	40.0	41.0	48.8	45.7	50.8	57.5	61.5	71.56
Goats	47.2	55.4	60.9	64.6	67.5	75.6	95.3	110.2	115.3	122.7	124.4	140.54
Horses & Ponies	1.5	1.5	1.3	1.1	0.3	0.9	0.9	0.8	0.8	0.8	0.8	0.61
Camels	0.6	0.8	0.9	1.0	1.1	1.1	1.1	1.0	1.0	0.9	0.6	0.52
Pigs	4.4	4.9	5.2	5.0	6.6	7.6	10.1	10.6	12.8	13.3	13.5	11.13
Mules	0.1	0.0	0.1	0.1	0.1	0.1	0.1	0.2	0.2	0.2	0.2	0.14
Donkeys	1.3	1.1	1.1	1.1	1.0	1.0	1.0	1.0	1.0	0.9	0.6	0.44
Total Livestock	292.8	306.6	335.4	344.1	353.6	369.0	419.6	445.3	470.9	485.4	485.0	529.70
Poultry	73.5	94.8	114.2	115.4	138.5	159.2	207.7	275.3	307.1	347.6	489.0	468.88
Dogs	NC	NC	NC	NC	NC	NC	18.5	18.0	21.8	25.5	29.0	19.09

ANSWERSHEET

CHAPTER 1: INTRODUCTION

1. (A); 2. (B); 3. (A); 4. (A); 5. (A); 6. (C); 7. (D); 8. (D); 9. (C); 10. (B); 11. (D); 12. (A); 13. (A); 14. (A); 15. (A); 16. (B); 17. (C); 18. (C); 19. (A); 20. (B); 21. (C); 22. (D); 23. (A); 24. (B); 25. (D); 26. (D); 27. (C); 28. (A); 29. (A); 30. (A); 31. (B); 32. (B); 33. (D); 34. (A); 35. (B); 36. (A); 37. (A); 38. (A); 39. (A); 40. (B); 41. (B); 42. (B); 43. (D); 44. (A); 45. (D); 46. (B); 47. (D); 48. (C); 49. (C); 50. (A); 51. (A); 52. (C); 53. (B); 54. (C); 55. (A); 56. (D); 57. (C); 58. (A); 59. (B); 60. (C); 61. (D); 62. (B); 63. (C); 64. (D); 65. (B); 66. (C); 67. (A); 68. (A); 69. (C); 70. (A); 71. (B); 72. (A); 73. (A); 74. (B); 75. (C); 76. (C); 77. (C); 78. (A); 79. (B); 80. (B); 81. (D); 82. (D); 83. (C); 84. (D); 85. (B); 86. (A); 87. (A); 88. (B); 89. (C); 90. (D); 91. (A); 92. (B); 93. (C); 94. (B); 95. (D); 96. (A); 97. (A); 98. (C); 99. (D); 100. (C); 101. (A); 102. (A); 103. (D); 104. (B); 105. (D); 106. (B); 107. (A); 108. (B); 109. (C); 110. (C); 111. (B); 112. (C); 113. (D); 114. (B); 115. (B); 116. (A); 117. (A); 118. (A); 119. (C); 120. (A); 121. (A); 122. (D); 123. (D); 124. (D); 125. (A); 126. (A); 127. (B); 128. (C); 129. (D); 130. (C); 131. (B); 132. (C); 133. (D); 134. (B); 135. (A); 136. (A); 137. (B); 138. (D); 139. (D); 140. (B); 141. (A); 142. (C); 143. (C); 144. (B); 145. (B); 146. (D); 147. (B); 148. (B); 149. (A); 150. (A); 151. (A); 152. (D); 153. (B); 154. (A); 155. (A); 156. (B); 157. (B); 158. (A); 159. (B); 160. (A); 161. (A); 162. (D); 163. (C); 164. (A); 165. (A); 166. (C); 167. (B); 168. (D); 169. (D); 170. (B); 171. (A); 172. (B); 173.(A); 174. (A); 175. (A); 176. (A); 177. (B); 178. (B); 179. (B); 180. (A); 181. (B); 182. (B); 183. (B); 184. (B); 185. (C); 186. (B); 187. (B); 188. (B); 189. (B); 190. (B); 191. (D); 192. (B); 193. (D); 194. (B); 195. (D); 196. (C); 197. (A); 198. (C); 199. (D); 200. (A); 201. (C); 202. (D); 203. (C); 204. (B); 205. (A); 206. (B); 207. (C); 208. (C); 209. (A); 210. (B); 211. (B); 212. (D); 213. (C); 214. (B); 215. (B); 216. (B); 217. (A); 218. (A); 219. (D); 220. (A); 221. (B); 222. (B); 223. (A); 224. (D); 225. (A); 226. (A); 227. (A); 228. (B); 229. (B); 230. (A); 231. (C); 232. A); 233. (A); 234. (C); 235. (A); 236. (C); 237. (D); 238 (B); 239. (A); 240. (C); 241. (C); 242. (D); 243. (A); 244. (B); 245. (B); 246. (B); 247. (A); 248. (A); 249. (A); 250. (B); 251. (A); 252. (C); 253. (B); 254. (B); 255. (C); 256. (D); 257. A); 258. (A); 259. (A); 260. (A); 261. (A); 262. (B); 263. (B); 264. (A); 265. (D); 266. (B); 267. (B); 268. (A); 269. (C); 270. (A); 271. (B); 272. (A); 273. (B); 274. (C), 275. (C); 276. (C); 277. (C); 278. (C); 279. (D); 280. (B); 281. (C); 282. (A); 283. (C); 284. (A); 285. (B); 286. (B); 287. (C); 288. (B); 289. (C); 290. C); 291. (A); 292. (B); 293. (C); 294. (A); 295. (B); 296. (C); 297. (A); 298. (B); 299. (B); 300. (D); 301. (D); 302. (A); 303. (D); 304. (C); 305. (B); 306. (D); 307. (C); 308. (B); 309. (A); 310. (C); 311. (A); 312. (B); 313. (A); 314. (D); 315. (A); 316. (A); 317. (A); 318. (A); 319. (A); 320. (A); 321. (B); 322. (B); 323. (D); 324. (C); 325. (A), 326. (T); 327. (F); 328. (F);

329. (T); 330. (T); 331. (F); 332. (F); 333. (T); 334. (F); 335. (F); 336. (F); 337. (T); 338. (F); 339. (F); 340. (F); 341. (F); 342. (F); 343. (T); 344. (T); 345. (F); 346. (T); 347. (T); 348. (T); 349. (T); 350. (F); 351. (F); 352. (T); 353. (T); 354. (T); 355. (F); 356. (T); 357. (T); 358. (F); 359. (T); 360. (F); 361. (T); 362. (T); 363. (F); 364. (F); 365. (F); 366. (T); 367. (T); 368. (F); 369. (F); 370. (T); 371. (F); 372. (T); 373. (T); 374. (T); 375. (F); 376. (F); 377. (F); 378. (F); 379. (F); 380. (F); 381. (T); 382. (T); 383. (T); 384. (T); 385. (T); 386. (F); 387.(F); 388. (T); 389. (F); 390. (F); 391. (T); 392. (T); 393. (F); 394. (T); 395. (1904); 396. (1948); 397. (1947); 398. (1948, S. K. Dubey); 399. (A. Mayer); 400. (1951); 401. (1978); 402. (Block); 403. (1977); 404. (1954); 405. (1970); 406.(1974. India); 407. (Doing); 408. (Dr. Seeman A. Kappa); 409. (Short); 410. (Long); 411. (Ministry of Food & Agriculture); 412. (Baster & Benor Scheme, Daniel Benor); 413. (USA); 414. (John Deway, 1966); 415. (Individual matter); 416. (Innovation, Diffusion, Motivation and Adoption); 417. (Mass communication); 418. (Learning call for); 419. (Spoken); 420. (Television); 421. (Doing); 422. (National demonstration); 423. (Communication); 424. (Latin); 425. (Need and interest of people); 426. (Helping the people who help themselves); 427. (Is development of people); 428. (Farmers); 429. (Professional leader); 430. (1978); 431. (ICAR); 432. (1959); 433. (Flexible); 434. (To educate the people); 435. (Demonstration); 436. (Collector); 437. (1975); 438. (1966-67); 439. (1959); 440. (Teaching, Display aids); 441. (Deputy commissioner/collector); 442. (1979, ICAR and IBRD); 443. (Towards betterment); 444. (Government of India); 445. (Mahatma Gandhi); 446. (Motivation); 447. (Overhead); 448. (2 lakhs); 449. (Rural development programme); 450. (Brayne); 451. (Flexible); 452. (Package programme); 453. ((SFDA, MFAL); 454. (Government); 455. (1997-98); 456. (Mc Leagan Committee); 457. (Prime Minister); 458. (District); 459. (Attitude); 460. (1953); 461. (Demonstration); 462. (5); 463. (1977-78); 464. (10); 465. (Status); 466. (VDO); 467. (Need); 468. (Ist result demonstration); 469. (Mass communication); 470. (Audio-visual); 471. (Programme); 472. (July, 1982); 473. (Birth); 474. (Edger Dale); 475. (Judgment); 476. (Primary group)

CHAPTER 2: HISTORICAL PERSPECTIVE OF AGRICULTURE AND EXTENSION

1. (A); 2. (C); 3. (B); 4. (D); 5. (B); 6. (C); 7. (A); 8. (C); 9. (D); 10. (B); 11. (D); 12. (B); 13. (B); 14. (D); 15. (A); 16. (B); 17. (B); 18. (C); 19. (B); 20. (B); 21. (B); 22. (C); 23. (D); 24. (C); 25. (D); 26. (C); 27. (A); 28. (D); 29. (D); 30. (A); 31. (B); 32. (C); 33. (C); 34. (B); 35. (D); 36. (C); 37. (A); 38. (A); 39. (A); 40. (D); 41. (B); 42. (D); 43. (A); 44. (A); 45. (C); 46. (B); 47. (D); 48. (D); 49. (B); 50. (D); 51. (B); 52. (D); 53. (A); 54. (A); 55. (C); 56. (C); 57. (D); 58. (D); 59. (D); 60. (D); 61. (C); 62. (B); 63. (D); 64. (C); 65. (D); 66. (A); 67. (A); 68. (A); 69. (C); 70. (C); 71. (B); 72. (C); 73. (B); 74. (D); 75. (A); 76. (D); 77. (D); 78. (D); 79. (A); 80. (A); 81. (A); 82. (C); 83. (D); 84. (B); 85. (C); 86. (B); 87. (A); 88. (B); 89. (A); 90. (B); 91. (D); 92. (C); 93. (B); 94. (A); 95. (C); 96. (D); 97. (B); 98. (C); 99. (A); 100. (B); 101. (A); 102. (A); 103. (C); 104. (B); 105. (D); 106. (A); 107. (A); 108. (A); 109. (A); 110. (B); 111. (B); 112. (A); 113. (D); 114. (C); 115. (A); 116. (A); 117. (C); 118. (C); 119. (C); 120. (B); 121. (D); 122. (D); 123. (D); 124. (B); 125. (A); 126. (B); 127. (C); 128. (C); 129. (D); 130. (C); 131. (B); 132. (B); 133. (A); 134. (C); 135. (D); 136. (A); 137. (D); 138. (B); 139. (C); 140. (A); 141. (C); 142. (A); 143. (B); 144. (A); 145. (B); 146. (B); 147. (B); 148. (A); 149. (A); 150. (A); 151. (C); 152. (C); 153. (B); 154. (C); 155. (B); 156. (C); 157. (A); 158. (C); 159. (C); 160. (A); 161. (A); 162. (A); 163. (C); 164. (B); 165. (D); 166. (C); 167. (C); 168. (A); 169. (C); 170. (B); 171. (C); 172. (A); 173.(B); 174. (D); 175. (A); 176. (B); 177. (D); 178. (D); 179. (B); 180. (C); 181. (B); 182. (B); 183. (D); 184. (D); 185. (A); 186. (C); 187. (C); 188. (A); 189. (D); 190. (D); 191. (A); 192. (A); 193. (A); 194. (C); 195. (A); 196. (A); 197. (C); 198. (A);

199. (D); 200. (B); 201. (B); 202. (C); 203. (A); 204. (A); 205. (C); 206. (A); 207. (B); 208. (B); 209. (B); 210. (B); 211. (B); 212. (B); 213. (A); 214. (D); 215. (C); 216. (D); 217. (A); 218. (D); 219. (C); 220. (C); 221. (B); 222. (B); 223. (A); 224. (A); 225. (A); 226. (A); 227. (B); 228. (B); 229. (C); 230. (A); 231. (B); 232. (D); 233. (B); 234. (C); 235. (B); 236. (A); 237. (A); 238. (C); 239. (B); 240. (A); 241. (B); 242. (C); 243. (C); 244. (A); 245. (B); 246. (C); 247. (C); 248. (C); 249. (D); 250. (C); 251. (A); 252. (B); 253. (B); 254. (C); 255. (C); 256. (A); 257. (C); 258. (C); 259. (B); 260. (A); 261. (B); 262. (A); 263. (C); 264. (C); 265. (A); 266. (C); 267. (B); 268. (B); 269. (B); 270. (A); 271. (B); 272. (D); 273. (B); 274. (C), 275. (B); 276. (C); 277. (B); 278. (B); 279. (A); 280. (A); 281. (A); 282. (B); 283. (A); 284. (A); 285. (C); 286. (C); 287. (C); 288. (C); 289. (B); 290. (A); 291. (C); 292. (C); 293. (B); 294. (A); 295. (D); 296. (C); 297. (C); 298. (C); 299. (D); 300. (C); 301. (B); 302. (C); 303. (C); 304. (C); 305. (C); 306. (C); 307. (C); 308. (B); 309. (C); 310. (C); 311. (C); 312. (C); 313. (A); 314. (A); 315. (C); 316. (B); 317. (B); 318. (B); 319. (B); 320. (B); 321. (A); 322. (A); 323. (B); 324. (B); 325. (C), 326. (C); 327. (D); 328. (D); 329. (A); 330. (A); 331. (B); 332. (A); 333. (A); 334. (C); 335. (B); 336. (B); 337. (D); 338. (A); 339. (B); 340. (C); 341. (B); 342. (B); 343. (A); 344. (C); 345. (B); 346. (C); 347. (B); 348. (A); 349. (A); 350. (D); 351. (B); 352. (B); 353. (A); 354. (A); 355. (B); 356. (C); 357. (A); 358. (A); 359. (A); 360. (A); 361. (C); 362. (A); 363. (B); 364. (A); 365. (C); 366. (B); 367. (C); 368. (C); 369. (B); 370. (B); 371. (C); 372. (B); 373. (A); 374. (C); 375. (C); 376. (A); 377. (A); 378. (C); 379. (B); 380. (C); 381. (D); 382. (D); 383. (A); 384. (B); 385. (D); 386. (B); 387.(A); 388. (D); 389. (B); 390. (C); 391. (C); 392. (B); 393. (A); 394. (B); 395. (B); 396. (C); 397. (C); 398. (D); 399. (B); 400. (A); 401. (B); 402. (A); 403. (C); 404. (B); 405. (C); 406.(B); 407. (B); 408. (A); 409. (B); 410. (D); 411. (C); 412. (A); 413. (B); 414. (B); 415. (A); 416. (D); 417. (A); 418. (B); 419. (D); 420. (A); 421. (A); 422. (D); 423. (C); 424. (B); 425. (A); 426. (B); 427. (B); 428. (C); 429. (D); 430. (D); 431. (B); 432. (A); 433. (B); 434. (D); 435. (A); 436. (D); 437. (A); 438. (C); 439. (A); 440. (C); 441. (B); 442. (B); 443. (C); 444. (D); 445. (A); 446. (B); 447. (D); 448. (C); 449. (C); 450. (C); 451. (A); 452. (B); 453. (A); 454. (B); 455. (C); 456. (D); 457. (C); 458. (B); 459. (A); 460. (A); 461. (A); 462. (B); 463. (C); 464. (C); 465. (A); 466. (D); 467. (B); 468. (D); 469. (D); 470. (B); 471. (D); 472. (A); 473. (B); 474. (C); 475. (B); 476. (C); 477. (A); 478. (A); 479. (B); 480. (D); 481. (C); 482. (B); 483. (B); 484. (A); 485. (B); 486. (D); 487. (A); 488. (A); 489. (C); 490. (B); 491. (B); 492. (A); 493. (B); 494. (A); 495. (A); 496. (D); 497. (A); 498. (D); 499. (B); 500. (B); 501. (B); 502. (D); 503. (B); 504. (C); 505. (A); 506. (D); 507. (B); 508. (C); 509. (B); 510. (B); 511. (C); 512. (B); 513. (A); 514. (C); 515. (A); 516. (D); 517. (D); 518. (C); 519. (B); 520. (B); 521. (B); 522. (D); 523. (A); 524. (A); 525. (C); 526. (A); 527. (C); 528. (C); 529. (C); 530. (A); 531. (B); 532. (B); 533. (D); 534. (A); 535. (C); 536. (D); 537. (C); 538. (C); 539. (D); 540. (A); 541. (C); 542. (C); 543. (C); 544. (A); 545. (D); 546. (B); 547. (D); 548. (D); 549. (C); 550. (A); 551. (A); 552. (B); 553. (D); 554. (C); 555. (C); 556. (A); 557. (A); 558. (C); 559. (B); 560. (A); 561. (B); 562. (C); 563. (A); 564. (A); 565. (C); 566. (A); 567. (A); 568. (C); 569. (A); 570. (C); 571. (A); 572. (D); 573. (A); 574. (C); 575. (D); 576. (D); 577. (A); 578. (A); 579. (B); 580. (C); 581. ()A; 582. (B); 583. (B); 584. (B); 585. (A); 586. (B); 587. (B); 588. (D); 589. (B); 590. (D); 591. (C); 592. (D); 593. (A); 594. (A); 595. (A); 596. (D); 597. (C); 598. (B); 599. (A); 600. (A); 601. (B); 602. (D); 603. (C); 604. (A); 605. (A); 606. (B); 607. (C); 608. (C); 609. (C); 610. (C); 611. (B); 612. (C); 613. (D); 614. (C); 615. (C); 616. (B); 617. (A); 618. (A); 619. (A); 620. (B); 621. (C); 622. (C); 623. (A); 624. (D); 625. (D); 626. (A); 627. (B); 628. (B); 629. (D); 630. (C); 631. (B); 632. (C); 633. (C); 634. (D); 635. (C); 636. (B); 637. (C); 638. (D); 639. (C); 640. (B); 641. (B); 642. (A); 643. (D); 644. (C); 645. (B); 646. (C); 647. (B); 648. (C); 649. (D); 650. (A); 651. (C); 652. (A); 653. (A); 654. (B); 655. (C); 656. (D); 657. (D); 658. (B);

659. (D); 660. (A); 661. (C); 662. (D); 663. (A); 664. (B); 665. (C); 666. (D); 667. (A); 668. (B); 669. (D); 670. (C); 671. (B); 672. (A); 673. (B); 674. (C); 675. (D); 676. (A); 677. (B); 678. (C); 679. (D); 680. (A); 681. (A); 682. (B); 683. (C); 684. (C); 685. (A); 686. (B); 687. (A); 688. (D); 689. (A); 690. (B); 691. (C); 692. (D); 693. (A); 694. (B); 695. (C); 696. (D); 697. (A); 698. (C); 699. (B); 700. (C); 701. (B); 702. (C); 703. (A); 704. (C); 705. (B); 706. (A); 707. (B); 708. (C); 709. (A); 710. (B); 711. (B)

CHAPTER – 3 EXTENSION PRINCIPLES, APPROACHES AND PROGRAMMING

1. (C); 2. (A); 3. (C); 4. (B); 5. (B); 6. (C); 7. (C); 8. (B); 9. (C); 10. (A); 11. (A); 12. (A); 13. (C); 14. (A); 15. (D); 16. (B); 17. (C); 18. (B); 19. (A); 20. (A); 21. (D); 22. (D); 23. (A); 24. (A); 25. (C); 26. (B); 27. (C); 28. (A); 29. (C); 30. (B); 31. (A); 32. (D); 33. (A); 34. (A); 35. (A); 36. (A); 37. (B); 38. (A); 39. (C); 40. (D); 41. (D); 42. (A); 43. (B); 44. (B); 45. (D); 46. (B); 47. (B); 48. (D); 49. (D); 50. (B); 51. (C); 52. (A); 53. (B); 54. (D); 55. (A); 56. (C); 57. (B); 58. (D); 59. (B); 60. (D); 61. (A); 62. (A); 63. (B); 64. (A); 65. (B); 66. (B); 67. (C); 68. (D); 69. (B); 70. (A); 71. (B); 72. (A); 73. (C); 74. (C); 75. (B); 76. (A); 77. (D); 78. (B); 79. (B); 80. (E); 81. (D); 82. (A); 83. (B); 84. (C); 85. (D); 86. (A); 87. (B); 88. (C); 89. (A); 90. (B); 91. (C); 92. (C); 93. (A); 94. (B); 95. (C); 96. (B); 97. (C); 98. (A); 99. (A); 100. (C); 101. (A); 102. (D); 103. (B); 104. (A); 105. (C); 106. (B); 107. (D); 108. (C); 109. (B); 110. (A); 111. (B)

CHAPTER 4 : EXTENSION EDUCATION, TEACHING AND METHODS

1. (C); 2. (A); 3. (A); 4. (A); 5. (D); 6. (C); 7. (C); 8. (C); 9. (C); 10. (C); 11. (A); 12. (C); 13. (D); 14. (B); 15. (D); 16. (C); 17. (B); 18. (B); 19. (A); 20. (A); 21. (C); 22. (C); 23. (C); 24. (D); 25. (B); 26. (A); 27. (B); 28. (A); 29. (C); 30. (C); 31. (A); 32. (A); 33. (C); 34. (B); 35. (C); 36. (C); 37. (A); 38. (C); 39. (B); 40. (D); 41. (A); 42. (B); 43. (B); 44. (D); 45. (A); 46. (A); 47. (A); 48. (A); 49. (A); 50. (C); 51. (C); 52. (B); 53. (D); 54. (B); 55. (A); 56. (A); 57. (A); 58. (C); 59. (D); 60. (B); 61. (C); 62. (B); 63. (D); 64. (C); 65. (B); 66. (C); 67. (C); 68. (D); 69. (D); 70. (C); 71. (D); 72. (A); 73. (E); 74. (B); 75. (C); 76. (A); 77. (C); 78. (D); 79. (F); 80. (A); 81. (C); 82. (D); 83. (C); 84. (C); 85. (A); 86. (C); 87. (D); 88. (C); 89. (D); 90. (D)

CHAPTER 5 : EXTENSION LEARNING AND EVALUATION

1. (A); 2. (A); 3. (B); 4. (A); 5. (C); 6. (D); 7. (D); 8. (B); 9. (B); 10. (B); 11. (A); 12. (C); 13. (A); 14. (A); 15. (B); 16. (A); 17. (D); 18. (B); 19. (B); 20. (A); 21. (C); 22. (B); 23. (C); 24. (A); 25. (C); 26. (B); 27. (A); 28. (A); 29. (C); 30. (A); 31. (A); 32. (D); 33. (C); 34. (B); 35. (D)

CHAPTER 6: COMMUNICATION AND COMMUNICATION TECHNOLOGY

1. (C); 2. (B); 3. (A); 4. (A); 5. (A); 6. (D); 7. (B); 8. (C); 9. (A); 10. (A); 11. (C); 12. (D); 13. (B); 14. (B); 15. (C); 16. (D); 17. (A); 18. (C); 19. (C); 20. (C); 21. (D); 22. (D); 23. (A); 24. (A); 25. (A); 26. (B); 27. (B); 28. (A); 29. (C); 30. (B); 31. (A); 32. (C); 33. (A); 34. (D); 35. (D); 36. (B); 37. (D); 38. (D); 39. (D); 40. (A); 41. (A); 42. (D); 43. (D); 44. (A); 45. (D); 46. (A); 47. (B); 48. (B); 49. (C); 50. (A); 51. (A); 52. (B); 53. (A); 54. (D); 55. (D); 56. (B); 57. (B); 58. (A); 59. (B); 60. (C); 61. (C); 62. (D); 63. (A); 64. (D); 65. (C); 66. (B); 67. (C); 68. (D); 69. (B); 70. (B); 71. (B); 72. (C); 73. (C); 74. (A); 75. (B); 76. (A); 77. (C); 78. (C); 79. (A); 80. (D); 81. (C); 82. (D); 83. (A); 84. (C); 85. (B); 86. (B); 87. (A); 88. (D); 89. (A); 90. (A); 91. (C); 92. (A); 93. (B); 94. (C); 95. (C); 96. (A); 97. (B); 98. (D); 99. (A); 100. (A); 101. (C); 102. (D); 103. (D); 104. (B); 105. (A); 106. (B); 107. (D); 108. (B); 109. (A); 110. (A); 111. (C); 112. (B); 113. (A); 114. (B); 115. (C); 116. (D); 117. (C); 118. (B); 119. (A); 120. (C); 121. (A); 122. (B); 123. (A); 124. (B); 125. (C); 126. (A); 127. (B); 128. (A); 129. (D); 130. (C); 131. (C); 132. (C); 133. (B); 134. (D); 135. (D); 136. (B); 137. (C); 138.

(A); 139. (B); 140. (C); 141. (B); 142. (C); 143. (A); 144. (B); 145. (D); 146. (C); 147. (A); 148. (B); 149. (C); 150. (D); 151. (C); 152. (D); 153. (C); 154. (C); 155. (A); 156. (B); 157. (C); 158. (D); 159. (B); 160. (C); 161. (C); 162. (D); 163. (C); 164. (C); 165. (C); 166. (D); 167. (C); 168. (A); 169. (C); 170. (C); 171. (C); 172. (E); 173. (E); 174. (C); 175. (C); 176. (C); 177. (D); 178. (D); 179. (D); 180. (E); 181. (C); 182. (E); 183. (D); 184. (E); 185. (E); 186. (D); 187. (E); 188. (C); 189. (A); 190. (B); 191. (C); 192. (A); 193. (A); 194. (D); 195. (D); 196. (A); 197. (D); 198. (A); 199. (D); 200. (B); 201. (A); 202. (D); 203. (C); 204. (B); 205. (D); 206. (A); 207. (A); 208. (C); 209. (A); 210. (B); 211. (A); 212. (D); 213. (B); 214. (C); 215. (A); 216. (B); 217. (D); 218. (C); 219. (D); 220. (C); 221. (C); 222. (C); 223. (D); 224. (D); 225. (C); 226. (A); 227. (B); 228. (A); 229. (A); 230. (C); 231. (A); 232. (C); 233. (D); 234. (E); 235. (A); 236. (C); 237. (B); 238. (A); 239. (A); 240. (D); 241. (D); 242. (D); 243. (C); 244. (A); 245. (D); 246. (D); 247. (C); 248. (C); 249. (D); 250. (B); 251. (A); 252. (C); 253. (A); 254. (D); 255. (D); 256. (C); 257. (A); 258. (B); 259. (C); 260. (D); 261. (A); 262. (B); 263. (C); 264. (D); 265. (D); 266. (F); 267. (E); 268. (E); 269. (E); 270. (A); 271. (C); 272. (C); 273. (C); 274. (C), 275. (C); 276. (A); 277. (C); 278. (A); 279. (A); 280. (B); 281. (D); 282. (C); 283. (B); 284. (B); 285. (A); 286. (E); 287. (A); 288. (B); 289. (B); 290. (C); 291. (A); 292. (C); 293. (C); 294. (A); 295. (C); 296. (C); 297. (A); 298. (D); 299. (D); 300. (B); 301. (A); 302. (A); 303. (C); 304. (C); 305. (B); 306. (C); 307. (A); 308. (A); 309. (B); 310. (C); 311. (C); 312. (E); 313. (F); 314. (D); 315. (D); 316. (C); 317. (D); 318. (E); 319. (C); 320. (D); 321. (A); 322. (C); 323. (B); 324. (B); 325. (A), 326. (B); 327. (D); 328. (E); 329. (A); 330. (D); 331. (B); 332. (C); 333. (A); 334. (B); 335. (C); 336. (D); 337. (C); 338. (A); 339. (B); 340. (C); 341. (D); 342. (A); 343. (A); 344. (D); 345. (B); 346. (C); 347. (C); 348. (B); 349. (C); 350. (C); 351. (C); 352. (A); 353. (A)

CHAPTER 7: AUDIO - VISUAL AIDS

1. (B); 2. (C); 3. (B); 4. (C); 5. (B); 6. (C); 7. (B); 8. (B); 9. (A); 10. (B); 11. (D); 12. (D); 13. (C); 14. (D); 15. (C); 16. (D); 17. (D); 18. (A); 19. (D); 20. (C); 21. (B); 22. (C); 23. (D); 24. (C); 25. (A); 26. (D); 27. (C); 28. (A); 29. (D); 30. (A); 31. (A); 32. (B); 33. (D); 34. (B); 35. (D); 36. (A); 37. (A); 38. (D); 39. (A); 40. (D); 41. (D); 42. (D); 43. (B); 44. (A); 45. (D); 46. (A); 47. (C); 48. (B); 49. (C); 50. (C); 51. (C); 52. (D); 53. (D); 54. (C); 55. (D); 56. (D); 57. (D); 58. (C); 59. (A)

CHAPTER 8 : INNOVATION, EVALUATION AND ADOPTION

1. (C); 2. (A); 3. (C); 4. (D); 5. (B); 6. (B); 7. (D); 8. (A); 9. (C); 10. (C); 11. (B); 12. (A); 13. (D); 14. (B); 15. (A); 16. (B); 17. (C); 18. (D); 19. (A); 20. (B); 21. (A); 22. (D); 23. (A); 24. (C); 25. (A); 26. (D); 27. (B); 28. (B); 29. (B); 30. (B); 31. (A); 32. (A); 33. (B); 34. (D); 35. (B); 36. (C); 37. (D); 38. (C); 39. (A); 40. (D); 41. (D); 42. (C); 43. (D); 44. (C); 45. (A); 46. (D); 47. (C); 48. (C); 49. (B); 50. (A); 51. (D); 52. (D); 53. (B); 54. (A); 55. (A); 56. (A); 57. (B); 58. (B); 59. (B); 60. (C); 61. (D); 62. (B); 63. (D); 64. (A); 65. (C); 66. (C); 67. (D); 68. (B); 69. (B); 70. (A); 71. (C); 72. (A); 73. (A); 74. (D); 75. (D); 76. (B); 77. (A); 78. (D); 79. (B); 80. (A); 81. (A); 82. (B); 83. (A); 84. (B); 85. (A); 86. (A); 87. (B); 88. (B); 89. (A); 90. (B); 91. (C); 92. (C); 93. (A); 94. (D); 95. (B); 96. (B); 97. (B); 98. (A); 99. (B); 100. (A); 101. (A); 102. (B); 103. (A); 104. (B); 105. (A); 106. (A); 107. (C); 108. (C); 109. (A); 110. (A); 111. (B); 112. (D); 113. (A); 114. (D); 115. (B); 116. (B); 117. (C); 118. (D); 119. (C); 120. (A)

CHAPTER 9: PARTICIPATORY RURAL APPRAISAL

1. (D); 2. (B); 3. (C); 4. (A); 5. (B); 6. (A); 7. (C); 8. (C); 9. (B); 10. (C); 11. (B); 12. (B); 13. (C); 14. (B); 15. (B); 16. (A); 17. (A); 18. (D); 19. (F); 20. (A); 21. (D); 22. (D); 23. (A); 24. (B); 25. (C); 26. (E); 27. (A); 28. (B); 29. (D); 30. (C); 31. (B); 32.

(A); 33. (B); 34. (A); 35. (E); 36. (A); 37. (C); 38. (C); 39. (A); 40. (D); 41. (D); 42. (B); 43. (B); 44. (D); 45. (A); 46. (B); 47. (A); 48. (A); 49. (D); 50. (A); 51. (A); 52. (D); 53. (A); 54. (B); 55. (A); 56. (B); 57. (C); 58. (B); 59. (A); 60. (D); 61. (C); 62. (D); 63. (C); 64. (F); 65. (F); 66. (G); 67. (D); 68. (D); 69. (C); 70. (A); 71. (A); 72. (A); 73. (D); 74. (A); 75. (D); 76. (A); 77. (C); 78. (D); 79. (D); 80. (E); 81. (E); 82. (C); 83. (D); 84. (C); 85. (C); 86. (D); 87. (D); 88. (D); 89. (D); 90. (D); 91. (C); 92. (A); 93. (D); 94. (C); 95. (D); 96. (A); 97. (B); 98. (D); 99. (D); 100. (C); 101. (A); 102. (D); 103. (B); 104. (D); 105. (A); 106. (D); 107. (A); 108. (D); 109. (A); 110. (D); 111. (F); 112. (F); 113. (B); 114. (C); 115. (A); 116. (C); 117. (E); 118. (D); 119. (A); 120. (A); 121. (B); 122. (C); 123. (A); 124. (C); 125. (B); 126. (C)

CHAPTER 10: RURAL DEVELOPMENT AND PANCHAYATI RAJ

1. (D); 2. (D); 3. (C); 4. (A); 5. (A); 6. (B); 7. (A); 8. (C); 9. (D); 10. (B); 11. (A); 12. (B); 13. (A); 14. (D); 15. (B); 16. (B); 17. (A); 18. (C); 19. (A); 20. (A); 21. (B); 22. (B); 23. (B); 24. (D); 25. (D); 26. (A); 27. (B); 28. (D); 29. (D); 30. (D); 31. (B); 32. (C); 33. (B); 34. (B); 35. (A); 36. (B); 37. (A); 38. (D); 39. (C); 40. (D); 41. (C); 42. (B); 43. (C); 44. (A); 45. (B); 46. (B); 47. (C); 48. (C); 49. (A); 50. (D); 51. (D); 52. (B); 53. (D); 54. (D); 55. (A); 56. (B); 57. (B); 58. (A); 59. (B); 60. (D); 61. (B); 62. (D); 63. (B); 64. (B); 65. (A); 66. (A); 67. (C); 68. (B); 69. (B); 70. (B); 71. (C); 72. (B); 73. (B); 74. (A); 75. (C); 76. (B); 77. (C); 78. (A); 79. (D); 80. (A); 81. (A); 82. (C); 83. (B); 84. (C); 85. (B); 86. (A); 87. (A); 88. (C); 89. (A); 90. (A); 91. (D); 92. (D); 93. (D); 94. (B); 95. (C); 96. (C); 97. (B); 98. (B); 99. (A); 100. (A); 101. (B); 102. (D); 103. (D); 104. (B); 105. (A); 106. (C); 107. (D); 108. (C); 109. (D); 110. (C); 111. (B); 112. (C); 113. (B); 114. (B); 115. (B); 116. (B); 117. (C); 118. (A); 119. (B); 120. (D); 121. (B); 122. (D); 123. (C); 124. (B); 125. (C); 126. (C); 127. (B); 128. (A); 129. (C); 130. (C); 131. (A); 132. (B); 133. (D); 134. (D); 135. (D); 136. (D); 137. (C); 138. (A); 139. (B); 140. (A); 141. (A); 142. (D); 143. (B); 144. (B); 145. (B); 146. (A); 147. (C); 148. (B); 149. (D); 150. (B); 151. (C); 152. (C); 153. (A); 154. (B); 155. (D); 156. (C); 157. (D); 158. (B); 159. (B); 160. (C); 161. (A); 162. (B); 163. (D); 164. (A); 165. (A); 166. (C); 167. (B); 168. (B); 169. (B); 170. (A); 171. (A); 172. (A); 173. (A); 174. (D); 175. (C); 176. (B); 177. (C); 178. (C); 179. (A); 180. (D); 181. (B); 182. (A); 183. (B); 184. (C); 185. (D); 186. (B); 187. (D); 188. (C); 189. (D); 190. (D); 191. (C); 192. (B); 193. (C); 194. (D); 195. (C); 196. (A); 197. (B); 198. (B); 199. (A); 200. (B); 201. (B); 202. (C); 203. (A); 204. (B); 205. (D); 206. (C); 207. (C); 208. (B); 209. (A); 210. (C); 211. (D); 212. (A); 213. (B); 214. (A); 215. (A); 216. (B); 217. (A); 218. (D); 219. (B); 220. (B); 221. (A); 222. (D); 223. (A); 224. (B); 225. (A); 226. (D); 227. (B); 228. (C); 229. (A); 230. (A); 231. (A); 232. (A); 233. (A); 234. (B); 235. (A); 236. (B); 237. (A); 238. (A); 239. (B); 240. (C); 241. (B); 242. (A); 243. (B); 244. (B); 245. (D); 246. (C); 247. (B); 248. (B); 249. (A); 250. (D); 251. (A); 252. (A); 253. (B); 254. (C); 255. (A); 256. (B); 257. (A); 258. (D); 259. (A); 260. (A); 261. (C); 262. (A); 263. (B); 264. (B); 265. (B); 266. (A); 267. (A); 268. (D); 269. (D); 270. (B); 271. (D); 272. (D); 273. (A); 274. (A), 275. (A); 276. (D); 277. (D); 278. (C); 279. (B); 280. (D); 281. (B); 282. (C); 283. (B); 284. (B); 285. (A); 286. (C); 287. (B); 288. (B); 289. (A); 290. (B); 291. (C); 292. (B); 293. (B); 294. (D); 295. (B); 296. (B); 297. (A); 298. (D); 299. (C); 300. (C); 301. (D); 302. (A); 303. (C); 304. (A); 305. (A); 306. (D); 307. (A); 308. (B); 309. (B); 310. (A); 311. (A); 312. (A); 313. (D); 314. (D); 315. (B); 316. (A); 317. (A); 318. (A); 319. (C); 320. (A); 321. (D); 322. (A); 323. (A); 324. (B); 325. (A), 326. (D); 327. (A); 328. (A); 329. (D); 330. (B); 331. (A); 332. (A); 333. (C); 334. (B); 335. (B); 336. (A); 337. (B); 338. (A); 339. (A); 340. (B); 341. (A); 342. (A); 343.(A); 344. (C); 345. (C); 346. (A); 347. (A); 348. (A); 349. (A); 350. (C); 351. (A); 352. (C); 353. (C); 354. (B); 355. (B); 356. (A); 357. (C); 358. (C); 359. (C); 360. (D); 361. (B); 362. (A); 363. (A); 364. (C); 365. (A); 366. (A); 367. (C); 368. (B); 369. (D); 370. (B)

CHAPTER 11: ECONOMIC PRINCIPLE

1. (D); 2. (A); 3. (B); 4. (A); 5. (D); 6. (C); 7. (B); 8. (B); 9. (B); 10. (C); 11. (B); 12. (A); 13. (B); 14. (B); 15. (A); 16. (C); 17. (A); 18. (C); 19. (A); 20. (C); 21. (D); 22. (D); 23. (D); 24. (C); 25. (A); 26. (C); 27. (C); 28. (C); 29. (B); 30. (A); 31. (B); 32. (C); 33. (A); 34. (A); 35. (B); 36. (B); 37. (D); 38. (B); 39. (B); 40. (B); 41. (B); 42. (A); 43. (A); 44. (A); 45. (A); 46. (D); 47. (A); 48. (B); 49. (A); 50. (B); 51. (A); 52. (C); 53. (B); 54. (B); 55. (C); 56. (A); 57. (B); 58. (D); 59. (D); 60. (D); 61. (A); 62. (B); 63. (A); 64. (C); 65. (A); 66. (D); 67. (C); 68. (A); 69. (B); 70. (C); 71. (B); 72. (A); 73. (D); 74. (B); 75. (A); 76. (B); 77. (C); 78. (D); 79. (B); 80. (B); 81. (C); 82. (D); 83. (D); 84. (C); 85. (B); 86. (C); 87. (D); 88. (C); 89. (A); 90. (B); 91. (B); 92. (A); 93. (B); 94. (B); 95. (C); 96. (A); 97. (B); 98. (D); 99. (C); 100. (C); 101. (B); 102. (A); 103. (C); 104. (D); 105. (A); 106. (B); 107. (A); 108. (A); 109. (B); 110. (A); 111. (D); 112. (B); 113. (B); 114. (A); 115. (A); 116. (B); 117. (A); 118. (A); 119. (C); 120. (C); 121. (B); 122. (B); 123. (C); 124. (D); 125. (C); 126. (C); 127. (B); 128. (A); 129. (B); 130. (A); 131. (B); 132. (B); 133. (C); 134. (C); 135. (B); 136. (D); 137. (C); 138. (A); 139. (C); 140. (B); 141. (B); 142. (B); 143. (A); 144. (D); 145. (B); 146. (A); 147. (B); 148. (D); 149. (B); 150. (B); 151. (D); 152. (B); 153. (C); 154. (A); 155. (A); 156. (D); 157. (D); 158. (A); 159. (A); 160. (A); 161. (D); 162. (D); 163. (B); 164. (B); 165. (C); 166. (A); 167. (B); 168. (D); 169. (A); 170. (B); 171. (D); 172. (A); 173. (C); 174. (A); 175. (C); 176. (B); 177. (B); 178. (A); 179. (C); 180. (B); 181. (C); 182. (D); 183. (B); 184. (); 185. (D); 186. (B); 187. (A); 188. (B); 189. (D); 190. (A); 191. (C); 192. (C); 193. (A); 194. (C); 195. (A); 196. (B); 197. (A); 198. (D); 199. (D); 200. (C); 201. (B); 202. (D); 203. (C); 204. (B); 205. (C); 206. (D); 207. (A); 208. (C); 209. (A) 210. (D); 211. (A); 212. (C); 213. (D); 214. (B); 215. (A); 216. (C); 217. (B); 218. (B); 219. (B); 220. (A); 221. (C); 222. (A); 223. (B); 224. (A); 225. (A); 226. (A); 227. (C); 228. (C); 229. (A); 230. (D); 231. (D); 232. (C); 233. (A); 234. (C); 235. (A); 236. (C); 237. (D); 238. (A); 239. (D); 240. (A); 241. (D); 242. (C); 243. (A); 244. (B); 245. (D); 246. (C); 247. (D); 248. (D); 249. (D); 250. (C); 251. (D); 252. (D); 253. (A); 254. (A); 255. (C); 256. (D); 257. (C); 258. (A); 259. (A); 260. (D); 261. (A); 262. (B); 263. (C); 264. (D); 265. (A); 266. (B); 267. (A); 268. (A); 269. (C); 270. (D); 271. (D); 272. (B); 273. (B); 274. (A), 275. (D); 276. (A); 277. (B); 278. (A); 279. (B); 280. (D); 281. (A); 282. (A); 283. (A); 284. (B); 285. (D); 286. (B); 287. (A); 288. (C); 289. (A); 290. (B); 291. (A); 292. (B); 293. (C); 294. (A); 295. (A); 296. (A); 297. (B); 298. (A); 299. (A); 300. (B); 301. (C); 302. (D); 303. (A); 304. (D); 305. (D); 306. (A); 307. (D); 308. (A); 309. (D); 310. (A); 311. (D); 312. (D); 313. (D); 314. (D); 315. (A); 316. (D); 317. (A); 318. (D); 319. (A); 320. (A); 321. (B); 322. (A); 323. (B); 324. (D); 325. (A), 326. (C); 327. (C); 328. (D); 329. (D); 330. (A); 331. (B); 332. (A); 333. (A); 334. (A); 335. (A); 336. (A); 337. (C); 338. (A); 339. (D); 340. (A); 341. (D); 342. (C); 343. (A); 344. (D); 345. (D); 346. (C); 347. (A); 348. (C); 349. (D); 350. (A); 351. (A); 352. (D); 353. (A); 354. (A); 355. (D); 356. (B); 357. (A); 358. (D); 359. (D); 360. (D); 361. (A); 362. (A); 363. (B); 364. (A); 365. (B); 366. (D); 367. (D); 368. (A); 369. (A); 370. (D); 371. (A); 372. (A); 373. (A); 374. (B); 375. (B); 376. (C); 377. (B); 378. (A); 379. (B); 380. (B); 381. (A); 382. (B); 383. (D); 384. (A); 385. (A); 386. (A); 387.(B); 388. (D); 389. (A); 390. (D); 391. (B); 392. (D); 393. (D); 394. (A); 395. (A); 396. (D); 397. (A); 398. (B); 399. (A); 400. (D); 401. (D); 402. (C); 403. (D); 404. (A); 405. (D); 406.(D); 407. (D); 408. (D); 409. (D); 410. (A); 411. (C); 412. (D); 413. (C); 414. (D); 415. (C); 416. (D); 417. (C); 418. (D); 419. (A); 420. (D); 421. (A); 422. (A); 423. (D); 424. (D); 425. (D); 426. (A); 427. (D); 428. (D); 429. (D); 430. (A); 431. (A); 432. (B); 433. (A); 434. (D); 435. (D); 436. (B); 437. (A); 438. (D); 439. (D); 440. (B); 441. (D); 442. (D); 443. (A); 444. (C); 445. (B); 446. (A); 447. (D); 448. (A); 449. (A); 450. (C); 451. (A); 452. (A); 453. (C); 454. (D); 455. (D); 456. (A); 457. (C); 458. (A);

459. (D); 460. (D); 461. (A); 462. (C); 463. (C); 464. (A); 465. (A); 466. (B); 467. (B); 468. (A); 469. (A); 470. (D); 471. (A); 472. (C); 473. (D); 474. (D); 475. (D); 476. (B); 477. (D); 478. (A); 479. (D); 480. (B); 481. (B); 482. (B); 483. (A); 484. (A); 485. (A); 486. (D); 487. (C); 488. (A); 489. (D); 490. (A); 491. (A); 492. (D); 493. (A); 494. (D); 495. (D); 496. (D); 497. (D); 498. (B); 499. (B); 500. (D); 501. (D); 502. (D); 503. (A); 504. (B); 505. (C); 506. (B); 507. (D); 508. (D); 509. (B); 510. (A); 511. (A); 512. (A); 513. (B); 514. (D); 515. (B); 516. (A); 517. (B); 518. (D); 519. (A); 520. (B); 521. (A); 522. (C); 523. (A); 524. (B); 525. (A); 526. (A); 527. (B); 528. (D); 529. (D); 530. (C); 531. (B); 532. (B); 533. (B); 534. (C); 535. (C); 536. (B); 537. (D); 538. (A); 539. (B); 540. (C); 541. (D); 542. (A); 543. (C); 544. (D); 545. (A); 546. (B); 547. (B); 548. (C); 549. (C); 550. (C); 551. (C); 552. (C); 553. (D); 554. (B); 555. (B); 556. (D); 557. (A); 558. (A); 559. (C); 560. (A); 561. (B); 562. (A); 563. (C); 564. (D); 565. (C); 566. (A); 567. (A); 568. (A); 569. (B); 570. (B); 571. (A); 572. (C); 573. (D); 574. (B); 575. (A); 576. (D); 577. (C); 578. (A); 579. (B); 580. (D); 581. (D); 582. (A); 583. (A); 584. (B); 585. (A); 586. (A); 587. (C); 588. (A); 589. (B); 590. (D); 591. (B); 592. (A); 593. (A); 594. (A); 595. (C); 596. (B); 597. (A); 598. (A); 599. (B); 600. (B); 601. (A); 602. (D); 603. (B); 604. (A); 605. (D); 606. (A); 607. (B); 608. (D); 609. (C); 610. (C); 611. (A); 612. (C); 613. (B); 614. (D); 615. (B); 616. (B); 617. (A); 618. (D); 619. (A); 620. (D); 621. (A); 622. (A); 623. (C); 624. (B); 625. (A); 626. (B); 627. (C); 628. (A); 629. (C); 630. (B); 631. (A); 632. (D); 633. (C); 634. (A); 635. (A); 636. (B); 637. (C); 638. (C); 639. (B); 640. (D); 641. (C); 642. (A); 643. (B); 644. (A); 645. (C); 646. (A); 647. (A); 648. (B); 649. (A); 650. (B); 651. (B); 652. (A); 653. (B); 654. (B); 655. (B); 656. (A); 657. (A); 658. (B); 659. (B); 660. (C); 661. (D); 662. (B); 663. (D); 664. (C); 665. (B); 666. (A); 667. (A)

CHAPTER 12: SOCIOLOGY AND SOCIAL WORK

1. (D); 2. (A); 3. (D); 4. (D); 5. (C); 6. (A); 7. (C); 8. (A); 9. (A); 10. (C); 11. (C); 12. (A); 13. (A); 14. (A); 15. (C); 16. (B); 17. (B); 18. (A); 19. (A); 20. (A); 21. (A); 22. (A); 23. (D); 24. (B); 25. (B); 26. (D); 27. (B); 28. (A); 29. (A); 30. (B); 31. (A); 32. (B); 33. (C); 34. (C); 35. (A); 36. (A); 37. (C); 38. (C); 39. (C); 40. (A); 41. (C); 42. (A); 43. (D); 44. (C); 45. (A); 46. (B); 47. (A); 48. (B); 49. (C); 50. (D); 51. (D); 52. (A); 53. (B); 54. (C); 55. (D); 56. (A); 57. (A); 58. (C); 59. (C); 60. (A); 61. (A); 62. (D); 63. (A); 64. (A); 65. (A); 66. (C); 67. (B); 68. (D); 69. (D); 70. (B); 71. (B); 72. (D); 73. (D); 74. (A); 75. (A); 76. (B); 77. (B); 78. (B); 79. (A); 80. (C); 81. (A); 82. (E); 83. (B); 84. (A); 85. (B); 86. (C); 87. (A); 88. (C); 89. (B); 90. (C); 91. (D); 92. (D); 93. (C); 94. (C); 95. (A); 96. (C); 97. (C); 98. (E); 99. (A); 100. (A); 101. (D); 102. (A); 103. (A); 104. (B); 105. (A); 106. (A); 107. (C); 108. (B); 109. (D); 110. (D); 111. (C); 112. (A); 113. (A); 114. (C); 115. (A); 116. (A); 117. (B); 118. (A); 119. (B); 120. (D); 121. (C); 122. (B); 123. (D); 124. (A); 125. (D); 126. (C); 127. (A); 128. (B); 129. (B); 130. (B); 131. (A); 132. (D); 133. (D); 134. (C); 135. (A); 136. (D); 137. (D); 138. (C); 139. (D); 140. (B); 141. (B); 142. (E); 143. (C); 144. (C); 145. (E); 146. (A); 147. (A); 148. (C); 149. (C); 150. (A); 151. (E); 152. (C); 153. (D); 154. (A); 155. (C); 156. (C); 157. (D); 158. (A); 159. (E); 160. (C); 161. (C); 162. (C); 163. (B); 164. (C); 165. (D); 166. (A); 167. (D); 168. (A); 169. (B); 170. (D); 171. (D); 172. (D); 173. (A); 174. (B); 175. (C); 176. (A); 177. (C); 178. (B); 179. (A); 180. (D); 181. (B); 182. (B); 183. (D); 184. (A); 185. (A); 186. (C); 187. (B); 188. (B); 189. (A); 190. (D); 191. (D); 192. (A); 193. (D); 194. (D); 195. (D); 196. (B); 197. (D); 198. (C); 199. (A); 200. (C); 201. (A); 202. (A); 203. (D); 204. (C); 205. (A); 206. (A); 207. (C); 208. (D); 209. (D); 210. (B); 211. (D); 212. (D); 213. (A); 214. (B); 215. (D); 216. (C); 217. (C); 218. (C); 219. (A); 220. (B); 221. (C); 222. (A); 223. (A); 224. (D); 225. (C); 226. (A); 227. (A); 228. (D); 229. (B); 230. (B); 231. (B); 232. (C); 233. (C);

234. (D); 235. (A); 236. (C); 237. (B); 238. (D);
239. (A); 240. (A); 241. (B); 242. (B); 243. (A);
244. (A); 245. (D); 246. (A); 247. (A); 248. (A);
249. (A); 250. (A); 251. (A); 252. (D); 253. (A);
254. (D); 255. (A); 256. (C); 257. (A); 258. (B);
259. (A); 260. (C); 261. (D); 262. (C); 263. (A);
264. (A); 265. (C); 266. (D); 267. (A); 268. (C);
269. (C); 270. (A); 271. (C); 272. (A); 273. (B);
274. (C), 275. (D); 276. (D); 277. (C); 278. (B);
279. (A); 280. (B); 281. (D); 282. (B); 283. (A);
284. (D); 285. (C); 286. (C); 287. (D); 288. (D);
289. (B); 290. (C); 291. (B); 292. (C); 293. (A);
294. (D); 295. (D); 296. (D); 297. (B); 298. (D);
299. (A); 300. (D); 301. (C); 302. (D); 303. (B);
304. (C); 305. (C); 306. (B); 307. (C); 308. (D);
309. (C); 310. (A); 311. (B); 312. (A); 313. (B);
314. (B); 315. (B); 316. (D); 317. (A); 318. (A);
319. (D); 320. (A); 321. (C); 322. (D); 323. (C);
324. (B); 325. (A), 326. (D); 327. (D); 328. (B);
329. (A); 330. (D); 331. (C); 332. (A); 333. (B);
334. (D); 335. (D); 336. (C); 337. (B); 338. (D);
339. (D); 340. (D); 341. (A); 342. (C); 343. (C);
344. (A); 345. (C); 346. (A); 347. (B); 348. (B);
349. (A); 350. (B); 351. (D); 352. (A); 353. (C);
354. (C); 355. (C); 356. (A); 357. (B); 358. (B);
359. (A); 360. (C); 361. (D); 362. (B); 363. (B);
364. (B); 365. (B); 366. (C); 367. (C); 368. (D);
369. (A); 370. (D); 371. (C); 372. (A); 373. (A);
374. (D); 375. (C); 376. (D); 377. (D); 378. (C);
379. (B); 380. (C); 381. (C); 382. (C); 383. (A);
384. (A); 385. (B); 386. (A); 387. (A); 388. (A);
389. (C); 390. (A); 391. (A); 392. (B); 393. (B);
394. (D); 395. (D); 396. (C); 397. (B); 398. (D);
399. (B); 400. (C); 401. (B); 402. (A); 403. (C);
404. (A); 405. (A); 406. (B); 407. (A); 408. (C);
409. (A); 410. (D); 411. (B); 412. (A); 413. (A);
414. (B); 415. (A); 416. (B); 417. (B); 418. (B);
419. (B); 420. (D); 421. (A); 422. (B); 423. (B);
424. (A); 425. (B); 426. (C); 427. (C); 428. (D);
429. (A); 430. (A); 431. (A); 432. (B); 433. (B);
434. (A 435. (A); 436. (B); 437. (A); 438. (B);
439. (A); 440. (B); 441. (A); 442. (C); 443. (A);
444. (A); 445. (A); 446. (D); 447. (D); 448. (A);
449. (A); 450. (A); 451. (D); 452. (A); 453. (C);
454. (A); 455. (B); 456. (D); 457. (C); 458. (C);
459. (C); 460. (D); 461. (C); 462. ()A; 463. (B);
464. (C); 465. (D); 466. (B); 467. (B); 468. (C);
469. (C); 470. (C); 471. (A); 472. (A); 473. (B);
474. (C); 475. (C); 476. (D); 477. (B); 478. (C);
479. (B); 480. (B); 481. (A); 482. (B); 483. (D);
484. (B); 485. (B); 486. (D); 487. (B); 488. (A);
489. (A); 490. (C); 491. (D); 492. (A); 493. (A);
494. (A); 495. (C); 496. (B); 497. (A); 498. (A);
499. (C); 500. (B); 501. (B); 502. (A); 503. (B);
504. (B); 505. (B); 506. (B); 507. (B); 508. (B);
509. (A); 510. (B); 511. (C); 512. (B); 513. (A);
514. (B); 515. (C); 516. (B); 517. (A); 518. (A);
519. (A); 520. (A); 521. (C); 522. (A); 523. (C);
524. (A); 525. (A); 526. (A); 527. (B); 528. (A);
529. (D); 530. (C); 531. (A); 532. (A); 533. (A);
534. (C); 535. (C); 536. (A); 537. (A); 538. (D);
539. (C); 540. (B); 541. (D); 542. (A); 543. (C);
544. (A); 545. (D); 546. (A); 547. (C); 548. (C);
549. (A); 550. (D); 551. (C); 552. (C); 553. (D);
554. (D); 555. (C); 556. (D); 557. (B); 558. (B);
559. (C); 560. (D); 561. (C); 562. (C); 563. (C);
564. (C); 565. (C); 566. (D); 567. (A); 568. (A);
569. (C); 570. (B); 571. (A); 572. (C); 573. (A);
574. (A); 575. (C); 576. (A); 577. (D); 578. (B);
579. (B); 580. (B); 581. (A); 582. (B); 583. (A);
584. (D); 585. (D); 586. (C); 587. (C); 588. (A);
589. (B); 590. (C); 591. (C); 592. (D); 593. (A);
594. (B); 595. (C); 596. (A); 597. (B); 598. (A);
599. (C); 600. (A); 601. (A); 602. (C); 603. (D);
604. (A); 605. (C); 606. (D); 607. (B); 608. (C);
609. (C); 610. (A); 611. (C); 612. (D); 613. (C);
614. (C); 615. (A); 616. (C); 617. (C); 618. (A);
619. (B); 620. (C); 621. (A); 622. (C); 623. (A);
624. (A); 625. (B); 626. (D); 627. (A); 628. (D);
629. (B); 630. (B); 631. (D); 632. (B); 633. (D);
634. (A); 635. (D); 636. (A); 637. (A); 638. (B);
639. (B); 640. (C); 641. (D); 642. (C); 643. (C);
644. (D); 645. (B); 646. (A); 647. (B); 648. (C);
649. (B); 650. (B); 651. (B); 652. (C); 653. (B);
654. (C); 655. (D); 656. (C); 657. (D); 658. (B);
659. (A); 660. (C); 661. (A); 662. (A); 663. (B);
664. (A); 665. (B); 666. (D); 667. (D); 668. (B);
669. (A); 670. (A); 671. (C); 672. (B); 673. (C);
674. (A); 675. (D); 676. (D); 677. (A); 678. (D);
679. (A); 680. (B); 681. (A); 682. (A); 683. (A);
684. (A); 685. (A); 686. (C); 687. (D); 688. (A);
689. (B); 690. (C); 691. (D); 692. (D); 693. (A);

694. (B); 695. (A); 696. (C); 697. (C); 698. (D);
699. (B); 700. (A); 701. (A); 702. (B); 703. (C);
704. (C); 705. (B); 706. (B); 707. (A); 708. (C);
709. (C); 710. (B); 711. (B); 712. (C); 713. (B);
714. (A); 715. (A); 716. (B); 717. (D); 718. (B);
719. (B); 720. (A); 721. (B); 722. (B); 723. (C);
724. (D); 725. (B); 726. (A); 727. (D); 728. (D);
729. (D); 730. (B); 731. (C); 732. (B); 733. (B);
734. (B); 735. (A); 736. (A); 737. (C); 738. (A);
739. (D); 740. (C); 741. (B); 742. (A); 743. (D);
744. (C); 745. (C); 746. (A); 747. (D); 748. (D);
749. (A); 750. (A); 751. (A); 752. (B); 753. (A);
754. (A); 755. (A); 756. (D); 757. (A); 758. (D);
759. (A); 760. (A); 761. (C); 762. (D); 763. (C);
764. (C); 765. (D); 766. (A); 767. (D); 768. (A);
769. (A); 770. (B); 771. (C); 772. (A); 773. (D);
774. (C); 775. (D); 776. (B); 777. (A); 778. (C);
779. (D); 780. (B); 781. (A); 782. (C); 783. (A);
784. (D); 785. (D); 786. (D); 787. (A); 788. [illegible]);
789. (A); 790. (D); 791. (C); 792. (A); 793. ([illegible]
794. (D); 795. (B); 796. (B); 797. (D); 798. (C);
799. (A); 800. (C); 801. (B); 802. (A); 803. (A);
804. (A); 805. (A); 806. (B); 807. (A); 808. (B);
809. (A); 810. (A); 811. (C); 812. (A); 813. (C);
814. (C); 815. (B); 816. (C); 817. (C); 818. (B);
819. (A); 820. (B); 821. (B); 822. (A); 823. (A);
824. (D); 825. (B); 826. (B); 827. (D); 828. (B);
829. (D); 830. (B); 831. (A); 832. (C); 833. (D);
834. (C); 835. (B); 836. (D); 837. (A); 838. (A);
839. (A); 840. (B); 841. (B); 842. (C); 843. (C);
844. (A); 845. (C); 846. (D); 847. (B); 848. (C);
849. (D); 850. (B); 851. (A); 852. (C); 853. (B);
854. (B); 855. (C); 856. (C); 857. (C); 858. (D);
859. (A); 860. (B); 861. (B); 862. (C); 863. (A);
864. (A); 865. (C); 866. (D); 867. (A); 868. (C);
869. (D); 870. (B); 871. (C); 872. (B); 873. (B);
874. (C); 875. (A); 876. (B); 877. (B); 878. (D);
879. (D); 880. (C); 881. (A); 882. (C); 883. (B);
884. (B); 885. (B); 886. (C); 887. (B); 888. (B);
889. (B); 890. (A); 891. (A); 892. (A); 893. (A);
894. (A); 895. (B); 896. (B); 897. (B); 898. (C);
899. (C); 900. (C); 901. (D); 902. (A); 903. (D);
904. (C); 905. (B); 906. (C); 907. (A); 908. (C);
909. (C); 910. (A); 911. (A); 912. (B); 913. (B);
914. (D); 915. (A); 916. (D); 917. (A); 918. (C);
919. (A); 920. (A); 921. (D); 922. (A); 923. (C);
924. (A); 925. (A); 926. (B); 927. (B); 928. (C);
929. (D); 930. (B); 931. (C); 932. (A); 933. (D);
934. (D); 935. (C); 936. (D); 937. (A); 938. (C);
939. (C); 940. (A); 941. (A); 942. (B); 943. (D);
944. (A); 945. (A); 946. (D); 947. (C); 948. (A);
949. (C); 950. (D); 951. (B); 952. (A); 953. (C);
954. (A); 955. (C); 956. (B); 957. (C); 958. (A);
959. (A); 960. (B); 961. (C); 962. (A); 963. (C);
964. (B); 965. (A); 966. (B); 967. (A); 968. (B);
969. (D); 970. (D); 971. (D); 972. (C); 973. (C);
974. (C); 975. (A); 976. (A); 977. (A); 978. (A);
979. (D); 980. (B); 981. (A); 982. (B); 983. (D);
984. (C); 985. (C); 986. (A); 987. (C); 988. (D);
989. (C); 990. (C); 991. (D); 992. (D); 993. (B);
994. (C); 995. (A); 996. (A); 997. (C); 998. (C);
999. (A); 1000. (A); 1001. (A); 1002. (D); 1003. (C); 1004. (C); 1005. (A); 1006. (B); 1007. (B); 1008. (C); 1009. (D); 1010. (B); 1011. (A); 1012. (A); 1013. (A); 1014. (D); 1015. (D); 1016. (D); 1017. (B); 1018. (C); 1019. (C); 1020. (B); 1021. (B); 1022. (A); 1023. (C); 1024. (C); 1025. (B); 1026. (A); 1027. (C); 1028. (D); 1029. (C); 1030. (C); 1031. (D); 1032. (C); 1033. (C); 1034. (B); 1035. (B); 1036. (C); 1037. (B); 1038. (C); 1039. (B); 1040. (D); 1041. (A); 1042. (B); 1043. (A); 1044. (B); 1045. (C); 1046. (D); 1047. (B); 1048. (A); 1049. (C); 1050. (C); 1051. (D); 1052. (A); 1053. (D); 1054. (C); 1055. (B); 1056. (C); 1057. (B); 1058. (A); 1059. (C); 1060. (B); 1061. (C); 1062. (D); 1063. (A); 1064. (A); 1065. (C); 1066. (D); 1067. (C); 1068. (D); 1069. (D); 1070. (C); 1071. (C); 1072. (D); 1073. (A); 1074. (C); 1075. (D); 1076. (A); 1077. (B); 1078. (A); 1079. (B); 1080. (D); 1081. (B); 1082. (C); 1083. (A); 1084. (B); 1085. (A); 1086. (A); 1087. (C); 1088. (B); 1089. (C); 1090. (D); 1091. (D); 1092. (B); 1093. (C); 1094. (A); 1095. (B); 1096. (B); 1097. (A); 1098. (C); 1099. (A); 1100. (D); 1101. (C); 1102. (C); 1103. (C); 1104. (B); 1105. (B); 1106. (C); 1107. (C); 1108. (D); 1109. (A); 1110. (B); 1111. (A); 1112. (A); 1113. (A); 1114. (A); 1115. (C); 1116. (A); 1117. (A); 1118. (C); 1119. (A); 1120. (D); 1121. (C); 1122. (D); 1123. (A); 1124. (A); 1125. (D); 1126. (A); 1127. (D); 1128. (D); 1129. (C); 1130. (C); 1131. (C); 1132. (A); 1133. (A); 1134. (A); 1135. (B); 1136. (C); 1137. (B); 1138. (D); 1139.

(A); 1140. (D); 1141. (A); 1142. (A); 1143. (D); 1144. (A); 1145. (C); 1146. (B); 1147. (D); 1148. (C); 1149. (B); 1150. (C); 1151. (D); 1152. (D); 1153. (D); 1154. (C); 1155. (A); 1156. (D); 1157. (D); 1158. (B); 1159. (C); 1160. (C); 1161. (D); 1162. (C); 1163. (C); 1164. (C); 1165. (C); 1166. (D); 1167. (C); 1168. (A); 1169. (C); 1170. (C); 1171. (D); 1172. (C); 1173. (D); 1174. (B); 1175. (D); 1176. (B); 1177. (B); 1178. (A); 1179. (E); 1180. (C); 1181. (D); 1182. (C); 1183. (D); 1184. (A); 1185. (D); 1186. (A); 1187. (D); 1188. (A); 1189. (D); 1190. (D); 1191. (C); 1192. (B); 1193. (D); 1194. (D); 1195. (A); 1196. (A); 1197. (B); 1198. (B); 1199. (C); 1200. (D); 1201. (A); 1202. (D); 1203. (A); 1204. (B); 1205. (B); 1206. (D); 1207. (B); 1208. (D); 1209. (B); 1210. (C); 1211. (C); 1212. (B); 1213. (B); 1214. (A); 1215. (A); 1216. (A); 1217. (C); 1218. (A); 1219. (A); 1220. (C); 1221. (D); 1222. (C); 1223. (B); 1224. (D); 1225. (A); 1226. (D); 1227. (A); 1228. (B); 1229. (C); 1230. (B); 1231. (A); 1232. (C); 1233. (C); 1234. (C); 1235. (C); 1236. (A); 1237. (C); 1238. (C); 1239. (B); 1240. (B); 1241. (A); 1242. (B); 1243. (C); 1244. (C); 1245. (B); 1246. (B); 1247. (B); 1248. (B); 1249. (A); 1250. (C); 1251. (A); 1252. (C); 1253. (A); 1254. (A); 1255. (B); 1256. (B); 1257. (B); 1258. (D); 1259. (D); 1260. (D); 1261. (A); 1262. (A); 1263. (B); 1264. (A); 1265. (A); 1266. (A); 1267. (B); 1268. (B); 1269. (B); 1270. (B); 1271. (A); 1272. (A); 1273. (A); 1274. (B); 1275. (D); 1276. (B); 1277. (A); 1278. (B); 1279. (A); 1280. (B); 1281. (A); 1282. (A); 1283. (A); 1284. (D); 1285. (B); 1286. (C); 1287. (B); 1288. (A); 1289. (A); 1290. (B); 1291. (C); 1292. (D); 1293. (C); 1294. (B); 1295. (C); 1296. (D); 1297. (C); 1298. (D); 1299. (C); 1300. (A); 1301. (B); 1302. (A); 1303. (C); 1304. (B); 1305. (A); 1306. (A); 1307. (D); 1308. (C); 1309. (B); 1310. (D); 1311. (A); 1312. (C); 1313. (A); 1314. (A); 1315. (A); 1316. (A); 1317. (A); 1318. (C); 1319. (C); 1320. (C); 1321. (B); 1322. (B); 1323. (C); 1324. (C); 1325. (C); 1326. (B); 1327. (A); 1328. (C); 1329. (C); 1330. (A); 1331. (A); 1332. (B); 1333. (B); 1334. (B); 1335. (A); 1336. (A); 1337. (A); 1338. (A); 1339. (B); 1340. (C); 1341. (C); 1342. (D); 1343. (D); 1344. (A); 1345. (A); 1346. (A); 1347. (B); 1348. (D); 1349. (D); 1350. (C); 1351. (A); 1352. (C); 1353. (C); 1354. (D); 1355. (B); 1356. (D); 1357. (B)

CHAPTER 13: TRAINING MANUAL

1. (B); 2. (B); 3. (A); 4. (B); 5. (D); 6. (A); 7. (C); 8. (C); 9. (A); 10. (A); 11. (B); 12. (D); 13. (A); 14. (B); 15. (A); 16. (A); 17. (A); 18. (D); 19. (C); 20. (C); 21. (D); 22. (C); 23. (B); 24. (D); 25. (A); 26. (D); 27. (A); 28. (B); 29. (A); 30. (B); 31. (B); 32. (D); 33. (D); 34. (D); 35. (D); 36. (D); 37. (D); 38. (D); 39. (A); 40. (D); 41. (C); 42. (B); 43. (A); 44. (A); 45. (D); 46. (B); 47. (D); 48. (C); 49. (B); 50. (D); 51. (D); 52. (B); 53. (C); 54. (D); 55. (B); 56. (B); 57. (D); 58. (C); 59. (B); 60. (C); 61. (D); 62. (D); 63. (B); 64. (A); 65. (C); 66. (B); 67. (D); 68. (B); 69. (C); 70. (A); 71. (C); 72. (D); 73. (D); 74. (C); 75. (D); 76. (D); 77. (B); 78. (A); 79. (B); 80. (B); 81. (D); 82. (D); 83. (B); 84. (B); 85. (B); 86. (A); 87. (B); 88. (A); 89. (B); 90. (C); 91. (A); 92. (B); 93. (A); 94. (B); 95. (B); 96. (D); 97. (D); 98. (A); 99. (A); 100. (A); 101. (A); 102. (C); 103. (D); 104. (B); 105. (D); 106. (B); 107. (B); 108. (D); 109. (A); 110. (D); 111. (C); 112. (C); 113. (D); 114. (B); 115. (B); 116. (C); 117. (D); 118. (C); 119. (A); 120. (A); 121. (D); 122. (C); 123. (C); 124. (C); 125. (A); 126. (A); 127. (D); 128. (A); 129. (B); 130. (B); 131. (B); 132. (B); 133. (B); 134. (C); 135. (D); 136. (B); 137. (C); 138. (D); 139. (C); 140. (B); 141. (C); 142. (B); 143. (C); 144. (D); 145. (A); 146. (B); 147. (B); 148. (C); 149. (A); 150. (D); 151. (D); 152. (B); 153. (D); 154. (B); 155. (A); 156. (B); 157. (B); 158. (A); 159. (A); 160. (B); 161. (D); 162. (B); 163. (D); 164. (C); 165. (C); 166. (B); 167. (B); 168. (B); 169. (D); 170. (A); 171. (A); 172. (B); 173.(C); 174. (B); 175. (C); 176. (A); 177. (C); 178. (B); 179. (D); 180. (B); 181. (B); 182. (C); 183. (C); 184. (C); 185. (C); 186. (C); 187. (A); 188. (A); 189. (D); 190. (C); 191. (D); 192. (C); 193. (B); 194. (A); 195. (B); 196. (C); 197. (D); 198. (C); 199. (B); 200. (A); 201. (D); 202. (C); 203. (D); 204. (C); 205. (C); 206. (D); 207. (C); 208. (B); 209. (C); 210. (C); 211. (D); 212. (D); 213. (A); 214. (B); 215. (A); 216. (B); 217. (B); 218. (A);

219. (A); 220. (C); 221. (A); 222. (A); 223. (A); 224. (D); 225. (D); 226. (A); 227. (B); 228. (B); 229. (A); 230. (C); 231. (B); 232. (A); 233. (A); 234. (A); 235. (C); 236. (B); 237 (A); 238. (C); 239. (A); 240. (D); 241. (B); 242. (B); 243. (B); 244. (A); 245. (D); 246. (A); 247. (B); 248. (C); 249. (A); 250. (A); 251. (D); 252. (A); 253. (D); 254. (C); 255. (D); 256. (B); 257. (B); 258. (C); 259. (B); 260. (C); 261. (C); 262. (D); 263. (A); 264. (C); 265. (B); 266. (C); 267. (A); 268. (B); 269. (C); 270. (C); 271. (A); 272. (B); 273. (B); 274. (A), 275. (D); 276. (D); 277. (A); 278. (C); 279. (B); 280. (B); 281. (A); 282. (C); 283. (A); 284. (A); 285. (C); 286. (B); 287. (B); 288. (C); 289. (A); 290. (D); 291. (B); 292. (A); 293. (B); 294. (D); 295. (D); 296. (D); 297. (D); 298. (D); 299. (D); 300. (D); 301. (A); 302. (D); 303. (D); 304. (B); 305. (D); 306. (D); 307. (D); 308. (B); 309. (B); 310. (D); 311. (B); 312. (C); 313. (C); 314. (D); 315. (D); 316. (B); 317. (D); 318. (D); 319. (A); 320. (C); 321. (B); 322. (C); 323. (B); 324. (A); 325. (C), 326. (D); 327. (B); 328. (B); 329. (B); 330. (A); 331. (D); 332. (B); 333. (C); 334. (B); 335. (A); 336. (A); 337. (A); 338. (D); 339. (A); 340. (B); 341. (C); 342. (D); 343. (A); 344. (A); 345. (C); 346. (B); 347. (C); 348. (B); 349. (D); 350. (B); 351. (A); 352. (C); 353. (C); 354. (A); 355. (A); 356. (A); 357. (A); 358. (C); 359. (A); 360. (A); 361. (A); 362. (C); 363. (D); 364. (C); 365. (D); 366. (D); 367. (A); 368. (B); 369. (C); 370. (A); 371. (A); 372. (B); 373. (C); 374. (A); 375. (B); 376. (A); 377. (A); 378. (A); 379. (D); 380. (D); 381. (D); 382. (C); 383. (C); 384. (B); 385. (A); 386. (B); 387. (C); 388. (D); 389. (D); 390. (B); 391. (A); 392. (B); 393. (C); 394. (A); 395. (A); 396. (A); 397. (A); 398. (D); 399. (C); 400. (B); 401. (A); 402. (C); 403. (A); 404. (D); 405. (A); 406. (A); 407. (A); 408. (C); 409. (D); 410. (D); 411. (C); 412. (A); 413. (A); 414. (C); 415. (D); 416. (A); 417. (D); 418. (B); 419. (A); 420. (A); 421. (A); 422. (A); 423. (A); 424. (A); 425. (A); 426. (D); 427. (D); 428. (B); 429. (A); 430. (B); 431. (D); 432. (B); 433. (D); 434. (A); 435. (A); 436. (A); 437. (A); 438. (A); 439. (A); 440. (B); 441. (A); 442. (A); 443. (A); 444. (D); 445. (A); 446. (A); 447. (D); 448. (D); 449. (B); 450. (D); 451. (C); 452. (A); 453. (C); 454. (A); 455. (B); 456. (C); 457. (B); 458. (B); 459. (B); 460. (C); 461. (A); 462. (A); 463. (A); 464. (B); 465. (B); 466. (A); 467. (B); 468. (B); 469. (D); 470. (A); 471. (D); 472. (B); 473. (B); 474. (B); 475. (C); 476. (A); 477. (C); 478. (B); 479. (D); 480. (D); 481. (A); 482. (C); 483. (A); 484. (A); 485. (C); 486. (A); 487. (A); 488. (C); 489. (A); 490. (A); 491. (D); 492. (B); 493. (C); 494. (A); 495. (C); 496. (A); 497. (A); 498. (A); 499. (D); 500. (C); 501. (A); 502. (D); 503. (B); 504. (D); 505. (C); 506. (A); 507. (A); 508. (B); 509. (B); 510. (D); 511. (D); 512. (B); 513. (B); 514. (A); 515. (D); 516. (D); 517. (A); 518. (C); 519. (C); 520. (B); 521. (D);

CHAPTER 14: AGRICULTURAL STATISTICS

1. (A); 2. (B); 3. (A); 4. (B); 5. (B); 6. (A); 7. (B); 8. (B); 9. (C); 10. (A); 11. (A); 12. (A); 13. (B); 14. (B); 15. (A); 16. (C); 17. (A); 18. (C); 19. (A); 20. (A); 21. (B); 22. (A); 23. (D); 24. (A); 25. (A); 26. (A); 27. (C); 28. (B); 29. (A); 30. (A); 31. (A); 32. (D); 33. (B); 34. (A); 35. (A); 36. (B); 37. (B); 38. (A); 39. (A); 40. (C); 41. (D); 42. (D); 43. (D); 44. (D); 45. (B); 46. (A); 47. (B); 48. (B); 49. (A); 50. (C); 51. (D); 52. (A); 53. (B); 54. (A); 55. (A); 56. (B); 57. (B); 58. (C); 59. (A); 60. (B); 61. (D); 62. (C); 63. (B); 64. (A); 65. (C); 66. (B); 67. (C); 68. (A); 69. (C); 70. (D); 71. (B); 72. (B); 73. (B); 74. (D); 75. (C); 76. (B); 77. (A); 78. (D); 79. (D); 80. (C); 81. (B); 82. (C); 83. (A); 84. (B); 85. (D); 86. (D); 87. (A); 88. (C); 89. (A); 90. (A); 91. (B); 92. (A); 93. (C); 94. (B); 95. (C); 96. (D); 97. (B); 98. (B); 99. (A); 100. (B); 101. (B); 102. (B); 103. (A); 104. (B); 105. (B); 106. (A); 107. (A); 108. (D); 109. (B); 110. (B); 111. (C); 112. (A); 113. (B); 114. (C); 115. (A); 116. (A); 117. (D); 118. (B); 119. (B); 120. (B); 121. (A); 122. (B); 123. (B); 124. (C); 125. (A); 126. (B); 127. (B); 128. (A); 129. (B); 130. (D); 131. (C); 132. (C); 133. (C); 134. (B); 135. (A); 136. (C); 137. (A); 138. (C); 139. (B); 140. (A); 141. (D); 142. (B); 143. (B); 144. (A); 145. (B); 146. (C); 147. (B); 148. (C); 149. (A); 150. (B); 151. (C); 152. (A); 153. (C);

154. (D); 155. (A); 156. (B); 157. (A); 158. (A);
159. (B); 160. (C); 161. (D); 162. (B); 163. (C);
164. (D); 165. (B); 166. (D); 167. (B); 168. (C);
169. (A); 170. (D); 171. (C); 172. (B); 173.(C);
174. (D); 175. (B); 176. (B); 177. (B); 178. (C);
179. (C); 180. (A); 181. (B); 182. (A); 183. (B);
184. (D); 185. (D); 186. (B); 187. (A); 188. (A);
189. (C); 190. (D); 191. (A); 192. (B); 193. (A);
194. (A); 195. (B); 196. (A); 197. (C); 198. (B);
199. (A); 200. (C); 201. (B); 202. (B); 203. (C);
204. (B); 205. (C); 206. (B); 207. (A); 208. (A);
209. (C); 210. (A); 211. (A); 212. (C); 213. (C);
214. (B); 215. (B); 216. (B); 217. (A); 218. (C);
219. (C); 220. (C); 221. (D); 222. (A); 223. (B);
224. (B); 225. (B); 226. (B); 227. (B); 228. (A);
229. (D); 230. (C); 231. (D); 232. (B); 233. (C);
234. (B); 235. (A); 236. (C); 237. (B); 238. (B);
239. (C); 240. (B); 241. (A); 242. (A); 243. (C);
244. (D); 245. (A); 246. (B); 247. (C); 248. (A);
249. (C); 250. (C); 251. (A); 252. (B); 253. (C);
254. (C); 255. (C); 256. (D); 257. (B); 258. (A);
259. (A); 260. (A); 261. (C); 262. (A); 263. (D);
264. (A); 265. (D); 266. (A); 267. (B); 268. (A);
269. (B); 270. (B); 271. (A); 272. (B); 273. (B);
274. (A), 275. (C); 276. (D); 277. (D); 278. (B);
279. (B); 280. (B); 281. (C); 282. (A); 283. (A);
284. (B); 285. (A); 286. (A); 287. (B); 288. (B);
289. (B); 290. (A); 291. (D); 292. (C); 293. (C);
294. (A); 295. (C); 296. (B); 297. (C); 298. (D);
299. (A); 300. (C); 301. (C); 302. (D); 303. (A);
304. (C); 305. (A); 306. (C); 307. (D); 308. (C);
309. (B); 310. (D); 311. (A); 312. (D); 313. (B);
314. (B); 315. (A); 316. (B); 317. (A); 318. (C);
319. (B); 320. (C); 321. (A); 322. (B); 323. (C);
324. (C); 325. (D), 326. (C); 327. (D); 328. (C);
329. (A); 330. (D); 331. (B); 332. (B); 333. (B);
334. (C); 335. (A); 336. (B); 337. (C); 338. (C);
339. (D); 340. (C); 341. (A); 342. (B); 343. (B);
344. (D); 345. (B); 346. (C); 347. (A); 348. (A);
349. (A); 350. (A); 351. (D); 352. (A); 353. (B);
354. (A); 355. (B); 356. (B); 357. (A); 358. (D);
359. (A); 360. (D); 361. (C); 362. (C); 363. (C);
364. (C); 365. (A); 366. (A); 367. (D); 368. (A);
369. (B); 370. (B); 371. (A): 372. (B); 373. (B);
374. (B); 375. (A); 376. (D); 377. (C); 378. (A);
379. (A); 380. (C); 381. (C); 382. (D); 383. (A);
384. (D); 385. (B); 386. (A); 387. (D); 388. (B);
389. (B); 390. (B); 391. (C); 392. (D); 393. (A);
394. (D); 395. (D); 396. (A); 397. (C); 398. (A);
399. (A); 400. (A); 401. (B); 402. (A); 403. (A);
404. (A); 405. (A); 406. (D); 407. (B); 408. (A);
409. (C); 410. (A); 411. (D); 412. (C); 413. (B);
414. (B); 415. (A); 416. (A); 417. (A); 418. (A);
419. (D); 420. (C); 421. (D); 422. (C); 423. (A);
424. (B); 425. (B); 426. (A); 427. (D); 428. (A);
429. (C); 430. (A); 431. (C); 432. (C); 433. (B);
434. (B); 435. (B); 436. (A); 437. (C); 438. (B);
439. (B); 440. (A); 441. (A); 442. (A); 443. (B);
444. (C); 445. (D); 446. (B); 447. (B); 448. (D);
449. (B); 450. (C); 451. (A); 452. (B); 453. (C);
454. (B); 455. (B); 456. (B); 457. (C); 458. (C);
459. (A); 460. (B); 461. (A); 462. (A); 463. (B);
464. (A); 465. (A); 466. (A); 467. (C); 468. (B);
469. (B); 470. (A); 471. (C); 472. (A); 473. (D);
474. (C); 475. (D); 476. (A); 477. (C); 478. (B);
479. (D); 480. (A); 481. (B); 482. (D); 483. (B);
484. (A); 485. (B); 486. (B); 487. (A); 488. (A);
489. (C); 490. (A); 491. (C); 492. (D); 493. (C);
494. (A); 495. (B); 496. (B); 497. (C); 498. (B);
499. (A); 500. (B); 501. (A); 502. (B); 503. (C);
504. (A); 505. (A); 506. (A); 507. (D); 508. (C);
509. (B); 510. (B); 511. (D); 512. (A); 513. (C);
514. (B); 515. (A); 516. (A); 517. (B); 518. (B);
519. (A); 520. (D); 521. (C); 522. (A); 523. (D);
524. (C); 525. (A); 526. (B); 527. (A); 528. (A);
529. (D); 530. (C); 531. (B); 532. (A); 533. (A);
534. (B); 535. (B); 536. (C); 537. (D); 538. (D);
539. (B); 540. (D); 541. (B); 542. (A); 543. (D);
544. (B); 545. (C); 546. (D); 547. (A); 548. (B);
549. (C); 550. (A); 551. (B); 552. (A); 553. (B);
554. (A); 555. (B); 556. (A); 557. (B); 558. (B);
559. (C); 560. (A); 561. (B); 562. (C); 563. (C);
564. (D); 565. (A); 566. (C); 567. (D); 568. (C);
569. (A); 570. (C); 571. (A); 572. (C); 573. (C);
574. (D); 575. (A); 576. (B); 577. (D); 578. (B);
579. (B); 580. (D); 581. (C); 582. (B); 583. (C);
584. (A); 585. (A); 586. (C); 587. (D); 588. (C);
589. (C); 590. (B); 591. (B); 592. (A); 593. (B);
594. (C); 595. (A); 596. (A); 597. (A); 598. (D);
599. (A); 600. (C); 601. (D); 602. (D); 603. (A);
604. (D); 605. (D); 606. (D); 607. (C); 608. (B);
609. (A); 610. (D); 611. (B); 612. (D); 613. (B);

614. (A); 615. (B); 616. (B); 617. (B); 618. (B); 619. (D); 620. (C); 621. (A); 622. (A); 623. (B); 624. (A); 625. (C); 626. (A); 627. (A); 628. (A); 629. (C); 630. (B); 631. (C); 632. (A); 633. (A); 634. (C); 635. (C); 636. (A); 637. (B); 638. (C); 639. (B); 640. (C); 641. (C); 642. (C); 643. (B); 644. (B); 645. (D); 646. (A); 647. (C); 648. (A); 649. (C); 650. (A); 651. (C); 652. (A); 653. (A); 654. (D); 655. (B); 656. (B); 657. (D); 658. (D); 659. (C); 660. (C); 661. (B); 662. (A); 663. (A); 664. (A); 665. (B); 666. (B); 667. (A); 668. (C); 669. (B); 670. (A); 671. (D); 672. (C); 673. (C); 674. (B); 675. (A); 676. (D); 677. (C); 678. (D); 679. (D); 680. (C); 681. (A); 682. (C); 683. (B); 684. (A); 685. (C); 686. (A); 687. (B); 688. (A); 689. (C); 690. (B); 691. (B); 692. (D); 693. (D); 694. (C); 695. (A); 696. (B); 697. (A); 698. (C); 699. (B); 700. (D); 701. (D); 702. (B); 703. (A); 704. (D); 705. (D); 706. (A); 707. (C); 708. (A); 709. (B); 710. (A); 711. (C); 712. (D); 713. (B); 714. (B); 715. (B); 716. (A); 717. (C); 718. (A); 719. (D); 720. (A); 721. (B); 722. (A); 723. (C); 724. (A); 725. (A); 726. (C); 727. (C); 728. (B); 729. (A); 730. (A); 731. (B); 732. (C); 733. (A); 734. (A); 735. (B); 736. (B); 737. (B); 738. (B); 739. (D); 740. (D); 741. (B); 742. (B); 743. (A); 744. (A); 745. (C); 746. (A); 747. (D); 748. (D); 749. (A); 750. (A); 751. (A); 752. (C); 753. (B); 754. (A); 755. (B); 756. (C); 757. (A); 758. (B); 759. (B); 760. (C); 761. (C); 762. (D); 763. (A); 764. (A); 765. (B); 766. (D); 767. (C); 768. (A); 769. (B); 770. (A); 771. (B); 772. (B); 773. (A); 774. (A); 775. (A); 776. (B); 777. (B); 778. (C); 779. (C); 780. (C); 781. (A); 782. (B); 783. (D); 784. (B); 785. (D); 786. (C); 787. (B); 788. (B); 789. (C); 790. (C); 791. (A); 792. (C); 793. (C); 794. (C); 795. (B); 796. (C); 797. (A); 798. (C); 799. (B); 800. (C); 801. (D); 802. (D); 803. (D); 804. (C); 805. (C); 806. (C); 807. (B); 808. (A); 809. (A); 810. (A); 811. (B); 812. (B); 813. (D); 814. (B); 815. (B); 816. (B); 817. (C); 818. (A); 819. (A); 820. (B); 821. (A); 822. (A); 823. (B); 824. (D); 825. (B); 826. (A); 827. (B); 828. (A); 829. (D); 830. (D); 831. (A); 832. (B); 833. (D); 834. (B); 835. (C); 836. (A); 837. (B); 838. (B); 839. (B); 840. (A); 841. (D); 842. (D); 843. (C); 844. (C); 845. (B); 846. (B); 847. (B); 848. (D); 849. (A); 850. (B); 851. (D); 852. (D); 853. (C); 854. (A); 855. (A); 856. (D); 857. (C); 858. (B); 859. (B); 860. (C); 861. (C); 862. (A); 863. (B); 864. (D); 865. (B); 866. (C); 867. (A); 868. (B); 869. (D); 870. (B); 871. (B); 872. (C); 873. (C); 874. (A); 875. (D); 876. (C); 877. (C); 878. (A); 879. (C); 880. (D); 881. (B); 882. (C); 883. (C); 884. (A); 885. (B); 886. (A); 887. (A); 888. (D); 889. (A); 890. (B); 891. (A); 892. (A); 893. (C); 894. (A); 895. (C); 896. (C); 897. (A); 898. (B); 899. (A); 900. (B); 901. (A); 902. (D); 903. (B); 904. (A); 905. (A); 906. (B); 907. (D); 908. (D); 909. (A); 910. (D); 911. (B); 912. (A); 913. (C); 914. (B); 915. (A); 916. (B)

Related Books

Titles	Authors
— A Technical Advisors Manual Managing Agricultural Irrigation Draining Water	*SWRCB*
— Agri-Entrepreneurship Training Manual	*NSA*
— Agricultural Extension A Glimpse	*S. Jananai*
— Agricultural Extension On A Refernce Manual	*H, Addison*
— Agriculture and Fertilizers	*O.C. Bockman*
— Agro Enterprises for Empowering Farm Women	*T. Rathakrishnan*
— Agro-Entrepreneurship	*S.R. Panigrahy*
— Beekeeping: A Compressive Guide to Bees and Beekeeping	*D.P. Abrol*
— Biofertilizer Manual	*FNCA*
— Cerial Seed Technology : A Manual of Cereal Seed Production Quality Control and Distribution	*W.P. Feistritrzer*
— Communication Skills and Personality Development	*J.R. Kadam*
— Communication Techniques in Farm Extension	*T. Rathakrishnan*
— Compensatory Afforestation in Rajasthan: An Evaluation	*G. Singh*
— Comprehensive Assessment of Soil Health *3rd Ed*	*SIPS*
— Educating Farmers About Agriculture	*Raghunath Singh*
— Entrepreneurship Development and Communication Skills	*R.R. Chole*
— Entrepreneurship in Farming	*D. Khan*
— Farmers Guidebook to Work Health and Safety	*SA Safework*
— Farmers Market Manual	*Kristin Brennan*
— Farmers Training Entrepreneurship Manual	*Francis Luyayi*
— Feeding and Management of Livestock During Drought and Scarcity	*N.V. Patil*
— Fertilizer Manual	*IFDC*
— Fundamentals of Human Ethics and Agriculture Extension	*D.K. Meena*
— Glimpse on General Agriculture	*P. Laxman Rao*
— Guide to Extension Training	*P. Oakley*
— Indiras Objective Agriculture	*R.L. Arya*
— Measuring Agricultural Sustainability	*Kaushalya Ramachandran*
— Operations Research	*P.K. Mohanty*
— Pioneer Research in Extension Education	*A.K. Singh*
— Plant Breeding With Farmers: A Technical Manual	*S. Ceccarelli*
— Text Book of Agricultural Heritage	*Utpal Giri*
— The Community Forestry Guidebook	*Jennifer Gunter*
— Transfer of Agricultural Technology	*R.R. Chole*
— गृह विज्ञान एवं प्रसार शिक्षा	*Renu Arya*